Introduction to Computer Science

Java Programming

by

Julie A. Anderson
Kathleen M. Austin
Lorraine N. Bergkvist

Publisher
The Goodheart-Willcox Company, Inc.
Tinley Park, Illinois
www.g-w.com

Introduction

Introduction to Computer Science: Java Programming presents a comprehensive approach to learning object-oriented programming with Java. Intended for first-time users, this engaging, hands-on text guides you to developing basic Java programs, troubleshooting, and debugging errors.

By studying this text, you will learn computer science concepts of computational thinking and encoding as well as information about variables, expressions, classes, repetition, and other important programming concepts. Extension opportunities provide cross-curricular connections to math, science, and language arts that help you relate to Java's widespread relevance in the real world.

In addition to learning programming concepts, practical coverage of computing and society is addressed. The importance of abiding by basic principles of right and wrong, respecting others' ideas, and behaving in an ethical manner are emphasized.

About the Authors

Julie A. Anderson has seventeen years of experience teaching, and is recently retired from Rollins College. She has coauthored college textbooks on Java and Python. She also has fifteen years of experience as a technology writer, editor, and columnist for computer publications with an emphasis on databases and programming. She began her computing career as a software developer. She holds a Master of Science degree in Computer Science from Johns Hopkins University.

Kathleen M. Austin was a senior lecturer in the School of Information Arts and Technologies at the University of Baltimore. She has participated in the development of many educational multimedia projects. She has authored, coauthored, or contributed to several textbooks, including *Introduction to Computer Science: Coding, Principles of Digital Information Technology,* and *Introduction to Microsoft Office 2019* published by Goodheart-Willcox Publisher as well as *Consumer Mathematics* and *Math for the World of Work.* She holds a Master of Science degree in Computer Science from Johns Hopkins University and a Doctor of Communications Design from the University of Baltimore as well as IC3 certification.

Lorraine N. Bergkvist was an Adjunct Professor at the University of Baltimore providing instruction in Visual Basic programming, database implementation, and web-page creation. She is the coauthor of *Introduction to Computer Science: Coding, Principles of Digital Information Technology,* and *Introduction to Microsoft Office 2019* published by Goodheart-Willcox Publisher. She is also the owner of Kingsville Résumé Center, which provides professional résumé-writing services as well as consulting and editing in the information technology field. She developed the curriculum and taught the Introduction to Technology course at the University of Baltimore and the College of Notre Dame of Maryland. She holds a Bachelor of Science degree from Trinity University, a Master of Education degree from Towson University, and IC3 certification. She has received several scholarships and grants in the technology field.

Reviewers

Goodheart-Willcox Publisher would like to thank the following instructors who reviewed selected chapters and provided input for the development of *Introduction to Computer Science: Java Programming.*

Troy Burns
Advanced Placement and International Baccalaureate
    and Computer Science Teacher
Marvin Ridge High School
Waxhaw, North Carolina

Charles Flynt
Teacher
Weaver Academy
Greensboro, North Carolina

R. Brent Greene
Computer Science and Autonomous Vehicles Instructor
Williamson County Schools
Franklin, Tennessee

Steven Klug
Computer Science Teacher
Hanford High School
Richland, Washington

Rachel Lawrence
Math and Computer Science Teacher
Woods Charter School
Chapel Hill, North Carolina

Erica Roberts
Department Chair
Computer Science
Northside College Preparatory High School
Chicago, Illinois

Stephanie B. Van Slyke
Computer Science Teacher
Lambert High School
Suwanee, Georgia

Prepare for Your Future

Introduction to Computer Science: Java Programming is designed to help prepare you for a future in the field of information technology (IT). By studying this text, you can take the first step toward learning how to program in the Java language. Developing foundational knowledge in computer science and programming can help prepare you for success in a future IT career.

Content is presented in an easy-to-comprehend and relevant format. Each chapter includes practical tools and activities that help enhance your learning experience and aid in your comprehension of the material.

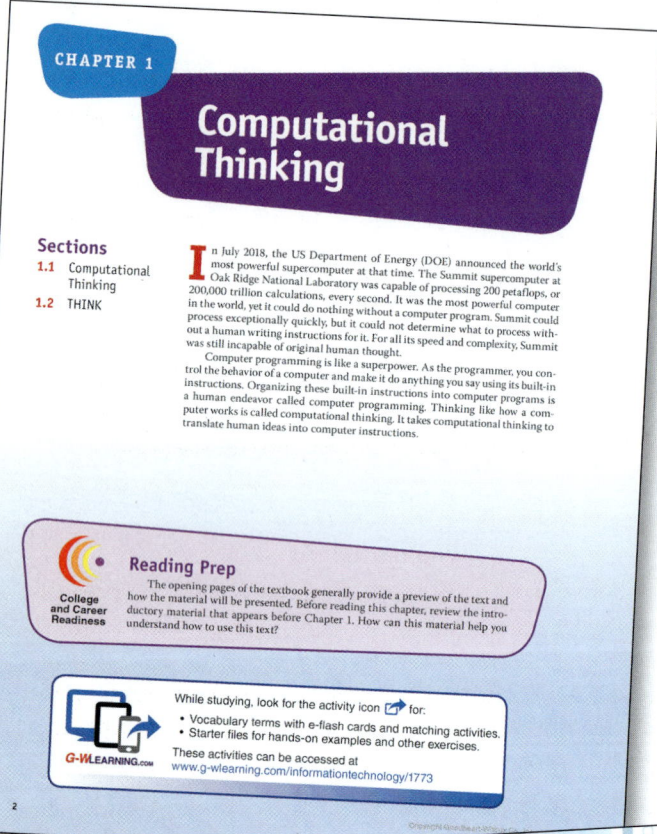

- Each chapter is broken into smaller **sections**. Chunking the content makes it easier for you to read and comprehend new material at your own speed.

- **Reading Prep** activities incorporate English/language arts standards for reading, writing, speaking, and listening to provide ways for you to demonstrate your literacy skills.

- A **chapter glossary** previews and defines each word presented in the chapter. This makes a convenient resource to prepare for new learning before diving into the content.

- An **Essential Question** at the beginning of each section will engage you to uncover the important points that are presented as the content develops.

- **Learning Goals** listed at the beginning of each section provide an overview of the concepts you will explore and can be used as personal learning benchmarks.

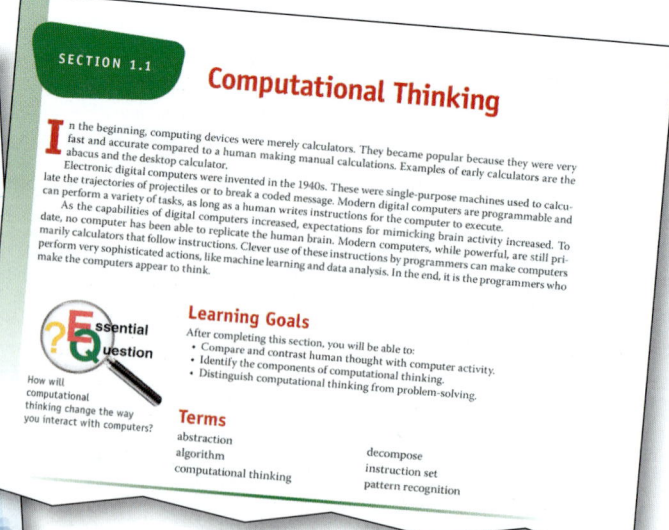

Amplify Your Learning

Applying what you learn reinforces and strengthens your skills. Each chapter teaches concepts of computer science related to Java programming. After learning a concept, you immediately apply it. Starter files to complete these activities are available for download from the companion website.

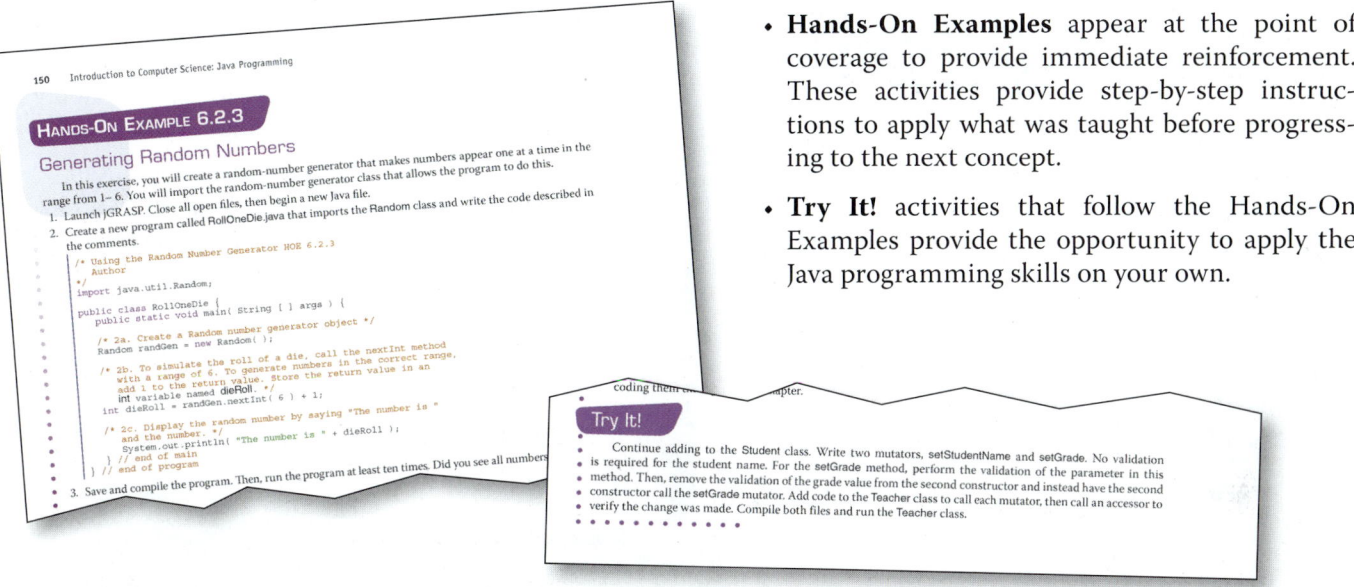

- **Hands-On Examples** appear at the point of coverage to provide immediate reinforcement. These activities provide step-by-step instructions to apply what was taught before progressing to the next concept.

- **Try It!** activities that follow the Hands-On Examples provide the opportunity to apply the Java programming skills on your own.

What about a challenge? When you learn to solve problems, you learn you can take on challenges presented to you. Working as an individual or part of a team, solving problems is a skill needed for life and the workplace.

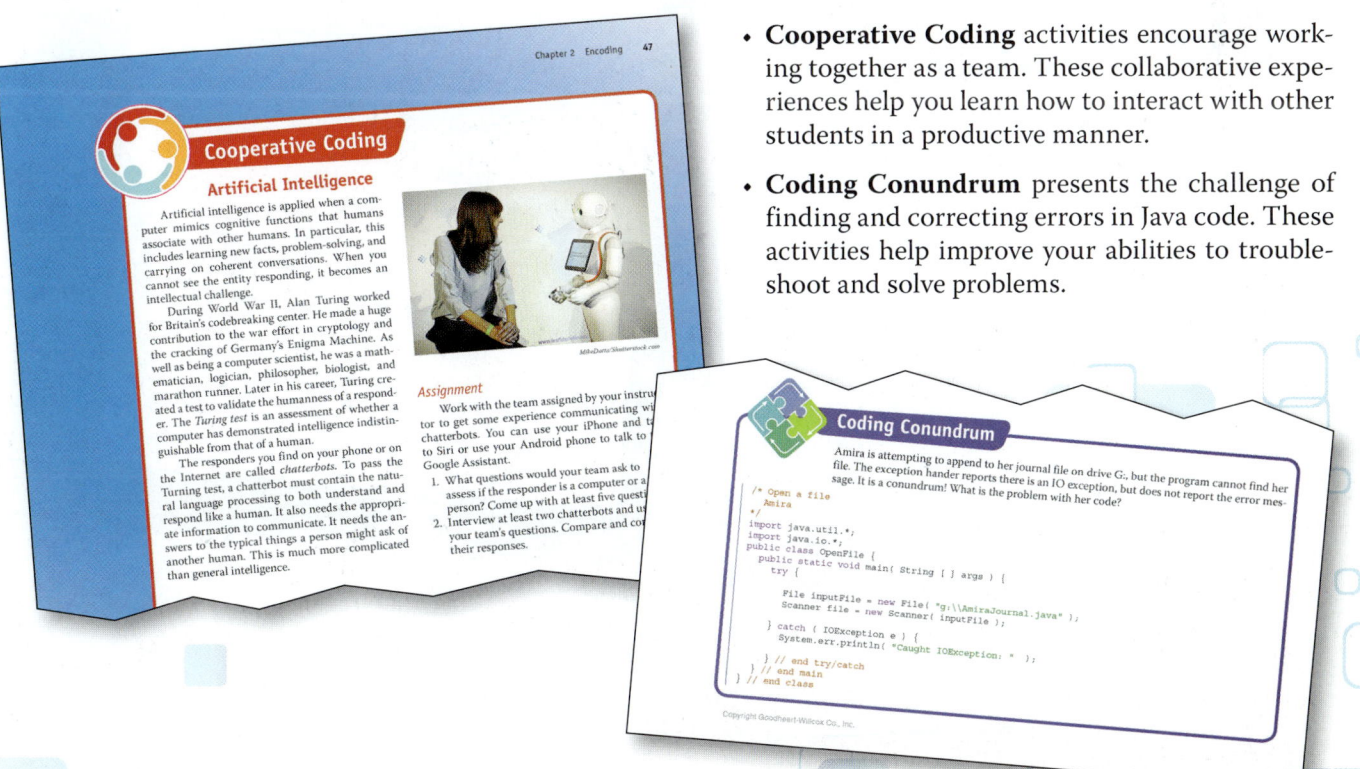

- **Cooperative Coding** activities encourage working together as a team. These collaborative experiences help you learn how to interact with other students in a productive manner.

- **Coding Conundrum** presents the challenge of finding and correcting errors in Java code. These activities help improve your abilities to troubleshoot and solve problems.

Maximize the Impact

Practical information helps prepare for your future. Throughout this text, special features highlight how computer programming is useful in a variety of subject areas and situations. These cross-curricular connections are demonstrated across other areas of study, such as math, language arts, and science.

- **Math and Java** features emphasize the importance of math in programming. This information provides opportunities for you to apply your math skills to Java coding.

- **Language Arts and Java** features are opportunities to demonstrate creating Java applications that illustrate language arts skills or apply computational thinking to everyday life.

- **Science and Java** features relate how programming can be applied to science concepts and projects.

Assess Your Progress

It is important to assess what you learn as you progress through the textbook. Multiple opportunities confirm learning as you explore the content. *Formative assessment* includes the following:

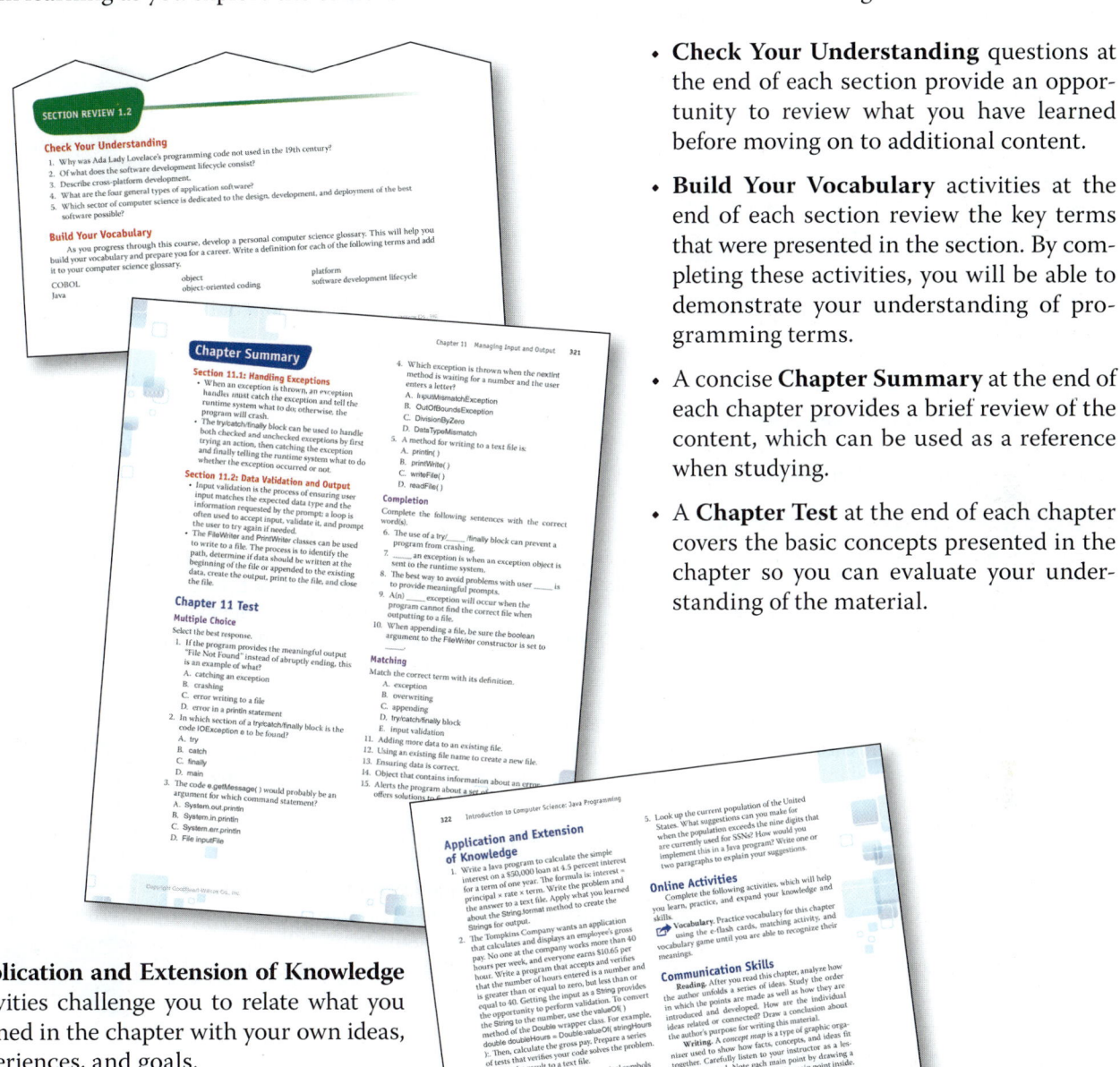

- **Check Your Understanding** questions at the end of each section provide an opportunity to review what you have learned before moving on to additional content.

- **Build Your Vocabulary** activities at the end of each section review the key terms that were presented in the section. By completing these activities, you will be able to demonstrate your understanding of programming terms.

- A concise **Chapter Summary** at the end of each chapter provides a brief review of the content, which can be used as a reference when studying.

- A **Chapter Test** at the end of each chapter covers the basic concepts presented in the chapter so you can evaluate your understanding of the material.

- **Application and Extension of Knowledge** activities challenge you to relate what you learned in the chapter with your own ideas, experiences, and goals.

- **Online Activities** on the student companion website provide a review of the key terms presented in each chapter.

- **Communication Skills** activities provide ways for you to demonstrate your literacy and career-readiness skills.

- **Portfolio Development** activities provide guidance to create a personal portfolio for use when exploring volunteer, education and training, and career opportunities.

Student Tools

Student Text

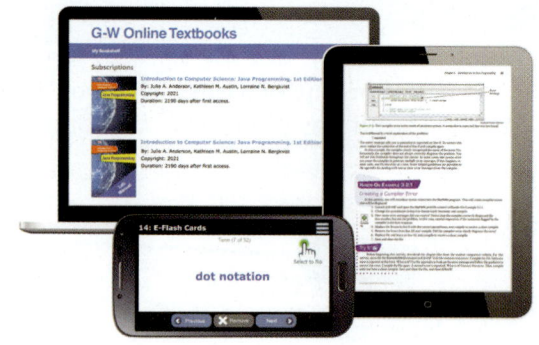

Introduction to Computer Science: Java Programming presents a comprehensive approach to learning object-oriented programming with Java. Intended for first-time users, this engaging, hands-on text guides students to developing basic Java programs, troubleshooting, and debugging errors.

By studying this text, students will learn computer science concepts of computational thinking and encoding as well as information about variables, expressions, classes, repetition, and other important programming concepts. Extension opportunities provide cross-curricular connections to math, science, and language arts that help students relate to Java's widespread relevance in the real world.

In addition to learning programming concepts, practical coverage of computing and society is addressed. The importance of abiding by basic principles of right and wrong, respecting others' ideas, and behaving in an ethical manner are emphasized.

Laboratory Manual

The student laboratory manual that accompanies *Introduction to Computer Science: Java Programming* provides hands-on practice with warm-up exercises, Java programming activities, and debugging challenges. Each chapter corresponds to the text to reinforce key concepts and applied knowledge.

Online Learning Suite (OLS)

The Online Learning Suite provides the foundation of instruction and learning for digital and blended classrooms. An easy-to-manage shared classroom subscription makes the OLS a hassle-free solution for both students and instructors. An online student text and laboratory manual, along with rich supplemental content, brings digital learning to the classroom. All instructional materials are found on a convenient online bookshelf and accessible at home, at school, or on the go.

G-W Companion Website

The G-W Learning companion website is a study reference that contains e-flash cards and vocabulary exercises. Starter files needed to complete selected programming and cross-curricular activities are available for download. The companion website is accessible from any digital device.

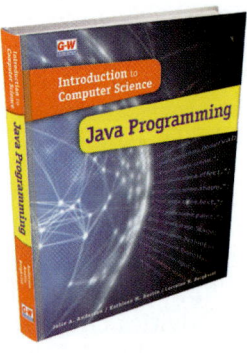

Online Learning Suite/Student Text Bundle

Looking for a blended solution? Goodheart-Willcox offers the Online Learning Suite bundled with the printed text in one easy-to-access package. Students have the flexibility to use the printed text, the Online Learning Suite, or a combination of both components to meet their individual learning styles. The convenient packaging makes managing and accessing content easy and efficient.

Instructor Tools

LMS Integration

Integrate Goodheart-Willcox content within your Learning Management System for a seamless user experience for both you and your students. LMS-ready content in Common Cartridge format facilitates single sign-on integration and gives you control of student enrollment and data. With a Common Cartridge integration, you can access the LMS features and tools you are accustomed to using and G-W course resources in one convenient location—your LMS.

To provide a complete learning package for you and your students, G-W Common Cartridge includes the Online Learning Suite and Online Instructor Resources. When you incorporate G-W content into your courses via Common Cartridge, you have the flexibility to customize and structure the content to meet the educational needs of your students. You may also choose to add your own content to the course.

QTI® question banks are available within the Online Instructor Resources for import into your LMS. These prebuilt assessments help you measure student knowledge and track results in your LMS gradebook. Questions and tests can be customized to meet your assessment needs.

Online Instructor Resources (OIR)

Online Instructor Resources provide all the support needed to make preparation and classroom instruction easier than ever. Available in one accessible location, the OIR includes Instructor Resources, Instructor's Presentations for PowerPoint®, and assessment software with question banks. The OIR is available as a subscription and can be accessed at school, at home, or on the go.

Instructor Resources One resource provides instructors with time-saving preparation tools such as answer keys, editable lesson plans, and other teaching aids.

Instructor's Presentations for PowerPoint® These fully customizable presentations for PowerPoint® provide a useful teaching tool for presenting concepts introduced in the text. Richly illustrated slides help you teach and visually reinforce the key concepts from each chapter.

Assessment Software with Question Banks Administer and manage assessments to meet your classroom needs. The question banks that accompany this textbook include hundreds of matching, true/false, completion, multiple choice, and short answer questions to assess student knowledge of the content in each chapter. Using the assessment software simplifies the process of creating, managing, administering, and grading tests. You can have the software generate a test for you with randomly selected questions. You may also choose specific questions from the question banks and, if you wish, add your own questions to create customized tests to meet your classroom needs.

G-W Integrated Learning Solution

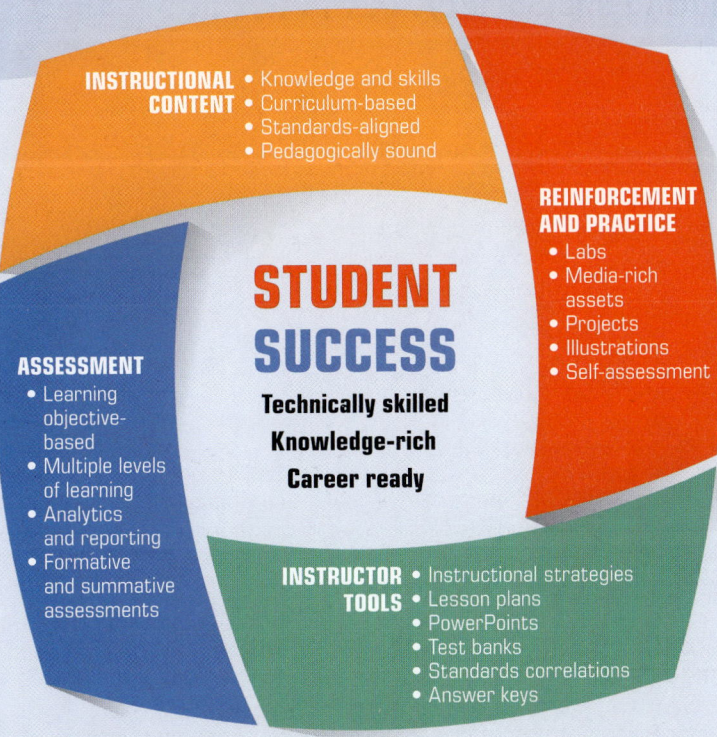

INSTRUCTIONAL CONTENT
- Knowledge and skills
- Curriculum-based
- Standards-aligned
- Pedagogically sound

REINFORCEMENT AND PRACTICE
- Labs
- Media-rich assets
- Projects
- Illustrations
- Self-assessment

ASSESSMENT
- Learning objective-based
- Multiple levels of learning
- Analytics and reporting
- Formative and summative assessments

INSTRUCTOR TOOLS
- Instructional strategies
- Lesson plans
- PowerPoints
- Test banks
- Standards correlations
- Answer keys

STUDENT SUCCESS

Technically skilled

Knowledge-rich

Career ready

The G-W Integrated Learning Solution offers easy-to-use resources that help students and instructors achieve success.

▶ **EXPERT AUTHORS**
▶ **TRUSTED REVIEWERS**
▶ **100 YEARS OF EXPERIENCE**

EMPLOYABILITY SKILLS · TECHNICAL SKILLS · ACADEMIC KNOWLEDGE · INDUSTRY RECOGNIZED STANDARDS

Brief Contents

Contents

HANDS-ON EXAMPLES

Special Features

Cooperative Coding

Language Arts and Java

Math and Java

Science and Java

Coding Conundrum

Computational Thinking

Sections

In July 2018, the US Department of Energy (DOE) announced the world's most powerful supercomputer at that time. The Summit supercomputer at Oak Ridge National Laboratory was capable of processing 200 petaflops, or 200,000 trillion calculations, every second. It was the most powerful computer in the world, yet it could do nothing without a computer program. Summit could process exceptionally quickly, but it could not determine what to process without a human writing instructions for it. For all its speed and complexity, Summit was still incapable of original human thought.

Computer programming is like a superpower. As the programmer, you control the behavior of a computer and make it do anything you say using its built-in instructions. Organizing these built-in instructions into computer programs is a human endeavor called computer programming. Thinking like how a computer works is called computational thinking. It takes computational thinking to translate human ideas into computer instructions.

College and Career Readiness

Reading Prep

The opening pages of the textbook generally provide a preview of the text and how the material will be presented. Before reading this chapter, review the introductory material that appears before Chapter 1. How can this material help you understand how to use this text?

While studying, look for the activity icon for:

- Vocabulary terms with e-flash cards and matching activities.
- Starter files for hands-on examples and other exercises.

These activities can be accessed at
www.g-wlearning.com/informationtechnology/1773

Chapter Glossary

abstraction: Generalization of patterns found in problem solving.

algorithm: Sequence of steps used to solve a problem.

COBOL: Stands for common business oriented language; a programming language written for business applications and one of the first programming languages that did not require the use of binary numbers for instructions.

computational thinking: Thinking like a computer operates.

decomposition: Breaking a problem into small doable steps.

instruction set: Collection of complex actions developed using the basic functions of calculation and decision.

Java: Object-oriented computer programming language with features that minimize confusion and errors and allow programmers to produce code that operates on a wide variety of platforms.

object: Self-contained unit of data and code to manage that data.

object-oriented coding: Writing code in a programming language that encapsulates code and data into objects.

pattern recognition: Identifying parts of one problem that are replicated within a problem or in other problems.

platform: Computer environment depending on a unique instruction set.

software development life cycle: Process that provides structure to the production and maintenance of software; software is written, installed, used, and maintained.

Computational Thinking

In the beginning, computing devices were merely calculators. They became popular because they were very fast and accurate compared to a human making manual calculations. Examples of early calculators are the abacus and the desktop calculator.

Electronic digital computers were invented in the 1940s. These were single-purpose machines used to calculate the trajectories of projectiles or to break a coded message. Modern digital computers are programmable and can perform a variety of tasks, as long as a human writes instructions for the computer to execute.

As the capabilities of digital computers increased, expectations for mimicking brain activity increased. To date, no computer has been able to replicate the human brain. Modern computers, while powerful, are still primarily calculators that follow instructions. Clever use of these instructions by programmers can make computers perform very sophisticated actions, like machine learning and data analysis. In the end, it is the programmers who make the computers appear to think.

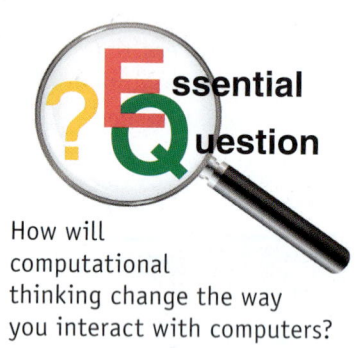

?EQ Essential Question

How will computational thinking change the way you interact with computers?

Learning Goals

After completing this section, you will be able to:
- Compare and contrast human thought with computer activity.
- Identify the components of computational thinking.
- Distinguish computational thinking from problem-solving.

Terms

abstraction
algorithm
computational thinking

decompose
instruction set
pattern recognition

Think like a Computer

Computers are often described using human characteristics. People speak of a computer "thinking." However, the computer does not create original thought. Even the most advanced video games and simulations are essentially just moving data around in memory. These programs decide which number is greater and make calculations. This may appear as if the computer is thinking, but it is just following instructions.

A video game program typically provides options for the computer player. This may include reacting to the human player's actions. Special programming techniques known as *artificial intelligence (AI)* are used to provide options for a computer program to execute when reacting to conditions during operation. AI makes it appear the game is thinking. In reality, it is executing a list of statements the programmers have provided.

Robotics is a field where AI is used to make a machine seem to be human. The robot Nadine created by Professor Nadia Thalman is an example of a humanlike robot that uses AI. In another example, **Figure 1-1** shows a robotic receptionist in use at a shopping center in Japan. These robots rely on their programming to remember visitors, their names, and when they were last at the reception desk.

To become great programmers, humans must learn how to think like a computer operates. This is *computational thinking,* which is discussed later. Programmers use computational thinking to translate great ideas into something the computer can accomplish.

Human Thought

Humans think quite differently from how computers operate. They can create ideas and imagine new scenarios. Humans build computers to do the things that humans are not good at doing. Calculations are one example of where computers are better than humans. Humans are slow and often inaccurate when performing calculations.

When humans think, they use their brains to make sense of their experiences. They can also analyze what they have seen, felt, heard, smelled, and tasted. Using information from their senses, humans can make decisions about the future. Human thought can use these decisions to make plans, as shown in **Figure 1-2.**

On occasion, a person will have a goal in mind. Recalling experiences, sorting through them, and looking for links provide a way to reach that goal. During this process, it is not unusual for a person to create a new idea. This new idea is the result of human thought and may be the key to reaching the goal.

Computer Processing

A computer without a program to direct its calculations is like a puppy dog jumping up and down trying to get your attention. An idle computer repeats the same steps as this loop:

1. Is there anything to do?
   A. Yes? Do it!
      i. Go to the first step.
   B. No? Go to sleep.
2. Wait a bit.
3. Wake up, and go to the first step.

These steps are hardwired into the computer. It is the only thing a computer can do without human coding. The only things a human can ask a computer to do are actions in its instruction set. The *instruction set* for a chip is the collection of complex actions developed using the basic functions of calculation and decision. The instruction set is also hard wired into the computer. A computer's programming can be changed, but its basic instruction set cannot. **Figure 1-3** shows part of the instruction set for an early modern computer. Typical actions are:

- add two numbers;
- move a number from one memory location to another; and
- decide if two numbers are equal or which is bigger.

It takes a human to write a program using these instructions. A human must also tell the computer to run the program for anything novel to happen in the computer.

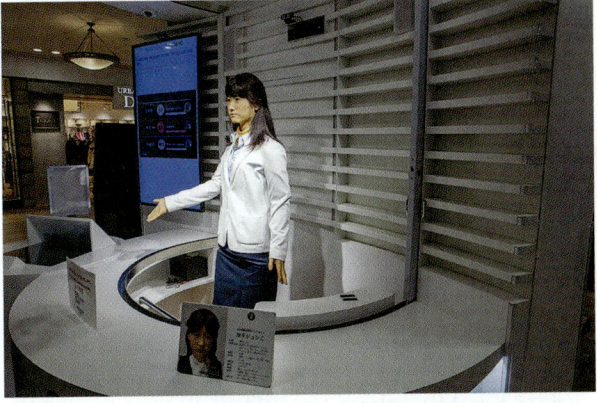

Yellow Cat/Shutterstock.com

Figure 1-1. Robot receptionists like this one in Japan may appear to act very lifelike, but they can do nothing without a human computer programmer telling them what to do.

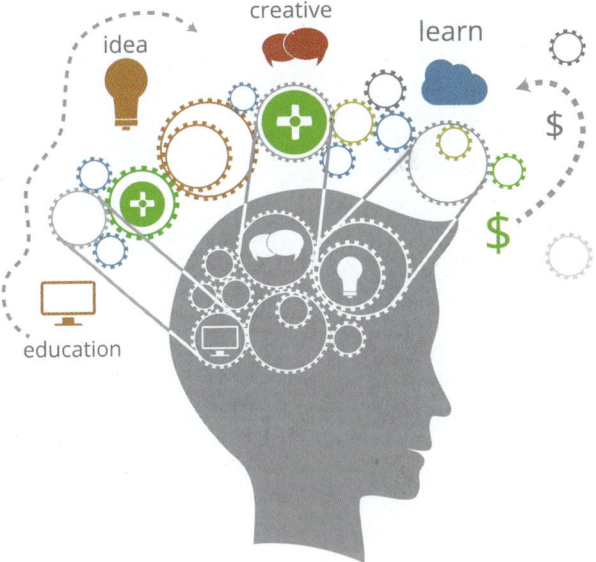

Marina Vector/Shutterstock.com

Figure 1-2. The human brain is capable of thought, assimilating past experiences, facts, and sensory information to form new ideas.

Science and Java

pH Algorithm

The lichen is an amazing plant. It is made up of two different types of plants: an alga and a fungus. They have a symbiotic relationship. The algal cells are photosynthetic and make food for the entire organism. The fungal partner protects the alga by retaining water and providing a large area for taking in mineral nutrients. Their close symbiotic relationship extends the ecological range of both partners. Lichens commonly grow on tree bark, as shown.

fibPhoto/Shutterstock.com

Certain species of lichens produce a dye called *litmus.* It was discovered in 1300 by a Spanish physician and then used more extensively in the Netherlands after the 16th century. The blue dye can be extracted from certain species found in South America, Madagascar, Norway, Sweden, and California.

Litmus has a useful property. It is made into paper that comes in two colors: red and blue. When a drop of a liquid is placed on litmus paper, the paper changes color. An acid, like vinegar, turns the blue litmus papers red. A base (or alkaline), like ammonia, turns the red litmus papers blue. Pure water turns both papers violet, which indicates a neutral solution. The same dye can be used to test solutions, as shown.

Litmus pH Indicator

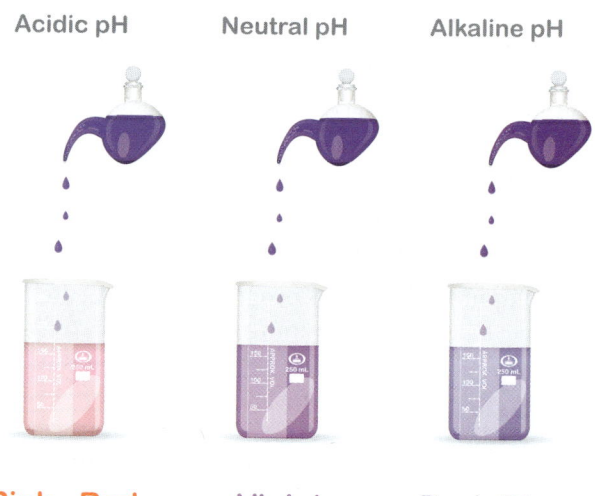

| Acidic pH | Neutral pH | Alkaline pH |

Pink - Red   **Violet**   **Dark Blue**

chromatos/Shutterstock.com

Assignment

You are presented with an unknown clear liquid. The problem is to determine if the liquid is an acid, base, or neutral solution. Your task is to write an algorithm to solve this problem using litmus paper.

Instruction	Action
MOV EAX, 100	Load the number 100 into the EAX register
SUB EAX, 5	Subtract 5 from the number in the EAX register
INC EAX	Add 1 to the number in the EAX register

Goodheart-Willcox Publisher

Figure 1-3. These are some of the instructions from the instruction set for an 8086 microprocessor.

Programming at this level is called *low-level programming* and uses assembly language. This textbook uses the Java programming language. Java is a high-level programming language. A *high-level programming language* is much easier for a human to read and write.

Actions of Computational Thinking

Mastering computational thinking is a key part of learning to be a great computer programmer. **Computational thinking** is expressing problems and solutions in a way that a computer can execute. There are four elementary actions in computational thinking:
- decompose
- recognize patterns
- generalize patterns
- write an algorithm

Each concept stands on its own as a life skill. Considered together, they lay the cornerstone of computer programming. These actions are subtasks needed to solve a problem.

Decompose

Any problem that requires a solution likely consists of several smaller problems. **Decompose** means to take apart. To decompose a large problem means to break it into smaller, doable steps. This is a skill that can be used for writing a computer program or solving a life problem. Decomposing also helps analyze the problem statement to be certain the programmer understands the problem that is being solved.

Suppose the problem is getting to school in the morning. Many people follow the same pattern every day. There are several smaller parts to the larger problem:

1. Wake up.
2. Clean up.
3. Get dressed.
4. Eat breakfast.
5. Go to school.

You can imagine that each of these smaller steps can be decomposed into even smaller parts. Wake up can be decomposed into:

1. Wake up to alarm.
2. Turn alarm off.
3. Stretch.
4. Get out of bed.

In terms of a *large-scale computational problem,* decomposing can be thought of as finding separate jobs that can be handed off to different people.

Consider the problem of maintaining statistics for major league baseball. One person could calculate the average number of runs per game. Another person could find the number of innings each player has played in the game in a season. Yet another person could calculate the ratio of hits to at-bats for each batter. Each person would use various productivity tools to gather authentic information for the application. In a computer program, these smaller jobs are handled by lines of instructions called *procedures* or *methods*.

Recognize Patterns

Problems that are good candidates for computer programs have patterns in the solution. **Pattern recognition** is the skill of identifying and isolating any repetitions. Repetitions can be inside the problem being solved. They also may be patterns recognized from earlier solutions. Either way, recognizing these patterns helps to create or reuse a solution.

Figure 1-4 shows a player taking a shot on goal in a soccer match. Suppose the problem is finding the average distance between a player and the goal line when a goal is scored. Finding an average distance to the goal is the same pattern as finding a grade average in math class. Recognizing this pattern helps you to see that individual data points need to be collected, the points totaled, and the total divided by the number of data points. Collecting the data points is a human action. Inputting them into the computer, totaling them, and dividing them are steps to use in a computer program.

Oleksandr Osipov/Shutterstock.com

Figure 1-4. This player is taking a shot on goal. Suppose the problem is finding the average distance between players and the goal line when a shot on goal is taken.

Generalize Patterns

Abstraction is the act of generalizing a pattern. It is taking specific patterns in a problem and relating them to other similar or identical patterns. For example, consider the following patterns.

- the date two days from today
- the number of people in a family after twins are born
- the next even number

These patterns can all be generalized with the algebraic expression x + 2. After all, that is the whole point of algebra, to generalize. Computers can solve problems if they can be generalized in an algebraic expression.

Once a pattern is identified and generalized, the solution of an equation is essentially the same for all algebraic equations: look at the equation and figure out how to isolate the variable. When you think of abstraction, think of algebra. One of the most-asked questions in algebra is, "When am I ever going to use this stuff?" Here is the answer: you will use it in computer programming.

In a *large-scale statistics* program, patterns can be identified. For example, sums of numbers are required, averages must be calculated, individual stats are tracked as well as team stats, and more. Generalizing these patterns into a sum method or an average method allows the problem-solver to reuse these methods for each specific case in the program.

Write an Algorithm

Once a problem is decomposed into small steps that a computer can perform, the computer program can be written. For example, to calculate an average of a set of data, the small steps are:

1. Enter the data.

FYI

Writing general expressions that can represent a specific situation and then applying the laws of algebra to find the value of the variable is the main purpose of algebra. Advanced algebra is called *analysis*.

2. Count the number of data points.
3. Total the data.
4. Divide the total by the number of data points.

After the problem is decomposed into smaller steps, then those steps can be translated into instructions for the computer to perform. Writing the program is the very last step in the process. Humans solve the problem. Computers calculate the result.

An **algorithm** is the sequence of steps used to solve a problem. Writing an algorithm is the last activity in computational thinking. The steps must be identified by decomposing the problem. The patterns must be recognized and then generalized using abstraction. Then abstractions must be organized into the proper sequence to solve the problem. This is the algorithm. The expression of an algorithm is called *pseudocode.* Pseudocode is discussed further in upcoming chapters.

Before there were computerized cash registers, cashiers had to make change using their own algorithms. The first step was to determine the change due the customer. They had to do subtraction in their heads! Then, they would compute how many quarters were needed, either one (.25 to .49), two (.50 to .74), or three (.75 to .99). The next step was the number of dimes, then nickels, then pennies, subtracting every time from the total due the customer. This same algorithm is used by computerized cash registers today.

Often, there are many algorithms that could solve a particular problem. One goal is to find the most efficient sequence of steps that produce the result required. The most efficient algorithm is often referred to as the *most elegant solution.* It is important to streamline the code. Extra steps take extra calculating time. While finding the most elegant solution may not make much difference for the programs you write in this course, programs that manage big data take many instructions. For those programs, wasted computer time means a longer wait for results.

FYI

Often a goal in making change is to use the fewest number of coins and bills. Certainly providing all pennies is not desired.

HANDS-ON EXAMPLE 1.1.1

Creating a Perfect Squares Algorithm

A perfect square is a number that has two equal factors. Perform the computational thinking required to find the first 10 perfect squares. The end result is the algorithm to solve the problem of finding perfect squares.

1. Analyze the problem to find the small, doable steps.
    1. Find out how to make a perfect square.
    2. Find the first perfect square.
    3. Find 10 of them.
2. Find the perfect squares in the following list. Identify the equal factors for those that are perfect squares.
    A. 16
    B. 10
    C. 49
    D. 5
3. Find the first perfect square. Answer: Start with the number 0.

$$0 \times 0 = 0$$

4. Abstract the process.

    Let x = 0. Its perfect square is x × x.

5. Make 10 of them.

> Add 1 to x, and find the next square. Keep going until x =10

6. Write the algorithm.

   1. let x = 0
   2. square = x × x
   3. add 1 to x
   4. if x = 10 stop; otherwise go to step 2

7. Follow the steps in the algorithm to verify the algorithm solves the problem. Check that there are 10 numbers generated. Check that each one is a perfect square. Check that no perfect squares were missed.

Try It!

Another method for generating perfect squares is to add the next odd number to the square. Notice that 0 + 1 = 1, 1 + 3 = 4, 4 + 5 = 9, and 9 + 7 = 16. Identify the pattern in these four equations. Use computational thinking to write an algorithm to generate the first 10 perfect squares.

Comparing Problem-Solving to Computational Thinking

Some problems do not require a computer calculation. Other problems are unsolvable. The process of problem-solving can be applied to any problem you encounter. Problem-solving is at its roots a perfectly human endeavor. Computational thinking is a subset of the actions of problem-solving. This section compares problem-solving and computational thinking.

General problem-solving can be used to answer questions. This is true whether or not a computer is used. Defining and specifying the purpose and goals of problem-solving involve steps or strategies like those for computational thinking, as shown in **Figure 1-5.**

1. Understand the problem statement.
2. Study the causes of the problem.
3. Brainstorm solutions.
4. Identify and implement one of the solutions.
5. Verify that the solution solved the problem.

Computational thinking results in an algorithm. Problems where the data can be numerically and textually analyzed are good candidates for computer solutions. Problems not solved by a computer are not considered in computational thinking. For example, a teacher makes a seating chart based on a variety

Problem-Solving

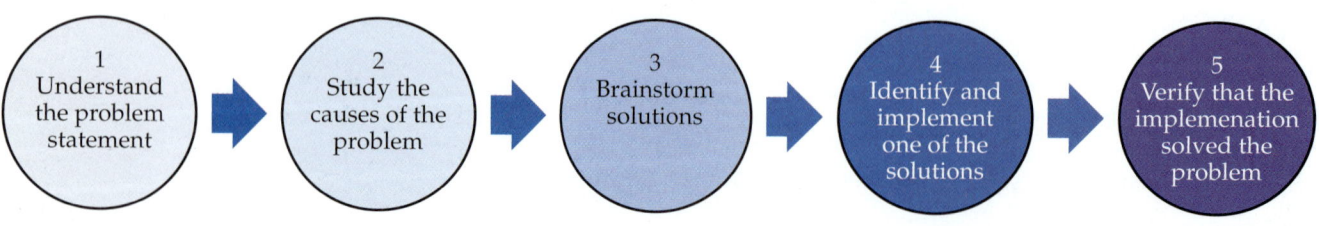

Goodheart-Willcox Publisher

Figure 1-5. Problem-solving is the process of choosing a course of action after assessing existing information and weighing the costs and benefits of different actions.

Math and Java

Sums of Powers Algorithm

Number theory is the branch of pure mathematics that studies the properties of the integers. Thinking about properties of integers can be recreational as well as performed by mathematicians. In the past, a common quest was to find formulas to calculate the sums of powers of integers. Because there were no computers to calculate these sums of powers, number theorists developed formulas to find sums of powers of integers indirectly. Bernoulli numbers were developed by Swiss mathematician Jakob Bernoulli in the late 17th century. These numbers were used in a complex formula to calculate the sums of powers of integers quickly; well, relatively quickly. It was still a labor-intensive calculation.

Today, computers can be used to solve this problem. A computer can calculate the sums of powers of integers directly and almost instantaneously without Bernoulli numbers. Examples of this task are shown below.

Example 1

This example shows the calculation of the sum of the first five powers of 3 directly.

Powers of 3	Sum of Powers of 3
$3^0 = 1$	1
$3^1 = 3$	4
$3^2 = 9$	13
$3^3 = 27$	40
$3^4 = 81$	121

Example 2

This example shows the calculation of the sum of the first four powers of 10.

Powers of 10	Sum of Powers of 10
$10^0 = 1$	1
$10^1 = 10$	11
$10^2 = 100$	111
$10^3 = 1000$	1111

Assignment 1

Apply what you have learned about computational thinking to write an algorithm for calculating the sums of powers of integers. Allow input of any integer. Calculate the sum of the first 100 powers of that integer. Follow the four actions of computational thinking.

Assignment 2

Study Example 2 above for calculating the sums of the powers of 10. Look for a pattern in the sums. Write a shortcut algorithm to find the sum of any number of powers of 10. Follow the four actions of computational thinking.

of conditions. The teacher thinks about accessibility issues, students who wear glasses, height of students, students with hearing deficiencies, chatty students, among other factors. It is not necessary to use a computer to make this seating chart, but it is indeed a complex problem to solve.

Computers can be programmed to follow rules for logistics or moving items from one place to another. They cannot be asked to solve world hunger or fix poverty in the United States. A human would have to solve the problem of world hunger. A computer could help by keeping track of the process and delivery. However, where to move items and how to move them are ultimately human decisions.

SECTION REVIEW 1.1

Check Your Understanding

1. Name three differences between the way humans think and computers think.
2. List the four actions for computational thinking.
3. Describe the process of abstraction.
4. What is an algorithm?
5. List the five steps or strategies to define and specify the purpose and goals in problem-solving.

Build Your Vocabulary

As you progress through this course, develop a personal computer science glossary. This will help you build your vocabulary and prepare you for a career. Write a definition for each of the following terms and add it to your computer science glossary.

abstraction   computational thinking   instruction set
algorithm   decompose   pattern recognition

THINK

IBM is the current version of a series of early computing companies and a world leader in computer software and hardware. It was created by Thomas J. Watson. Watson's famous slogan THINK has encouraged workers to imagine and create new ideas in computing since 1911. In 1992, IBM's Personal Computer Division based the name of the ThinkPad notebook on the slogan. An adaptation of the slogan is Apple's motto, "Think different." The THINK motto and its various forms acknowledges a fundamental difference between humans and computers. It is humans who must do the thinking for computers and then tell the computer what to do. This is the embodiment of computational thinking.

Software is the result of computer programing. Some people refer to the instructions in a computer programs as *code.* No matter how it is referred to, it is software that makes the computer do something. Many key advancements have been made by great thinkers that resulted in today's rich computing environment. This section follows the development of the programming side of computing.

Learning Goals

After completing this section, you will be able to:
- Describe the development of programming languages.
- Diagram the software-development lifecycle.
- List seven traits of successful programmers.

Terms

COBOL

Java

object

object-oriented coding

platform

software development lifecycle

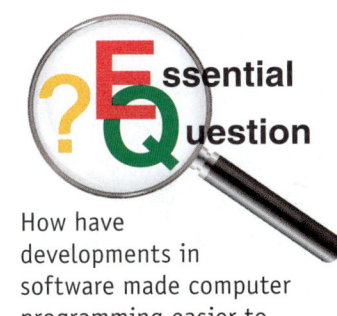

Essential Question

How have developments in software made computer programming easier to perform?

Brief History of Software

Many computing machines have been created over the years to perform a single task. Each of these early machines required the hardware be modified to change the task performed. It was not until the mid-19th century that a set of external instructions was written for a computing device. That is often marked as the beginning of computer programming. Incremental improvements advanced the endeavor to what is being used today to program computers.

Ada Lady Lovelace

In the mid-1800s, Charles Babbage designed what is considered to be the first computer with an instruction set. He called the machine the Analytical Engine. Ada Lady Lovelace prepared a program of instructions for this device to calculate Bernoulli Numbers. Sadly, machine tools in the 19th century were not able to mill the gears of this machine precisely enough. The engine was not completed at that time.

A century later, manufacturing technology had sufficiently advanced. Engineers followed Babbage's design and produced a working version of the Analytical Engine. The code written by Lovelace was used. The engine ran perfectly. Many name Lovelace as the first computer programmer.

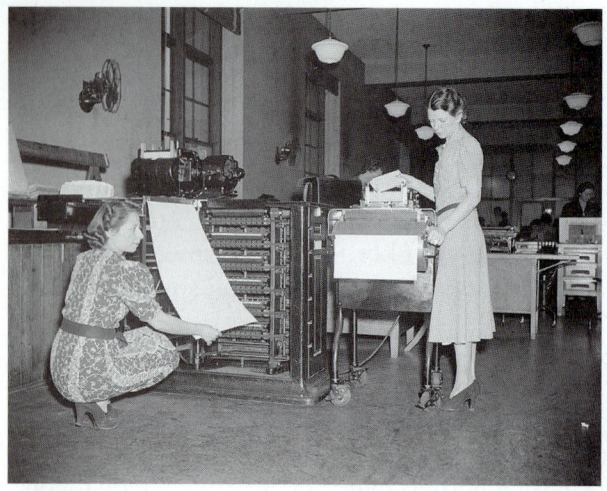

Everett Historical/Shutterstock.com

Figure 1-6. In this 1940 photograph of the Census Bureau, Herman Hollerith's "unit tabulator" can be seen on the left. This machine processed census information on punched cards at the rate of 4800 statistics per minute.

FYI

The scientific community needed a more mathematical language than COBOL to solve its problems. FORTRAN, named for formula translator, was created to fit this need.

Herman Hollerith

The 1880 US census resulted in a great number of responses. There were so many that it would have been impossible to analyze all responses before the 1890 census was to be taken. Herman Hollerith had a solution. He devised a tabulating machine that read punched cards. He programmed this device to collect the data from the census and sort it in just three days. It was so effective that it was used well into the 1950s when computers replaced the tabulating machine. **Figure 1-6** shows some of the tabulators used at the Census Bureau in 1940.

Hollerith's device was a programmed device. However, it was not programmed with software. The process was hard-wired into the tabulating machine.

Grace Hopper

In the 1940s, Grace Murray Hopper and other pioneering women were the programmers on the first digital computers. Software programming at that time involved rewiring the computer. These wires were organized into hardware units called *wireframes.* Hopper showed how the programs could be changed by rearranging the wires in the program section of the computer. She also noticed that the code could be relocated to different parts of the wireframes. The robust computer languages used today have their origins in the wireframes of Grace Hopper's era.

Hopper realized that not everyone would be able to reprogram a computer by rewiring it. She and her team created COBOL and compilers in 1959 to make it easier to program computers. Her attempt was to create a language that could be used on many different types of computers. COBOL stands for common business oriented language. As its name suggests, **COBOL** was a programming language written for business applications, such as collecting data and writing reports. It was one of the first programming languages that did not require the use of binary numbers for instructions. Programmers wrote human-readable instructions using COBOL. Another piece of software called a *compiler* translated the COBOL code into the machine-readable instructions. COBOL was originally seen as a stopgap measure, but the Department of Defense promptly forced computer manufacturers to provide it, resulting in its widespread adoption.

Hopper is shown in **Figure 1-7** working on a UNIVAC system in the early 1960s. She was a member of the United States Navy and reached the rank of Rear Admiral.

Computer Scientists

Each type of computer has its own unique instruction set. Each new programming language was written to create programs for a specific computer. Many talented computer programmers wrote their own programming languages to suit the work they were doing. Most of these languages were cryptic and very hard to use. Over time, the number of computer applications increased rapidly, and languages were created to suit the applications. With the rapid growth of the computing industry, programmers were concerned with getting the project to work. They were not concerned with the readability or maintenance of their code. Another programmer looking at the code rarely could follow the logic.

Figure 1-7. Grace Murray Hopper was an early pioneer in computer science. One of her contributions was helping design the UNIVAC I system. She is shown here with her team working on the UNIVAC.

As applications became very large, the development of applications took the collaboration of many people. The programming process became error-prone and confusing. A crisis developed in the computer world. Errors were causing loss of money and eventually loss of life. In the case of the "Patriot missile error," a rounding error caused an inaccurate quantity that ended up causing the death of US service personnel.

With the increased use of computers, computer science was created as a field of study. The first computer science degree program in the United States was created at Purdue University in 1962. Computer scientists have created many programming languages over the years. These were designed to reduce errors and provide new functionality. Programming procedures have also been developed to improve readability and accuracy of the code. The field of software development emerged to manage the problems created by lack of communication and management of these projects.

A report on the "Patriot missile error" can be found on the US Government Accountability website (www.gao.gov). Search for report number IMTEC-92-26.

Software Development

Software development is the process of identifying a need, specifying the components for a solution, formulating a design, writing the programs, developing documentation, writing test plans, executing the test plans, and repairing anomalies related to creating and maintaining software applications, data collection, or other software components. Notice that coding, or writing the programs, is only one of a series of steps in the process. Major considerations in software development involve the overarching demands of the project and the people who make up the software development team.

Software Development Lifecycle

The software development lifecycle emerged as the guiding principle for software development. The **software development lifecycle** describes how a computer program moves from concept through development, release, maintenance, and eventual removal. This cycle acknowledges that development is not over once a program is written and placed into production, as shown in **Figure 1-8.** Over time, modifications may be required. When this occurs, the process of software development is repeated to create a new version. The cycle may be repeated several times before the end of the program's life.

The accuracy of the original code must be maintained throughout these modifications. The original programmer likely will not make the revisions. Therefore, communication between team members is critical. A maxim developed to "write the program for the next programmer." This means using self-documenting names for data. It also means including comments in the code to guide the next programmer.

Software Platform

A big problem in software development is that each version of a computer has its own instruction set. These instructions will not match those of a different computer. Each computer environment is called a **platform.** Programs needed to be written in different languages to suit each platform. This was a headache of major proportions. Often a different programming team was used for each platform. Keeping the software functions similar over each platform took a massive coordination and communication effort. Programmers sought a cross-platform approach to coding: write it once and run it everywhere. Object-oriented programming emerged to meet this need.

There are two basic types of software: system software and application software. System software works to help the CPU find programs, assign memory, run devices, and provide utility programs. Application software is the software that performs the user's work, such as writing term papers, sending e-mail, paying taxes, editing photos, playing games, and taking online courses. There are five general types of application software: productivity, entertainment, educational, utility, and development. The type of application software you will use in this course to help you write Java programs is development software.

Object-Oriented Programming

With many people working on the same project, the code and data were often not secure. This meant the code and data were open to use by the entire team, but they could also be misused. To solve this problem, both the code and

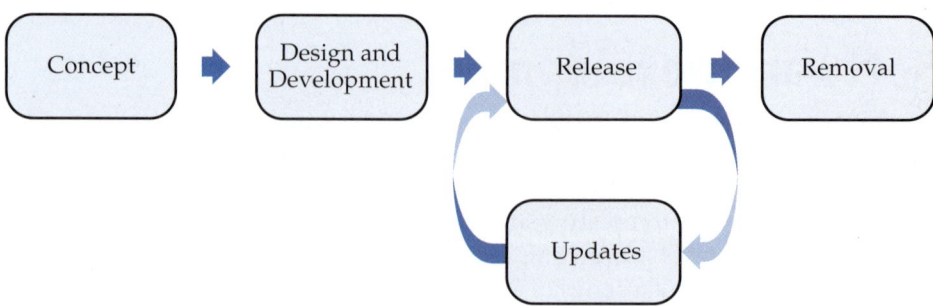

Goodheart-Willcox Publisher

Figure 1-8. The software development lifecycle includes not only creating the software, but maintaining it over time.

Language Arts and Java

Algorithm Exposition

You are going with a group of students on a cruise of the Baltic Sea. You will stop in many of the countries from Copenhagen, Denmark, heading counterclockwise eventually to Norway. These countries use different forms of currency. The expression of an algorithm is called pseudocode. The following is the pseudocode for a program that tells you what form of currency is used in each country.

```
/* Application that tells what currency is used when a country name is clicked by the user
Student Name
Date
*/
//Set up the program
Display a list of countries that can be clicked with a mouse.
Code a message box to display the name of the currency.
// Output the results
If Denmark is clicked
   Then change the message to Danish krone
      Else if Germany is clicked
         Then change the message to euro
            Else if Poland is clicked
               Then change the message to Polish zloty
                  Else if Estonia is clicked
                     Then change the message to euro
                        Else if Russia (St. Petersburg) is clicked
                           Then change the message to ruble
                              Else if Finland is clicked
                                 Then change the message to euro
                                    Else if Sweden is clicked
                                       Then change the message to Swedish krona
                                          Else if Norway is clicked
                                             Then change the message to Norwegian krone
```

Claudio Divizia/Shutterstock.com

Assignment

Examine the pseudocode shown above. Describe this algorithm in paragraph form. Assume the reader is not a computer programmer. Write in complete sentences, and follow grammar rules for Standard English. Be sure to introduce the topic.

FYI

In the 1990s, interactive television was a promising new technology. Java was invented for use in the digital cable industry. However, the industry was not up to the challenge, and Java found many other homes.

data were encapsulated into software units named objects. An **object** is a self-contained unit of data and code to manage that data. Using objects helps protect the data for large scale programs written by a large team of different people. Most of today's programmers work in object-oriented coding environments. **Object-oriented coding** is writing code in a programming language that encapsulates code and data into objects. Examples of modern programming languages that support object orientation are Java, C++, C#, and Python.

Java is an object-oriented computer programming language with features that minimize confusion and errors and allow programmers to produce code that operates on a wide variety of platforms. It was created in the 1990s under the leadership of James Gosling at Sun Microsystems. The language was named Oak during development, but was renamed Java, which is another name for coffee. The logo used with Java appears like a cup of coffee. Java is the computer programming language used in this textbook.

Characteristics of Successful Programmers

Many people are drawn to computer programming because of its creative nature. Damien Filiatraut is the CEO of Scalable Path. He conducted a study of more than one thousand computer programmers. He noticed seven traits that great programmers have, as shown in **Figure 1-9.** These are very important traits. Employers are eager to find people with a strong work ethic with traits like Filiatraut lists. In addition, there are three more qualities that should be mentioned: basic beliefs, computational thinkers, and relentless checkers.

- Positive attitude or "can do" approach to problem-solving
- Great communicator who can understand the problem, decompose it, and communicate solutions to the development team
- Great time- and task-manager; gets work accomplished on time
- Quick learner and adapts to new technologies with ease
- Broad and rich technical experiences
- Works well with a team
- Puts importance on the end user, and incorporates a superior user experience in all code

nd3000/Shutterstock.com

Figure 1-9. A survey of programmers found great programmers have these seven characteristics.

Basic Beliefs

A notable theory of problem-solving holds that the bulk of a person's ability to solve a problem rests in two fundamental beliefs:

- the problem has a solution
- he or she can find the solution

If a problem-solver wavers in either of these two beliefs, he or she gives up way too soon. Successful computer programmers stick to the problem. They try a variety of strategies to decompose and develop an algorithm.

On a recent tour of a game company, the tour guide mentioned that visitors often walk by cubicles and see programmers sitting perfectly still, staring at the computer display. The guide pointed out that they are quite busy. They are thinking. They are trying to find the solution to a problem.

Computational Thinkers

Good programmers understand that computers "think" differently from humans. They are good computational thinkers. Programmers spend time looking for tasks a computer can do to solve a stated problem. All computer programs are a series of small steps. Even advanced computer simulations simply complete small steps and follow the rules the programmer set up to model a condition or process.

Good programmers notice patterns in everything they do. They are good at writing expressions that describe a problem. They are great at logical thinking. They are good at finding relationships between things and making connections that are key to the solutions. Many programmers are also very musical. The ability to recognize and follow patterns in music can transfer to programming.

Relentless Checkers

Good programmers check, recheck, and double-check. It is easy to lose track of the fine points of a problem statement during the computational-thinking process. Programmers need to go back and verify that they actually solved the problem correctly and completely. They check the algorithm. They check each line of code. They check input. They check output. They verify that the problem presented was solved.

An entire segment of computer science is dedicated to the design, development, and deployment of the best software possible. This sector is known as *software quality assurance.* Those working in this sector are involved with the project from the decomposition stage though the final testing. They check every aspect of the program to ensure an excellent product.

Coding Conundrum

This coding conundrum feature is your chance to check the results to identify an error in the process. Use computational thinking to find the mishap in the following example.

Problem: Generate the first 10 natural numbers.

Decompose: Natural numbers are the positive counting numbers: 1, 2, 3 ...

1. Start at the first natural number.
2. Add one to the first natural number to get to the second.
3. Keep counting until the first 10 natural numbers are shown.

Pattern: Each natural number is 1 more than the prior natural number.

Abstract: Use n + 1 as the general notation for the next natural number where n is the current natural number.

Algorithm:

1. Let n = 0. This is the first natural number.
2. Show the natural number.
3. Ask if we have 10 numbers?

    A. Yes, stop.
    B. No, Let n = n + 1. Go to step 2.

Find the error in this computational thinking.

HANDS-ON EXAMPLE 1.2.1

Exploring Software Programming Languages

Computer software programming has a long history. In this activity, you will explore some of this history and report on your findings.

1. Launch a web browser, and navigate to a search engine. Enter the search string: history of programming languages.
2. Evaluate the validity of the search results, and select a website to visit.
3. How would you characterize the number of programming languages that have been developed: a few, several, or many?
4. Select one of the languages listed in the article, and find its date of creation, target platform if available, and developer.
5. Identify five major advances in computing languages from 1842 to present day. Visit other websites if necessary to find additional information.
6. Prepare to share your results with the class.

Try It!

Research the early computer language FLOW-MATIC. Locate information on its developer, date of creation, target platform, and uses. Find at least one advantage and one disadvantage of this language.

SECTION REVIEW 1.2

Check Your Understanding

1. Why was Ada Lady Lovelace's programming code not used in the 19th century?
2. Of what does the software development lifecycle consist?
3. Describe cross-platform development.
4. What are the four general types of application software?
5. Which sector of computer science is dedicated to the design, development, and deployment of the best software possible?

Build Your Vocabulary

As you progress through this course, develop a personal computer science glossary. This will help you build your vocabulary and prepare you for a career. Write a definition for each of the following terms and add it to your computer science glossary.

COBOL   object   platform
Java   object-oriented coding   software development lifecycle

Cooperative Coding

Determine a Win in Tic-Tac-Toe

Almost everyone has played the classic game *Tic-Tac-Toe*, as shown. In other cultures, this game may be known as *Noughts and Crosses.* The rules are:

- Two players use a grid like a hashtag or pound sign (#).
- One player uses X and the other player uses O.
- Players must agree on who goes first.
- Players take turns placing their letter in a cell.
- The first player to have three letters in a row horizontally, vertically, or diagonally wins the game; the game is over.
- If all the cells are full and neither player has three in a row, there is no winner; the game is over.

Golubovystock/Shutterstock.com

A game is over when someone has played three in a row or all cells are filled with Xs or Os.

Not every game ends in a win. When the game ends with no winner, it is called a draw.

An X or O can only be played in an empty cell. It is easy for humans to determine how to play an X or O, and it is easy for humans to see when a player has won. They use their sense of sight to see three Xs or Os in a row. A computer cannot "see" without a massive program for machine vision. Self-driving cars have this type of setup. How can a computer figure out who won? An algorithm is needed for a computer to judge whether there is a win or if the game is a draw. Computational thinkers must take several things into consideration. Some questions to consider:

- What constitutes a win?
- How does the computer know someone won?
- What are the smaller steps in the problem?
- Is there a pattern to consider that can be used for X play versus O play?
- Is there an algebraic equation to express a winning pattern?
- Does every winning path have to be checked every time?
- What is the most efficient algorithm?

Assignment

Work with the team assigned by your instructor. Go through the four steps of computational thinking to generate an algorithm that checks for a win. Be ready to share your algorithm with the rest of the class. Select a team member to explain the algorithm and the decisions the team made.

1. What is your team's algorithm?
2. What decisions did you find your team had to make?

Chapter Summary

Section 1.1: Computational Thinking

- Computers cannot create original thought as humans can, and to program a computer a human must think in the same way a computer processes information.
- Computational thinking is expressing actions in a way a computer can execute and consists of decomposing, recognizing patterns, generalizing patterns, and writing an algorithm.
- Computational thinking is a subset of the problem-solving process, which involves understanding the problem statement, studying the causes of the problem, brainstorming solutions, identifying and implementing a solution, and verifying the solution solved the problem.

Section 1.2: THINK

- The first external instructions for a computing device were written in the mid-19th century, which is often marked as the beginning of computer programming, and software was developed through the years by individuals such as Ada Lovelace, Herman Hollerith, Grace Hopper, and today's computer scientists.
- Software has a lifecycle that begins with the original concept and continues through design and development, release and all following upgrades, and removal or retirement; most software written today is based in an object-oriented programming language to allow cross-platform compatibility.
- Successful programmers have a positive attitude, are great communicators, are good time managers, are quick learners, have broad technical experience, work well with a team, and put importance on the end user as well as have the basic beliefs a problem has a solution that can be found, are computational thinkers, and are relentless checkers.

Chapter 1 Test

Multiple Choice

Select the best response.

1. Which type of processing can only be done by humans?
   A. artificial intelligence
   B. original thought
   C. robotics
   D. converting number systems

2. All of the following are characteristics of computers except _____.
   A. storing numbers in memory locations
   B. calculating very quickly
   C. encoding text data
   D. writing algorithms

3. What was the primary mechanism used by Herman Hollerith to analyze the census?
   A. smartphones
   B. on-off switches
   C. Intel processors
   D. punched cards

4. One of the first programming languages developed that was able to be easily read by humans was:
   A. COBOL
   B. assembly language
   C. C++
   D. Python

5. Which event is *not* a major phase of the software development lifecycle?
   A. research
   B. compilation
   C. maintenance
   D. deployment

Completion

Complete the following sentences with the correct word(s).

6. Computers can solve problems if they can be generalized in a(n) _____ expression.

7. Analyzing large-scale computational problems requires identifying generalizable _____.

8. One goal of a(n) _____ is to define the most efficient sequence of steps for solving a problem.

9. _____ led a team of programmers to create COBOL.

10. Writing a program that runs on many types of computers is described as the cross-_____ approach.

Matching

Match the correct term with its definition.

    A. decomposition

    B. pattern recognition

    C. algorithm

    D. abstraction

    E. computational thinking

11. Steps to follow when solving a problem.

12. Finding repeating schemes in problem-solving.

13. Skill needed to write a successful program.

14. Breaking apart a problem into small parts.

15. Generalization of patterns; algebra.

Application and Extension of Knowledge

1. The software development lifecycle can be represented in a diagram, as shown in **Figure 1-8.**

    A. Concept

    B. Design

    C. Development

    D. Release

    E. Updates

    F. Removal

Research the Agile software development model. What does *release* mean? What does *monitor* mean? Make a diagram of the Agile model including the meanings of those two terms.

2. Assume you are in a room with 10 other people. Everyone must shake hands with every person, but only once each. Use decomposition, pattern recognition, and an algorithm to solve the following problems. How many times will you shake hands? How many handshakes will occur in the room? No matter how many people are in the room, how can you express this problem algebraically? Use n to equal the number of people and h to equal the total number of handshakes.

3. Apply what you learned in Section 1.1 about writing algorithms. Write an algorithm for making change where the goal is to return as few coins and bills as possible to the customer. Apply the technique of counting up from the amount paid to the amount tendered. Provide an example of your algorithm for making change for $5 when the amount paid is $3.42.

4. The Florida manatee is a native species found in many of the state's waterways. Various research, management, and educational efforts have been undertaken to bring back this species that was on the verge of extinction. Due to these efforts, the Florida manatee population has grown to more than 6,600 animals. Manatees are considered one of Florida's keystone species whose behavior can alert researchers to the environmental and habitat changes that may otherwise go unnoticed in Florida's waterways. The following figure is a partial list of counties from the Florida Fish and Wildlife Conservation Commission that details the causes of mortality of manatees in 2016. Use various productivity tools to gather authentic data related to manatee populations. In paragraph form, write an algorithm that reads the data and identifies the county with the largest number of fatalities of manatees.

County	Watercraft	Other Human	Perinatal	Cold Stress	Natural	Undetermined	Unrecovered	Total
Bay	0	0	0	0	0	0	1	1
Brevard	10	1	45	4	3	41	6	110
Broward	5	0	1	0	3	9	0	18
Charlotte	2	0	2	0	11	5	0	20
Citrus	3	0	6	1	3	3	0	16

5. A common quest in mathematics is to find the greatest common divisors of two integers. The greatest common divisor (GCD) is the largest integer that divides two other integers with no remainder. For example, the GCD of 28 and 35 is 7. This calculation is used in programming a computer to simplify fractions. Use computational thinking to devise an algorithm for finding the greatest common divisor of two integers. Remember that a computer will not have the insight gained by looking at the numbers first as a human can do. This must be a general algorithm to work on any two integers. There are several approaches to this problem. After you have found your original algorithm, look up Euclid's Algorithm for the Greatest Common Divisor. Try your algorithm and Euclid's algorithm on these two integers: 168 and 32. Which algorithm is more efficient? Euclid is considered one of the three best mathematicians of all time, so do not feel bad if your algorithm takes more steps. As long as it gets the job done, it is a successful algorithm.

Online Activities

Complete the following activities, which will help you learn, practice, and expand your knowledge and skills.

Vocabulary. Practice vocabulary for this chapter using the e-flash cards, matching activity, and vocabulary game until you are able to recognize their meanings.

Communication Skills

Reading. *Sight words* are those words you recognize just by seeing them. They are words that appear on almost every page of text. Examples of sight words are *the*, *he*, and *me*. Identify 10 sight words in this chapter.

Writing. *Note taking* is the process of writing key information from a lecture, text, or other source on paper or a digital device. Taking notes can help you recall information, and the notes can serve as a resource when studying. Reread this chapter and take notes on the content. When you come to the end of a section, write a brief summary in your own words. To summarize, identify the most important ideas in the material and retell them in your own words. Be selective about what you include in your notes.

Speaking. Effective speaking requires individuals to use correct pronunciation. This is especially true when using words you learned recently because you may be less familiar with how the word is pronounced. Identify three key terms you learned in this chapter. Practice pronouncing them by saying them aloud to yourself until you can pronounce each one correctly.

Listening. *Hearing* is a physical process. *Listening* combines hearing with evaluation. Listen carefully to your instructor as a lesson is presented. Pay attention to the sounds and patterns of the words used. How would you rate your listening skills?

Portfolio Development

College and Career Readiness

Portfolio Overview. When you apply for a job, community service, or college, you will need to tell others why you are qualified for the position. To support your qualifications, you will need to create a portfolio. A *portfolio* is a selection of related materials that you collect and organize to show your qualifications, skills, and talents to support a career or personal goal. For example, a certificate that shows you have completed lifeguard and first-aid training could help you get a job at a local pool as a lifeguard. An essay you wrote about protecting native plants could show that you are serious about ecofriendly efforts and help you get a volunteer position at a park. A transcript of your school grades could help show that you are qualified for college. A portfolio is a *living document*, which means it should be reviewed and updated on a regular basis.

Artists and other communication professionals have historically presented portfolios of their creative work when seeking jobs or admission to educational institutions. However, portfolios are now used in many professions.

Two types of portfolios commonly used are print portfolios and digital portfolios. A digital portfolio may also be called an *e-portfolio.*

1. Use the Internet to search for *print portfolio* and *digital portfolio.* Read articles about each type of portfolio.
2. In your own words, compare and contrast a print portfolio with a digital one.

CTSOs

Student Organizations. Career and technical student organizations (CTSOs) are national student organizations, with local school chapters, that are related to career and technical education (CTE) courses. There is a variety of organizations from which to select, depending on the goals of your educational program. CTSOs are a valuable asset to any educational program. These organizations support student learning and the application of the skills learned in real-world situations.

Competitive events sponsored by CTSOs recognize outstanding student performance. Competing in various events enables students to show mastery of specific content. These events also measure the use of decision-making, problem solving, and leadership skills. Competitive events may be written, oral, or a combination of both. To prepare for any competitive event, complete the following activities.

1. Go to the website of your organization to find specific information for the events. Visit the site often as information changes quickly. If the organization has an app, download it to your digital device.
2. Read all the organization's guidelines closely. These rules and regulations must be strictly followed or disqualification can occur.
3. Communication plays a role in all the competitive events, so read which communication skills are covered in the event you select. Research and preparation are important keys to a successful competition.
4. Select one or two events that are of interest to you. Print the information for the events and discuss your interest with your instructor.

Encoding

Sections

The physical parts of a computer are called the hardware. A computer with its attached devices is called computer system. Most computer systems have additional hardware components to provide extended functions. These may be attached externally or internally. Each hardware component has a specific job in digitizing, displaying, storing, and retrieving information. Computer systems have very few moving parts.

Computer systems require specific directions to function. These directions or instructions are known as the software. These instructions are very specific to the type of computer. This presents a problem for software developers who want to write programs for a variety of computers. However, as you learned in Chapter 1, there are software programs that are cross-platform. Java is one example.

Software programs must be encoded for the computer to use them. Knowledge of number systems is essential to understanding how computers encode data. Binary, decimal, and hexadecimal notation contain the language that hardware uses. Text and decimal numbers need to be encoded into binary so the computer will understand them. Fortunately, programmers have software tools that perform that translation.

Reading Prep

A table of contents is a list of each section in a book and the page where it can be found. Before reading this chapter, review the table of contents for this text. Trace the content from simple to complex ideas. What does this tell you about what you will be learning?

College and Career Readiness

While studying, look for the activity icon for:

- Vocabulary terms with e-flash cards and matching activities.
- Starter files for hands-on examples and other exercises.

These activities can be accessed at
www.g-wlearning.com/informationtechnology/1773

Chapter Glossary

base: How many numbers are used in a number system.

binary number system: Positional number system that has a base of two.

bytecode: Machine language–like instructions designed to be interpreted at runtime.

central processing unit (CPU): Fetches coded instructions, decodes them, and then runs or executes them.

class: Contains the instructions for creating objects.

compiled: Conversion of code written in a high-level programming language into machine code or bytecodes *before* it is run.

hexadecimal notation: Positional number system that has a base of 16.

high-level language: Programming language that has a limited set of recognizable English words.

input: Data that is typed, scanned, or otherwise sent to a computer system.

instantiation: Creation of an object from a class.

interpreted: Conversion of code written in a high-level programming language bytecode into machine code *as* it is run.

Java Virtual Machine: Software that interprets Java bytecode into machine code at run time.

machine language: Code that the CPU uses consisting of 0s and 1s.

memory: The part of the computer that stores information for immediate processing.

method: Behaviors of an object.

operating system (OS): Specific set of software persistently stored on the computer that manages all of the devices as well as locates instructions and provides them to the CPU.

output: Data that has been digitized into a useful format and provided to the user.

peripheral: External computer device not critical to basic operation.

runtime engine: Software that interprets bytecode into machine code.

storage: Where information is kept by the computer so that it can be saved to view, play, or be reused.

syntax: Rules of grammar that specify how code is to be written in a programming language.

Unicode encoding system: Encoding standard for representing international text characters.

user interface (UI): How the user enters data.

Hardware

Hardware components of a computer are easy to visualize because all the parts are visible. The keyboard and monitor are the most obvious. Opening the computer case reveals the other parts. Four components are necessary for hardware to be categorized as a computer. The system must contain a device that allows the user to input data in programs. It must have a means to output data. It must include a mechanism to process the data between the input and output. It also must have a mechanism to store information. If one of these components is missing, the device is not considered a computer system. All four of these components are present in computers from smartphones all the way to supercomputers.

Understanding the operation of a computer makes users more effective. In this section, you will study how information is entered, how it is stored, how it is processed, and how it is displayed.

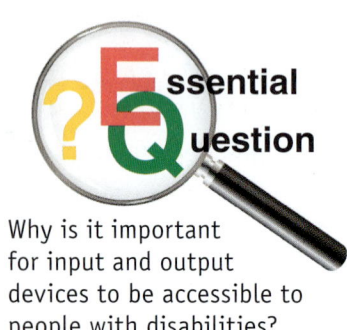

ssential Question

Why is it important for input and output devices to be accessible to people with disabilities?

Learning Goals

After completing this section, you will be able to:
- List input devices and their functions.
- Identify output devices and their functions.
- Differentiate between primary and secondary memory.
- Describe the role of the central processing unit (CPU).

Terms

central processing unit (CPU)
input
memory
output
peripheral
storage
user interface (UI)

Input

The four major components of a computer system are input, memory, processing, and output. *Input* can be described as data that is entered, scanned, or otherwise sent to a computer system. The data can originate from a person, the environment, or another computer. Using a keyboard, mouse, and touchpad are common methods of inputting data, as shown in **Figure 2-1.** External computer devices not critical to basic operation are called *peripherals.* For example, sensors can detect changes in the environment, such as an increase in temperature. This data can be sent to a computer system that regulates the heat of a room. Messages in the form of e-mails or tweets are also types of input. These are data sent to the computer.

Input devices make it possible for the user to communicate with the computer. Many devices are available to the user at one time. The set of choices the user has for entering data and receiving output is called the *user interface (UI).* UI is a general term used for this. The user interface in Java is called the *integrated development environment (IDE).*

The most common input device is the keyboard. Pointing devices, such as an optical mouse, mechanical mouse, trackball, and touchpad, are also input devices. There are also touchscreen monitors, scanners, cameras, microphones, and special input devices for users with disabilities.

Output

Output is data that has been digitized into a useful format and provided to the user. The most familiar form of output is the video display on a computer monitor, as shown in **Figure 2-1.** Other forms of output are printed pages; recorded words, music, and video; and 3D-printed objects.

The most common type of flat-panel monitor is a liquid crystal display (LCD). They are light, inexpensive, and energy-efficient. A newer type of monitor is based on light-emitting diode (LED) technology. These monitors are even more efficient and have better color representation than LCD monitors.

A data projector is similar to a monitor. It collects video data from a computer or other media player and projects the images onto a large screen. Data projectors are most commonly used in educational and business settings. They are used to show images, lessons, and presentations.

The printer is a popular output device for personal computers. Two types of printers are currently widespread: inkjet and laser. These are used to create printed pages or images. Another type of printer that is becoming more common is the 3D printer. This printer creates a physical object from plastic or other materials.

All personal computers come equipped with internal speakers. The computer system uses sounds to alert the user about the status of the system. The system comes with an audio jack in the unit to allow the user to plug in headphones or earbuds.

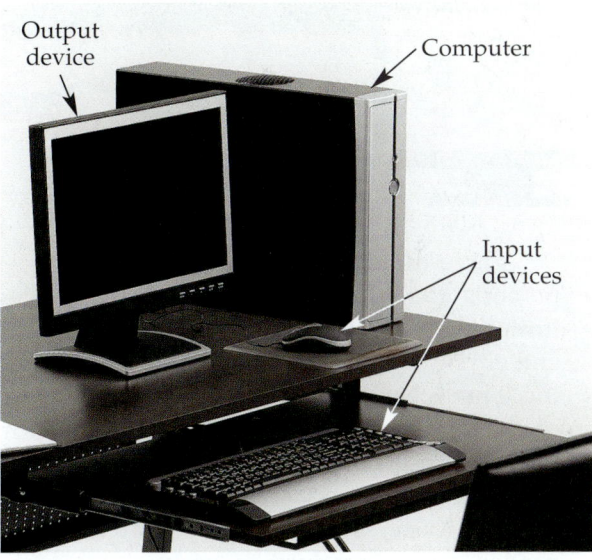

Output device — Computer — Input devices

Venus Angel/Shutterstock.com

Figure 2-1. A mouse and keyboard are common computer input devices. A monitor is a common computer output device.

Storage

Storage is where information is kept by the computer so that it can be saved to view, play, or be reused. Storage may be inside the computer or on external devices. The most familiar storage devices are the computer system's memory and hard disk drive. Thumb or flash drives and other forms of external storage devices are also common, as shown in **Figure 2-2.**

Storage devices are considered *secondary memory.* They are needed to keep semipermanent copies of the user's data in digital format for reuse. The most popular types of storage media are hard drives, USB drives (such as flash drives), CDs, and DVDs. The cloud is another popular way to store data.

Broewnis Photo/Shutterstock.com

Figure 2-2. External storage devices like flash drives are common computer peripherals.

Another type of storage is called memory. *Memory* is the part of the computer that stores information for immediate processing. This is considered *primary memory.* There are two types of primary memory.

- Memory that can be changed is called *random-access memory (RAM).*
- Memory that cannot be changed is called *read-only memory (ROM).*

The results of any processor action can be saved in a RAM location.

Science and Java

3D Printing

Three-dimensional (3D) printing has the potential to revolutionize manufacturing. It was invented in the 1980s by Chuck Hull, an engineer and physicist. It is a type of additive manufacturing. That is because the process deposits material one thin layer at a time. Three-dimensional printing allows users to design and produce complex products in new ways. It can also help reduce material waste, save energy, and shorten the time needed to bring products to market.

The process starts by creating a 3D blueprint, or model, using computer-aided design software. The 3D model is then converted into a format the 3D printer software can manage. The printer uses raw materials such as plastics, metal powders, or binding solutions to create the object. When the operator hits the print button, the machine takes over, automatically building the desired object, as shown.

Monkey Business Images/Shutterstock.com

Material extrusion is the most common process used in desktop 3D printers. The printing material, most often a plastic filament, is heated until it melts. Then it is pushed through the print nozzle. Using information from the digital file, the design is split into cross sections. The nozzle deposits the polymer in thin layers, often no more than 0.1 millimeter thick. The polymer solidifies quickly, attaching to the layer below before the print head adds another layer. Depending on the size and complexity of the object, the entire process can take anywhere from minutes to days.

As the price of desktop 3D printers continues to drop, some innovators are experimenting with different materials like chocolate and other food items, wax, ceramics, and biomaterial like human cells. Though the possibilities for additive manufacturing are endless, today 3D printing is mostly used to build small, relatively costly components using plastics and metal powders. When new or replacement parts are required on the International Space Station, it is not easy to send one up. Sending the print instructions allows astronauts to print the part on demand.

Three-dimensional printers have been used to manufacture robots, prosthetic limbs, custom shoes, and musical instruments. Some medical researchers have begun creating human organs. There are applications in veterinary medicine as well. In one example, Dr. Michelle Oblak, a veterinarian at the Ontario Veterinary College, used a 3D-printed titanium implant to treat a dog that had part of her skull removed in order to treat a tumor.

Assignment

Conduct a search using the Internet for 3D printing innovations. Evaluate the validity of your search results. Identify at least five innovative items that have been created using 3D printing. Be ready to present a description of each item to your class.

Computers use an on-off pattern of signals to work with data. The on-off pattern can be set by optical and magnetic means as well as electronically. The types of storage are magnetic, optical, and solid state. Flash drives use solid-state technology and come in a variety of forms. They are small, extremely portable, and can plug into a USB port. Hard drives may use magnetic or solid-state technology. CD and DVD drives use optical technology.

When the computer is turned off, all data and instructions stored in RAM disappear.

Coding Conundrum

A task required many people many hours to manually input data from stacks and stacks of paper forms into a computer program. The programmer wanted to save the data to a safe place, so it was stored in RAM. The computer display verified the data was in the computer. To the dismay of everyone, when the program was reloaded the next day to view the data, all data was gone. It is a conundrum! Can you spot the error?

Processing

Processing of the data takes place in between the input and the output. The hardware where this takes place is in electronic parts that cannot be seen unless the user looks inside the computer's metal or plastic case. This case also houses other important pieces of hardware, such as an electrical power source and fan to keep the unit cool.

The actual transformation of keystrokes on a keyboard or clicks of a mouse into data that can be processed by the computer takes place internally. A small device called the *central processing unit (CPU)* fetches coded instructions, decodes them, and then runs or executes them. The CPU is where programs are run and memory is managed. It is also known as the *microprocessor* or *chip* and is located on the main computer board (motherboard), as shown in **Figure 2-3.**

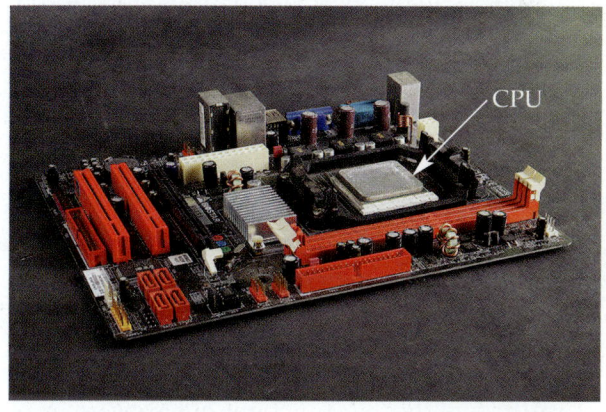

Spotters/Shutterstock.com

Figure 2-3. The CPU is the heart of a computer system. It runs programs and handles memory.

The CPU uses many electronic switches to digitize the input. It sets switches to either on (1) or off (0). The internal structure of the processor determines which switches to turn on and off as it receives instructions from the computer's program.

The CPU controls all the jobs performed by the computer's other parts. Modern CPUs can perform as many as 3.5 billion tasks per second without error. That means in one second a modern computer can locate, decode, and execute up to three and a half billion instructions. Many of these processors contain multiple CPUs and other technologies that increase their speeds.

In principle, the processing unit is very simple. It quickly performs calculations and makes decisions. Chip designers use these two capabilities to create a set of instructions to load programs and data, process the data, and produce output. As defined in Chapter 1, the instruction set for a chip is the more complex

In the future, artificial intelligence will be more like the human brain. Circuits, like neural synapses, will both process and store information.

actions developed using the basic functions of calculation and decision. The basic instruction set is different for each chip. It is software's job to take these few steps and produce the elaborate outcomes that computers generate today.

HANDS-ON EXAMPLE 2.1.1

Determining RAM in Your PC

RAM is an important part of computer performance. Many people do not know how much RAM is installed in their computers. Follow the steps to see the installed RAM in your PC. This activity is based on Windows 10.

1. Click in the search box on the Windows task bar, and enter settings. Click the Settings app in the search results.
2. In the **Settings** window, click **System**.
3. In the **System** window that is displayed, click **About** on the lower left. The right-hand side of the window displays various information about your system.
4. Locate the Installed RAM entry. How much RAM is installed on your computer?

Try It!

Find out how much storage you have. The **System** window offers a choice called **Storage**. Click that choice, and see how many gigabytes (GB) of storage you have available on the hard disk as well as on any peripheral storage devices that are currently attached. The view will also show you how much storage is being used.

SECTION REVIEW 2.1

Check Your Understanding

1. Which of the four functions of hardware involves data being sent to the computer?
2. What is the most common type of computer monitor?
3. Compare and contrast the two types of primary memory.
4. What is the most important hardware component for computer processing?
5. What is the term for the complex actions performed by the CPU?

Build Your Vocabulary

As you progress through this course, develop a personal computer science glossary. This will help you build your vocabulary and prepare you for a career. Write a definition for each of the following terms and add it to your computer science glossary.

central processing unit (CPU)	output	storage
input	peripheral	user interface (UI)
memory		

Software

Software makes the hardware work. Although the user cannot see, hear, or feel it, the software contains all of the instructions to operate the hardware. Without software, a computer does nothing but wait for instructions. A software program is a set of directions that tells the computer what to do. Programs tell a computer to do specific jobs. Programs are also known as software applications, sometimes referred to as apps. Software applications target specific tasks. These tasks are as varied as writing a document, scanning a photo, editing a video, or browsing the web.

Programmers have a wide choice of languages in which to write software applications. There are over 10,000 programming languages that exist today. The type of hardware connected to a computer determines, in part, which programming language must be used. Other factors depend on what tasks the program is to perform. The programmer's personal preference is also a factor in which programming language is used to create a program.

Learning Goals

After completing this section, you will be able to:
- Compare and contrast high-level and low-level programming languages.
- Describe the difference between procedural and object-oriented programming.
- Discuss the functions of operating system software.

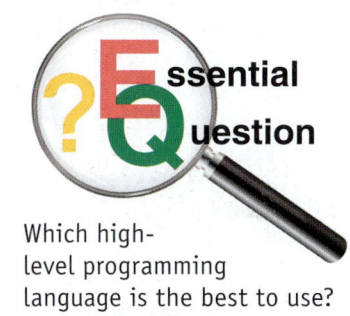

Which high-level programming language is the best to use?

Terms

bytecode	Java Virtual Machine
class	machine language
compiled	method
high-level language	operating system (OS)
instantiation	runtime engine
interpreted	syntax

Levels of Languages

A programming language is a set of words, symbols, and codes that enables a programmer to communicate instructions to a computer. Writing a program is called *coding*. The instructions written are called computer *code.*

Programs are the sets of instructions that carry out the tasks of the user. Programs are written in languages. Just as humans speak different languages, a computer programmer can select from a multitude of programming languages. The two basic types of programming languages are low level and high level.

Low-Level Language

The CPU uses *machine language* to handle instructions. Machine language consists of 0s and 1s. It is a low-level language and the only language understood by computers. However, machine language is very hard for humans to read. For a computer to understand the commands that programmers write, the commands must be converted into machine language.

Language Arts and Java

Machine Learning

Robots are with us. They function based on artificial intelligence. You can find robots working as vacuum cleaners, performing surgery, folding laundry, flipping hamburgers, responding to chats for technical support, and manufacturing cars.

You can even use robots to play games. The website 20Q (20 questions) communicates with a neural network–based artificial intelligence (AI). It learns by repetition. The more visitors use the site, the smarter the AI gets. The player thinks of an object or a concept and the AI attempts to guess what the object is. In addition to objects or concepts, you can choose from a variety of people from different topics including sports, movies, TV, and music.

At the end of the questions, the AI will list contradictions, or questions that it thinks you answered incorrectly. The AI is merely giving you a peek into its state of mind with regard to the object you played. Since it learns everything it knows from the people who play, it is working with opinions, not facts. The responses come from social knowledge and are used in everyday speech.

When what the player has thought of is an object the AI already knows, it adjusts the weightings within the neural network. If the AI performs poorly with a particular object or concept and you play that object multiple times, you might recognize that the system is learning. It will get better at guessing that object. It may also help to play similar objects to reinforce the differences between objects.

Assignment

Go to the 20Q website (www.20q.net). Since the application is multilingual and multicultural, you need to choose "Think in American" or whichever language or culture you wish to use. Answer the demographic questions. Start with an object like a sunflower. Let the software try to guess it. Follow the directions as to the types of responses you can give. Also, try one of the suggested topics.

Write a paragraph describing your experience. Was it amusing? Could the application read your mind? Include what you selected as the object or concept, the contradictions, and how many players have been to the site. Use Standard English in writing your paragraph, and be sure to check for spelling and grammar errors.

Goodheart-Willcox Publisher

High-Level Languages

Most programmers use high-level languages, as shown in **Figure 2-4.** A *high-level language* has a very limited set of recognizable English words. It contains commands like Print, If, Next, and Get. The particular set of rules or grammar that specifies how the instructions are to be written is called the *syntax.*

An early high-level language was invented at Dartmouth College that became very popular in the 1980s. It is called Beginners All-purpose Symbolic Instruction Code (BASIC). It was found in many forms on the first personal computers. Many computers loaded it into the ROM. Others had cartridges that allowed users access to it. BASIC was the first programming language that permitted line numbers.

There are thousands of high-level programming languages. Examples of modern high-level programming languages include Visual Basic, C++, Java, Swift, Python, PHP, and COBOL. C++ and C# are useful in high-performance programs such as rocket science. Interpreted BASIC is good for rapid editing and execution when prototyping is required in a short amount of time. COBOL is useful in moving data around, such as keeping data for business use. PHP has uses for interactions with a web server. Java is good for general-purpose applications.

Language	Year Introduced	Description
FORTRAN	1957	Procedural language for scientific and engineering applications; dominated early programming and used continuously for over 60 years.
COBOL	1959	Compiled, English-like procedural language; primarily used in business, finance, and administrative systems for companies and governments and still used in legacy systems.
Microsoft BASIC	1975	Procedural language that became the *de facto* programming language for all home computer systems.
C	1973	Still used in applications that were previously coded in assembly language including operating systems as well as application software for supercomputers.
C++	1985	Procedural and object-oriented programming language that is a complex extension of C for e-commerce, web search, and SQL servers as well as performance-critical applications such as telephone switches and space probes.
Python	1991	Emphasizes simplification of code and significant white space that is used by many large companies such as Wikipedia, Google, Yahoo, NASA, Amazon, and Facebook.
Java	1995	Object-oriented language with syntax largely influenced by C++ and when compiled, can run on all platforms without the need for recompilation.
C#	2000	Part of Microsoft .NET Framework; similar to Java.

Goodheart-Willcox Publisher

Figure 2-4. Most programmers use a high-level language to write computer code. While there are many high-level languages, these are some of the most popular.

HANDS-ON EXAMPLE 2.2.1

Exploring BASIC

This is a sample of a BASIC program that a user could enter. The command prompt says LIST. The next two lines are the program. There is another command prompt at the end for running the program.

```
]LIST
10 INPUT "What is your name:"; U$
20 PRINT "Hello "; U$
]RUN
```

The **LIST** command readies the computer for accepting the program. The first line is 10 and has the command **INPUT**. It tells the computer the user is going to enter something that should be saved. The text inside the quotation marks is displayed to the user on the screen when the program runs. The U$ is a place in memory to hold the user's name. The second line is 20 and prints Hello and a space and then the contents of U$, which in this case is the name the user entered. The last command tells the computer to execute, or run, the program. When the program runs, it first asks the user to enter a name and then says hello to that name.

Try It!

The following is a program written in BASIC code. Analyze this code. Then, write a description of what happens when the program is run.

```
]LIST
10 INPUT "What is your name:" U$
20 PRINT "Hello "; U$
30 PRINT "You are a five-star student";
40 S$ = " *****"
50 PRINT S$
]RUN
```

Translating High-Level Languages

High-level languages are converted to machine language by being compiled, interpreted, or some blend of the two. It is an issue of timing. If the code is *compiled*, the entire program is converted to machine language before it is run. Examples of languages that are compiled include C, C++, Fortran, and COBOL.

Some languages are interpreted instead of compiled. *Interpreted* means the instructions are converted to machine language as the program is executed, one line at a time. This has two basic effects. One is that the execution of interpreted code is slower than compiled code. Time must be taken to do the conversion to machine code for each instruction *every* time the program is run. The second effect is the code for an interpreted program can be modified rapidly. Examples of interpreted languages include Smalltalk, Python, Perl, and some versions of BASIC.

A happy medium between fully compiled and interpreted is bytecode. *Bytecode* is machine language–like instructions designed to be interpreted at runtime. Java and C# are two languages that are partially compiled so that the code is more portable between operating systems and CPUs. At runtime, another software program called the *runtime engine* or the virtual machine interprets the bytecode file for the target CPU. The advantage is that higher-level programs can be written for a wide range of CPU instruction sets and the interpretation does not take as long as fully interpreted code. In the case of Java, this step is performed every time the application is run. The *Java Virtual Machine* interprets the bytecode file at runtime.

Procedural versus Object-Oriented Programming

Procedural programming breaks down a task into a collection of subroutines. A *subroutine* is a block of code within a program that performs a specific task. This block or unit can be used in programs whenever that particular task is needed to run. For example, a program could call a subroutine to display the current date or print the text "This page is intentionally left blank." Examples of this type of programming language are C and FORTRAN.

Object-oriented programming (OOP) allows programmers to use objects. An object is a software representation of anything that can be seen, touched, heard, or used. Examples of objects in Windows programming are check boxes and buttons. A payroll program could also be written with OOP. A timecard could be an object, an object could be created for each employee, and an object could be used to issue paychecks. Examples of object-oriented programming languages are Java, C++, and Visual Basic.

Every object is created from a class. The *class* contains the instructions on how the object is to look and behave. It contains the blueprint for the instantiation of the object. *Instantiation* is the creation of an object. The best analogy is a baking comparison, as shown in **Figure 2-5.** The cookie cutter is the class. The cookies that result are the objects. The object not only includes what the object looks like, but also what it can do. The behaviors are called the object's *methods.* Attributes of objects are called *fields.* Java is an object-oriented programming language created in 1995.

OOP was invented in the 1970s.

nata_nytiaga/Shutterstock.com

Figure 2-5. Think of a class as a cookie cutter. Each object created by the class will be identical in form, just like each shape cut out by a given cookie cutter. However, depending on the dough used, each will have different data.

Operating Systems

The *operating system (OS)* is a specific set of software persistently stored on the computer that manages all of the devices as well as allocates memory to programs and manual execution of programs. General-use computers such as a PC or tablet must have an OS to work. Single-use computers, such as those found on satellites, may not have an OS.

For example, one function of an operating system is to tell the printer how to work. Whatever application is running sends the OS a print command. The OS checks to see if the selected printer is online and ready to accept output from the computer. Next, the OS sends the output to the printer and either tells the application the output is printed or sends an error message to the application. Each printer has a separate driver program that tells the OS how it works. However, to the user, all printers appear to work the same.

A proprietary operating system is one owned by a company or organization. The code for these OSs are secret. An open-source operating system is one for which the code is available to anyone for change. Examples of proprietary operating systems are Microsoft Windows, Apple's macOS (formerly OS X) and iOS, Chrome OS, Android, and BlackBerry Tablet OS. An example of an open-source OS is Linux.

SECTION REVIEW 2.2

Check Your Understanding

1. Rules of grammar that specify how code is to be written is called _____.
2. Which type of programming language has a limited set of recognizable English words?
3. Compiling code turns the program into _____ language.
4. What is the creation of an object from a class called?
5. Name one function provided by an operating system.

Build Your Vocabulary

As you progress through this course, develop a personal computer science glossary. This will help you build your vocabulary and prepare you for a career. Write a definition for each of the following terms and add it to your computer science glossary.

bytecode	instantiation	method
class	interpreted	operating system (OS)
compiled	Java Virtual Machine	runtime engine
high-level language	machine language	syntax

Binary Code

Electronic computers date from 1945 when the Electronic Numerical Integrator and Calculator (ENIAC) was in development at the University of Pennsylvania. The difference between mechanical and electronic computers is that the processors in an electronic computer have no moving parts. In an electronic computer, each on and off piece of information is kept in the form of a tiny amount of electricity. The electricity is on (5 volts) or off (0 volts). These on and off signals are represented in the computer programming by 1 and 0.

These two digits, 0 and 1, represent all information in a digital computer. The computer programs are developed to decode programs into 1s and 0s. All text, sounds, videos, and images that are input, stored, or output by a computer are composed as 1s and 0s. Ultimately, all numbers are nothing more than combinations of 1s and 0s. This section investigates the basics for this encoding.

Learning Goals

After completing this section, you will be able to:
- Compare and contrast number systems in computing.
- Convert values between number systems.
- Describe how to encode characters in Unicode.

Terms

base

binary number system

hexadecimal notation

Unicode encoding system

Essential Question

How is a strong understanding of math important to computer programmers?

Number Systems

The number or counting system used most widely in the world by humans is the decimal system. It is based on ten digits, 0 through 9. The *base* denotes how many digits are used in the number system. Other number systems exist in addition to the decimal system. Number systems used in computing include binary, octal, and hexadecimal. These have bases of two, eight, and sixteen digits.

Decimal Numbers

The decimal number system was invented by Indian mathematicians between the 1st and 4th centuries. Arabian mathematicians adopted the system in the 9th century. The digits became known as Arabic or Hindu-Arabic numerals.

The decimal system is a *positional* system. The position that a digit holds in a number indicates its value, as shown in **Figure 2-6.** In the number 2,587, the 5 indicates 500. However, in the number 2,857, the 5 indicates 50. Ten is the base for the decimal system, which is why it is called a base-10 system. This means counting and arithmetic are done in groups of ten. It also means that the individual digits of a decimal number are represented by a power of 10. Recall from math class that the number 2,587 as an expanded number is:

$$2587$$

$$2000 = 2 \text{ thousands} = 2 \text{ times } 10^3$$
$$500 = 5 \text{ hundreds} = 5 \text{ times } 10^2$$
$$80 = 8 \text{ tens} = 8 \text{ times } 10^1$$
$$7 = 7 \text{ ones} = 7 \text{ times } 10^0$$

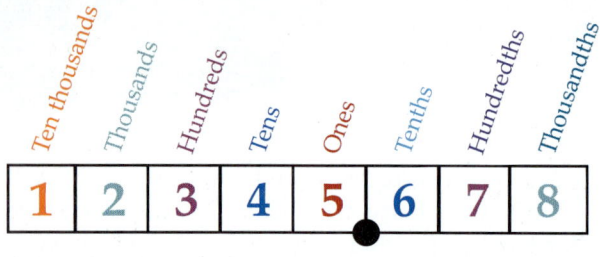

Figure 2-6. In the decimal number system, the position of a digit determines its value.

Binary Numbers

It became clear early in the development of computers that trying to represent ten different digits could make the computer too slow and bulky. The developers decided to use binary numbers to represent everything. The basic building block for the electronic computer is the binary digit, or *bit*. The smallest addressable unit in computer memory is a *byte*. Each byte holds eight bits.

The *binary number system* is a base-2 positional system. This means there are only two choices for the value of each digit. The base is 2, so each individual digit of a binary number is represented by a power of 2. This is a sample binary number: 1111. As an expanded number, it is:

$$
\begin{array}{rcll}
1111 & & & \\
1000 & = & 1 \text{ eights} & = \quad 1 \text{ times } 2^3 \\
100 & = & 1 \text{ fours} & = \quad 1 \text{ times } 2^2 \\
10 & = & 1 \text{ twos} & = \quad 1 \text{ times } 2^1 \\
1 & = & 1 \text{ ones} & = \quad 1 \text{ times } 2^0 \\
\end{array}
$$

The binary number 1111 is the same value as 15 in the decimal system. Verify this by adding 1 eight + 1 four + 1 two + 1 one. The sum is 15.

The decimal number 2,587 converted to binary is 1010 0001 1011. The conversion is carried out in the same way as illustrated above. Note that the binary number equivalent of a decimal number usually has more digits. However, the computer arithmetic is much faster using binary numbers than decimal numbers.

Hexadecimal Number System

Some of the numbers needed for output diagnostic procedures, memory management, and identification of network adapters and other hardware are very large. Large binary numbers are difficult for people to read because there are so many digits. For example, the binary number 1010 0001 1011 is hard for most people to quickly interpret. Hexadecimal notation came to the rescue as a way to balance the ability for a human to read the number with a number system that could be easily read by a computer.

Hexadecimal notation, or hex, is a base-16 positional number system. When counting in base-16, new symbols are required to represent digits higher than 9. The 16 digits in the hexadecimal number system are: 0, 1, 2, 3, 4, 5, 6, 7, 8, 9, A, B, C, D, E, and F. Each place represents a power of 16. The contents of any memory location can be represented by a pair of hex numbers.

In hexadecimal notation, binary digits are grouped into units of four, each represented by a hexadecimal digit. Four binary digits hold 16 different numbers, 0 to 15 or 0000 to 11111. Therefore, it is easy to write each group of four binary digits as one hex digit. This means fewer digits are needed to represent the number. This is why hex was adopted in computing.

Conversion between Number Systems

Figure 2-7 compares initial values of the decimal, binary, and hexadecimal number systems. The binary number 1010 0001 1011 can be converted to hex using this table. Because each hex digit represents four binary digits, first break

Decimal	Binary	Hex
0	0000	0
1	0001	1
2	0010	2
3	0011	3
4	0100	4
5	0101	5
6	0110	6
7	0111	7
8	1000	8
9	1001	9
10	1010	A
11	1011	B
12	1100	C
13	1101	D
14	1110	E
15	1111	F
16	1 0000	10

Goodheart-Willcox Publisher

Figure 2-7. A comparison of values in the decimal, binary, and hexadecimal number systems.

the number into groups of four digits starting from the right. Next, look up the corresponding hex digit for each group.

$$1010 \quad\quad 0001 \quad\quad 1011$$
$$1010 = A$$
$$0001 = 1$$
$$1011 = B$$

The hex conversion of 1010 0001 1011 is A1B. To avoid confusion with other number systems, it is a convention to write hex numbers with the 0x prefix. Therefore, the hex representation of 1010 0001 1011 is 0xA1B.

It is important to understand how numbers can be converted between number systems. However, technology can also be used to automatically perform the conversion. There are handheld calculators that can be used for this. There are also computer-based calculators. The calculator utility in the Windows operating system has a programmer view that allows conversions. Using this utility, converting between number systems is as easy as clicking buttons. **Figure 2-8** shows the Windows calculator in the programmer view. Notice there are options for hex, decimal, octal, and binary. Oct is octal. This is a base-8 number system. Octal is used in programming, but not commonly.

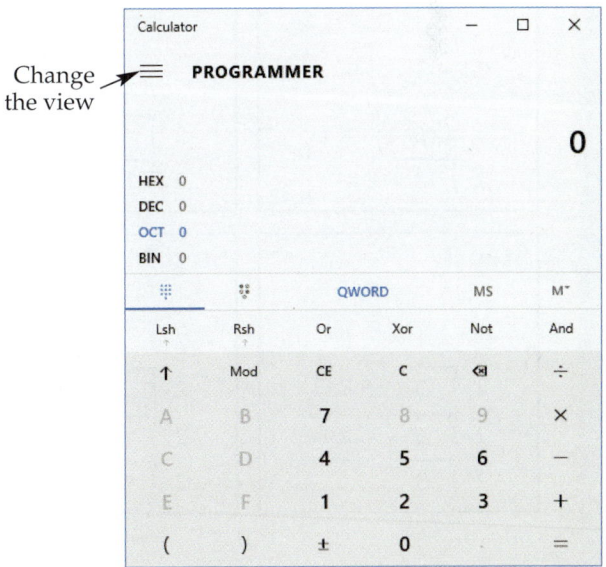

Goodheart-Willcox Publisher

Figure 2-8. The Windows calculator utility can be used to convert between number systems.

Math and Java

Algorithms for Converting Decimal to Binary

Binary numbers existed before computers were developed. Gottfried Leibniz, a German philosopher and mathematician, published the article "Explanation of Binary Arithmetic" in the early 18th century. He was the first to create a number system with only two digits, 0 and 1. A handheld or software-based calculator can be used to convert decimal numbers to binary. However, someone had to write the algorithm to make that calculator program. There is more than one possible algorithm for how that is done.

Division Algorithm. Use the long division symbol. Place the decimal number as the dividend under the symbol. Place 2 as the divisor to the left of the symbol. Compute the quotient and record the remainder. Divide the quotient by 2, and record the remainder. Continue repeating the division until the quotient is zero. The remainders are combined in reverse order to form the binary equivalent. An example of this algorithm using the decimal number 13 is shown. The binary equivalent is 1101.

Subtraction Algorithm. Notice in the algorithm shown previously that four 2s are used. They represent the first four powers of 2: 2^0, 2^1, 2^2, and 2^3. This means the largest power of 2 in the number 13 is $2^3 = 8$. In the subtraction algorithm, find the largest power of 2 in the decimal number and subtract it from the decimal number. Then, subtract the next lower power of 2, if possible, from the remainder. If this subtraction is possible, write a 1 in the binary number. If not, write a 0. Repeat until the remainder is zero. An example of this algorithm using the decimal number 13 is shown. Notice the binary equivalent calculated is the same as with the division algorithm (1101).

$13 - 2^3 = 5$ $8 = 2^3 \times 1$

$5 - 2^2 = 1$ $4 = 2^2 \times 1$

2^1 is too big $2 = 2^1 \times 0$

$1 - 1 = 0$ $1 = 2^0 \times 1$

Assignment

Apply both algorithms to the decimal number 55. What is the binary equivalent of this number? Which algorithm do you prefer to perform by hand? Is one algorithm more efficient than the other? Explain your answer.

Step 1:

```
        6    R 1
   2 ) 13
```

Step 2:

```
        3    R 0
   2 )  6    R 1
   2 ) 13
```

Step 3:

```
        1    R 1
   2 )  3    R 0
   2 )  6    R 1
   2 ) 13
```

Step 4:

```
        0    R 1
   2 )  1    R 1
   2 )  3    R 0
   2 )  6    R 1
   2 ) 13
```

Goodheart-Willcox Publisher

HANDS-ON EXAMPLE 2.3.1

Converting between Number Systems

Calculators make converting between number systems quick and easy. The Windows calculator is one example of a utility that can be used for conversions.

1. In Windows 10, launch the calculator by clicking the **Start** button followed by **All apps** and then **Calculator**.
2. In Windows 10, click the button in the upper-left corner of the calculator, and then click **Programmer** in the drop-down menu. The display should look like the one shown in **Figure 2-8** if you are using Windows 10.
3. Click the **Dec** label or radio button. This specifies the entry will be in decimal units.
4. Enter the number 365 using the keyboard or by clicking the number buttons.
5. Notice the **Hex**, **Oct**, and **Bin** labels display the hex, octal, and binary equivalents of decimal 365.

Try It!

Set the calculator to the binary option. Enter this binary number: 1110 0111 0110. What are the equivalent numbers in decimal, octal, and hex?

Code

Binary values are used to represent more than numbers in the computer. They are used to convert media instructions, make programming codes, identify memory locations, identify peripherals, and many others tasks. It is important for an IT professional to be able to understand and use these number systems. Besides using binary numbers for calculations, computers use them to represent data and run computer programs.

Encoding Instructions

Different CPUs may have different ways of interpreting code. This makes it difficult for programmers to write programs that work on all computers. An example of code that tells a certain CPU to add two numbers is:

```
Eb Gb 00
```

Fortunately, most programmers do not need to worry about this. There are many layers between what the CPU understands and what programmers write. When the programmer's code is compiled or interpreted, the result is binary code for a given computer platform.

Encoding Data

Besides writing instructions, programmers must know how computers translate data the user may enter. Some agreements have been made by industry professionals to make sharing data among different computers easier than sharing programs. These agreements are called *standards*. One early standard was called ASCII. The American Standard Code for Information Interchange (ASCII) was a standard for representing text that most computers support. The acronym ASCII is usually pronounced *askee*.

ASCII is a system for encoding characters in the English alphabet and the digits 0–9. It also includes basic punctuation symbols and a blank space. There are 32 nonprinting characters used to control processes, such as a hard return for the end of a paragraph.

A disadvantage to ASCII is that it encodes the English alphabet only. To address this limitation, the Unicode Consortium was founded in 1991. Its corporate, university, and governmental members developed the Unicode encoding system. The *Unicode encoding system* covers not only the English alphabet, numbers, and punctuation, but also international alphabets, number systems, and punctuation. Even emojis are covered by Unicode. Most modern programming languages, including Java, use Unicode for encoding text.

Unicode uses two bytes (16 bits) for each character. This allows the encoding of as many as 65,636 characters. Recognizing the widespread use of ASCII, the Unicode Consortium defined the first 256 Unicode characters to be compatible with the ASCII characters. **Figure 2-9** shows the printable characters of the Unicode code chart for the decimal numbers 32 to 126. All of the charts can be seen on the Unicode Consortium website (unicode.org).

A curiosity that the character table points out is that the number 1 and the text character 1 are represented two different ways in a computer although they look the same when displayed on the monitor. The 16 bits that represents the binary number 1 is 0000 0000 0000 0001. However, the 16 bits that represents the text character 1 in Unicode is 0000 0000 0011 0001.

Note that the table shows the decimal and hexadecimal code for each symbol in the table. To store the characters in the word *Book*, these hex codes are used:

B	0042
o	006F
o	006F
k	006B

That means the word "Book" is encoded as 0042006F006F006B. No calculations are performed on these hex numbers. They are used only to represent the letters.

HANDS-ON EXAMPLE 2.3.2

Encoding Literals

Encode the text string *Hello World!* using hex notation. Use the chart shown in **Figure 2-9** to locate the correct symbols.

1. Locate the codes for the characters in the word *Hello*.
2. Locate the code for a space.
3. Locate the codes for the characters in *World*.
4. Locate the code for the exclamation point.
5. Write the codes in order from left to right to form the code for the entire text string. Use the 0x prefix.

Binary	Dec	Hex	Symbol	Binary	Dec	Hex	Symbol	Binary	Dec	Hex	Symbol
0000 0000 0010 0000	0032	0020	(space)	0000 0000 0100 0000	0064	0040	@	0000 0000 0110 0000	0096	0060	`
0000 0000 0010 0001	0033	0021	!	0000 0000 0100 0001	0065	0041	A	0000 0000 0110 0001	0097	0061	a
0000 0000 0010 0010	0034	0022	"	0000 0000 0100 0010	0066	0042	B	0000 0000 0110 0010	0098	0062	b
0000 0000 0010 0011	0035	0023	#	0000 0000 0100 0011	0067	0043	C	0000 0000 0110 0011	0099	0063	c
0000 0000 0010 0100	0036	0024	$	0000 0000 0100 0100	0068	0044	D	0000 0000 0110 0100	0100	0064	d
0000 0000 0010 0101	0037	0025	%	0000 0000 0100 0101	0069	0045	E	0000 0000 0110 0101	0101	0065	e
0000 0000 0010 0110	0038	0026	&	0000 0000 0100 0110	0070	0046	F	0000 0000 0110 0110	0102	0066	f
0000 0000 0010 0111	0039	0027	'	0000 0000 0100 0111	0071	0047	G	0000 0000 0110 0111	0103	0067	g
0000 0000 0010 1000	0040	0028	(	0000 0000 0100 1000	0072	0048	H	0000 0000 0110 1000	0104	0068	h
0000 0000 0010 1001	0041	0029	)	0000 0000 0100 1001	0073	0049	I	0000 0000 0110 1001	0105	0069	i
0000 0000 0010 1010	0042	002A	*	0000 0000 0100 1010	0074	004A	J	0000 0000 0110 1010	0106	006A	j
0000 0000 0010 1011	0043	002B	+	0000 0000 0100 1011	0075	004B	K	0000 0000 0110 1011	0107	006B	k
0000 0000 0010 1100	0044	002C	,	0000 0000 0100 1100	0076	004C	L	0000 0000 0110 1100	0108	006C	l
0000 0000 0010 1101	0045	002D	-	0000 0000 0100 1101	0077	004D	M	0000 0000 0110 1101	0109	006D	m
0000 0000 0010 1110	0046	002E	.	0000 0000 0100 1110	0078	004E	N	0000 0000 0110 1110	0110	006E	n
0000 0000 0010 1111	0047	002F	/	0000 0000 0100 1111	0079	004F	O	0000 0000 0110 1111	0111	006F	o
0000 0000 0011 0000	0048	0030	0	0000 0000 0101 0000	0080	0050	P	0000 0000 0111 0000	0112	0070	p
0000 0000 0011 0001	0049	0031	1	0000 0000 0101 0001	0081	0051	Q	0000 0000 0111 0001	0113	0071	q
0000 0000 0011 0010	0050	0032	2	0000 0000 0101 0010	0082	0052	R	0000 0000 0111 0010	0114	0072	r
0000 0000 0011 0011	0051	0033	3	0000 0000 0101 0011	0083	0053	S	0000 0000 0111 0011	0115	0073	s
0000 0000 0011 0100	0052	0034	4	0000 0000 0101 0100	0084	0054	T	0000 0000 0111 0100	0116	0074	t
0000 0000 0011 0101	0053	0035	5	0000 0000 0101 0101	0085	0055	U	0000 0000 0111 0101	0117	0075	u
0000 0000 0011 0110	0054	0036	6	0000 0000 0101 0110	0086	0056	V	0000 0000 0111 0110	0118	0076	v
0000 0000 0011 0111	0055	0037	7	0000 0000 0101 0111	0087	0057	W	0000 0000 0111 0111	0119	0077	w
0000 0000 0011 1000	0056	0038	8	0000 0000 0101 1000	0088	0058	X	0000 0000 0111 1000	0120	0078	x
0000 0000 0011 1001	0057	0039	9	0000 0000 0101 1001	0089	0059	Y	0000 0000 0111 1001	0121	0079	y
0000 0000 0011 1010	0058	003A	:	0000 0000 0101 1010	0090	005A	Z	0000 0000 0111 1010	0122	007A	z
0000 0000 0011 1011	0059	003B	;	0000 0000 0101 1011	0091	005B	[	0000 0000 0111 1011	0123	007B	{
0000 0000 0011 1100	0060	003C	<	0000 0000 0101 1100	0092	005C	\	0000 0000 0111 1100	0124	007C	|
0000 0000 0011 1101	0061	003D	=	0000 0000 0101 1101	0093	005D	]	0000 0000 0111 1101	0125	007D	}
0000 0000 0011 1110	0062	003E	>	0000 0000 0101 1110	0094	005E	^	0000 0000 0111 1110	0126	007E	~
0000 0000 0011 1111	0063	003F	?	0000 0000 0101 1111	0095	005F	_				

Figure 2-9. Unicode characters (symbols) and their binary, decimal, and hexadecimal codes.

SECTION REVIEW 2.3

Check Your Understanding

1. What does it mean to say a number system is positional?
2. What is the smallest addressable unit in computer memory?
3. The hexadecimal system is a base-_____ number system.
4. The binary number 0000 0001 is equivalent to the decimal number 1. What binary number is one greater and equivalent to decimal number 2?
5. Encode the text string *I love you* in hex using the chart in **Figure 2-9.**

Build Your Vocabulary

As you progress through this course, develop a personal computer science glossary. This will help you build your vocabulary and prepare you for a career. Write a definition for each of the following terms and add it to your computer science glossary.

base

binary number system

hexadecimal notation

Unicode encoding system

Cooperative Coding

Artificial Intelligence

Artificial intelligence is applied when a computer mimics cognitive functions that humans associate with other humans. In particular, this includes learning new facts, problem-solving, and carrying on coherent conversations. When you cannot see the entity responding, it becomes an intellectual challenge.

During World War II, Alan Turing worked for Britain's codebreaking center. He made a huge contribution to the war effort in cryptology and the cracking of Germany's Enigma Machine. As well as being a computer scientist, he was a mathematician, logician, philosopher, biologist, and marathon runner. Later in his career, Turing created a test to validate the humanness of a responder. The *Turing test* is an assessment of whether a computer has demonstrated intelligence indistinguishable from that of a human.

The responders you find on your phone or on the Internet are called *chatterbots.* To pass the Turning test, a chatterbot must contain the natural language processing to both understand and respond like a human. It also needs the appropriate information to communicate. It needs the answers to the typical things a person might ask of another human. This is much more complicated than general intelligence.

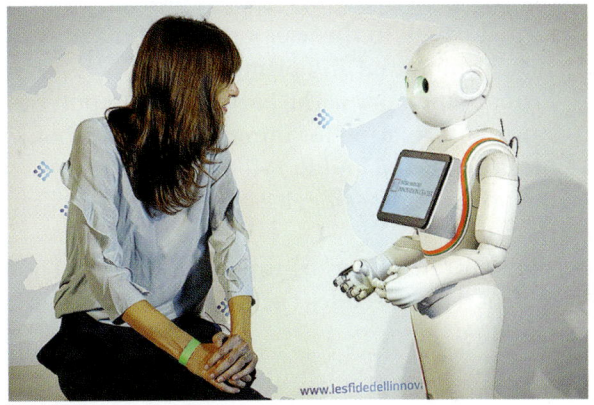

MikeDotta/Shutterstock.com

Assignment

Work with the team assigned by your instructor to get some experience communicating with chatterbots. You can use your iPhone and talk to Siri or use your Android phone to talk to the Google Assistant.

1. What questions would your team ask to assess if the responder is a computer or a person? Come up with at least five questions.
2. Interview at least two chatterbots and use your team's questions. Compare and contrast their responses.

Chapter Summary

Section 2.1: Hardware

- Input devices allow data to be sent to the computer and include devices such as keyboards and touchpads.
- Output devices digitize data to provide it in a format useful to the user and includes devices such as monitors and printers.
- Primary memory is the part of the computer that stores information for immediate processing, including RAM and ROM, while secondary memory is storage devices such as hard disks and flash drives.
- The central processing unit (CPU) fetches coded instructions, decodes them, and then runs or executes them.

Section 2.2: Software

- A low-level programming language is known as machine language and consists of 1s and 0s that the computer can understand, while a high-level programming language contains a limited set of recognizable English words and must be compiled or interpreted for the computer to understand.
- Procedural programming breaks down a task into a collection of subroutines, which are blocks of code that performs specific tasks, while object-oriented programming allows the use of prebuilt objects, which are software representations of anything that can be seen, touched, heard, or used.
- General-use computers must have an operating system to work, which is a specific set of software persistently stored on the computer that manages all of the devices as well as allocates memory and executes programs.

Section 2.3: Binary Code

- The decimal number system is a positional base-10 system and the system most widely used by humans in the world, while the binary number system is a base-2 system used in computers; the hexadecimal number system is a base-16 system also used in computers.
- Values can be converted between the decimal, binary, and hexadecimal number systems using a conversion table or chart or a handheld or software-based calculator.
- American Standard Code for Information Interchange (ASCII) is a system for encoding characters in the English alphabet, the digits 0–9, and basic punctuation symbols, while the Unicode encoding system extends the ability to encode international alphabets, number systems, and punctuation.

Chapter 2 Test

Multiple Choice

Select the best response.

1. What are the four major components of a computer system?
   A. Computer programs, input, processing, and joysticks.
   B. Phones, Internet, apps and social networks.
   C. Input, memory, processing, and output.
   D. Personal computers, tablets, smartphones, and video cameras.

2. Which type of memory is a storage device?
   A. Primary memory
   B. Secondary memory
   C. Random-access memory
   D. Read-only memory

3. Which of the following is a procedural programming language?
   A. C
   B. Java
   C. C++
   D. Visual Basic

4. Of these programming languages, which would most likely be used for a rocket-control system?
   A. COBOL
   B. PHP
   C. BASIC
   D. C++

5. What is the standard that represents English alphabets, numbers, and punctuation, but also international alphabets, number systems, and punctuation?
   A. Unicode
   B. ASCII
   C. exabyte
   D. instantiation

Completion

Complete the following sentences with the correct word(s).

6. External computer devices not critical to basic operation are called _____.

7. The two kinds of memory chips, or primary memory, in a computer are RAM and _____.

8. Each CPU has its own instruction set written in _____.

9. In an object-oriented programming language, the instructions for creating an object are found in a(n) _____.

10. The binary number 1001 1110 is written as _____ in decimal notation.

Matching

Match the correct term with its definition.

A. base-2
B. high-level language
C. interpreted
D. instantiation
E. compiled

11. Creation of an object.

12. Examples include C++, Java, and Visual Basic.

13. Binary number system.

14. Converts code to machine language before being run.

15. Converts the bytecode file at runtime.

Application and Extension of Knowledge

1. The following program is written in BASIC, which is a high-level programming language. The program line numbers are incremented by 10s. This allows the programmers to add code between the initial lines if needed. Examine this code. What do you think each line means? Can you guess the overall shape that is drawn?

```
]LIST
10 HCOLOR = 7
20 LINE (100,100)-(400,100)
30 LINE (100,400)-(400,400)
40 LINE (100,100)-(100,400)
50 LINE (400,100)-(400,400)
]RUN
```

2. COBOL is a procedural programming language. The following is an example of a COBOL program. Examine this code. What is the result of running the program? What command does COBOL use instead of the **PRINT** command used in BASIC? What is missing from the COBOL code that is found in BASIC? How does the program know when to end?

```
---- hello.cob ------------------------
 * Sample COBOL program
 IDENTIFICATION DIVISION.
 PROGRAM-ID. hello.
 PROCEDURE DIVISION.
 DISPLAY "Hello World!".
 STOP RUN.
----------------------------------------
```

3. FORTRAN is a procedural programming language. FORTRAN was developed by IBM in the 1950s for scientific and engineering applications. It dominated programming then and has been used continuously for over 50 years. There are many versions. FORTRAN's design was the basis for many other programming languages, including BASIC. The following is an example of a FORTRAN program. Examine this code. What is the result of running the program? What command does this FORTRAN code use instead of the **PRINT** command used in BASIC? What is missing from the FORTRAN code that is found in BASIC? How does the program know when to end?

```
program hello_world
implicit none

character*32 text

text = 'Hello World'
write (*,*) text

end hello_world
```

4. Human beings have attempted to predict the weather for thousands of years. In the 1800s, records began to be kept. Weather forecasts are made by collecting quantitative data about the current state of the atmosphere at a given place and using meteorology to project how the atmosphere will change. Making mathematical models of the atmosphere and oceans to predict the weather was first attempted in the 1920s. It was the invention of computer-simulation programs in the 1950s that allowed numerical weather predictions to produce more realistic results. Weather satellites and supercomputers make it possible. The supercomputers manipulate the huge datasets and perform complex calculations. Even with the increasing power of supercomputers, the ability to accurately forecast weather extends to only about six days. The limitation is based on the fact that it is impossible today to get a complete representation of the entire atmosphere at any given time. Once this problem is solved, weather prediction will be perfect. Each model makes assumptions about the beginning conditions of the atmosphere. Write a paragraph about the effect accurate weather forecasting can have on a region. Why are there many mathematical models when predicting weather, such as the path of a hurricane?

5. The metric prefixes used to describe memory configurations are based on powers of ten. However, when applied to computer memory, the sizes are represented in powers of 2. This introduces a margin of error. For example, 1 kilobyte is roughly 1,000 bytes, as the metric prefix K indicates. However, it really is 2^{10} bytes, or 1,024 bytes. A megabyte is roughly 1,000,000 bytes, M representing millions. But, a megabyte is actually $2^{20} = 1,048,576$ bytes. Complete the following table to find the power of two that matches each of the metric prefixes.

Metric Prefix	Power of 10	Power of 2	Value of Power of 2
kilo (K)	thousand	2^{10}	1,024
mega (M)	million	2^{20}	1,048,576
giga (G)	trillion	2^{30}	
tera (T)	quadrillion		

Online Activities

Complete the following activities, which will help you learn, practice, and expand your knowledge and skills.

Vocabulary. Practice vocabulary for this chapter using the e-flash cards, matching activity, and vocabulary game until you are able to recognize their meanings.

Communication Skills

Reading. *Imagery* is descriptive language describing how something looks, feels, smells, sounds, or tastes. Using the information in this chapter, find an example of how the author uses imagery to describe a concept. What mental picture did you create? Note the page number and paragraph where you found the example. Why do you think this is a good example?

Writing. Standard English means that word choices, sentence structures, paragraphs, and the narrative follow conventions used by those who speak English. Well-written paragraphs are usually the product of editing. *Editing* means to rewrite sentences and paragraphs to improve their content and organization. Using standard English, write several paragraphs to describe your reasons for taking this class. Edit and revise your work until the ideas are refined and clear to the reader.

Speaking. *Circumlocution* is the use of many words to convey an idea when fewer would do. It can be used to communicate an idea when the exact word is not known. For example, suppose you do not know the word "democracy." Instead, you say, "the form of government in which each person has a say." Use circumlocution to describe to a classmate how to establish goals without using the word "goal." Was your classmate able to determine your meaning?

Listening. *Informative listening* is the process of listening to gain specific information or instructions from the speaker. Ask a classmate to give you directions to the school cafeteria. Take notes as the directions are given. If necessary, ask the speaker to slow down or repeat a point. Summarize your notes and retell the directions to your classmate to confirm your understanding of them. Follow the directions as they were given. Did you arrive at the cafeteria?

Portfolio Development

College and Career Readiness

Objective. Before you begin collecting information for your portfolio, write an objective for the finished product. An *objective* is a complete sentence or two that states what you want to accomplish.

The language in your objective should be clear and specific. Include enough details so you can easily judge when it is accomplished. Consider this objective: "I will try to get into college." Such an objective is too general. A better, more detailed objective might read: "I will get accepted into the computer science program at one of my top three colleges of choice." Creating a clear objective is a good starting point for beginning to work on your portfolio.

1. Decide the purpose of the portfolio you are creating, such as short-term employment, career, community service, or college application.
2. Set a timeline to finish the final product.
3. Write an objective for your portfolio.

CTSOs

Performance. Some competitive events for CTSOs have a performance component. Performance events provide an opportunity to demonstrate verbal communication skills, as well as decision-making and problem-solving abilities.

Depending on the organization, this event can be for individual or team participation. The activity could potentially be a presentation, role-play, or decision-making scenario for which the participants provide a solution and present to the judges. To prepare for the performance component of a presentation, complete the following activities.

1. On the website of your CTSO, locate a rubric or scoring sheet for the event.
2. Confirm if visual aids may be used in the presentation and the amount of setup time permitted.
3. Review the rules to confirm the type of activity and if questions will be asked or if a case or situation will be defended.
4. Make notes on index cards about important points to remember. Use these notes to study. You may also be able to use these notes during the event.
5. Practice the performance. You should introduce yourself, review the topic to be presented, defend the topic, and conclude with a summary.
6. After the practice performance is complete, ask for feedback from your instructor. You may also consider having a student audience listen and give feedback.

Introduction to Java Programming

Sections

3.1 Java Program Structure

3.2 Understanding Errors

Programmers have a choice of many programming languages. The Java programming language is used in this textbook. Java is used professionally in many fields and runs on countless computers and devices. With Java, a programmer can create applications for Windows, MacOS, Linux, and other operating systems. If you have an Android phone, your apps were most likely written in Java. Java is the official programming language for Android devices. Java is also used in other smart devices, such as kitchen appliances, televisions, automobiles, and the Google Home assistant. The best part about Java is its large library of code that makes writing applications faster.

The Java compiler, library, and runtime system are free to download and use in creating applications. Java applications may be distributed or sold without charge to the Java creators. Java applications may be found in many areas, including banking, health care, games, weather prediction, and countless other areas.

Reading Prep

Before reading this chapter, turn to the opening pages of the chapter and read the section titles. By previewing these, you can be prepared for the topics presented in the chapter. What do the titles reveal about what you will be learning?

College and Career Readiness

While studying, look for the activity icon for:

- Vocabulary terms with e-flash cards and matching activities.
- Starter files for hands-on examples and other exercises.

These activities can be accessed at
www.g-wlearning.com/informationtechnology/1773

Chapter Glossary

block: Subsection of code; begins with { and ends with }.

block comment: Comment that starts with /* and continues over multiple lines until ended by */.

class: Unit of code that works together; every program is made up of one or more classes.

comment: Text that the compiler ignores; intended to communicate information to the programmer.

compiler error: Error in Java syntax detected by the compiler.

integrated development environment (IDE): Application that provides a source code editor and an interface to write, compile, and run programs.

Java Development Kit (JDK): Set of tools and code for writing Java programs.

keyword: Code or word that has special meaning in the Java language.

line comment: Comment that begins with // and continues to the end of the line; does not extend beyond the line.

logic error: Error in the program's algorithm or in the implementation of the algorithm that causes incorrect results; not detected by the compiler.

portability: Ability of a program to run on multiple computers.

runtime error: Error detected by the JVM while the program is executing; not found by the compiler.

statement: Performs the work of the program; can extend over multiple lines and ends with a semicolon.

user error: Incorrect action by the user.

white space character: A space, tab, or new line character; used as a separator.

Java Program Structure

The Java language was developed by Sun Microsystems more than two decades ago. Fifteen years after Java's creation, Oracle Corporation bought Sun Microsystems and took over the Java language. Oracle periodically releases improvements to the language. In addition, Oracle is working with the Java community to make Java freely available in an open-source version as OpenJDK. This is released under the GNU General Public License. Oracle's Java website is a rich source of information as you explore programming using the Java language. There you will find software, tutorials, blogs, and reference material related to the Java language and Java software.

Learning Goals

After completing this section, you will be able to:
- Explain how a Java program is compiled and executed.
- Diagram the basic structure of a Java program.
- Create a simple Java program.

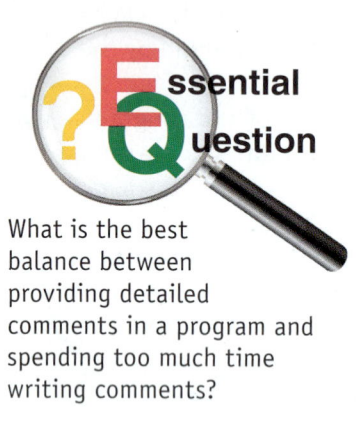

Essential Question

What is the best balance between providing detailed comments in a program and spending too much time writing comments?

Terms

block	Java Development Kit (JDK)
block comment	keyword
class	line comment
comment	portability
Integrated Development Environment (IDE)	statement
	white space character

Java Compiler and Virtual Machine

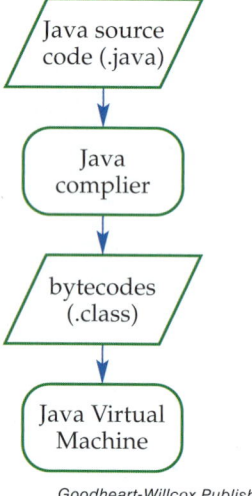

Goodheart-Willcox Publisher

Figure 3-1. This flowchart shows the execution of a Java program.

Chapter 2 introduced compilers and interpreters as two different ways to convert source code into executable code. Java uses a combination of a compiler and an interpreter to run programs. **Figure 3-1** shows how a Java program travels from source code to execution. The programmer writes Java source code in a plain-text file that is saved with a .java file extension. This is then fed into the Java compiler. The compiler converts the source code into binary bytecodes. The compiler stores the bytecodes into a file with a .class file extension.

To run the program, the Java Virtual Machine (JVM) interprets the bytecodes in the .class file. As the program executes, the JVM translates the bytecodes into the machine language of the computer on which the program is running. To run a Java program on any computer, all that is needed is a .class file and a JVM for that computer. Fortunately, JVMs exist for practically every computer platform. The ability to run on multiple computers is called *portability.* Because of the portability of code, Java is often referred to as a "write once, run many" programming language.

Before writing a program in Java, some software needs to be set up. First, the Java Development Kit (JDK) needs to be downloaded and installed. The *Java Development Kit (JDK)* is a set of tools for creating programs in Java. It includes the Java compiler, Java Virtual Machine, and lots of code that can be used to make your programs fun.

To help in writing programs, an integrated development environment (IDE) also needs to be downloaded and installed. An *integrated development environment (IDE)* is a program that makes it easy to write, compile, and run programs. Many integrated development systems are available for Java. This textbook uses jGRASP. It was developed at Auburn University and is available without charge. It is ideal for students who will be writing many small programs.

Java Program Structure

Java programs follow a pattern. **Figure 3-2** shows an example of this pattern. Creating a new program usually means starting with this shell and making any changes needed for the application at hand. Although the lines are numbered in the figure, the numbers are not part of the program. They are only used to allow references to specific lines as the program is discussed. Refer to this figure as you read the following sections.

Class

Each program is written as one or more classes. A *class* is a unit of code that works together. On line 5 in **Figure 3-2,** the class for the program is defined. The programmer has said it is public, which means anyone can run the program, and it is the class for this program. The words public and class are keywords. A *keyword* is a code or word that has special meaning in the Java language. This is discussed in more detail in Chapter 4.

On the same line, the program is then given a name. The name should indicate what the program does. For example, if creating a program that adds a list of numbers, the name might be Calculator or SumNumbers. Java's syntax for naming classes calls for an initial letter followed by a combination of letters, numbers, underscores (_), or dollar signs ($). The name *cannot* contain spaces.

A good convention to follow in naming classes is to start with a capital letter and to capitalize any internal words, such as SumNumbers. This is often referred to as *CamelCase*. The capital letters in the middle of the name resemble the humps on a camel's back.

Which name is selected is important. Java requires the source code to be saved in a file with the same name and capitalization as the class name. The file extension of .java must be used as well. The compiler will then save its bytecodes in a file with the same name as the class, but with a .class file extension.

FYI

Refer to the appendix for instructions on downloading and installing the JDK and the jGRASP IDE. If your class is using a different IDE, the procedures for compiling and running a Java program will likely be different, but similar.

```
1  /* what the program does
2     who wrote it
3  */
4
5  public class NameOfProgram {
6     public static void main( String [ ] args ) {
7
8        // code to perform
9
10    }
11 }
```

Goodheart-Willcox Publisher

Figure 3-2. This example shows the standard pattern to use when writing a Java program.

Main Method

On line 6 in **Figure 3-2,** the main method is defined. As defined in Chapter 2, a method is code that performs one or more operations. The main method indicates where the code for the program starts. When the JVM starts interpreting the program, it looks for the code defining the main method as:

```
public static void main( String [ ] args )
```

Notice that public keyword is repeated on this line and two new keywords appear, static and void. Each program needs to include a main method as written here. Why each piece of the main method is needed and what the code means are explained as you progress through this text.

Statements and Blocks

Blocks are subsections of code. On lines 5 and 6 in **Figure 3-2,** there are open curly brackets. On lines 10 and 11, there are closing curly brackets. These curly brackets, or braces, mark the beginning and end of blocks. In this case, there are two blocks in the shell program:

- the class itself (lines 5 through 11)
- the main method (lines 6 through 10)

Statements perform the work of the program, such as performing a calculation or outputting some information. The code in **Figure 3-2** is just a program shell, so it has no statements. The comment on line 8 indicates where to insert any statements appropriate to an application. Most applications have many statements that may extend over multiple lines. A semicolon (;) is used to indicate the statement is finished. Think of the semicolon as similar to the period in English.

Java is often referred to as a punctuation-driven language because punctuation is used to indicate the beginning and end of blocks and the end of statements.

Comments

A *comment* is text ignored by the compiler and not part of the program function. In **Figure 3-2,** lines 1 through 3 and line 8 are comments. Comments are optional, but are a good way to communicate to yourself and to other programmers what the code is doing. Java allows for two types of comments:

- block comments
- line comments

A *block comment* is a comment that can span multiple lines. Lines 1 through 3 make up a block comment. A block comment starts with /* and can continue over multiple lines until ended by */. It is good practice to write a block comment at the beginning of each program. This comment should say what the program does and who wrote it.

The second format for a comment in Java is a line comment. A *line comment* is a comment limited to one line. It does not span lines. A line comment begins with two forward slashes (//) anywhere on a line. The compiler ignores everything from the slashes to the end of the line. Line comments are often used to explain what is happening in a specific line of code. Line 8 shows a line comment.

White Space and Indentation

Java requires that at least one white space character be used to separate keywords from other code. A *white space character* may be a space, tab, or the new line character. The new line character is created by pressing the [Enter] or [Return] keyboard key. It is good programming practice to use white space liberally in source code. White space makes the code easier for the programmer to read. Inserting blank lines between logical sections of a program is also a good idea for the same reason: improving readability.

Notice in **Figure 3-2** that the end brace for the class block on line 11 vertically aligns with the public class definition on line 5. Also, the main method on line 6 is indented a few spaces and its end brace on line 10 vertically aligns with the main method. This alignment makes it easier to see where each block ends. The placement of the comment on line 8 indicates that any statements added to the main method should be indented one more level. None of this indentation is required to make the program execute. It is, nevertheless, good programming practice to indent code for readability. **Figure 3-3** summarizes good programming practices for writing programs.

Good Programming Practices
• Include a comment at the top of each program saying what the program does and the programmer's name.
• Start a class name with a capital letter and capitalize internal words.
• Insert blank lines between logical sections of a program.
• Indent the code inside blocks.

Goodheart-Willcox Publisher

Figure 3-3. Strive to follow these good programming practices when writing code.

First Java Program

Starting with the shell, it is easy to create a Java program that outputs the message, "Hello World!" **Figure 3-4** shows the modified program. The block comment on lines 1 through 3 now tells you what the program will do: say hello. Remember that comments are ignored by the compiler. They are used only to communicate to other programmers what the program is doing.

On line 5, the program has been named SayHello. Java requires the source code to be saved in a file that matches this name (SayHello.java). The compiler will automatically create a file named SayHello.class to contain the bytecodes.

For this program, the main method contains one statement. This appears on line 8. The operation this statement performs is outputting the message, "Hello World!" The code System.out.println is used to create output from applications. The output text is surrounded with quotation marks and the quoted text is placed within parentheses. Notice that the rest of the program is the same as shown in **Figure 3-2.**

This program is an example of a sequential algorithm for a problem that does not branch or iterate. The entire program is processed in linear fashion, line by line, from beginning to end. No part of the program repeats. Branching and iterative applications are discussed later in this text.

Did you notice that the IDE applies colors to some of the words in the source code? The keywords public, class, static, and void are purple. The comments are orange, and the text to output is green. This color coding will be helpful. For

It is a tradition in computer science to write the Hello World program as the first exercise in any new language you are learning.

To keep your programs for this course organized and easily located, create a folder for each chapter, and save programs by chapter.

```
1  /* Program to say hello
2     Your Name Here
3  */
4
5  public class SayHello {
6     public static void main( String [ ] args ) {
7
8        System.out.println( "Hello World!" );// output message
9
10    }
11 }
```

Goodheart-Willcox Publisher

Figure 3-4. The basic Java shell is modified to create a program that says hello.

example, if a keyword is misspelled, perhaps as classs with an extra S, the word will not be purple. This would be obvious that something is wrong.

Different IDEs use different colors for each category of code. Most IDEs, including jGRASP, make it possible to select specific colors for keywords, punctuation, text, and comments. This is only a preference and has no effect on the program.

HANDS-ON EXAMPLE 3.1.1

Writing a First Program

For your first program, you will write the simple application shown in **Figure 3-4** to output the message, "Hello World!" In this activity and throughout the text, jGRASP is used as the IDE. If you are using a different IDE, the instructions will be similar, but you may have different icons and menus.

1. Launch jGRASP.

2. Using the pane on the left, navigate to the folder where you will store the programs for Chapter 3. The buttons on the toolbar at the top of the pane can be used to navigate.

3. Click **File**>**New**>**Java** in the pull-down menu to begin a new document for writing your source code. The document will appear in the source code pane in the upper-right of the jGRASP window.

4. Click **View**>**Line Numbers** in the pull-down menu so the menu item is checked. This turns on the display of line numbers in the source code document. The number 1 should now appear in the top-left corner of the source code pane. Remember that line numbers are not part of a program's code, but are helpful in keeping track of the code. jGRASP will automatically add line numbers as you enter code.

5. In the source code pane, enter the code shown in **Figure 3-4.** Enter all code exactly as you see it, including white space characters, blank lines, capitalization, and punctuation, except include your name on line 2. Note that in println the second-to-last character is a lowercase letter L rather than the number 1.

6. Click **View**>**Generate CSD** in the pull-down menu. Blue lines appear indicating the program structure. This is a useful feature for verifying that the indentation of the program is correct. CSD stands for control structure diagram.

7. Click **View**>**Remove CSD** in the pull-down menu. The blue lines are removed and the code is properly indented.

8. Click **File**>**Save As** in the pull-down menu. A save dialog box is displayed. Save the file as SayHello.java, which should be suggested by jGRASP. If not, enter the correct name matching capitalization. Also note there are no spaces in the file name.

Compile file

9. The next step is to compile the code. Click **Build**>**Compile** in the pull-down menu or simply click the **Compile file** button on the toolbar. In response, jGRASP saves any changes you made to the file, and then starts the Java compiler.

10. Click the **Compile Messages** tab at the bottom-right of the jGRASP window. If the message "operation complete" is displayed without any compiler errors, good job! You have what is called a *clean compile.* If errors are displayed, these will need to be corrected before going on. Carefully check the code to be sure you have entered the words and punctuation exactly as shown. Remember, capitalization is important. Fix any problems you find, then compile again.

11. Look in the "browse" pane (left-hand pane). You should now see the SayHello.class file. The compiler created this file, which contains the bytecodes that the JVM will interpret. You are now ready to run the program.

Find and run main method or applet

12. Click **Build**>**Run** in the pull-down menu or simply click the **Find and run main method or applet** button. This invokes the JVM to interpret the .**class** file. You should be rewarded with the message Hello World! in the **Run I/O** tab at the bottom-right of the jGRASP window, as shown. Use the scroll bars as needed to see the entire message.

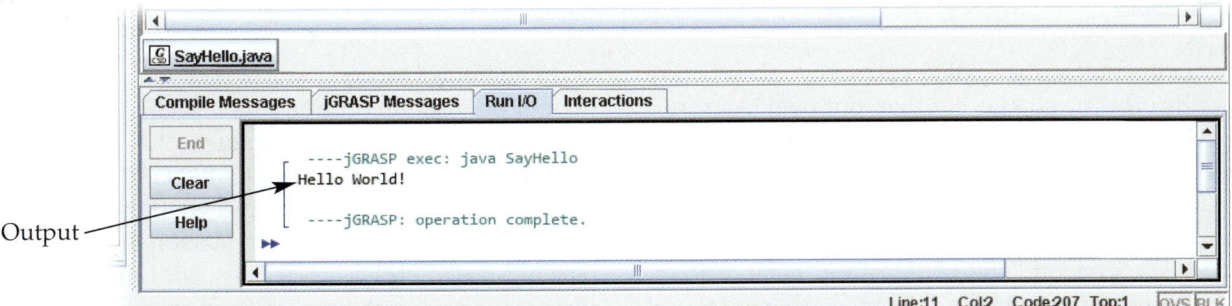

Output

Goodheart-Willcox Publisher

13. Click **File**>**Close** in the pull-down menu to close the current program.
14. Click **File**>**Exit** in the pull-down menu to close jGRASP.

Try It!

To illustrate that much of a Java application is reusable, you will modify the program you just wrote. Open the SayHello.java file in jGRASP. Change the name of the program to MyName. Then, save the program as MyName.java in your Chapter 3 folder. Modify the program to output your name rather than "Hello World!" Compile and run the program. Save your work, close the program, and close jGRASP.

SECTION REVIEW 3.1

Check Your Understanding

1. Explain the purpose of the Java compiler.
2. What does the Java Virtual Machine (JVM) do?
3. Describe the structure of a basic Java program.
4. What is the advantage of inserting blank lines between logical sections of source code?
5. How is a program for a sequential algorithm that does not branch or iterate processed?

Build Your Vocabulary

As you progress through this course, develop a personal computer science glossary. This will help you build your vocabulary and prepare you for a career. Write a definition for each of the following terms and add it to your computer science glossary.

block
block comment
class
comment

Integrated Development
  Environment (IDE)
Java Development Kit (JDK)
keyword

line comment
portability
statement
white space character

Science and Java

Exoplanets

Exoplanets, also called extrasolar planets, are worlds orbiting stars other than our own. Astronomers estimate that there could be trillions of planets around other stars. Within the last few years, astronomers have started to dedicate much more time to the search of other worlds in orbit around other stars beyond the realm of our solar system. Thousands of possible exoplanets have been found through ground-based and space-based observatories. The current number of confirmed exoplanets discovered is more than 500, and new planets are being discovered regularly. You can keep up to date with the latest count at NASA's PlanetQuest site.

Dotted Yeti/Shutterstock.com

The Kepler Mission was launched to search out distant planets. It continues its search today. Other missions that have found distant worlds include the Hubble Space Telescope, the CoROT mission from the European Space Agency, the WISE mission, the TESS space telescope, and the Herschel spacecraft. Ground-based observatories also continue to be an important part of the search for distant worlds.

The oldest exoplanet that has been discovered is PSR B1620-26b. This exoplanet is almost 13 billion years old. It was first discovered in 1993, but its discovery was only confirmed in 2003. The densest exoplanet discovered is Janssen (55 Cancri e). This planet also has a year of just less than 18 hours. It circles its star very quickly. The planet's mass, which is thought to be almost nine times that of the Earth, is almost as dense as pure lead.

Assignment

You have been tasked with creating better names for the oldest exoplanet and densest exoplanet. Come up with two names to assign to these exoplanets. Be creative. Then, write a Java program that displays on two separate lines your new names. Save the program as Exoplanets.java.

Understanding Errors

It would be great if you could write a program, compile it, and run it with perfect output. Alas, that ideal situation almost never happens. Humans tend to make errors. The Java syntax, like the syntax in other programming languages, requires special punctuation and keywords. It is easy for programmers to use the wrong characters or keywords. Perhaps a semicolon is omitted from the end of a statement. Perhaps a closing brace is used where an open brace is required. Perhaps inadvertently you have programmed an equation that attempts to divide by 0.

Do not get discouraged when errors happen. The compiler and JVM will help find some of these errors. Careful inspection of your program and its output can find the others. In addition, some programming techniques can help recognize and recover from mistakes that have been made.

Learning Goals

After completing this section, you will be able to:
- Identify compiler errors.
- Recognize runtime errors.
- Assess logic errors.
- Describe user errors.

Terms

compiler error

logic error

runtime error

user error

Essential Question

Why is it important to fully understand what the program is doing in order to correct errors?

Compiler Errors

It may be impossible to write a program without errors. Complex programs may contain hundreds or thousands of lines of code. It is very easy for mistakes to be made when writing this much code. However, even simple programs of just a few lines may contain errors. Four types of errors can occur when programming:
- compiler errors
- runtime errors
- logic errors
- user errors

A *compiler error* occurs when the compiler finds incorrect Java syntax in the source code. Java is strict with its syntax rules. In jGRASP, if the compiler finds problems with the syntax of a program, an error message is displayed in the **Compiler Message** tab.

For example, in the SayHello program, try deleting the semicolon after the statement on line 8 and compiling again. You will receive the compiler error shown in **Figure 3-5.** This message conveys some helpful information. First, the message lists the file name and the line number (8) where the compiler found the incorrect syntax.

Math and Java

Patterns in Unicode Character Sets

Since 1991, Unicode has lightened the load for programmers around the globe. Before Unicode, there were many encoding systems to support different languages, symbols for math and science, and symbols for other technologies. The Unicode standard contains more than 65,000 codes for characters from alphabets and symbols used around the world. Unicode makes it simple to share files because consistent coding of characters saves time otherwise spent encoding for a different character set. It also reduces errors that come along with data manipulation. Unicode is also used for emojis.

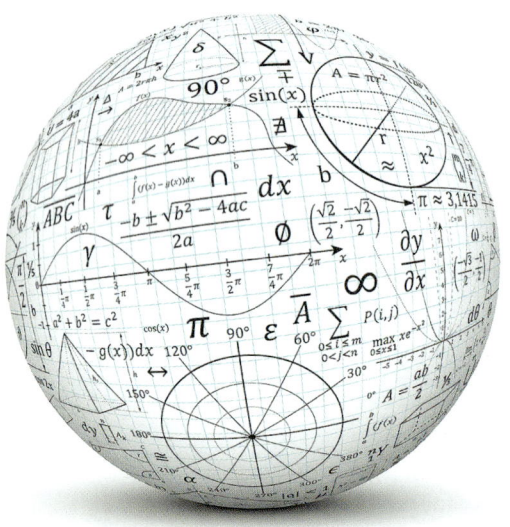

Senoldo/Shutterstock.com

Within the standard are many patterns that make character manipulation easy. For example, consider the codes for the uppercase and lowercase English alphabet. They are found in the range of 0000–007F in the Basic Latin chart. View the Unicode character charts at www.unicode.org/charts.

Before looking for patterns, complete this warmup exercise to ensure you can read the chart. To find the Unicode for the E character, look below the letter in the chart. The code is 0045. Find the Unicode for the characters e and 5. Then, express these hex codes in binary.

Character	Unicode in Hex	Unicode in Binary
E	0045	0000 0000 0100 0101
e		
5		

Examine the hex codes. What pattern do you see? The three codes are the same except for the third hex digit. Examine the binary codes. What pattern do you see? The three codes are the same except for the third set of four digits.

Assignment

Examine the numeric and lowercase and uppercase alphabetic characters. Can you suggest a reason why they start at 30, 61, and 41 respectively? Examine the uppercase and lowercase alphabetic character codes. Find the explicit difference between them. Compare the assignment of the codes with the positions of the characters on the keyboard. Find other patterns in this chart and be prepared to share with your fellow students.

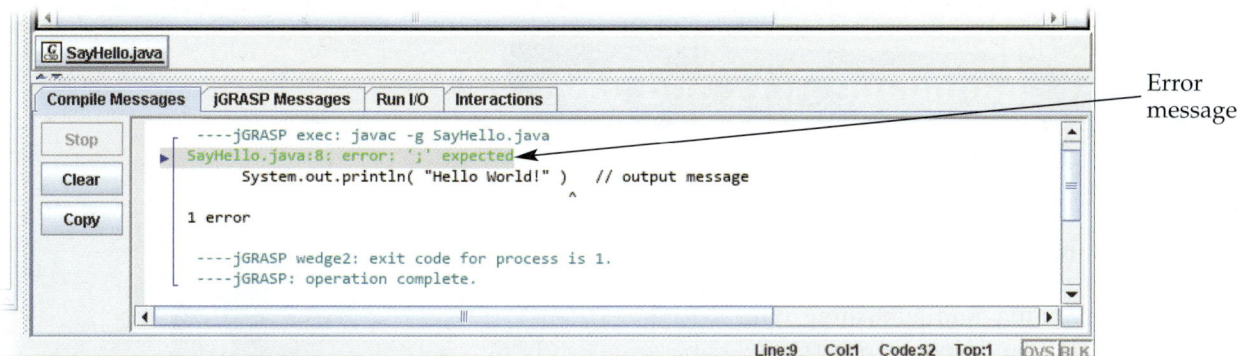

Figure 3-5. This compiler error is the result of incorrect syntax. A semicolon is expected, but was not found.

This is followed by a brief explanation of the problem:

';' expected

The entire message tells you a semicolon is expected on line 8. To correct this error, replace the semicolon at the end of line 8 and compile again.

In this example, the compiler clearly recognized the cause of the error. Unfortunately, the compiler does not always correctly diagnose the problem. You will see this firsthand throughout the course. In some cases, one syntax error can cause the compiler to generate multiple error messages. If this happens, remain calm, and fix one error at a time. Some helpful guidelines are provided in the appendix for dealing with not-so-clear error messages from the compiler.

HANDS-ON EXAMPLE 3.2.1

Creating a Compiler Error

In this activity, you will introduce syntax errors into the SayHello program. This will create compiler errors that will be displayed.

1. Launch jGRASP, and open the SayHello.java file created in Hands-On Example 3.1.1.
2. Change the parentheses in line 8 to braces (curly brackets), and compile.

Compile
file

3. How many error messages did you receive? Notice that the compiler correctly diagnosed the line number, but not the problem. In this case, careful inspection of the statement flagged by the compiler is the best response.
4. Replace the braces in line 8 with the correct parentheses, and compile to receive a clean compile.
5. Remove the brace from line 10, and compile. Did the compiler error clearly diagnose the error?
6. Replace the end brace on line 10, and compile to receive a clean compile.
7. Save and close the file.

Try It!

Before beginning this activity, download the chapter files from the student companion website. For this activity, open the file SayHelloWithErrors.java in jGRASP. This file contains two errors. Compile the file. Only one error is reported at this time. What is it? Use the appendix to look up the error message and follow the guidance to correct the error. Compile the file again. A second error is reported. What is it? Correct this error. Then, compile until you have a clean compile. Save and close the file, and close jGRASP.

Language Arts and Java

Describing Error Types

The writing process is a set of sequential steps for each writing task. These steps are prewriting, writing, postwriting, and publishing. The writing process can be used to create reports and papers. It can also be helpful in programming, especially if the program output is written material.

Prewriting is the planning stage. This involves defining the purpose and topic of the writing as well as analyzing the audience. Then, information is gathered, ideas are researched, thoughts are organized, and the publishing medium is selected.

Writing involves creating a first draft. After the first draft is created, it is revised to improve the message. Then, it is edited for grammar, mechanics, and spelling. Writing also involves soliciting feedback from others.

Postwriting involves a final edit of the work. Proofread the work for spelling, punctuation, and typographical errors. Check for grammar errors as well.

Publishing is preparing the work for distribution. This involves making the work available for the intended user. In the case of writing a message to be included in a program, publishing involves coding the message into the final program.

WAYHOME studio/Shutterstock.com

Assignment

For each of the four types of programming errors, write an example of the mistake. Follow the writing process to develop each example. Then, write a Java program that outputs a list of the error types and your examples. Follow the model program. Output each error and example on separate lines. Since there are four types of errors, the program should have eight System.out.println commands. Save the program as CopingWithErrors.java.

Runtime Errors

A *runtime error* occurs when the program stops when an attempt is made to run, but the compiler found no errors. One example of something that would cause a runtime error is attempting to divide by 0.

To force a runtime error in the SayHello program, delete the word static from line 6. This line defines the main method. Then, compile the program. You should not receive any compiler errors. However, now try to run the program. Instead of the Hello World! message being output, the dialog box shown in **Figure 3-6** is displayed. This indicates a runtime error. The pertinent information in this error message is that the JVM does not know where to begin the program. The main method, as originally written in line 6, indicates to the JVM where the code for the program starts. To correct the error, the word static must be reinserted.

Logic Errors

A *logic error* is the result of incorrect algorithms or faulty code. For example, a logic error would occur if a program is designed to calculate a sum, but the final sum is incorrect. If this occurs, you know that either the algorithm on which the program is based is faulty or the source code incorrectly implemented the algorithm.

In a simple program such as the SayHello program, a logic error would be present if "Hello World" is misspelled. If the output line were omitted, a logic error would also occur because the program would never say Hello World.

Catching logic errors early is a good practice. Rather than waiting until the entire program is written, coders use a technique called a *unit test.* To perform unit testing on a completed section of the program, the coder chooses an input for which the correct outcome can be predicted. Then, the coder runs the program, enters the chosen value, and verifies the output against the predicted value. This can then be repeated with known invalid data.

User Errors

The people who run a program, the users, can make mistakes, too. A *user error* is the result of an incorrect action by the person using the program. For example, consider a game that asks the user to guess a secret number between 1 and 10. The user might misunderstand what is requested and enter 11, −1, or even the letter A. Obviously, the user's entry will not match the secret number. The programmer cannot control user errors. However, a well-designed program will notify the user of the error. In this example, the game can prompt the user to enter a valid value and continue.

Java has some programming techniques to use for recognizing an error like this. Sometimes, however, the user may enter incorrect information that is difficult, if not impossible, to detect. For example, assume that the user makes a

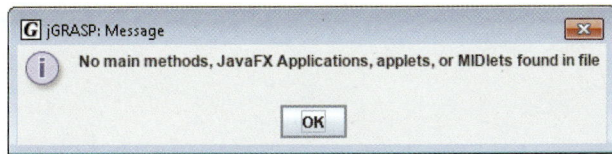

Figure 3-6. A runtime error is one that occurs when an attempt is made to run the program, but the compiler found no errors.

Coding Conundrum

You have just begun your career as a programmer. One of your first tasks is to identify the errors in several snippets of code. Look at each of the following snippets. Identify the error based on the message, explain the problem, describe the consequence, and propose a fix.

1. When this statement is compiled:

```
system.out.println( "Hello World!" );
```

this compiler message appears:

```
SayHello.java:8: error: package system does not exist
   system.out.println( "Hello World!" );  // output message
   ^
```

2. When this statement, which is saved in the file named Program 2.java, is compiled:

```
public class Program2 {
  // lots of good code here
}
```

this compiler message appears:

```
SayHello.java:5: error: class Program2 is public, should be declared in a file named Program2.java
   public class Program2 {
                ^
```

3. When this program is compiled:

```
1 /* Program to say hello
2    your name here
3
4
5 public class SayHello {
6    public static void main( String [ ] args ) {
7
8       System.out.println( "Hello World!" );  // output message
9    }
10  }
```

this compiler message appears:

```
SayHello.java:1: error: unclosed comment
   /* Program to say hello
   ^
```

4. When this program is compiled:

```
5 public class SayHello {
6    public static void main( String [ ] args ) {
8       System.out.println( "Hello World!" );
9
10  }
```

this compiler message appears:

```
SayHello.java:10: error: reached end of file while parsing
     }
     ^
```

typo and enters the wrong ZIP code for an address. If an error is not detectable, the program may simply continue with incorrect data and consequently produce incorrect output. This is a common problem. It is so common, in fact, that there is a phrase used in computing for this: garbage in, garbage out. Techniques for recognizing and handling errors are covered later in this text.

SECTION REVIEW 3.2

Check Your Understanding

1. What is a compiler error?
2. How is a runtime error different from a compiler error?
3. What would cause a logic error?
4. What type of test would be performed with known valid and invalid data to verify correctness of the program?
5. Which error type is beyond the control of the programmer, but can be planned for by the programmer?

Build Your Vocabulary

As you progress through this course, develop a personal computer science glossary. This will help you build your vocabulary and prepare you for a career. Write a definition for each of the following terms and add it to your computer science glossary.

compiler error	runtime error
logic error	user error

Cooperative Coding

Devices Using Java

The portability of Java is one of its greatest advantages. The slogan associated with its creator, James Gosling, is "write once, run anywhere," or WORA. With the correct Java Virtual Machine and integrated development environment, programmers can choose to use any hardware that they have. Or, programmers can select the computer platform that is familiar to them.

Java was originally designed for interactive television. However, it was too advanced for the digital television industry at the time Java was developed. So, other applications were found for it. Today, Java can be found on millions of devices, including digital televisions, Android-based smartphones, and automobile systems.

Hadrian/Shutterstock.com

Assignment

Work with the team assigned by your instructor. Research devices that run Java applications or Android apps. This task can be divided into smaller tasks by distributing the platforms to research:

- desktop, laptop, and tablet computers
- cell phones
- wearable devices
- home appliances
- automobiles
- supercomputers

Remember, all a platform needs to run Java applications or apps is a JVM.

1. List the devices your team found by category.
2. Which devices were you most surprised used Java? As a team, prepare a brief statement explaining why.

Chapter Summary

Section 3.1 Java Program Structure

- Java source code is written in a .java file, which is then compiled into a .class file that can be executed by the Java Virtual Machine (JVM).
- A class is a unit of code that works together, a method is code that performs one or more operations consisting of statements that perform the work of the program; use white space and indentation to help make the code more readable.
- The jGRASP integrated development environment uses color coding to identify keywords, comments, and other parts of the code, which can help identify errors and typos.

Section 3.2 Understanding Errors

- The four types of programming errors are compiler, runtime, logic, and user.
- Compiler errors occur when the compiler finds incorrect syntax.
- Runtime errors are not found by the compiler, but occur when the program stops when an attempt is made to run it.
- Logic errors are the result of an incorrect algorithm or faulty code and are indicated when the output does not make sense.
- User errors are the result of incorrect action by the user; while the programmer cannot control user errors, programming can be designed to handle what happens when a user makes an error.

Chapter 3 Test

Multiple Choice

Select the best response.

1. What is the function of the main method?
   A. signals to the JVM where the program code starts
   B. turns code into machine language
   C. interprets the code line-by-line
   D. compiles the coded punctuation

2. Which program or function is used to convert source code into bytecodes?
   A. Java Virtual Machine (JVM)
   B. Java compiler
   C. Microsoft Word
   D. Google Chrome

3. If the source code includes this line:

   ```
   public class SayGoodbye
   ```

   what must be the name of the source code file?
   A. sayGoodbye.java
   B. SayGoodbye.java
   C. sayGoodbye.class
   D. SayGoodbye.class

4. The types of programming errors are:
   A. Programming, inputting, and processing.
   B. Networking, methods, and binary.
   C. Compiler, runtime, logic, and user.
   D. Statement, white space, and decimal.

5. What does this error message indicate is the problem?

   ```
   MyName.java:15: error: ';' expected
   ```

   A. Line 15 is missing from the source code.
   B. Line 15 in the source code is missing a semicolon (;).
   C. The user has made an incorrect data entry.
   D. The Java Virtual Machine is unable to run the program.

Completion

Complete the following sentences with the correct word(s).

6. A(n) _____ exists for every type of computer platform to turn Java bytecodes into machine language.

7. The _____ is an application that provides a source code editor for the programmer to enter Java code.

8. The punctuation to signal the start of a block is _____.

9. If the program's output is incorrect, a(n) _____ error has occurred.

10. The programmer cannot control _____ errors, but code can be included to handle what happens if this type of error occurs.

Matching

Match the correct term with its definition.

   A. public, class, void, static
   B. compiler error
   C. logic error
   D. block comment
   E. portability

11. Caused by a faulty algorithm.

12. Caused by a typo or incorrect punctuation.

13. Capability to run a programming language on multiple computers.

14. Keywords in Java code.

15. Notation found at the beginning of the code including the programmer's name and the date the code was written.

Application and Extension of Knowledge

1. Some computer applications display a splash screen for the user to view while waiting for the program to load. For example, the University of Baltimore might use text characters that make large versions of the letters U and B, as shown below. Before beginning these activities, download the chapter files from the student companion website. The UB.java file will be used as a starting point for this activity. Open this file in jGRASP. Include block comments for your name, the date, and the program's purpose. Extend your knowledge of printing a message by including a tab character in your message. The code for a tab is \t. Examine the file to see how \t is used to build the U.

```
U     U
U     U
U     U
U     U
U     U
 UUUUU

BBBBBB
B    B
BBBBBB
B    B
B    B
BBBBBB
```

2. A friend is starting an orchid nursery and wants you to write a program that displays the company name, tag line, and e-mail address. Launch jGRASP, and open the SayHello.java program you created in this chapter. Change the class name to Orchids, and save the program as Orchids.java. Edit the comment to reflect the new use for this file. Then, modify the file to print the information your friend has requested.

```
Orchids by Kate
New Varieties and Old Standbys
kate@OrchidsByKate.com
```

3. The following is an example of a flowchart, which is a tool that programmers use to solve problems. Write a program that follows the chart. Launch jGRASP, and open the SayHello.java program you created in this chapter. Change the class name to Textbook, and save the program as Textbook.java. Edit the comment to reflect the new use for this file. Then, modify the program as needed.

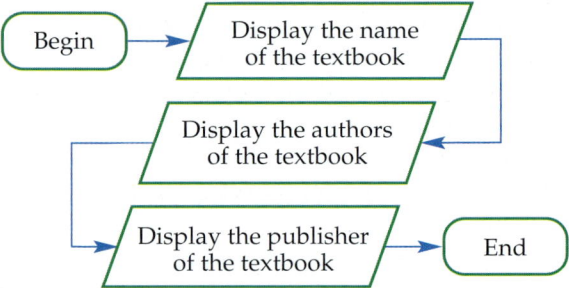

4. Write a Java program to print contact information for a fictional person. Launch jGRASP, and open the SayHello.java program you created in this chapter. Change the class name to ContactInformation, and save the program as ContactInformation.java. Edit the comment to reflect the new use for this file. Then, modify the program to print a line for each of these data points for the fictional person: name; occupation; phone number; Twitter handle; address including city, state, and ZIP code; and e-mail address. Include a title for the information.

5. The code below contains two errors of different types. Identify the errors, and state which type they are.

```
1 /* Program to demonstrate error types
2    your name here
3 */
4
5 public class NamingErrorTypes {
6
7    public static void main( String [ ] args ) {
8
9      System.our.println( "The quick brown fox" );
10     System.out.println( "jumped over the lazey dog." );
11
12   } // end main
13 } // end class
```

Online Activities

Complete the following activities, which will help you learn, practice, and expand your knowledge and skills.

Vocabulary. Practice vocabulary for this chapter using the e-flash cards, matching activity, and vocabulary game until you are able to recognize their meanings.

Communication Skills

Reading. A *sentence* is a group of words that expresses a complete thought. In the English language, a complete sentence has a subject and a predicate. The *subject* is the person speaking or the person, place, or thing a sentence describes. The *predicate* describes the action or state of being for the subject. Select three sentences in this chapter. Identify the subject and predicate in each. Exchange your sentences with a partner to check each other's work.

Writing. Select three key terms from the next chapter that you have not encountered before. Write a definition for each key term in your own words by drawing on your prior knowledge. *Prior knowledge* is experience and information you already possess about the term. Compare the definitions you wrote to those that appear in the glossary of this text.

Speaking. Developing effective communication skills requires individuals to be able to participate in and contribute to discussions. A *discussion* is a speaking situation in which two or more individuals share their ideas about a subject and talk about them with one another. Contribute thoughtful, relevant comments when participation is invited during class discussion. Ask questions if you need help determining or clarifying the meaning of the topics discussed during the lesson.

Listening. *Contextual clues* are hints that can help you define an unfamiliar word by considering the surrounding words or sentences. For example, consider how these sentences help explain the meaning of the word *clout*: "Celebrities often believe they have clout. Many think they have the power to get whatever they want." Using contextual clues can help you understand unfamiliar words you hear as well as read. When you hear an unfamiliar word during class, listen for contextual clues to help you understand its meaning.

Portfolio Development

College and Career Readiness

Checklist. Once you have written your portfolio objective, consider how you will achieve the objective. It is helpful to have a checklist of components that will be included in your portfolio. The checklist will be used to record ideas for documents and other items that you might include. Starting with a checklist will help you brainstorm ideas that you want to pursue.

The elements that you select to include in your portfolio will reflect your portfolio's purpose. For example, if you are seeking acceptance into college as a computer programming major, create a portfolio that includes your best Java programs.

1. Ask your instructor for a checklist. If one is not provided, use the Internet and research student portfolio checklists. Find an example that works for your purpose.

2. Create a checklist. This will be your road map for your portfolio.

CTSOs

Objective Test. Some competitive events for CTSOs require that entrants complete an objective component of the event. This event will typically be an objective test that includes terminology and concepts related to a selected subject area. Participants are usually allowed one hour to complete the objective test component of the event. To prepare for an objective test, complete the following activities.

1. Read the guidelines provided by your organization.

2. Visit the organization's website and look for objective tests that were used in previous years. Many organizations post these tests for students to use as practice for future competitions.

3. Look for the evaluation criteria or rubric for the event. This will help you determine what the judge will be looking for in your presentation.

4. Create flash cards for each vocabulary term with its definition on the other side. Ask a friend to use these cards to review with you.

5. Ask your instructor to give you practice tests for this chapter of the text that would prepare you for the subject area of the event. It is important that you are familiar with answering multiple choice and true-false questions. Have someone time you as you take a practice test.

Variables

Sections

4.1 Identifiers and Data Types

4.2 Variable Values

The job of a program is to input data, process that data, and output results. For each program, the data, the processing, and the output is different. For example, for a program written to calculate the circumference of a circle, the input data would be the circle's diameter and the value of pi. The calculation would be pi multiplied by the diameter. The output would be the result of the calculation, which is the circle's circumference.

Java provides ways to keep the data in memory while a program executes. Each data item is given a name. In doing so, what kind of data the item is must be specified as well as how the data is intended to be used. In the case of calculating a circumference, the names diameter, pi, and circumference might be used. Java can then be told these will be real numbers. Eventually, the programmer will tell Java what to do with the numbers after they have been identified and stored.

College and Career Readiness

Reading Prep

Before reading this chapter, read the essential question for each section. Write a short paragraph in response to the question. Share your answer with a classmate. Discuss how each other's answers relate to the chapter topic.

While studying, look for the activity icon for:

- Vocabulary terms with e-flash cards and matching activities.
- Starter files for hands-on examples and other exercises.

These activities can be accessed at
www.g-wlearning.com/informationtechnology/1773

Chapter Glossary

call: To use a method.

concatenation operator: Joins data values and String literals into one message for output; represented by the plus sign (+).

constant: Value that cannot change during the execution of the program.

data type: Format in which the data will be stored and the size the variable will be given in memory.

exception: Error detected by a method from which it cannot recover.

hard-coding: Defining a variable value in the code.

identifier: Name the programmer chooses for a variable in the program; also used as names for programs and for any methods the program defines.

literal value: Textual representation of a value.

primitive data type: One of eight data types built into Java.

prompt: Message displayed to the user indicating what to input or what action to take.

real number: Number with a fractional part.

String literal: Sequence of characters enclosed in quotation marks.

variable: Named memory location containing data that can change from one execution of the program to another.

Identifiers and Data Types

Choosing names for data serves several purposes. Named data can be stored in memory while the program executes so the value can be retrieved as needed. Also, naming data allows a generalized algorithm to be created that works with any data. To use other data for a program, those other data values are simply stored in the names that have been defined. Using the circumference example, code can be written that multiplies the diameter by pi to calculate the circumference. The diameter is a variable in this case. Then, for each execution of the program, a different value can be stored for the diameter. Without changing the code, the program will calculate the correct circumference for the current value of the diameter.

What level of importance should be placed on selecting identifiers for a Java program?

Learning Goals

After completing this section, you will be able to:
- Select identifiers for data.
- Identify correct data types.
- Illustrate the process for defining variables in Java.

Terms

data type
identifier
primitive data type

real number
variable

Identifiers

A *variable* is a named memory location containing data that can change from one execution of the program to another. In some cases, a program might change the value of a variable as the program executes. In other cases, a variable may store data input by the user.

Identifiers are names the programmer chooses for the variables in programs. Identifiers are also used as names for programs and for any methods the program defines. For example, the main method was used in the previous chapter. The identifier is main.

Java's rules for creating identifiers are shown in **Figure 4-1.** As you can see, a name must start with a letter, underscore, or dollar sign. Numbers are allowed after the first character, but a name cannot start with a number. One important rule is that spaces are not allowed in names. Also important is that names are case-sensitive, meaning that number1, Number1, and NUMBER1 are all considered to be different identifiers. Although an underscore is permitted anywhere in a name, a name consisting of only an underscore is not allowed.

Always remember that spaces are not allowed in Java names.

Rules for Creating Java Identifiers
• Must start with a "Java letter," which is any lowercase (a–z) or uppercase (A–Z) letter, an underscore (_), or a dollar sign ($); letters in other languages represented by Unicode characters are also allowed.
• After the first letter, can contain Java letters or digits (0–9).
• Cannot contain any spaces.
• Is case-sensitive.
• Cannot be a Java keyword or reserved word.

Goodheart-Willcox Publisher

Figure 4-1. Certain rules must be followed when creating an identifier.

Java's keywords and reserved words also cannot be used as names. Java's keywords and reserved words are shown in **Figure 4-2.** These words have special meaning in the Java language. You may notice familiar words from the SayHello program in Chapter 3: public, class, static, and void. Other words may be unexpected, such as new, this, and continue. **Figure 4-3** shows examples of valid and invalid identifiers.

Data Types

In addition to naming each piece of data in the program, programmers also need to tell Java the kind of data that will be assigned to the name. The *data type* is the format in which the data will be stored and the size the variable will be given in memory. You must use the data type that matches the value of the

Java's Keywords and Reserved Words				
(underscore) _	continue	for	new	switch
abstract	default	goto	package	synchronized
assert	do	if	private	this
boolean	double	implements	protected	throw
break	else	import	public	throws
byte	enum	instanceof	return	transient
case	extends	int	short	try
catch	final	interface	static	void
char	finally	long	strictfp	volatile
class	float	native	super	while
const				

Goodheart-Willcox Publisher

Figure 4-2. These are keywords and reserved words in Java. They cannot be used as identifiers.

Valid		Not Valid	
Identifier	**Explanation**	**Identifier**	**Explanation**
george	all letters	Janette Jones	spaces are not allowed
Alisha_Jones	underscores are allowed	2small	numbers cannot start an identifier
big$Cost	dollar signs are allowed	public	public is a Java keyword
my3rdAttempt	numbers are allowed after the first character	im@school	the @ character is not allowed

Goodheart-Willcox Publisher

Figure 4-3. These examples show valid and invalid identifiers.

data. *Primitive data types* are Java's eight built-in data types. **Figure 4-4** shows the size for each primitive data type as well as the minimum and maximum values each can hold.

Java Primitive Data Types

Four of the primitive data types can hold integers: byte, short, int, and long. These four data types differ in their memory size. Because values are stored in binary, the number of bytes allocated to a data type determines its maximum and minimum values. In a program that manages large amounts of data, execution time and storage space can be saved by allocating the data type with the fewest bytes required to hold the largest or smallest value. If the value is an integer, choose and either use the byte, short, int, or long data type. To simplify, in this text, the int data type will be used when the data is an integer.

The float and double data types hold real numbers. A *real number* is a number with a fractional part. That fractional part can be expressed with a decimal point, as in 415.67, or with scientific E notation, 4.1567E2. Notice that $4.1567E2 = 4.1567 \times 10^2$. If the value is a real number, choose and use either the float and double data type.

The float and double data types differ in the number of digits of precision that can be stored in that data type. The float data type holds what is called single

FYI

Real numbers are also called floating-point numbers because of the decimal points. The name of the data type float comes from this concept.

	Data Type	Number of Bytes	Minimum Value	Maximum Value
Integers	byte	1	−128	+127
	short	2	−32,768	+32,767
	int	4	−2,147,483,648	+2,147,483,647
	long	8	approximately −9 quintillion	approximately +9 quintillion
Real Numbers	float	4	+1.4E-45 (minimum positive nonzero value)	+3.4028235E38
	double	8	+4.9E-324 (minimum positive nonzero value)	+1.7976931348623157E308
Characters	char	2	0000 (hex)	FFFF (hex)
Boolean	boolean	undefined	N/A	N/A

Goodheart-Willcox Publisher

Figure 4-4. Java contains eight built-in data types called primitive data types.

precision and the double holds what is called double precision. To simplify, we will use double when the data is a real number.

The char data type holds one Unicode character. This may be a letter or number. It may also be a punctuation character. If the value is a character, choose and use the char data type.

The boolean data type can be either of the reserved words true or false. This data type has no specified size. A boolean variable might be used to test whether a condition is true or false. Note that boolean, true, and false all start with a lowercase letter.

Choosing the correct type for each data item is important to finding the solution to a problem. The data type must be large enough to store all possible values for the data, but not too large as to waste memory and CPU time.

For example, when YouTube set up a counter for the number of views for a video, it defined the counter as an int. Given that an int can store a value of more than two billion, it seemed to be a good choice. However, YouTube did not anticipate the overwhelming popularity of "Gangnam Style," which soon exceeded two billion views. As a result, YouTube needed to redefine the view counter as a long. This data type can store a value of more than nine quintillion (9E18)! The new value should be sufficiently large for the near future, at least.

Another example of choosing the wrong size for data is what was called the Y2K (Year 2000) Problem. In the early days of computing, memory was limited and expensive. COBOL programmers cleverly chose two digits to hold the year. For example, 86 meant 1986. What the programmers did not foresee was that their programs would still be used many years later when the first two digits of the year would change. When the year changed to 2000, the two-digit value would become 00. This made it seem that 1999 (99) was 99 years later than 2000. The problems this could cause were many.

For example, it was predicted the computers that controlled elevators would calculate that the time for maintenance was long past and, therefore, permanently hold the elevators on the bottom floor. Also, bus, train, and airplane schedules that depend on dates would be upended because January 1, 2000 would appear to be January 1, 1900.

Because of these potential problems, some people predicted a disaster when the year would change to 2000. Companies called COBOL programmers out of retirement to redefine the year and update the mountains of data already stored in two digits. It was a large undertaking, but because of those efforts, on January 1, 2000, the Time Square New Year's ball fell into a calm and orderly business-as-usual new day.

Typed and Untyped Languages

A *typed language* is any one that uses explicit data types. An *untyped language* is a language that does not have a type system or all variables are considered one type. Most high-level languages have a typing system. A type system is viewed as a set of constraints that are verified at compile time. Some typed languages include Java, C++, Visual Basic, Python, Perl, and Ruby. Untyped languages do not make you define the type of a variable. JavaScript is an example of an untyped language. This means that a JavaScript variable can hold a value of any data type. For the most part, untyped languages, such as most assembly languages, allow any operation to be performed on any data, which are generally considered to be sequences of bits of various lengths.

The current programming philosophy is type-driven languages are preferred. They are more efficient and correct. If the data is structured, that infers a structure for the program.

FYI

The boolean data type is named after the English mathematician George Boole who invented Boolean logic. Boolean logic is the foundation of digital computer logic.

Math and Java

Two's Complement

Apply what you have learned to write the Java statements that assign 5 to a variable named plusFive with a byte data type and −5 to a variable named minusFive, also a byte data type. These are the required statements:

```
byte plusFive = 5;
byte minusFive = -5;
```

It is the nature of a higher-level programming language that you do not need to know what is happening at the machine-code level. Nevertheless, it is good to understand the storage formats of data. Storing positive numbers as binary numbers is straightforward. Negative numbers are a different story.

CPU designers most often employ a scheme called Two's Complement to represent both positive and negative integers. The leftmost bit is reserved for a sign: 0 means the number is positive, 1 means the number is negative. This is called the *sign bit.* The remaining bits are used for the value, as shown. One notable impact is that because the leftmost bit cannot be used as a value, the number of bits that can be used for the value is reduced. In a byte, for example, instead of having all eight bits available to represent the binary value, there are only seven bits available for the data value. The largest binary number that fits in seven bits is 127.

0	1	1	1	1	1	1	1

The Two's Complement scheme is not obvious, but it is simple, and it works! First, write the positive number. Second, flip each of the bits in that number. If the bit is a 1, make it a 0. If the bit is a 0, make it a 1. Last, add 1 to the rightmost bit. Do not forget to carry if a 1 is already in the rightmost bit.

This binary addition table can be used for your work.

+	0	1
0	0	1
1	1	10

To use this table, find the number at the intersection of the two numbers you are adding. For example, 0 + 1 = 1 and 1 + 1 = 10.

Example 1

Show the binary representations of the two variables defined above.

plusFive

0	0	0	0	0	1	0	1

First, write the positive number.   0000 0101
Second, flip the bits.   1111 1010
Third, add 1 to the rightmost bit.   1111 1011

minusFive

1	1	1	1	1	0	1	1

If the Two's Complement scheme worked, the sum of 5 and −5 will be 0. Add the two binary numbers to verify that the sum is 0.

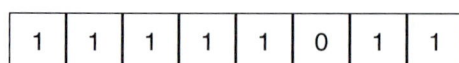

```
      0000  0101
  +   1111  1011
  1  0000  0000
```

(continued)

Math and Java

(Continued)

Use the eight rightmost bits for the byte. The value is indeed 0. Two's Complement delivers the correct result.

Example 2

Show the binary representations of 32 and −32.

32

0	0	1	0	0	0	0	0

First, write the positive number.   0010 0000
Second, flip the bits.   1101 1111
Third, add 1 to the rightmost bit.   1110 0000
Remember, 1+1 = 10, so write the 0 and carry the 1, as shown.

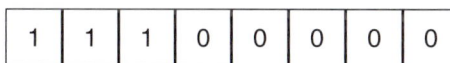

	1	1		1	1	1		Carry row
1	1	0	1		1	1	1	1	Bits-flipped row
+								1	Add 1 row
1	1	1	0		0	0	0	0	Two's Complement

−32

1	1	1	0	0	0	0	0

Assignment

1. Show the content of the following variables in binary format.
   A. byte bakersDozen = 13
   B. byte temperature = −11
   C. byte centuryAgo = −100
   D. byte minNegative = −128

2. The above examples used the byte data type to conserve space. The Two's Complement process is similar for the int, short, and long data types. Show the content of the following variables in binary format.
   A. int halfDozen = 6
   B. short temperature = −10
   C. long centuryAgo = −100

3. Explain why the largest value for the byte data type is 127 while the lowest is −128.

4. Speculate why the Two's Complement scheme always puts a 1 in the leftmost bit of a negative number.

5. Can you think of another way to represent negative numbers in a memory location? Explain how the sum of this number and its opposite will be 0.

Defining Variables

To define a variable, provide a data type and a name followed by a semicolon. It is a good programming practice to choose names that clearly indicate what data will be stored in the variable. For example:

```java
int numberOfPets;   // an integer

double price;       // a real number

char letterQ;       // a single letter

boolean isRaining;  // true or false
```

From the names, you can deduce what data these variables will store. A rule of thumb is to define identifiers for the next programmer. Self-documenting identifiers make it easy for the next programmer to understand your code. Often, the next programmer is you. If you return to edit a program even a few days later, clear, meaningful identifiers will immediately remind you of what you were trying to accomplish.

It is also possible to define multiple variables of the same data type by separating each name with a comma. For example:

```java
int numberOfBedrooms, numberOfBathrooms;   // 2 ints

double itemPrice, salesTax, finalPrice;    // 3 doubles

char letterX, letterY, letterZ, nine;      // 4 chars

boolean isCloudy, isSnowing, isWinter;     // 3 booleans
```

Notice a form of CamelCase is used in the names: the initial letter is lowercase and internal words are capitalized. The Java naming convention for variables is to start the name with a lowercase letter and capitalize any internal words. It is strongly recommended to follow this standard to improve the readability of the code. An advantage of following this practice is that with case-sensitive names, if you know the name of the variable, you will know which letters are capitalized and which letters are lowercase.

Hands-On Example 4.1.1

Naming Data and Identifying Data Types

Suppose you want to paint your bedroom. You decide to write a program to calculate the square footage of the bedroom walls, the amount of paint you will need, and the total cost of the paint. Assuming that your bedroom is rectangular, this is the data needed for the program:

- height of the room in feet
- width of the room in feet
- length of the room in feet
- number of square feet one gallon of paint will cover
- number of gallons of paint needed
- price of a gallon of paint in dollars and cents
- total cost to paint the bedroom

1. Write down a meaningful identifier for each data item using Java's standard naming convention. The first two data items are as follows. The other identifiers you must create.

   heightInFeet

   widthInFeet

2. Give a sample value for each data item. Identify and choose an appropriate data type for each data item, int or double. The first two data items may be typed as follows.

   heightInFeet might be 8 and would be an int

   widthInFeet might be 11.5 and would be a double

3. Write the Java statements to define the variables. The first two data items are defined as follows.

   int heightInFeet;

   double widthInFeet;

Try It!

As part of your bedroom renovation, you decide to lay carpet on the floor. You want to write a program to calculate the total cost. The square footage is calculated by multiplying the number of feet in the width by the number of feet in the length. Carpet is sold by the square yard. There are nine square feet in one square yard. Identify the data items needed for this program. For each data item, select an appropriate identifier and data type, then give the Java statement to define the variable. A table like the following can help keep the information organized.

Data Item	Identifier	Data Type	Variable Definition

SECTION REVIEW 4.1

Check Your Understanding

1. A Java identifier must begin with a(n) _____, _____, or _____.
2. Why is the identifier static invalid in Java?
3. List the eight primitive data types in Java.
4. Which type of programming language is Java, strongly typed or strongly untyped?
5. When defining a variable, which comes first, the data type or the identifier?

Build Your Vocabulary

As you progress through this course, develop a personal computer science glossary. This will help you build your vocabulary and prepare you for a career. Write a definition for each of the following terms and add it to your computer science glossary.

data type

identifier

primitive data type

real number

variable

Variable Values

After the data items for a program have been identified, meaningful names assigned, and the appropriate data types declared, values need to be stored in variables. Java provides several ways for storing values in data items. One way is to assign a value directly by entering a textual version of the value. This is useful for some data, especially if the data value does not change from one execution of program to the next. When using this method, if the data value needs to change, the code would need to be edited and recompiled to take advantage of the new value.

A second method is to ask the user to input a value. In this case, the user will enter a value using the keyboard or other input device and that value is assigned to the variable. This makes for more flexible execution of programs. Each execution of the program can process new input values without needing to recompile.

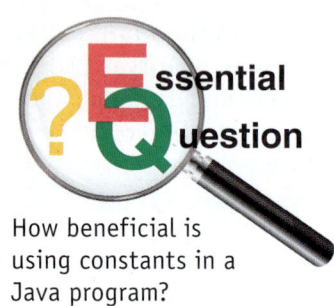

Essential Question

How beneficial is using constants in a Java program?

Learning Goals

After completing this section, you will be able to:
- Assign a literal value to a variable.
- Set up proper constants.
- Obtain input from the user.

Terms

call	hard-coding
concatenation operator	literal value
constant	prompt
exception	String literal

Assigning Values

Once an identifier is defined for a variable, the programmer must provide values for that variable. Numbers, single characters, or strings can be provided. A constant value can be defined that will remain the same for the entire program or input can be obtained from the user. Java will convert the values to binary to store in memory. Therefore, there are rules for describing the input to Java.

Assigning Literal Values

A *literal value* is simply a textual representation of a value. To assign a value directly to a variable, Java provides the assignment operator, which is the equals sign (=). The equals sign should be familiar from creating initializations in math class, such as let x = 6. In a Java assignment statement, the target variable is given first, then the equals sign, then the value to be assigned. Thus, the assignment statement works from right to left. The Java syntax is:

```
targetVariable = value; // assign value to targetVariable
```

Variable definitions and assignments are statements, so they must end with a semicolon. The data type and value can be combined in the same statement, if

FYI

Take care: assignment is different from an equation in algebra. It is *not* a statement of equality. It is an active storage of information in the program.

desired. The data type of the variable is provided only once—when the variable is initially defined. After that, only the name of the variable is used. The compiler remembers the data type. Also, once the data type is assigned, it cannot be changed: once an int, always an int. Once a double, always a double. This is true for all data types. Some sample assignment statements are as follows.

```
int numberOfPets;  // define the variable
numberOfPets = 5;  // then assign a value

double price = 4.99; // define and assign value together

boolean isRaining; // define the data type
isRaining = false; // then assign a value

char letter = 'b';   // define and assign value together; a char
//value is enclosed in single quotes
```

Programmers often refer to the assignment operator as "gets" because the variable is receiving, or "getting," a value. So the last statement above would be read as "letter gets b." The above values assigned to the variables (5, 4.99, false, and b) are called literals. Assigning values in this manner is called hard-coding. *Hard-coding* is defining a variable value in the code.

Java's rules for literals that can be assigned to variables are summarized in **Figure 4-5.** Note that a literal for a long variable must be terminated with an l or an L. Because a lowercase l can easily be confused with the number 1, it is recommended to use the uppercase L, as in:

```
long population = 923456670L;
```

Although two data types for real number are available, you may find it easier to use double by default. One of the reasons is that to define a float variable, you need to remember to add an F at the end:

```
float taxRate = .15F;
```

whereas a double can be defined without any terminating character:

```
double taxRate = .15;
```

Data Types	Valid Literal Values	Examples
int, byte, short	Only numbers with an optional sign, no commas, dollar signs, or percentage signs (%).	int numberPeople = 2534; int depth = −86;
long	Same as for other integer types except the number must be terminated with a lower- or uppercase L.	long countryNo = 8325678990L;
double	Decimal or scientific format with an optional sign; no commas, dollar signs, or percentage signs.	double price = 34.76; double gdp = 19.39E12
float	Same as for doubles except the number must be terminated with a lower- or uppercase F.	float discount = .20f;
char	Any printable keyboard character enclosed in single quotes. A decimal value between 0 and 65,535. Any two-character escape sequence begun with a backslash; for example, '\n' represents a new line character and '\t' represents a tab character.	char wow = '!' char omega = 937; char tab = '\t';
boolean	One of the values true or false.	boolean isPassing = true; boolean isFailing = false;

Figure 4-5. Java has rules for how literals are formed.

A variable can be assigned multiple values over the course of the program's execution. However, a variable can have only a single value at any one time. These statements show that values can be assigned when the variable is defined and new values assigned later in the program:

```
int reynaAge = 11;    // reynaAge gets 11
int suriAge = 14;     // suriAge gets 14
suriAge = 15;         // suriAge is changed to 15
int benAge = suriAge; // benAge gets 15
```

Also, any variable can be assigned the value of a variable that can hold a smaller value. For example, the value of a short variable can be assigned to an int or a long, or even to a float or double.

Assigning String Literals

Remember from Chapter 3 that the statement used for output was:

```
System.out.println( "Hello World!" );
```

The message to output was placed inside quotation marks. The message in this statement is actually a String literal. A *String literal* is a sequence of characters enclosed in quotation marks ("). String is not a primitive data type. Rather, String is a Java class. Classes are discussed in detail in Chapter 6.

When a String literal is placed inside parentheses in an output statement (System.out.println), the characters are outputted exactly as coded. When we put a variable name inside the parentheses, the value of the variable is outputted. With these statements:

```
int numberOfBooks = 2000;
System.out.println( numberOfBooks );
```

the output is:

```
2000
```

Although outputting the value of a variable is helpful, it might leave the user wondering what the 2000 means. Labels can be provided for data values using String literals in combination with the concatenation operator (+). The *concatenation operator* joins data values and String literals into one message for output. For example, with these statements:

```
int numberOfBooks = 2000;
System.out.println( "I have " + numberOfBooks + " books." );
```

the output is:

```
I have 2000 books.
```

Notice the concatenation operator is used twice. The concatenation operator is needed each time the message switches between a String literal and a variable. To insert spaces before and after the variable value, place them into the String literal, as shown in **Figure 4-6.**

```
System.out.println( "I have " + numberOfBooks + " books." );
```
                              ↑   ↑
                            Space   Space

Figure 4-6. To have spaces appear before and after the variable's value, they must be added to the code.

One last rule is that String literals must start and end on the same line. If you have a long message, you can break it into two String literals. Then, concatenate the two literals.

Defining Constants

Some data values are known and do not change: number of inches in a foot, number of days in a week, ounces in a quart, and so on. Known values like these are called constants. A *constant* is a value that cannot and should not change during the execution of the program. To define a constant, use the keyword final in front of the definition. Here are some examples:

```
final int DAYS_IN_WEEK = 7;
final double CENTIMETERS_PER_INCH = 2.54;
```

Notice that the Java naming convention for constants is different from other variables. All letters are capitalized, and words are separated by underscores.

An advantage to defining constants is to avoid what are called "magic numbers." These are numbers that "magically" appear in a program and lead another programmer to ask, "Where did that number come from?" For example, 32 is a magic number in this statement:

```
double totalOunces = quarts * 32;  // 32 is a magic number
```

Instead, it would be much clearer to define 32 as a constant:

```
final int OUNCES_IN_QUART = 32;
double totalOunces = quarts * OUNCES_IN_QUART;
```

Another advantage to defining a value as a constant is that Java will monitor the use of that variable. It will stop any attempt by the program to change the constant's value after it has been assigned a value.

HANDS-ON EXAMPLE 4.2.1

Defining Variables

In this exercise, you will define variables and output their values. Before beginning this exercise, download the chapter files from the student companion website. The starter file VariableSnippet.java is used in this exercise.

1. Launch jGRASP, and open the VariableSnippet.java file.
2. Add your name to the header comment.
3. Examine the code. You will see five places where comments direct your activity. The work for comment 1 is completed for you. The variable doubleBurgerCalories has been defined and its value outputted in a sentence.

```
/* VariableSnippet -- define variables of primitive types
   your name here
*/

public class VariableSnippet {
    public static void main( String [ ] args ) {
        /***** 1. Define a variable named doubleBurgerCalories that has
            the value 563 and output its value.
        */
        int doubleBurgerCalories = 563;
        System.out.println( "The calories in a double hamburger are "
          + doubleBurgerCalories + "." );
```

(Continued)

```
        // more code here

    }
}
```

4. Compile and run the program to see the output.

5. Locate comment 2, and follow the instructions. The value 6745 is an integer, so the data type should be int. Enter the following code below the comment. Remember to insert a space after walk and before steps. Then, compile and run the program.

```
/***** 2. Define a variable named targetSteps that has the
      value 6745 and output its value.
*/

int targetSteps = 6745;
System.out.println( "Today, I want to walk " + targetSteps + " steps." );
```

6. Locate comment 3, and follow the instructions. The value .5 is a real number, so the data type should be double. Enter the following code below the comment. Then, compile and run the program.

```
/***** 3. Define a variable named oneHalf that has the value .5
      and output its value.
*/

double oneHalf = .5;
System.out.println( "One-half as a decimal is " + oneHalf + "." );
```

7. Locate comment 4, and follow the instructions. The asterisk (*) is a character, so the data type should be char. Enter the following code below the comment. Then, compile and run the program.

```
/***** 4. Define a variable named star that has the value *
      and output its value.
*/
char star = '*'; // note the apostrophes around the asterisk
System.out.println( "I earned a " + star + " for my work today." );
```

8. Locate comment 5, and follow the instructions. The number of inches in a foot is an integer and does not change. So the data type should be an int set as a constant. For a constant, insert the keyword final before the data type. Enter the following code below the comment. Then, compile and run the program.

```
/***** 5. Define a constant for the number of inches in a foot
      and output its value.
*/

final int INCHES_IN_FOOT = 12;
System.out.println( "There are " + INCHES_IN_FOOT
            + " inches in one foot." );
```

9. Check the final output. Does it match this text?

```
The calories in a double hamburger are 563.
Today, I want to walk 6745 steps.
One-half as a decimal is 0.5.
I earned a * for my work today.
There are 12 inches in one foot.
```

Try It!

In the VariableSnippet file, add code to define each of the following variables using a meaningful name and appropriate data type. Output a sentence with the value of each variable. Compile and run the program to check your work.

- number of puzzle pieces in a jumbo puzzle; assign the value 1000
- price of your favorite fruit snack; assign the value .89 and include a dollar sign in the output

Inputting Values from the User

Another way to assign a value to a variable is to ask the user for input. The advantage to user input is that the program is more flexible. If using hard-coded values and the data values need to change, the code must be edited and recompiled before the program is run again. With user input, the program can be run multiple times, and the user can enter different values each time.

To input values from the user, the Scanner class is used from the Java Class Library. The Java Class Library is a collection of prewritten code that programmers use repeatedly. It is better to write these code elements once and make them available to programmers whenever they need them. This library is divided into packages that keep similar classes together. In order to use this code, programmers simply tell the compiler to look in the library for it. The Scanner class is explored in detail in Chapter 6. For now, simply follow three steps to get input.

1. Import the Scanner class.
2. Create a Scanner object.
3. Ask the user for input, and read the inputted value into a variable.

The first step is to import the Scanner class using the following statement. This statement is placed after the identification block comment and *before* the definition of the class name.

```
import java.util.Scanner;
```

The compiler needs to know where the Scanner class is defined. The import statement tells the compiler where to find it in the Java Class Library. In the above example, the import statement specifies the Scanner class is in the package java.util. All classes in the Java Class Library are downloaded with the Java JDK.

The second step is to create a Scanner object to use for input. Put this statement inside the main method:

```
Scanner input = new Scanner( System.in );
```

The System.in reference means the keyboard. So, this statement creates a Scanner object named input that will be used to input the data the user enters with the keyboard.

The third step is to ask the user for the input and read the input into a variable. This is done for each data item to input. To ask the user for input, a message called a prompt is outputted. A *prompt* is a message to the user indicating what the user needs to input or what action to take. To read the user input, call one of the Scanner methods shown in **Figure 4-7.** To *call* means to use a method. To call these methods, prefix the method name with the name of the Scanner object created (input) and a period or dot (.). The empty parentheses after the method name are required. The Scanner class provides the two methods in **Figure 4-7** for reading integers and real numbers. The Scanner class also has methods for reading the other primitive data types, except for chars.

Method	Explanation
nextInt()	Read an integer from the user
nextDouble()	Read a real number from the user

Goodheart-Willcox Publisher

Figure 4-7. Two of the methods available with the Scanner class.

For example, suppose a program calculates the area of a circle based on a radius the user enters. After importing Scanner and creating a Scanner object, these statements could be used:

```
System.out.println( "Enter the radius of the circle:" ); // prompt
double radius = input.nextDouble( );                      // read value
```

When this code runs, the prompt is displayed in the output window. Then, the program pauses and waits for the user to input a value and press the [Enter] key.

Science and Java

Scuba Diving

The word scuba comes from the acronym for *self-contained underwater breathing apparatus*. It allows divers to breathe underwater with no connection to the surface. Since the divers are not hooked to a tube, they can swim almost anywhere. The divers carry diving cylinders containing a gas mixture at high pressure that is supplied to the diver through a regulator for breathing underwater. The open-circuit scuba systems discharge the exhaled gas into the environment.

Dudarev Mikhail/Shutterstock.com

The normal atmospheric pressure at the surface is 14.7 pounds per square inch (psi). As a diver descends, the water exerts increasing hydrostatic pressure of approximately 14.7 psi for every 33 feet of depth. If you have ever swum to the bottom of the deep end of a pool to retrieve an object, you can feel the pressure in your ears. The pressure of the inhaled breath must balance the surrounding increased pressure to allow inflation of the lungs.

It becomes impossible to breathe air at normal atmospheric pressure through a tube below 30 feet under the water.

The depth range for scuba diving depends on the application and training. The major worldwide recreational diver certification agencies consider 130 feet to be the limit for recreational diving. For most recreational diving, ordinary atmospheric air can be used (21 percent oxygen, 78 percent nitrogen, 1 percent trace gases). If the diver is experienced and competent, he or she may attempt dives deeper than 130 feet. In that case, the diver needs more oxygen and less nitrogen. The most commonly used mixture is Nitrox. Nitrox is air with extra oxygen (between 32 percent to 36 percent). Therefore, it will contain less nitrogen by percentage. This reduces the risk of decompression sickness, which causes nitrogen gas in the blood to become bubbles as the diver ascends. Blood does not flow if it contains bubbles.

Assignment

Define a set of variables and constants to measure certain aspects of preparing to dive. You need the following data: a constant for the amount of pressure per square inch, a constant for the percentage of oxygen in air, a constant for the percentage of nitrogen in air, a variable for the depth of the dive in feet, a variable input by the user for the percentage of oxygen in a Nitrox gas mixture, and a variable that stores whether or not that is enough oxygen. Percentages are integers. These variables and constants will be used in a later activity.

Whatever value the user enters is then assigned to the variable radius. Refer to **Figure 4-8.** In this case, the user has input 23.5 and pressed the [Enter] key. The variable radius now contains the value 23.5.

The user can be prompted and input accepted on the same line. To do this, a different version of the output statement is used. Use print, rather than println. The difference between the two is that println adds a new line character to move the cursor to the next line. The cursor remains on the same line with print. Using print with the same prompt and user input as in the previous example:

```
System.out.print( "Enter the radius of the circle: " ); // using print
double radius = input.nextDouble( );
```

the output (prompt) and user input are:

```
Enter the radius of the circle: 23.5
```

Note that a space needs to be added to the end of the prompt to separate the message from the input.

It is important to use the correct Scanner method for the data type being entered. For example, if the program attempts to read an integer using the nextInt method, but the user enters a real number or even a letter, the operation will fail. The nextInt method will generate a runtime error called an exception, and the program will stop. An *exception* is an error detected by a method from which it cannot recover. In this case, the exception will be InputMismatchException. Because the input does not match the data type expected by the method, the method cannot continue. With what you have learned to this point, there is not much to do if the user enters data that does not match the data type the program is attempting to read. Later in the course, ways to avoid this exception or to recover from it gracefully will be discussed.

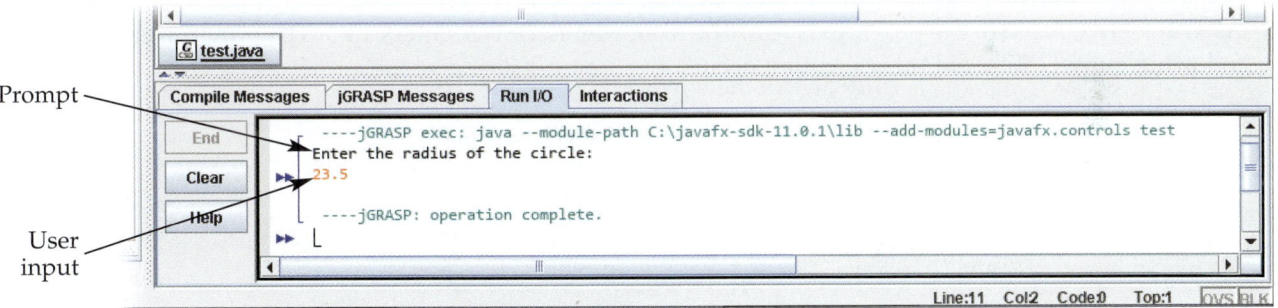

Goodheart-Willcox Publisher

Figure 4-8. The program issues a prompt, and the user inputs a number as data.

Coding Conundrum

The following code has been written to prompt the user for his or her age. However, when the program is compiled, an error is reported. It is a conundrum!

```
1  /* Inputting an age
2  */
3
4
5  public class Conundrum {
6    public static void main( String [ ] args ) {
7       Scanner input = new Scanner( System.in );
8
9       // prompt and read the user's age
10      System.out.print( "Enter your age " );
11      int age = input.nextInt( );
12
13      System.out.println( "You are " + age + " years old." );
14
15    }
16 }
```

Error Report

```
Conundrum.java:7: error: cannot find symbol
Scanner input = new Scanner( System.in );
^
symbol: class Scanner
location: class Conundrum
Conundrum.java:7: error: cannot find symbol
Scanner input = new Scanner( System.in );
                    ^
symbol: class Scanner
location: class Conundrum
2 errors
```

1. What is the problem, and how can you fix the error?
2. After the error is fixed, the program is run. The user, who is 15 1/2, enters 15.5. The program stops with the following output. Why did this exception occur?

```
Enter your age 15.5
Exception in thread "main" java.util.InputMismatchException
at java.base/java.util.Scanner.throwFor(Scanner.java:939)
at java.base/java.util.Scanner.next(Scanner.java:1594)
at java.base/java.util.Scanner.nextInt(Scanner.java:2258)
at java.base/java.util.Scanner.nextInt(Scanner.java:2212)
at Conundrum.main(Conundrum.java:11)
```

Language Arts and Java

Scuba Diving

Safety is of critical importance to the sport of scuba diving. Scuba divers must know the equipment associated with the sport. Look at the photograph. Find at least four pieces of gear that are necessary to safely perform scuba diving. If you do not know the names of the equipment, research the correct names. You will create a program that lists the equipment as an introduction for the user to see.

blue-sea.cz/Shutterstock.com

A critical facet of programming is to make the output easy to understand for the user. Writing output using proper English, punctuation, spacing, and sentence structure is essential. Telling the user what is in the program as well as how to enter information are very important. This communication must be clear. Unclear instructions or output can hinder the usability of the program. Typos and other grammatical errors will make the program look unprofessional to the user.

Assignment

In the Science and Java feature in this chapter, you created constants and variables for future calculation. Review that activity for the explanations of the identifiers. Now, you will create a program to display all constants while describing what they are. Launch jGRASP and begin a new file. Enter the following code into the editor, adding your name to the header comment:

```
/** Describing scuba equipment
   your name here
*/

public class ScubaCompleted {
   public static void main( String [ ]
args ) {
      // write your code here

   }
}
```

1. Add code to display a sentence listing the gear that a scuba diver needs. The output should be presented using proper sentence construction and grammar.
2. Define each constant and use its value in a descriptive sentence. Use concatenation to list the values. Pay careful attention to spaces between words and punctuation
3. Save the file as *ScubaCompleted.java*. Compile and run the program.

HANDS-ON EXAMPLE 4.2.2

Asking for User Input ⤤

In this exercise, you will ask the user for input and then output what the user entered. Before beginning this exercise, download the chapter files from the student companion website. The starter file InputSnippet.java is used in this exercise.

1. Launch jGRASP, and open the InputSnippet.java file.

2. Add your name to the header comment.

3. Examine the code. You will see three places where comments direct your activity. The work for comment 1 is completed for you. The variable booksRead has been defined, a prompt issued to request input from the user, and a message output to the user.

```java
/* Practice with input
   your name here
*/

import java.util.Scanner; // this imports the Scanner class

public class InputSnippet {
    public static void main( String [ ] args ) {
        Scanner input = new Scanner( System.in ); // create the input object

        /***** 1. Prompt the user for the number of books they have read
               this year, read the value into a variable of the appropriate
               type, and output the value in a message.
        */
        System.out.print( "How many books have you read this year? " );
        int booksRead = input.nextInt( );

        System.out.println( booksRead + " is a lot of books!" );

        // more code here

    }
}
```

4. Compile and run the program to see the output.

5. Locate comment 2, and follow the instructions. The number of people in your family is an integer, so the data type should be int and the Scanner method should be nextInt( ). Enter the following code below the comment. Then, compile and run the program.

```java
/***** 2. Prompt the user for the number of people in their family,
       read the value into a variable of the appropriate type,
       and output the value in a message.
*/

System.out.print( "How many people are in your family? " );
int peopleInFamily = input.nextInt( );

System.out.println( "You have " + peopleInFamily
                + " people in your family." );
```

6. Locate comment 3, and follow the instructions. The sum of two real numbers is a real number, so the data type should be a double and the Scanner method should be nextDouble( ). Enter the following code below the comment. Then, compile and run the program.

```java
/***** 3. Prompt the user for the sum of 1.55 and .20,
       read the value into a variable of the appropriate type,
       and output the value in a message.
*/

System.out.print( "What is 1.55 plus .20? " );
double sum = input.nextDouble( );

System.out.println( "1.55 plus .20 is " + sum + "." );
```

Try It!

In the InputSnippet file, add code to prompt the user for two more values, read those values into variables, and write a message using each value in a sentence. Create comments 4 and 5 to describe what you are doing. Identify and choose the correct data types to use. Compile and run the program to check your work.

- input the sales tax rate for the user's state
- input the number of courses the user is taking

SECTION REVIEW 4.2

Check Your Understanding

1. Which symbol is used as the assignment operator in Java?
2. A String literal is surrounded by _____.
3. Describe the function of the concatenation operator.
4. How is a variable in Java set to be a constant?
5. What does this statement do, and when would it be used?

```
Scanner input = new Scanner( System.in );
```

Build Your Vocabulary

As you progress through this course, develop a personal computer science glossary. This will help you build your vocabulary and prepare you for a career. Write a definition for each of the following terms and add it to your computer science glossary.

call	exception	prompt
concatenation operator	hard-coding	String literal
constant	literal value	

Cooperative Coding

Inputting and Outputting Data

You and your team are chaperoning a group of 4th graders on a trip to your local zoo. Each member of your team will choose one animal that is found at the zoo. So, if your team has four members, you will select four animals. As a team, research each animal's average lifespan, the average weight of the animal, how many pounds of food it eats each day, and how many minutes of daily exercise it needs. This data will be used as input for a Java program.

Assignment

Launch jGRASP, start a new file, and save it as Zoo.java. List all programmers (team members) in the initial comments. Write a program that asks the user to input the data for each animal that you have selected. You can copy and paste code to minimize the amount of code that needs to be written. Use the data types described in the table and use Java's standard naming conventions for the variable names. After the program has accepted the last of the user input, display the data to the user in sentences. User proper grammar and punctuation in the sentences.

Susan Schmitz/Shutterstock.com

Compile and run the program. Thoroughly test the program. If there are errors, work as a team to find the source and correct the error. Errors may include misspellings, lack of semicolons at the end of lines, and missing braces.

1. Why are lifespan and minutes of exercise set as int data types while weight and pounds of food set as double data types?
2. What process did your team use to find any reported errors and correct them?

Data	Data Type
Lifespan	int
Weight	double
Pounds of food	double
Minutes of exercise	int

Chapter Summary

Section 4.1: Identifiers and Data Types

- Identifiers are names for variables, programs, and methods and must start with a letter, underscore, or dollar sign; numbers are allowed after the first character, but spaces are not allowed.
- The data type is the format in which the data will be stored and the size the variable will be given in memory; Java contains eight built-in data types called primitive data types.
- To define a variable, provide a data type and a name followed by a semicolon; it is a good practice to choose names that clearly indicate what data will be stored in the variable.

Section 4.2: Variable Values

- A literal value is a textual representation of a value and is set in Java with the assignment operator (=); a String literal is a sequence of characters enclosed in quotation marks ("), and the concatenation operator (+) can be used to join data values and String literals into one message for output.
- A constant is a value that cannot change during the execution of the program and is created by placing the final keyword at the beginning of the variable definition.
- The Scanner class is used to input values from the user; the three-step process is to import the Scanner class, create the Scanner object, and ask the user for input with a prompt.

Chapter 4 Test

Multiple Choice

Select the best response.

1. Which of these identifiers does not follow standard Java naming conventions?
   A. testDate
   B. cleanupTime
   C. winterChill
   D. timechange

2. Numeric data types are chosen based on the _____ of the value that will be placed in them.
   A. size
   B. method
   C. identifier
   D. capitalization

3. A boolean data type can have values of _____.
   A. integers
   B. floating decimals
   C. true and false
   D. large (astronomical) numbers

4. The concatenation operator (+) is used to:
   A. Combine text and variable values.
   B. Add two numbers.
   C. Eliminate the use of double quotes.
   D. As a shortcut for defining variables.

5. Which class is imported when desiring user input?
   A. Scanner
   B. Prompt
   C. Literal
   D. Char

Completion

Complete the following sentences with the correct word(s).

6. A(n) _____ is the punctuation used at the end of a statement that defines a variable.

7. The four primitive data types that can hold integers are byte, short, _____, and long.

8. The data types that can hold real number values are _____ and float.

9. The _____ keyword is used to create a constant.

10. A data mismatch is a(n) _____ runtime error, which can be avoided by using the correct data type.

Matching

Match the correct term with its example.
   A. int
   B. double
   C. long
   D. square feet
   E. public

11. Cost of a gallon of gas
12. Invalid identifier
13. Reserved word
14. Number of eggs in a dozen
15. Width of the Milky Way galaxy

Application and Extension of Knowledge

1. Write a program to ask the user how many hours were worked in the month and how much he or she is paid per hour. Output both values in a sentence.

2. Astronomical distances are measured in light years. A light year is the distance light travels in one year. It is *not* a measure of time. One light year is about six trillion miles. Define and assign a value to a variable to express the distances in miles.

Distant Object	Description	Light Years Away	Miles Away
Proxima Centauri	Nearest Star	4.22	
Sirius	Brightest star in the sky	8.6	
A0620-00	Nearest known black hole	3,000	

3. Write a program that prompts the user to enter if this year is a leap year. The user should enter true or false. Hint: use the Scanner class nextBoolean method to input a boolean value.

4. Punctuation is important in most programming languages. Exact placement of semicolons, parentheses, and braces is essential. Write a short paragraph that describes the errors you have made so far and what you are going to look for if your program does not compile.

5. Refer to **Figure 4-4.** Compare the maximum values that can be stored based on the integer data types. For example, using two bytes instead of one byte increases the maximum value from 127 to 32,767. Two bytes provide a capacity of 32,767, which is 258 times that provided by one byte ($32,767 \div 127 = 258$). Make this comparison for each of the integer data types. Provide an explanation of why there is such a big jump in capacity.

Online Activities

Complete the following activities, which will help you learn, practice, and expand your knowledge and skills.

Vocabulary. Practice vocabulary for this chapter using the e-flash cards, matching activity, and vocabulary game until you are able to recognize their meanings.

Communication Skills

Reading. *Print awareness* is the understanding that printed text is organized in a methodical way. For example, texts written in English contain individual letters that make up words, which are separated by spaces. Words are read from left to right and top to bottom. Print awareness helps readers understand text can be separated by images, pages, or other divisions. These breaks may or may not indicate a break in thought. Review the pages of this chapter. How does your understanding of print awareness help you read and understand the text?

Writing. *Rhetoric* is the study of writing or speaking as a way of communicating information or persuading someone. When you *persuade,* you convince a person to take a proposed course of action. There will be many instances in which you will be required to write paragraphs to persuade a reader. Write a script you could use as the basis for a conversation with a classmate about whether a clean learning environment is important. Use your best persuasive techniques to argue for or against the value of a clean school. Use solid reasoning that will influence your classmate's understanding of the topic.

Speaking. What role do you think ethics and integrity have in decision-making? Think of a time when your ideals and principles helped you make a decision involving ethics. Do you think you made the correct decision? What was the impact of your decision? Deliver an informal speech to your class to narrate the situation that led to the decision. Describe your thoughts and opinions. Explain how ethics and integrity influenced your decision.

Listening. *Active listening* is fully participating as you process what a person says. Listen actively to a classmate as he or she is talking with you and focus on what the person is saying. Provide feedback by asking for more explanation about anything you do not understand.

Portfolio Development

College and Career Readiness

Hard Copy Organization. As you collect material for your portfolio, you will need an effective strategy to keep the items clean, safe, and organized for assembly at the appropriate time. Structure and organization is important when working on an ongoing project that includes multiple pieces. A large manila envelope works well to keep hard copies of documents, photos, awards, and other items. A three-ring binder with sleeves is another good way to store your materials.

Plan to keep similar items together and label the categories. For example, store sample documents that illustrate your writing or coding skills together. Use notes clipped to the documents to identify each item and state why it is included in the portfolio. For example, a note might say, "Printout of first Java program."

1. Select a method for storing hard copy items you will be collecting.

2. Create a master spreadsheet to use as a tracking tool for the components of your portfolio. You may list each document alphabetically, by category, date, or other convention that helps you keep track of each document that you are including.

3. Record the name of each item and the date that you stored it.

CTSOs

Role-Play and Interview. Some competitive events for CTSOs require that entrants complete a role-play or interview. Some competitive events require a role-play or interview. Role-play is representing a situation by acting. An interview is a formal conversation during which one person asks questions and the other person answers them. Those who participate will be provided information about a situation and given time to practice. A judge or panel of judges will review the presentations or conduct the interview. To prepare for the role-play or interview event, complete the following activities.

1. Read the guidelines provided by your organization.

2. Visit the organization's website and look for role-play and interview events that were used in previous years. Many organizations post these events for students to use as practice for future competitions. Also, look for the evaluation criteria or rubric for the event. This will help you determine what the judge will be looking for in your presentation.

3. Practice in front of a mirror. Are you comfortable speaking without reading directly from your notes?

4. Ask a friend or an instructor to listen to your presentation or conduct an interview. Give special attention to your posture and how you present yourself. Concentrate on the tone of voice. Be pleasant and loud enough to hear, but do not shout. Make eye contact with the listener. Do not stare, but engage the person's attention.

5. After you have made your presentation, ask for constructive feedback.

Java Expressions

Sections

5.1 Arithmetic Operators

5.2 Operators and Expressions

5.3 Output Results

One of the advantages of a computer program is that it can be run, or executed, many times with different values. Chapter 4 shows how variables are created to accommodate this. Just as in algebra, a programming variable holds values that can change over time. Expressions describe how the values are combined to get a new value. Programmers create variables and write expressions to calculate results using constants and values stored in variables.

Values for variables can be defined by the programmer or obtained from the user via input to the application. Calculations are made with variables that have the same data type. Digital design supports the combination of values having the same number of bytes. Java can accommodate mixed-type expressions using techniques called casting and promoting. This chapter explores the use of different variable types and expressions to make calculations and output results.

Reading Prep

College and Career Readiness

Before reading this chapter, review the learning goals for each section. Based on this information, write down two or three items you think are important to note while you are reading. How can this help you prepare to understand the content?

While studying, look for the activity icon ⤴ for:

- Vocabulary terms with e-flash cards and matching activities.
- Starter files for hands-on examples and other exercises.

These activities can be accessed at
www.g-wlearning.com/informationtechnology/1773

Chapter Glossary

binary operator: Combines only two values at a time.

bug: Anomaly or error in a program.

casting: Creation of a temporary value for the duration of an operation; also *type casting*.

debug: Remove bugs from a program.

decrement operator: Subtracts 1 from the value of a specified variable.

divide equals operator: Divides the value of a specified variable by a given amount.

echo: To say back what was input so it can be verified the input was correct.

escape character: Indicates the following character has special meaning.

explicit type casting: Occurs when the programmer tells the compiler to convert a value to a different type, which applies for that operation only.

expression: Combination of operators, numbers, constants, and variables that results in a single value.

implicit type casting: Java automatically promotes the operands.

increment operator: Adds 1 to the value of a specified variable.

minus equals operator: Decreases the value of a specified variable by a given amount.

mixed-type arithmetic: Use of an operator with two differently typed variables.

mod equals operator: Divides the value of a specified variable by a given amount and stores the remainder.

modulus operator: Finds the remainder in division.

operand: Each value in an operation.

operator precedence: Order in which operations are performed in Java.

plus equals operator: Increases the value of a specified variable by a given amount.

shortcut operator: Streamlines common expressions to simplify the code and make it easy to read.

test plan: Outlines a systematic approach to evaluating the success of a program.

test set: Combination of sample inputs and outputs for a single test.

times equals operator: Multiplies the value of a specified variable by a given amount.

Arithmetic Operators

Java's arithmetic operators are known as binary operators because only two values can be combined at a time using the operators. Some formulas require multiple calculations, such as converting Fahrenheit temperatures to Celsius. The following formula requires three separate calculations:

$$C = \frac{5(F - 32)}{9}$$

The operations required are multiplication, subtraction, and division. Java supports what you learned in algebra class and allows the combination of multiple arithmetic operators in a calculation. Java follows the order of operations rules to evaluate these formulas. Because the formula is so complex, conversion between temperature scales is a common app for smartphones.

You will recognize most of the arithmetic operations from elementary school. Adding, subtracting, multiplying, and dividing are the fundamentals of combining values in arithmetic. A new operator covered in this section is the modulus, which saves the remainder when two numbers are divided.

?Essential Question

How can your knowledge of the order of operations learned in math class be applied to Java programming?

Learning Goals

- Identify the binary operators in Java.
- Analyze a Java expression for the correct operator precedence.
- Compare a shortcut operator to the full-length version in Java code.

Terms

binary operator
decrement operator
divide equals operator
expression
increment operator
minus equals operator
mod equals operator

modulus operator
operand
operator precedence
plus equals operator
shortcut operator
times equals operator

Binary Operators

In Java, an *expression* is a combination of operators, numbers, constants, and variables that results in a single value. This value is then assigned to a variable in the application. Care must be taken when writing expressions. A misplaced operator or variable will cause the result to be incorrect and could cause syntax errors.

Java has binary operators and shortcut operators. Values are combined in Java using binary operators. A *binary operator* combines only two values at a time. Each value in an operation is called an *operand.* The binary operators in Java are addition, subtraction, multiplication, division, and modulus. **Figure 5-1** summarizes the binary operators for integer and real number data types. Java follows the order of precedence to perform calculations.

FYI

Break down complex algebraic expressions into binary operations to make Java expressions easier to code.

Addition Operator

The addition operator in Java is the plus sign (+). Place this operator between two values and the computer will add the two numbers. Recall the assignment operator (=) is used to save a value to a variable. For example, height + 5 is an expression using the addition operator. The code newHeight = height + 5; adds 5 to the value of the variable height and assigns the result to a variable named newHeight.

Subtraction Operator

The subtraction operator in Java is the minus sign (–). Place this operator between two values and the computer will subtract the second value from the first value. Use the assignment operator to save the new value to a variable. For example, height – 5 is an expression using the subtraction operator. The code newHeight = height – 5; subtracts 5 from the value of the variable height and assigns the result to a variable named newHeight.

Operator	Meaning	Example
+	addition	x = x + 5;
–	subtraction	x = x – 10;
*	multiplication	x = y * c
/	division	x = 12 / a;
%	modulus (remainder after division)	x = 12 % a

Goodheart-Willcox Publisher

Figure 5-1. Binary operators for integer and real number data types.

Multiplication Operator

The multiplication operator in Java is the asterisk (*). Place this operator between two values and the computer will multiply the two numbers. For example, height * 5 is an expression using the multiplication operator. The code newHeight = height * 5; multiplies the value of the variable height by 5 and assigns the result to a variable named newHeight.

In arithmetic, the times sign (×) is used for multiplication. However, Java thinks this is the letter *x* and considers this a variable identifier. Instead, the asterisk character is used to indicate multiplication.

Programmers say *star* when referring to the multiplication operator.

Division Operator

The division operator in Java is the forward slash character (/). Place this operator between two values and the computer will divide the two numbers. For example, height / 5 is an expression using the division operator. The code newHeight = height / 5; divides the value of the variable height by 5 and assigns the result to a variable named newHeight.

In arithmetic, the division sign (÷) is used for division. Sadly, the ÷ character does not appear on a computer keyboard. Instead, Java uses the forward slash character to indicate division. Think of this as the sign used to write a fraction on one line, such as 4/5.

Modulus Operator

In long division, there is a quotient and a remainder. The *modulus operator* finds the remainder in division. The modulus, or mod, operator in Java is the percent sign (%). Place this operator between two values and the computer will divide the first number by the second number and return only the remainder. For example, height % 5 is an expression using the modulus operator. The code newHeight = height % 5; divides the value of the variable height by 5 and assigns the remainder after the division to a variable named newHeight.

The percent sign is the closest keyboard character to the / used for division. Because modulus is also a division operation, with the added step of saving the remainder, % is used to indicate modulus.

HANDS-ON EXAMPLE 5.1.1

Using Binary Operators

In this exercise, you will define variables and output their values. Apply what you have learned about defining variables with proper data types to suit the activity. Before beginning this exercise, download the chapter files from the student companion website. The starter file ArithmeticOperatorsSnippet.java is used in this exercise. The file is called a *snippet* because it contains partial code that you will complete in this exercise.

1. Launch jGRASP, and open the ArithmeticOperatorsSnippet.java file.

2. Add your name to the header comment.

3. Examine the code, and look for comment 1. This block comment begins with /***** 1 and ends with */. After this comment, add code to prompt the user for two costs and output the sum. First, prompt the user for the cost of a jacket and get the input. Money has decimals, so use a double variable. For readability, enter a blank line after the input. Notice that the line comments entered do not address the syntax of the code but state what the code is doing.

```
System.out.print( "Enter the cost of the jacket you bought: " ); // prompt
double costJacket = input.nextDouble(); // read value
```

4. Now, prompt the user for the cost of a tee shirt and get the input.

```
System.out.print( "Enter the cost of the tee shirt: " ); // prompt
double costTee = input.nextDouble(); // read value
```

5. Add the two prices together, store the sum in a double variable, and print the sum.

```
double costPurchases = costJacket + costTee; // evaluate expression
System.out.println( "The cost of purchases is $" + costPurchases );
```

6. Compile and run the code.

7. Locate comment 2. After the comment, define a variable for the difference in the prices and print the result of the subtraction. To separate this output from the sum, first print a blank line.

```
System.out.println( ); // skip a line

double difference = costJacket - costTee; // evaluate expression

System.out.println( "The jacket cost $" + difference + " more than the tee." );
```

8. Compile and run the code.

9. Locate comment 3. After the comment, calculate the cost of two jackets and three tee shirts. Reuse the costPurchases variable for this calculation. To separate this output from the difference, first print a blank line.

```
System.out.println( ); // skip a line

costPurchases = 2 * costJacket + 3 * costTee; // evaluate expression

System.out.println( "The cost of two jackets and three tees is $" + costPurchases );
```

10. Compile and run the code.

11. Locate comment 4. After the comment, calculate how many tee shirts the user could buy with $30. To separate this output from the multiplication, first print a blank line.

```
System.out.println( ); // skip a line

final double MONEY_ON_HAND = 30.00;
double numberTees = MONEY_ON_HAND / costTee; // evaluate expression

System.out.println( "The number of tee shirts you could purchase for $" + MONEY_ON_HAND
                    + " is " + numberTees );
```

12. Compile and run the code.

13. Check your output. The number of decimal points may not be correct. You will learn how to format output later in this chapter. Make edits as needed and save the final version of your file.

Try It!

Operator Precedence

No programmer deals exclusively with binary operations. Often, expressions may combine values and operators to result in a single value. To isolate binary operations, parentheses can be added to let the compiler know what operations must be performed first, second, and so on. However, overuse of parentheses to create operations between two values is not very easy for the programmer to read or follow.

Suppose you need to average four numbers. You would add all four numbers and then divide by four. With binary operations, you could only add two at a time, get a result, and then add two more. Because of the associative and commutative laws of math, the order or associations could vary. This is one way to group the numbers:

$$(((\text{n1} + \text{n2}) + \text{n3}) + \text{n4}) \div 4$$

The compiler would add n1 + n2 and get a result. Then it would add n3 to the result and get a new result. Then it would add n4 to the new result and get the full sum. Then that sum would be divided by 4 to get the average. Observe the individual steps:

$$(((\text{n1} + \text{n2}) + \text{n3}) + \text{n4}) \div 4$$
$$((\text{result} + \text{n3}) + \text{n4}) \div 4$$
$$(\text{newResult} + \text{n4}) \div 4$$
$$\text{sum} \div 4 = \text{average}$$

Apply what you learned in algebra about the order of operations to Java expressions. Java follows the same order of operations. In Java, the order in which operations are performed is referred to as *operator precedence.* The rules are the same as in algebra. These are outlined in **Figure 5-2.** In the Java expression below, see that only one set of parentheses is required.

```
( n1 + n2 + n3 + n4 ) / 4
```

The compiler will apply operator precedence to add n1 and n2 first, then add n3, and then add n4. Finally, it will divide by 4. The only parentheses needed are the ones to say do the additions first. Without the parentheses, following the order of operations, n4 would be divided by 4 first and then the additions would occur.

Java style requires ample spaces to separate the elements in an expression. Paying close attention to operator precedence can eliminate many extra parentheses and makes the expressions easier to read. The code is much cleaner with the proper application of the order of precedence.

Step	Operation	Action
1	( )	Perform all the operations inside a set of parentheses first. With nested parentheses, start with the innermost set of parentheses. Within each pair of parentheses, follow the next three rules.
2	* / %	Scan the expression from left to right. Perform each of these three operations as they are encountered.
3	+ –	Scan the expression from left to right. Perform each of these two operations as they are encountered.
4	=	After the expression on the right of the assignment is evaluated, store the value in the variable on the left of the equals sign.

Goodheart-Willcox Publisher

Figure 5-2. These are the rules for order of operations or operator precedence.

HANDS-ON EXAMPLE 5.1.2

Applying Operator Precedence

In this assignment statement, apply the order of operations to the expression on the right and assign the final value to the variable on the left. Perform the calculation by hand first, then write a snippet to calculate it in Java. Before beginning this exercise, download the chapter files from the student companion website. The starter file TemperatureFormulaSnippet.java is used in this exercise.

$$C = \frac{5\,(F - 32)}{9}$$

1. Let F = 68°.
2. Perform what is inside the parentheses first:

$$(F - 32) = (68 - 32) = 36$$

$$C = \frac{5\,(36)}{9}$$

3. Perform the *, / operations as encountered from left to right:

$$5 * 36 / 9 = 180 / 9 = 20$$
$$C = 20$$

4. Launch jGRASP, and open the TemperatureFormulaSnippet.java file. Add your name to the header comment.
5. Locate comment 1, and follow the instructions to define a double variable for fahrenheit, and assign 68 to it. This way, you can check your code with the hand calculation above. Notice that the ending braces are pushed down in the file as you enter more statements. Be very careful not to delete them.

```
double fahrenheit = 68;
```

6. Write an assignment statement for the formula. Java style requires ample spaces to separate the elements in an expression. Define the celsius variable here, just as it is needed.

```
double celsius = 5 * ( fahrenheit - 32 ) / 9; // convert Fahrenheit to Celsius
```

7. Write an output statement for the user.

```
System.out.println( "The temperature in Celsius for " + fahrenheit
        + " degrees Fahrenheit is " + celsius + " degrees." );
```

8. Compile, run, and verify the output of the snippet. Make edits as required, and save the file.

Shortcut Operators

A favorite instructor of the authors often said, "It is very fashionable to be lazy in computer programming." In the case of shortcut operators, this is true. A *shortcut operator* streamlines common expressions to simplify the code and make it easy to read. These operators also save time entering and reading the code. Shortcut operators in Java include increment, decrement, plus equals, minus equals, times equals, divide by equals, and mod equals. **Figure 5-3** shows the operators.

The *increment operator* adds 1 to the value of a specified variable. The operator is represented by two plus signs (++). For example, to add 1 to a variable called puppies using full code:

```
puppies = puppies + 1;
```

Notice the variable puppies appears twice in that statement. The increment shortcut removes some of the redundancy:

```
puppies++;
```

If the value of puppies was 6 to start, after this statement is executed, the value of puppies is 7.

Shortcut	Operator	Operation
++	increment	Adds 1 to the value of a specified variable.
--	decrement	Subtracts 1 from the value of a specified variable.
+=	plus equals	Increases the value of a specified variable by a given amount.
−=	minus equals	Decreases the value of a specified variable by a given amount.
*=	times equals	Multiplies the value of a specified variable by a given amount.
/=	divided by equals	Divides the value of a specified variable by a given amount.
%=	mod equals	Divides the value of a specified variable by a given amount and returns the remainder.

Goodheart-Willcox Publisher

Figure 5-3. These are the operators in Java.

Language Arts and Java

Health Insurance

Someone who wants health insurance must pay a certain amount every month. The payment is called a premium. Many employers fund health insurance for their employees. Depending on the company one works for, employees may need to pay part of the monthly premium while the employer pays the rest. People who work for themselves must pay the entire premium. Some people cannot afford health insurance or choose not to insure themselves.

Monkey Business Images/Shutterstock.com

A program called Medicare was created in 1965 under President Johnson. It came about because people over 65 found it virtually impossible to get private health insurance coverage. Older people tend to have more health problems than younger people. Medicare has made access to health care a universal right for Americans once they reach age 65. Eligibility for Medicare does not discriminate on a person's income, location, ethnicity, social standing, or gender. This has helped improve the health and longevity of older Americans. The federal government pays the premiums for Medicare through taxes on worker earnings. Seniors also contribute to the costs through premiums deducted from Social Security earnings. Medicare pays approximately 80 percent of hospital and physician bills.

Assignment

Before beginning this exercise, download the chapter files from the student companion website. Launch jGRASP, and open the HealthSnippet.java file. Write a paragraph that summarizes the benefits of Medicare for senior citizens. Do you think this program should apply to American children? Should it apply to all Americans, even those under age 65? Include your reasons, why or why not, in the paragraph. Then, write a Java program that asks the user for his or her birth year. Have the program calculate in what year the user will be eligible for Medicare. Use integer variables for the years.

In evaluating an expression, you can specify if you want to increment the value and then use it (pre-increment) or use the value and then increment it (post-increment). Placement of the increment operator indicates how that should be done, as shown in **Figure 5-4.**

The *decrement operator* subtracts 1 from the value of a specified variable. The operator is represented by two dashes or minus signs (– –). For example:

```
puppies = puppies - 1;
```

is the same as:

```
puppies--;
```

Pre-increment Example	Post-increment Example

int a = 42;
int b = ++a;
• a is incremented and then assigned to b
• both a and b are 43
	int a = 42;
	int b = a++;
	• a is assigned to b and then incremented
	• a is 43 and b is 42

Goodheart-Willcox Publisher

Figure 5-4. Placement of the operator affects the result.

The decrement operator supports pre-decrement and post-decrement.

A shortcut for increasing a value by any number is the plus equals operator. The *plus equals operator* increases the value of a specified variable by a given amount. The operator is represented by a plus sign and an equals sign with no space between (+=). For example:

```
puppies = puppies + 5;
```

is the same as:

```
puppies += 5;
```

This shortcut operator and the following ones do not have pre-versions and post-versions.

A shortcut for decreasing a value by any number is the minus equals operator. The *minus equals operator* decreases the value of a specified variable by a given amount. The operator is represented by a dash or minus sign and the equals sign (–=). For example:

```
puppies = puppies - 5;
```

is the same as:

```
puppies -= 5;
```

A shortcut for multiplying by a number is the times equals operator. The *times equals operator* multiplies the value of a specified variable by a given amount. The operator is represented by an asterisk and an equals sign (*=). For example:

```
puppies = puppies * 5;
```

is the same as:

```
puppies *= 5;
```

A shortcut for dividing by a number is the divide equals operator. The *divide equals operator* divides the value of a specified variable by a given amount. The operator is represented by a forward slash and an equals sign (/=). For example:

```
puppies = puppies / 5;
```

is the same as:

```
puppies /= 5;
```

A shortcut for finding the modulus of a number is the mod equals operator. The *mod equals operator* divides the value of a specified variable by a given amount and stores the remainder. The operator is represented with a percentage sign and an equals sign (%=). For example:

```
puppies = puppies % 5;
```

is the same as:

```
puppies %= 5;
```

Be careful using shortcut operators. Do not confuse the reader of the code. Just because it looks cool, does not mean the code will be readable. A good rule of thumb for readability is, "Just because you could, doesn't mean you should." Always consider the readability of your code. Program with the next programmer in mind; it may very well be you.

SECTION REVIEW 5.1

Check Your Understanding

1. In Java, what are combinations of operators, numbers, constants, and variables that result in a single value?
2. The arithmetic operators are called _____ operators because only two values can be used with one operator.
3. List the symbols for the five arithmetic operators in Java.
4. What is the order of operations known as in Java?
5. If the value of puppies is 6, the result of puppies++; is _____.

Build Your Vocabulary

As you progress through this course, develop a personal computer science glossary. This will help you build your vocabulary and prepare you for a career. Write a definition for each of the following terms and add it to your computer science glossary.

binary operator	minus equals operator	operator precedence
decrement operator	mod equals operator	plus equals operator
divide equals operator	modulus operator	shortcut operator
expression	operand	times equals operator
increment operator		

Science and Java

Physics of Roller Coasters

Roller coasters can be thrilling. The very first sensation is the car being pulled to the top of the first hill. The occupants cannot see what lies beyond the crest. The initial hill is the tallest in the entire ride. At the top, the cars are released. They roll freely along the track without any external mechanical assistance for the remainder of the ride. The purpose of the ascent of the first hill is to build up potential energy that will take the cars on their way to the end of the ride. The potential energy is turned into kinetic energy when the cars go down.

VIAVAL/Shutterstock.com

Energy is measured in joules in the metric system. At the apex of the first hill, the cars have more potential energy than in the rest of the journey. The formula for measuring potential energy is:

$$\frac{\text{potential}}{\text{energy}} = \text{mass} \times \frac{\text{acceleration}}{\text{of gravity}} \times \text{height}$$

Briefly, it is PE = m*g*h. The acceleration of gravity is a constant. It is always 9.5 m/sec^2. Since different roller coasters have different masses and the tracks have different initial heights, two roller coasters generate varying potential energy. The two variables are the mass of the cars (in kilograms) and the height of the roller coaster (in meters).

PitK/Shutterstock.com

Assignment

Before beginning this exercise, download the chapter files from the student companion website. Launch jGRASP, and open the CoasterSnippet.java file. Using the above formula, write a program to calculate the potential energy from two roller coaster rides. The first roller coaster is the Mako in SeaWorld in Orlando, Florida. Its initial height is 61 meters (m) and the mass of the cars is 75 kilograms (kg). The second roller coaster is the Kingda Ka in Six Flags Great Adventure in Jackson, New Jersey. Its initial height is 139 m and the cars have a mass of 80 kg. Use the Scanner class in the program to have the user input the data. Then, the program should display the name of the first roller coaster and its potential energy. Test and debug the program. Once this is error-free, copy and modify the code to display the data for the second roller coaster.

Operators and Expressions

The arithmetic operators are designed to perform one calculation using two values of the same data type. The computer CPU works with integers and floating-point numbers in very different ways. Registers are memory locations within the CPU. Registers are set up to handle integers *or* floating-point numbers, but not both. There is an arithmetic logic unit (ALU) and a floating-point unit (FPU) in the CPU. The default is that integers are combined with integers and floating-point numbers are combined with floating-point numbers.

On occasion, a programmer wants to combine values that are stored in different-sized or different-typed variables. Java provides ways to accommodate this need, but it is important to understand what Java is doing. Meaningless results may occur if the consequences are disregarded.

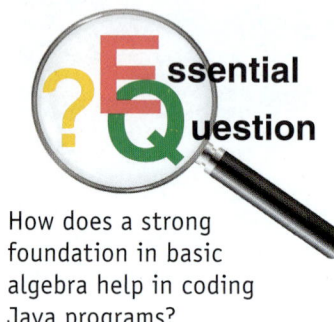

Essential Question

How does a strong foundation in basic algebra help in coding Java programs?

Learning Goals

- Compare and contrast division operations in Java for integer and floating-point numbers.
- Evaluate Java expressions for common expression errors.

Terms

mixed-type arithmetic

casting

implicit type casting

explicit type casting

Data Types and Operations

Data types in Java are important for many reasons. The type assigned to a variable determines how that value is stored in memory and how it is combined with other values. Generally, each data type can only be combined with a value of the same type. Whenever two values that are not of the same type must be combined, Java will promote one of the values to match the higher-precision type. The original variable is not changed, just the value used for the calculation. Because promoting is done automatically and it may not be what the coder wants to happen, the programmer is able to explicitly state how to promote the values. The following topics explain how data types are combined and how to manage the process.

Division for Integer and Floating-Point Variables

Division is carried out differently for integer and floating-point variables. Floating-point values can contain a decimal place, so division is straightforward. However, because integers have no decimal places and division potentially will generate decimal places, a problem pops up: what to do with the remainder? Handling the remainder means division for integers must be approached differently than for floating-point numbers.

Integer Division

The result of dividing an integer by an integer is an integer. Any remainder is truncated or lopped off. The result is not rounded to the nearest integer. The remainder simply disappears. If a programmer is interested in the remainder, the mod operator provides the remainder.

Figure 5-5 shows a pile of loose bagels. Suppose these bagels need to be sorted into boxes by the dozen. Given the number of bagels, how many boxes are needed? This problem involves two values, the number of bagels and the number of boxes. Both values are integers. The following code snippet demonstrates the actions in integer division.

Mattburchell/Shutterstock.com

Figure 5-5. Sorting these bagels into boxes by the dozen requires division of two integers.

```
final int BAGELS_PER_DOZEN = 12;
int bagels = 246;
System.out.println( "There are " + bagels + " bagels." );

int boxes = bagels / BAGELS_PER_DOZEN; // result is 20
System.out.println( "We need " + boxes + " boxes to hold " +
   boxes * BAGELS_PER_DOZEN + " bagels." );

int bagelsLeftOver = bagels % BAGELS_PER_DOZEN; // result is 6
System.out.println( "There are " + bagelsLeftOver +
   " bagels left over." );
```

The output from this snippet is:

```
There are 246 bagels.
We need 20 boxes to hold 240 bagels.
There are 6 bagels left over.
```

Floating-Point Division

When dividing a floating-point number by another floating-point number (double or float), the result is a floating-point number. Suppose you want to calculate miles per hour. Both values, miles and hours, are floating-point numbers. The following code snippet demonstrates the actions in floating-point division.

```
double hours = 10.5;
double miles = 583.8;

double milesPerHour = miles / hours;
System.out.println( "The MPH is " + milesPerHour  + "." );
```

The output from this snippet is:

```
The MPH is 55.6.
```

HANDS-ON EXAMPLE 5.2.1

Solving Integer and Floating-Point Division

What happens in memory as a result of the code below? Create the following table on paper or in a word-processing document. Perform the calculations and assignments like a snippet would do. Record the value of the assigned variable for each row in the table. Show your work. The first row is completed for you. This is an "unplugged" exercise.

Step	Statement	Assigned Variable	Substitute Values for Variables	Value of the Assigned Variable
Step 1	int k = 9;	k	9	9
Step 2	int j = k / 2;	j	9 / 2	
Step 3	int m = k % j;	m	9 % 2	
Step 4	m = k + 6 / 7;	m	9 + 0	
Step 5	m = (k + 6) / 7;	m	(9 + 6) / 7	
Step 6	double a = 3.0;	a	3.0	
Step 7	double b = 10.0;	b	10.0	
Step 8	double c = a / 2.0;	c	3.0 / 2.0	
Step 9	double d = a * (b / 2.5);	d	3.0 * (10.0 / 2.5)	

When your paperwork is complete, jGRASP can be used to check your answers. Use the **Interactions** tab of jGRASP to check your answers, as shown. The **Workbench** tab on the left displays the answers. Enter all nine assignments. Compare the compiler's answers to your answers in the table above.

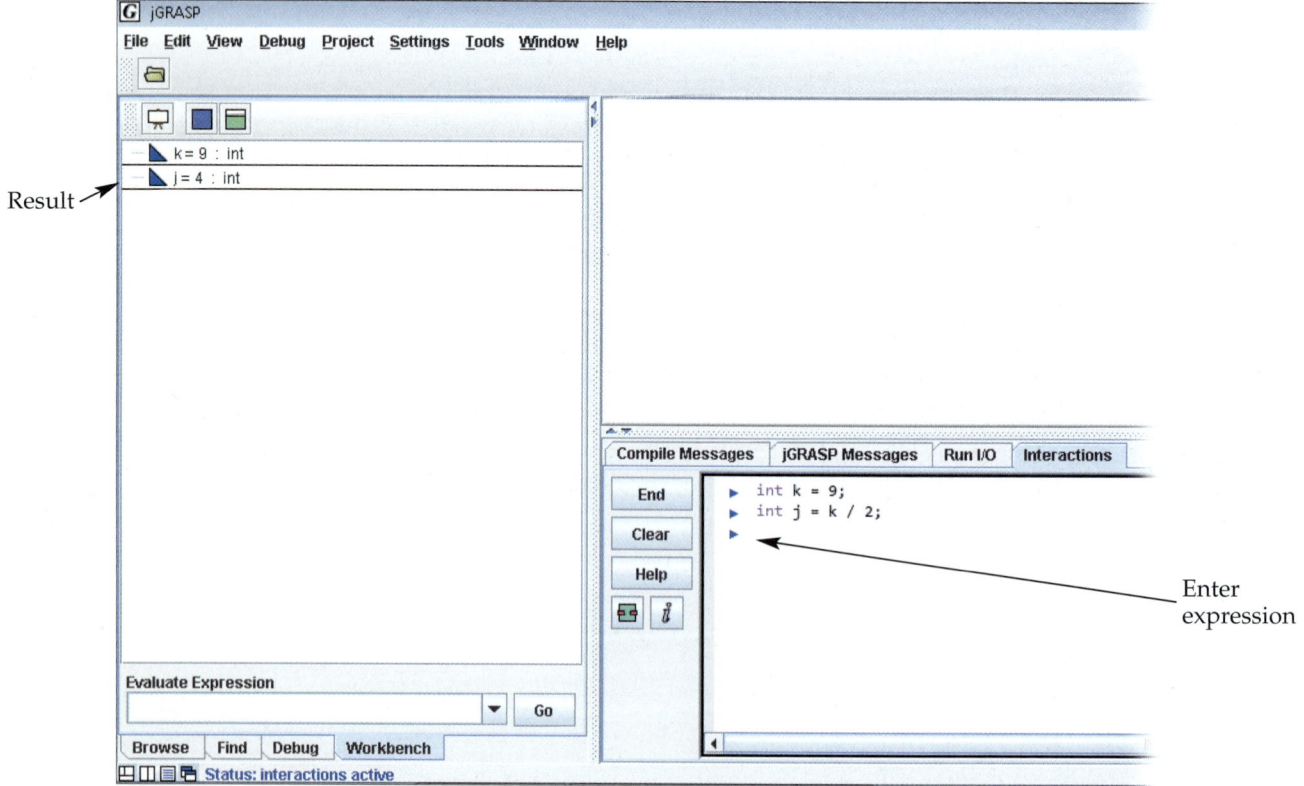

Try It!

Simplify the following expression first using a set of integer values and then using a set of double values. Follow the rules for integer and floating-point division. Use int a, b, c, d for the first set and double a, b, c, d for the second set. Use the **Interactions** tab in jGRASP to check your results. To reinforce the differences, use a = −5, b = 3, c = 4, d = 10 and then delete the first definitions of a, b, c, and d. Redefine a = −5.0, b = 3.0, c = 4.0, d = 10.0 as doubles. Write an observation in a sentence.

$$\frac{-a * (-b - c)}{d^2}$$

Mixed-Type Arithmetic and Type Casting

When a program requires that variables of different types be combined, where should the calculation occur? Which of the registers should be used? If integers are mixed with floating-point numbers, what will happen to the decimal places? These are questions that mixed-type arithmetic and type casting answer.

Mixed-type arithmetic is the use of an operator with two differently typed variables. For example, suppose an expression called for the following combination of values:

3 * 6.75

The 3 is an int type and the 6.75 is a double. This calls for mixed-type arithmetic.

Casting, or type casting, is the creation of a temporary value for the duration of an operation. Type casting occurs when an operation or assignment involves values of two different types. Lower-precision types are *promoted* to, or changed into, the higher-precision types. Type casting can be implicit or explicit.

Implicit Type Casting

In *implicit type casting,* the compiler automatically promotes the operands. Any mixed-type operation with a floating-point value and an integer value will ultimately be performed as a combination of floating-point values. Java provides a set of rules for promotion. There is a hierarchy to the rules. The first rule encountered in the list that applies to a mixed-type operation is the one applied.

1. If one of the operands is a double, the other is promoted to a double.
2. If one of the operands is a float, the other is promoted to a float.
3. If one of the operands is a long, the other is promoted to a long.
4. If one of the operands is an int, the other is promoted to a int.
5. If none of the operands is a double, float, long, or int, both are promoted to int.

In the following example, 2.0 is a double and a is an integer.

```
int a = 7;
double b = a / 2.0;
```

Rule 1 applies: a is promoted to a double for the operation. Its value in the expression becomes 7.0. The operation becomes 7.0 / 2.0 with the result of 3.5 assigned to b.

Explicit Type Casting

At times, a programmer may not want to follow the automatic-promotion scheme. *Explicit type casting* occurs when the programmer tells the compiler to convert a value to a different type, which applies for that operation only. When a floating-point value is explicitly cast as an integer, its decimal value is truncated. Explicit type casting is performed using the new type written in parentheses before the variable. The syntax is:

```
(datatype) expression
```

Consider the earlier bagels example. You may wish to know what fraction of a box is left over. To do this, explicitly cast the integers to double for the division to preserve the decimal remainder.

```java
final int BAGELS_PER_DOZEN = 12;
int bagels = 246;

double boxes = (double) bagels / (double) BAGELS_PER_DOZEN;
System.out.println( "We need " + boxes
            + " boxes to hold " + bagels + " bagels." );
```

The output is:

```
We need 20.5 boxes to hold 246 bagels.
```

This example also illustrates that the variable bagels is not permanently promoted to double. It remains an integer, as displayed in the output.

It is enough to explicitly cast just one of the operands. Then, implicit type casting will promote the other automatically by following the rules for promotion. The following example shows this.

```java
double boxes = (double) bagels / BAGELS_PER_DOZEN;
System.out.println( "We need " + boxes
            + " boxes to hold " + bagels + " bagels." );
```

The output is the same as with the previous snippet in which both operands are explicitly cast:

```
We need 20.5 boxes to hold 246 bagels.
```

Implicit casting is subtle. This presents a drawback in that it may be unnoticed by a future programmer. If casting is important, the best practice is to be very clear about what is happening by using *explicit* casting.

Common Expression Errors

There are several common errors that novice programmers repeat. The compiler will flag some errors for you. Others simply produce incorrect results. Examine the following topics and keep them in mind for future coding. A good rule is to hand check a calculation using a sample of similar data with a known outcome to verify the result is what you anticipated.

Division by Zero Errors

A famous problem in arithmetic is the undefined nature of a division by zero. If 6 / 3 means divide six things so there are three in each pile, what does 6 / 0 mean? Try to divide six things so that there are 0 in each pile. It is meaningless. Likewise, division by zero is meaningless in Java. However, the result depends on the data type.

For integer division, an exception runtime error occurs. The program terminates, and the following error is reported.

```
Exception in thread "main" java.lang.ArithmeticException: / by zero
```

For floating-point division, the program does not terminate, but the results are not numbers:

- If the dividend is nonzero, the result is infinity.
- If both dividend and divisor are 0, the result is NaN (not a number).

```
double result1 = 4.3 / 0.0;
double result2 = 0.0 / 0.0;
System.out.println( "Result 1 is " + result1 + " and Result 2 is "
   + result2 );

// Output: Result 1 is Infinity and Result 2 is NaN
```

Order of Precedence Errors

If there is an unbalanced number of parentheses, the compiler will flag this error. However, balanced but misplaced parentheses may cause a calculation error. For example, a baseball pitcher's earned run average (ERA) is calculated by dividing the number of earned runs allowed by the number of innings pitched. Then, that value is multiplied times nine innings per game. The following example shows an incorrect approach.

```
double earnedRuns = 5.0;
double inningsPitched = 5.33;
double era = earnedRuns / ( inningsPitched * 9 );

System.out.println( "The ERA is " + era );

// Output: The ERA is 0.10
```

The actual ERA is 8.44, but the output is incorrect. Fix this error by removing the parentheses and allowing order of precedence to work.

Mixed-Type Errors

Coders should always check that the promotion is performed at the proper time in the program. Timing can be a problem. The following example shows an incorrect approach. Integer division is performed first. Then, an implicit type cast occurs as assignment is made to a **double**.

```
int sum = 260;
int numItems = 8;
double average = sum / numItems;
System.out.println( "The average is " + average );

// Output: The average is 32.0
```

The integer division occurs before the cast, and truncation occurs. The integer quotient 32 becomes 32.0 when assigned to average which is a double. The average should be 32.5.

The following is another example of an incorrect approach. Integer division is performed first. Then, the explicit type cast occurs before the assignment.

```
double average = ( double ) ( sum / numItems );
System.out.println( "The average is " + average );

// Output: The average is 32.0
```

Again, the cast is too late. To ensure the decimal places are saved, cast *before* division:

```
average = ( double ) sum / ( double ) numItems;
System.out.println( "The average is " + average );

// Output: The average is 32.5
```

Math and Java

Structural Loads

A trapezoid is a quadrilateral with at least two parallel sides. The trapezoid is a very important shape in engineering, particularly in calculating the trapezoidal load for building bridges. The *trapezoidal load* is the sum of all forces acting on the bridge. This includes the weight of the bridge itself, forces due to weather, and the maximum weight of all the vehicles on the bridge at any given time.

Mapleman13/Shutterstock.com

The *centroid* is the center of gravity of a figure. It is important to the calculation of trapezoidal load. Refer to the figure to identify the two parallel bases *a* and *b*, the height of the trapezoid *h*, and the coordinates of its centroid (*Cx*, *Cy*). The centroid is located at the intersection of the two diagonals of the trapezoid. Three formulas are frequently used when working with trapezoids. These formulas, shown in the figure, calculate the coordinates of the centroid and the area of the trapezoid.

The *polar moment of inertia* is another measure of stress on a structure. The formula for this calculation is:

$$\text{polar moment of inertia} = \frac{h(4h^2b + 12h^2a + 7b^3 + 7ab^2 + ba^2 + a^3)}{48}$$

Notice that this formula requires the use of exponents. In Java, a method of the Math class is used to raise a value to a power. The Math class is discussed in depth in the next chapter. For now, use a direct method to raise these values to a power. For example, a^3 can be calculated using a * a * a. When writing the code in Java, replace a^3 in the above formula with a * a * a, and a^2 with a * a.

Assignment

Before beginning this exercise, download the chapter files from the student companion website. Launch jGRASP, and open the TrapezoidSnippet.java file. Write an application to accept input of the lengths of the bases, *a* and *b*, and the height (*h*) of a trapezoid. The program should calculate the centroid and the area of the trapezoid and output these values. Use this data to test the output of your snippet.

a = 10, b = 30, h = 6, (Cx, Cy) = (15, 2.5), area = 120

Save, compile, and run your snippet. Make edits if required, and compile again. For a challenge, try adding code to your snippet to calculate the polar moment of inertia.

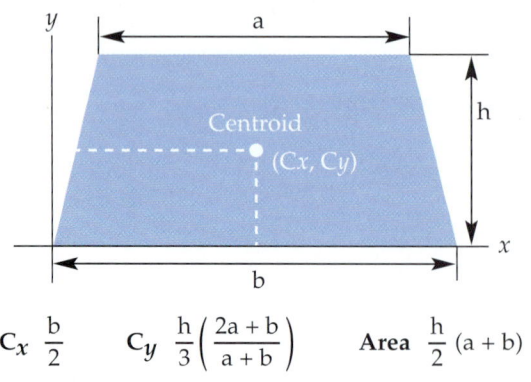

$$C_x \quad \frac{b}{2} \qquad C_y \quad \frac{h}{3}\left(\frac{2a + b}{a + b}\right) \qquad \textbf{Area} \quad \frac{h}{2}(a + b)$$

Goodheart-Willcox Publisher

Overflow Errors

In the previous chapter, data types were introduced to provide information to Java about the maximum and minimum values a variable might hold. Whenever an operation produces a value that exceeds or falls below those limits, and the code tries to store that excessive number into the limited space allotted, an "overflow" condition occurs. The number is simply too large for the reserved space and rolls over into an incorrect value with positive numbers becoming negative and negative numbers becoming positive.

It is the responsibility of the coder to check for overflow in an operation. In the following example, the sum of two integers is larger than the maximum value allowed for the int data type. The total is not correct because the sign bit got stepped on.

```
int bigInt = 2147483600;
int smallInt = 2000;

int sumInts = bigInt + smallInt;
System.out.println( "The sum of " + bigInt + " plus " +
            smallInt + " equals " + sumInts );

// Output: The sum of 2147483600 plus 2000 equals -2147481696
```

The coder must ensure the operation does not cause this condition by checking the result of a calculation against the size of the storage area before attempting to save the new value.

Coding Conundrum

Probably the most challenging part of Java programming is writing expressions correctly. For example, use the metal-cutting formula below for calculating a cutting speed as the specification for a program. The square brackets indicate the greatest integer in the expression. The code for the program is shown below.

$$V_c = \left[\frac{RPM \times 2\pi}{60} \right] \times 25$$

```
/*   Ch 5 Conundrum code
   your name here
*/

public class CH05ConundrumSnippet {
   public static void main( String [ ] args ) {
      final int RPM = 1200; // RPM is revolutions per minute

      double cuttingSpeed = int ( RPM * 2 * 3.1415926 / 60 ) * 25;
      System.out.println( "The cutting speed for " + RPM + " is " +
              cuttingSpeed );
   }
}
```

Download the chapter files from the student companion website. Launch jGRASP, open the Ch05ConundrumSnippet.java file, and compile it. The compiler indicates three errors. Find the error in the code, and propose a correction. When properly edited, compiled, and run, the program output is:

```
The cutting speed for 1200 RPM is 3125.0
```

SECTION REVIEW 5.2

Check Your Understanding

1. What happens to the decimal places in integer division?
2. Temporarily changing the data type of a value is called _____.
3. When the compiler automatically promotes the data type of a variable in an expression, it is called _____.
4. What is it called when a programmer chooses to override the automatic promotion of a variable?
5. The result of division by zero depends on the _____.

Build Your Vocabulary

As you progress through this course, develop a personal computer science glossary. This will help you build your vocabulary and prepare you for a career. Write a definition for each of the following terms and add it to your computer science glossary.

mixed-type arithmetic   implicit type casting
casting   explicit type casting

Output Results

Programmers are judged by the professional look of their output. Output of your programs so far has been determined by Java and, in some cases, has been rather clunky. Too many decimal places have been reported. Money has been displayed without a dollar sign, with extra digits, or with only one decimal place. Numbers larger than 1,000 have been displayed without commas. Java provides a feature to improve the looks of output. Use String.format( ) to format the data. However, misspellings and poor grammar are still the responsibility of the programmer.

Once the output has been generated, a test must be conducted to verify the validity of the results. A structured test plan helps to confirm the output.

Learning Goals

- Format output for clear presentation of results.
- Design a software test plan.

Terms

echo	test set
escape character	bug
test plan	debug

Essential Question

What effect does the format of outputted results have on the usefulness of a computer program?

Output Format

So far, System.out.print and System.out.println have been used to display information on the screen for your programs. As you have seen, the output format has not always been the best. The result of division may have far more decimal places than appropriate. Large numbers have been displayed without commas. Values that represent currency have been displayed without dollar signs. Java provides a method from the String class for formatting the output: String.format( ). To use this method, both a format string and a list of values to format are required. The following example creates a formatted string and prints it.

```java
int age = 16;
String birthdayGirl = "Meg";

String output = String.format( "Happy %dth Birthday, %s!", age,
    birthdayGirl );
System.out.println( output );
// Output: Happy 16th Birthday, Meg!
```

The String.format() method creates the string to print. Then, use System.out.println() to display the formatted string. The characters %d and %s in the above code are called specifiers. They refer to the data types of the variables after the quotation marks. The variables are listed in the same order as the specifiers are inserted. The formatter uses these characters to specify the type of variable and how each should be output. Here, %d designates an integer and %s means a string.

One advantage to using String.format( ) in this simple example is that the code does not have all those concatenation operators splitting up the output string. This statement creates the same output using concatenation:

```
System.out.println( "Happy " + age + "th Birthday, " + birthdayGirl +
    "!" );
```

Ultimately, with fewer concatenations, the output string will be easier to read and less prone to errors when coded. That is not the only benefit. Using String.format( ), the programmer can specify the exact number of decimal places of numeric values. There are many more advantages that will be illustrated as you progress through this course.

System.format( ) Syntax

Formatted output requires a format string and a list of variables, constants, or expressions. The creation of the string format variable follows this pattern:

```
String s = String.format( "text and specifiers", list of values );
```

The values arguments are variables separated by commas. The specifiers describe how the variables are to be formatted.

Arguments and Specifiers

The format specifier always begins with the % character. The pattern that specifiers follow is:

```
%[argumentIndex$] [flags] [width] [.decimalPlaces] conversion
```

Arguments are values output in the String.format( ) method. The argument index tells where in the list the value occurs. For example, the third value is referenced with 3$. This is optional. However, if you want to use a value more than once in a format string, it is not necessary to add it multiple times to the list of values. Simply reference the argumentIndex.

The flags are a set of characters that add to the format of the value, as shown in **Figure 5-6.** For example, the flag (,) specifies that the number is displayed with grouping symbols. In the United States, this grouping is a comma every three digits.

The width specifies how many spaces the value takes up in the output display. For example, 4 specifies a field width of four characters. If the number has fewer digits, spaces are added to widen the field to four characters. In the following example, a sample format specifier instructs the String.format( ) method to construct output for an argument with value 32767 to be displayed with commas, and in ten characters. Notice the extra spaces added to provide ten total characters.

```
String output = String.format( "The max short is %,10d.", 32767 );
System.out.println( output );
```

The output is:

```
The max short is     32,767.
```

The decimal places specify how many characters are to be displayed to the right of the decimal point. For example, 3.1415926 can be restricted to 3.14 by using a .2f specifier.

The conversion specifies how a data type is displayed. For example, d is used for integers and f is used for numbers with decimals. Care must be taken when using the String.format( ) method. It is very easy to match the wrong conversion

FYI

Notice that this is a new use for the % character. The compiler will know the proper use of the % by the context in which it is used. This % specifier character will always be used inside a string. The % modulus character will always be used within an expression.

Specifiers, Flags, or Escape Sequences	Symbol	Explanation
Floating-Point Conversion	%#.#f	Specify number of characters wide with # and number of decimal places, #.
Decimal-Integer Conversion	%#d	Specify number of characters wide with #.
Percent Conversion	%%	Display the % sign.
Line-Separator Conversion	%n	Go to the next line.
Left-Justified Flag	%-	Default is right-justified. This flag causes number to be left-justified.
Zero-Padded Flag	%0	Fill empty characters in field with 0s.
Comma-Separators Flag	%,	Insert commas in number every three digits.
Sign Flag	%+	Show the sign of a positive or negative number.
Parentheses Flag	%(	Show negative numbers in parentheses.
Escape Strings	\n	Go to the next line.
	\t	Insert a tab.

Figure 5-6. Specifiers, flags, and escape sequences for the System.format( ) method.

with a data type and the result throws an UnknownFormatConversionException exception. If this happens, check that the specifier is correctly formatted.

Some of the more commonly used specifiers are explained below. For all the examples, assume s is a String data type.

Floating-Point Conversion

An f specifies the argument should be displayed as a decimal number. The field width and number of decimal places are optionally specified. Examine the following example. Notice that the dot after the specifier provides the period for the end of the sentence.

```
s = String.format( "A common approximation for PI is %,10.2f.",
   3.14159 );
System.out.println( s );
```

The output is:

```
A common approximation for PI is       3.14.
```

Decimal-Integer Conversion

A d specifies that the argument should be displayed as a decimal integer. For example:

```
s = String.format( "The max short is %d.", 32767 );
System.out.println( s );
```

The output is:

```
The max short is 32767.
```

Percent Conversion

For a percentage, outputting the % character is desired. Suppose you want to output the % sign? The String.format( ) method will think you are starting a specifier. So, a special specifier is created for the % sign. Use two percent signs, one for the specifier and one for the sign.

Examine the following example. Notice you need the first % character to indicate a conversion and the second to indicate it is the percent sign itself.

```
s = String.format( "My grade average in coding class is %d%%", 95 );
System.out.println( s );
```

The output is:

```
My grade average in coding class is 95%
```

Line-Separator Conversion

Use %n to force a new line in the output. For example:

```
s = String.format( "Here is line 1 %nand here is line 2" );
System.out.println( s );
```

The output is:

```
Here is line 1
and here is line 2
```

Left-Justified Flag

By default, a number is right-justified in a field. Use of the − flag will left-justify the number in a field. For example:

```
s = String.format( "The answer to life, the universe, and everything "
+ "is %10d.", 42 );
System.out.println( s );
s = String.format( "The answer to life, the universe, and everything "
+ "is %-10d.", 42 );
System.out.println( s );
```

The output is:

```
The answer to life, the universe, and everything is        42.
The answer to life, the universe, and everything is 42       .
```

Zero-Padded Flag

By default, spaces are used to fill the field width if there are not enough characters in the number or string. To use zeros instead of spaces to fill, or pad, the specified field width, use the 0 flag. For example:

```
s = String.format( "The binary numbers for 1, 2, and 3 are %04d, "
+ "%04d, %04d.", 1, 10, 11 );
System.out.println( s );
```

The output is:

```
The binary numbers for 1, 2, and 3 are 0001, 0010, 0011.
```

Sign Flag

Numbers can be negative or positive. To show the sign of a positive or negative number, use the + flag. For example:

```
s = String.format( "One Kilobyte (1 K) is %+6d bytes.", 1024 );
System.out.println( s );
```

The output is:

```
One Kilobyte (1 K) is + 1024 bytes.
```

Parentheses Flag

In some applications, negative numbers should be displayed inside parentheses instead of with a negative sign. To display negative numbers in this format, use the ((left parenthesis) flag. For example:

```
s = String.format( "%(6.2f", -8.45 );
System.out.println( s );
```

The output is:

```
(8.45)
```

Escape Strings

Some simple formatting can be inserted directly into a System.out.println( ) argument string. The backslash character (\), called the escape character, can be inserted with another character into the output string. The *escape character* indicates the following character has a special meaning. It "escapes" being printed. The escape sequence of characters for a new line is \n. The escape sequence \t inserts a tab space. The tab space is useful for centering a title or formatting a list. For example:

```
System.out.println( "\t\tJava Expressions Game\n\t\t\tBy Team Coders"
    ); //title
```

The output is:

```
        Java Expressions Game
            By Team Coders
```

All these format specifiers are described in Appendix B. You can refer to this appendix throughout the text for formatting help.

HANDS-ON EXAMPLE 5.3.1

Formatting Output ➦

The output of the programs you have created to this point have been formatted with default settings. In this exercise, you will apply formatting specifications to customize the look of the output.

Before beginning this exercise, download the chapter files from the student companion website. The starter file StringFormatterSnippet.java is used in this exercise.

1. Launch jGRASP, and open the StringFormatterSnippet.java file. The first part of this example displays the results of the code in the section above.

2. Locate comment 1, and follow the instructions to define a string and format it to display the number PI to two decimal places. Provide a field of 10 characters for this number. Print the string.

```
String s = String.format( "A common approximation for PI is %10.2f.", 3.14159 );
System.out.println( s );
```

3. Compile and run the program. Do this after adding each bit of code to verify the output.

4. Locate comment 2, and follow the instructions to assign a new format to s. Do not redefine s, just reuse the variable. This time, display the largest number that can be stored in a short type.

```
s = String.format( "The max short is %d.", 32767 );
System.out.println( s );
```

5. Locate comment 3, and follow the instructions to edit s to display a percent conversion.

```
s = String.format( "My grade average in coding class is %d%%", 95 );
System.out.println( s );
```

6. Locate comment 4, and follow the instructions to edit s to show a line separator conversion.

```
s = String.format( "Here is line 1 %nand here is line 2" );
System.out.println( s );
```

7. Locate comment 5, and follow the instructions to edit s to show the default justification of a number and the same number left justified.

```
s = String.format( "The answer to life, the universe, and everything is %10d.", 42 );
System.out.println( s );
s = String.format( "The answer to life, the universe, and everything is %-10d.", 42 );
System.out.println( s );
```

8. Locate comment 6, and follow the instructions to edit s to show the zero-padded output.

```
s = String.format( "The binary numbers for 1, 2, and 3 are %04d, %04d, %04d", 1, 10, 11 );
System.out.println( s );
```

9. Locate comment 7, and follow the instructions to edit s to show the comma separator output.

```
s = String.format( "%,d", 1234567890 );
System.out.println( s );
```

10. Locate comment 8, and follow the instructions to edit s to show the sign of a number (positive or negative).

```
s = String.format( "One Kilobyte (1 K) is %+6d bytes.", 1024 );
System.out.println( s );
```

11. Locate comment 9, and follow the instructions to edit s to show a negative number in parentheses.

```
s = String.format( "%(6.2f", -8.45 );
System.out.println( s );
```

Alina McCullen/Shutterstock.com

12. Locate comment 10, and add another formatting example. Suppose a local gas station advertises prices for three grades of gasoline, as shown. Add the definition of a double variable named costOfGas, and assign the price of the mid-range gas to the variable. The price is given with a fractional cent, so first manually convert the price to a decimal reflecting the total cost in cents per gallon (269.9).

```
// real cost of gas
double costOfGas = 269.9; // cents per gallon
```

13. Show the price of gas.

```
// string format specifiers and arguments
String s = String.format( "%nThe cost of gas touted at %.1f cents.", costOfGas );
System.out.print( s );
```

14. Convert the value of costOfGas to dollars by dividing by 100.0. Format this output as currency with two decimal places. The String.format( ) method will round the answer.

```
costOfGas /= 100.0;
s = String.format( " is closer to $%.2f per gallon.", costOfGas );
System.out.println( s );
```

15. Save the file in your working folder.

Try It!

Open your version of the OnBasePercentageSnippet.java file in jGRASP. Format the output so only three decimal places are reported. Save the file as OnBasePercentageSnippetFormatted.java in your working folder.

Test Validity of Output

Too many novice programmers feel they are finished when the IDE produces a clean compile. They will run the program because the instructor asks for output. However, the job is still not done at this point in development. Great programmers check inputs and outputs and match the results with the requirements of the program. A test plan helps with these tasks. If problems surface, the programmers fix the program and test again.

Verify Output

It may sound simple, but an important part of testing is to verify the output. Compare the output with the problem statement. Refer to the specification or problem statement to ensure the assigned problem has been solved.

The output statements should be written in proper English. Inputs should be echoed in complete sentences. *Echo* means to say back what was input so it can be verified the input was correct. An echo is a replay of something. The results should be announced in complete sentences. Proper grammar and punctuation should be used.

Test Plan

A sure way to verify a program works correctly is to build a test plan. A *test plan* outlines a systematic approach to evaluating the success of a program. In this plan, include the problem statement and sample data you will use to test the program. Writing the test plan before any coding begins can verify you truly understand the problem. This will help produce a program that delivers the correct result to solve the intended problem.

A *test set* is a combination of sample inputs and outputs for a single test. Always know in advance what to expect for a test set. There are several types of test sets to include in a test plan.

Known Results

The preliminary stage of programming involves understanding the problem and breaking it into small, doable steps. While doing this, create a test that can be worked out by hand to show a result you know is a correct result. Then, create a test set using these values for input to check for the known results. For example, the area of a 10-inch circle can be manually calculated as 78.54 square inches. If the program input is 10 inches for diameter, the program should output 78.54.

Boundary Values

If the input falls within a certain range of numbers, create test sets for the minimum values and the maximum values. These set the boundaries of possible input. Again, perform the calculation by hand to generate a known result. Create test sets for numbers above and below the boundary values to verify that the program does not allow those inputs. For example, suppose an application is calculating the average of letter grades. A grade of *G* would not be a proper input value. Testing should verify that only *A* through *F* are accepted. Or, if a test grade is specified in the range 0 to 100, test the valid boundary values 0 and 100 to verify these values are properly handled. Also, test the invalid boundary values −1 and 101 to verify these values are not accepted. In each case, predict the output so an incorrect output is readily recognized.

Inputs That May Throw an Exception

If the program is taking integer inputs, create a test set with zero, floating-point inputs, or other inputs that may cause a problem. When these test sets are run, the predicted output is an exception.

Debug the Program

An anomaly or error in a program is called a *bug*. To remove bugs is to *debug* the program. Once an improper result is identified, the programmer must determine the cause of the incorrect result. Once the root of the problem is identified, the programmer edits the code, recompiles, and tests again. This process is called *debugging*. While this type of rigor may seem excessive for the type of problems solved in this chapter, practicing writing test plans for these sequential algorithms will make it easier to make test plans for programs that have many paths or branches.

Reference materials can be used to help debug and solve problems. The Oracle Docs website provides a lot of information on the proper use of classes, methods, and objects. Additionally, there are many online forums where coders help other coders solve problems and provide advice on the quality and accuracy of their programs. When first learning to code, initiate conversations by posting messages about problems you are trying to solve or bugs you are trying to fix in your programs. Respond to the feedback you receive with follow-up questions if needed. Then, implement the suggestions as appropriate. These electronic communities are a great place to extend your learning environment beyond the classroom. They enable you to participate not only as a learner, but as a contributor and even a teacher and mentor to others as you gain experience in coding. Note that this is *not* a place where you can ask someone else to write your coding homework for you. Be sure to comply with your instructor's guidelines on how much help you can seek on homework.

HANDS-ON EXAMPLE 5.3.2

Exploring Online Coding Forums

An effective way to extend your learning beyond the classroom is through participation in online forums. Professional coders as well as those just learning to code participate in these forums as both teachers and learners. Participating on these forums allows you to receive feedback from others and seek help in debugging programs and solving problems.

1. Launch a web browser, and navigate to the Stack Overflow website (www.stackoverflow.com).
2. Locate the About link at the bottom of the page, and click it.
3. Read about Stack Overflow. What is its purpose? How many users does it have each month?
4. Use the site's search function to search for Java debugging help.
5. Scroll through the list of results. Do any of the topics sound familiar to you?
6. Navigate to Wikipedia (www.wikipedia.org), and search for source code playgrounds.
7. What is a source code playground? Are there any available for Java?

SECTION REVIEW 5.3

Check Your Understanding

1. Using the String.format( ) specifier(s), what is the correct format to output money?
2. Write the correct specifier(s) to format an integer in seven places, including commas and its sign.
3. What is included in a test plan?
4. List the three types of test sets.
5. What is an anomaly or error in a program called?

Build Your Vocabulary

As you progress through this course, develop a personal computer science glossary. This will help you build your vocabulary and prepare you for a career. Write a definition for each of the following terms and add it to your computer science glossary.

echo	test plan	bug
escape character	test set	debug

Cooperative Coding

Comparing Java Expressions

Coding is a very creative endeavor. Everyone tends to write code just a bit differently from other coders. Yet, almost magically, the results are the same.

Assignment

Form a team of coders from your class. Separately, each coder should write a program to encode the following set of expressions provided in paragraph form. Each program should provide meaningful output.

A local senior-assistance charity is preparing its meal deliveries for the week. When a client requests a day of meals, he or she is provided with three meals. There are 87 elderly clients on the list. Six of the clients have said they do not need a delivery this week. Two-thirds of the remaining clients have requested a full seven days of meals. Twenty of the clients need meals only on weekdays. The remainder of the clients have requested meals only on the weekends. How many meals must be prepared and loaded into the truck for this week's deliveries?

Monkey Business Images/Shutterstock.com

When satisfied with your program, meet with your team and compare the code each of you wrote. Were the comments from your team members helpful? Were each person's expressions correctly coded? Were the calculations correct? Were the outputs identical? Prepare a summary of what you and your team discovered.

1. What seemed to be similar in each program? How are the programs different?
2. Suppose you had to follow a teammate's code. Would you understand what they had done?

Chapter Summary

Section 5.1 Arithmetic Operators

- Binary operators combine only two values at a time, and Java includes addition, subtraction, multiplication, division, and modulus operators.
- Java expressions must follow the standard algebraic order of operations, which is called operator precedence in Java.
- Shortcut operators streamline common expressions to simplify the code and make it easy to read; Java includes several shortcut operators, such as the increment, decrement, plus equals, minus equals, times equals, divide by equals, and mod equals operators.

Section 5.2 Operators and Identifiers

- When floating-point numbers are divided, the decimal remainder is retained, but when integers are divided, the remainder is truncated and must be obtained with the mod operator.
- Common expression errors in Java include divide by zero, incorrect order of precedence, mixed data types, and overflow errors.

Section 5.3 Output Results

- The String.format( ) method is used to format the output and requires a format string and an argument list, which may contain several specifiers.
- A software test plan is used to evaluate the success of a program and consists of the problem statement and sample data, which is called the test set; any errors that are discovered must be debugged.

Chapter 5 Test

Multiple Choice

Select the best response.

1. Which symbol represents the operator used to derive the remainder in division?
   A. /
   B. %
   C. \
   D. {

2. Which of the following arithmetic operations is calculated first in an arithmetic expression?
   A. within the parentheses
   B. multiplication
   C. division
   D. exponentiation

3. In this example, the value of numY will have a decimal. Which form of type casting does this represent?

   ```
   int numX = 9;
   double numY = numX / 4.0;
   ```

   A. explicit
   B. implicit
   C. inherited
   D. hard-coded

4. The specifier %n sends the output to the next line when using which method?
   A. System.out.println
   B. System.format
   C. System.out.print
   D. String.format

5. If a variable long numX has the value of 56903, the output of String.format("%d", numX); is:
   A. 0056903
   B. +56903
   C. 56903
   D. 56903.0

Completion

Complete the following sentences with the correct word(s).

6. The result of dividing an integer by an integer is an integer. Any remainder is _____.

7. Since the double data type takes up more memory and is more precise than the int data type, changing from int to double is called _____.

8. When the coder uses _____ type casting before a division calculation, it ensures the decimal places are saved.

9. A(n) _____ error occurs when the data type is not large enough to hold the value.

10. When putting together a test plan, create a(n) _____ of data that can be used to calculate output by hand.

Matching

Match the correct term with its definition.

A. operator precedence
B. explicit type casting
C. Java expression
D. test plan
E. implicit type casting

11. Problem statement and sample data.

12. Combination of operators, numbers, constants, and variables that result in a single value.

13. Hierarchy of how arithmetic calculations are performed in Java.

14. If one of the operands is a double, the other is automatically promoted to a double.

15. Coder tells the compiler to convert a value to a different data type.

Application and Extension of Knowledge

1. The table that follows summarizes the month's activity of three checking accounts. Write a program that displays the account number and the end-of-month balance for each account. The program should then display the total amount of money in the three accounts. Develop an algorithm and a test plan. Then, write the program. Save the program as EOMbalance.java in your working folder.

Account Number	JA123	KA456	LB789
Beginning Monthly Balances	1234.56	987.54	2345.78
Deposits	300.00	100.00	1000.00
Withdrawals	100.00	350.00	500.00

2. One of the most frequent tasks in information technology is to produce a report. In each case, tables of descriptors and numbers are printed. The data must all align. Write a program that creates a table of occupations and salaries. Store the values of salaries in variables. Write String.format() expressions that print each of the following job titles and align their values in a list with the salaries. As an extension to what you have learned about specifiers, Strings can be formatted using the conversion s. For example, %-10s prints a string in ten spaces, left-justified.

Occupation	Median Pay
Computer Network Architects	$104,650
Computer Programmers	$82,240
Computer Support Specialists	$52,810
Software Developers	$103,560

3. A grower of oranges in California sends out 150 oranges in each crate. She has harvested 3,478 oranges this day. She donates the extra oranges to the homeless shelter rather than send a partial crate. Write a program that displays how many oranges the shelter will receive today. Develop an algorithm and a test plan. Then, write the program. Save it as Oranges.java in your working folder. After that, write a paragraph explaining why you think this is an important gesture on the grower's part.

4. The population of three New England states in thousands is as follows: Maine: 1,275; Massachusetts: 6,349; and Connecticut: 3,406. Write a program that calculates the average population of these states. Develop an algorithm and a test plan. Then, write the program. Save it as NEstates.java in your working folder.

5. One of the conversions for String.format( ) output is e. It formats the decimal number in scientific notation. The Andromeda galaxy is 2,537,000 light years from Earth. Choose an appropriate data type and write a String.format( ) specifier to express this number in scientific notation. Refer to Appendix B for the use of e in a specifier.

Online Activities

Complete the following activities, which will help you learn, practice, and expand your knowledge and skills.

Vocabulary. Practice vocabulary for this chapter using the e-flash cards, matching activity, and vocabulary game until you are able to recognize their meanings.

Communication Skills

Reading. *Scanning* is done when you know the information you need is in a document, you just have to find it. Scan this chapter for information on type casting. Did scanning work for you? How long did it take you to find the information?

Writing. When taking notes, it is common to write down only key information, rather than every word. Select several pages of notes you have taken during class. Rewrite your notes as complete sentences and paragraphs. Use *transition words* to make smooth connections as your writing moves from one idea to the next.

Speaking. *Impromptu speaking* is talking without advance notice to plan what to say. Turn to the person next to you and explain what you did over the weekend. Clarify anything the other person finds confusing or does not understand. Were you able to hold an impromptu conversation on this topic?

Listening. Active listening requires you to *show* you are listening. To do this, face the speaker and pay attention. Engage in eye contact, but avoid staring, which can be intimidating and distracting. Nod when you understand a point. Practice showing you are listening by asking a friend or classmate about a familiar topic, such as favorite movies. Identify important details of what your friend or classmate is saying.

Portfolio Development

College and Career Readiness

Digital Presentation Options. Before you begin collecting items for a digital portfolio, you will need to decide how you are going to present the final product. For example, you could create an electronic presentation with slides for each section. The slides could have links to documents, videos, graphics, or sound files. This will dictate file naming conventions and file structure.

Websites are another option for presenting a digital portfolio. You could create a personal website to host the files and have a main page with links to various sections. Each section page could have links to pages with your documents, videos, graphics, or sound files. Be sure you read and understand the user agreement for any site on which you place your materials.

Another option is to place the files on a CD or flash drive. There are many creative ways to present a digital portfolio. The method you choose should allow the viewer to easily navigate and find items.

Establish the types of technology available for you to create a digital portfolio. Will you have access to cameras or studios? Do you have the skill to create videos? Decide the type of presentation you will use. Research what will be needed to create the final portfolio product.

CTSOs

Teamwork. Some competitive events for CTSOs have a teamwork component. If the event is a team event, it is important that the team prepares to operate as a cohesive unit. Effective team members are individuals who contribute ideas and personal effort. To prepare for teamwork activities, complete the following activities.

1. Review the rules to confirm if questions will be asked or if the team will need to defend a case or situation.

2. Practice performing as a team by completing the Cooperative Coding feature in each chapter of this text. This will help members learn how to interact with each other and participate effectively.

3. Locate a rubric or scoring sheet for the event on your organization's website to see how the team will be judged.

4. Confirm whether visual aids may be used in the presentation and the amount of setup time permitted.

5. Make notes on index cards about important points to remember. Team members should exchange note cards so each member evaluates the others' notes. Use these notes to study. You may also be able to use these notes during the event.

6. Assign each team member a role for the presentation. Practice performing as a team. Each team member should introduce himself or herself, review the case, make suggestions for the case, and conclude with a summary.

7. Ask your instructor to play the role of competition judge as your team reviews the case. After the presentation is complete, ask for feedback from your instructor. You may also consider having a student audience to listen and give feedback.

Classes

Sections

6.1 Introduction to Classes

6.2 Java Class Library

A class makes it possible to combine data and the code needed to manipulate that data into one unit. The Java Class Library has many classes available for use. For example, to generate a random number for a game, you can use the Random class. To calculate mathematical functions, you can use the Math class. Although the Java Class Library is full of useful classes, any programmer can also write a class and make it available to other programmers without putting that class in the Java Class Library.

This chapter concentrates on using classes that have already been written. In Chapter 12, you will explore writing your own classes.

College and Career Readiness

Reading Prep

Before reading this chapter, review the key terms. They are listed at the beginning of each section and highlighted within the body of the chapter. As you read, determine the meaning of each key term.

While studying, look for the activity icon for:

- Vocabulary terms with e-flash cards and matching activities.
- Starter files for hands-on examples and other exercises.

These activities can be accessed at
www.g-wlearning.com/informationtechnology/1773

Chapter Glossary

application programming interface (API): Documentation on how to create objects of a given class, what methods are available, and how to call those methods.

buffer: Storage area in memory.

constructor: Special method in a class that creates (instantiates) an object, making sure the object has valid values for all its data.

dot notation: Placing a period (.) between the object or class name and the method name when calling a method; also used between the object or class name of a static constant or public variable.

immutable: Unable to be changed.

instance: An object of a class.

null: Keyword indicating an object reference has no value.

object reference: Name of the object; holds a location in memory used to find the object.

overloaded method: Two or more methods that share a name but take different arguments.

package: Collection of related classes in one location.

pixel: Point on the display that has an x and y location and a color; picture element.

pseudorandom number: Number that belongs to a set of a sequence of numbers long enough to appear to be random, but is not.

static method: Called with a class name instead of an object reference.

token: Unit of characters collected via tokenizing.

tokenizing: Skipping white space and collecting non-white-space characters.

Introduction to Classes

A class is a framework from which objects can be created. The class describes the data each object will have. The class also provides methods (code) to store, obtain, and manipulate the data in the object.

In effect, the class is like instructions for building birdhouses. The instructions tell you to build a birdhouse with walls, a door, and a roof. Each birdhouse has walls, a door, and a roof, but each birdhouse also is unique in that you can mount it individually and it will house its own family of birds. An object made from a class would be like one birdhouse created following those instructions. For example, one birdhouse could have a rounded door with a shingle roof, while another birdhouse could have a square door with a wooden roof. Each object made from a class is also unique. Each object has its own data and can be manipulated independently from other objects.

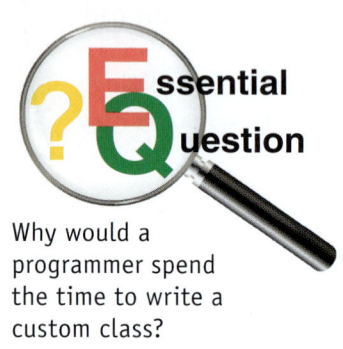

Essential Question

Why would a programmer spend the time to write a custom class?

Learning Goals

- Explain the use of a class API to create objects and call methods.
- Construct graphics using the custom Turtle class.

Terms

application programming interface (API)

constructor

dot notation

instance

null

object reference

overloaded method

pixel

Class API

An advantage to using classes is that some programmers can write a class that many programmers can reuse. Using classes written by other programmers makes it easier to develop new programs more quickly. Another advantage of using prewritten classes is the methods have been tested and are most likely free of bugs. Using classes, therefore, can make programs more robust and reliable.

A programmer who has written a class tells other programmers how to use the class by publishing the API. The *application programming interface (API)* tells programmers wishing to use the class how to create objects of that class, what methods are available, and how to call those methods. The beauty of using an API is you do not need to know how the class is written or look at the code of the class.

For example, if the class is in the Java Class Library, you just need to import the class. You have seen how to do this with the Scanner class. Other classes not in the Java Class Library can be put into the same folder as the program. The compiler will automatically find any classes in the same folder. In this case, an import is not needed.

Using Constructors and Passing Arguments

Typically, the first order of business when using a class is to create an object. This is accomplished by calling a special method of the class called a constructor. The function of the *constructor* is to create the object, making sure the object has valid values for all its data. As discussed in Chapter 2, creating an object using a constructor is called instantiating an object. The object is called an *instance* of the class.

By convention, class names are capitalized and use CamelCase. A constructor has the same name as the class and returns an object reference to the newly created object. The *object reference* is the name of the object and holds a location in memory used to find the object. An object reference has the same data type as the class. Yes, a class is another data type. The difference between a reference type and a primitive type is the primitive type holds its value directly, while the reference type contains the memory address of the object. From that address, the JVM can access the object's data and methods.

The constructor is called using the new keyword. The syntax is:

```
ClassType reference = new ClassType( arguments );
```

The arguments are usually initial values for the data in the object the client of the class specifies. This is called *passing arguments.* The arguments to be passed vary from constructor to constructor. Arguments are listed in the API as comma-separated pairs of:

```
dataType argumentName
```

So far, you have used a constructor to create a Scanner object. The API for the constructor to create a Scanner object looks like this:

```
Scanner ( InputStream source )
```

The API shows this constructor takes only one argument, which is an object of the InputStream class and is the source of the input. To call this constructor and name the Scanner object reference input, use this now-familiar statement:

```
Scanner input = new Scanner( System.in );
```

This statement instantiates the input object, as shown in **Figure 6-1.** In this case, the argument passed as the input source is System.in, which is the name assigned by default to the keyboard.

The data type of each argument is specified in the API so the programmer knows what data type is expected. However, do not include that data type when calling methods. Send only the value for the argument. In other words, this statement is incorrect:

```
// INCORRECT! InputStream is a data type; do not include this
Scanner input = new Scanner( InputStream System.in );
```

This statement, without the data type for the argument, is correct:

```
// CORRECT, only the argument value is given
Scanner input = new Scanner( System.in );
```

To follow the Java naming convention for classes, capitalize the class name and use CamelCase.

Reference   Class type again

```
Scanner input = new Scanner( System.in );
```

Class type   Keyword   Argument
that "makes"   of type
the object   InputStream

Goodheart-Willcox Publisher

Figure 6-1. This is how the constructor for a Scanner object is formed.

Remember *not* to include the data type of the argument when calling a method.

Be careful *not* to call a method using a null object reference.

When a reference variable is first defined, but has not been used to instantiate an object, as in:

```
Scanner input;
```

the reference has a special value called null. ***Null*** means no value. If a null reference is then used to call a method, such as in this code:

```
int age = input.nextInt( );
```

the nextInt method will generate a NullPointerException at runtime. In some circumstances, the compiler will recognize the reference variable's value is null and flag the statement as an error. In any case, it is best to remember to instantiate an object using the object reference so it is no longer null.

Calling Methods and Handling Return Values

The API for methods of the class other than the constructors is in the format shown here:

```
returnValueType methodName( datatype1 arg1, dataType2 arg2, …)
```

A method can take from zero to many arguments.

When a program calls a method, the JVM suspends executing the current code and starts executing the code in the method. When the method finishes, the JVM resumes executing the program's code. A method may or may not return a value. If the method returns a value, that value replaces the method call in the expression. For example, in this statement that calls the nextInt method of the Scanner class:

```
int age = input.nextInt();
```

the nextInt method returns an int. The value entered by the user replaces the method call, and that value is assigned to the variable age.

If the method returns a value, the returnValueType shown in the API format above is the data type of the value the method returns. If the method does not return a value, the returnValueType will be the keyword void. The void keyword means the method does not return a value. As explained in Chapter 3, main is a method. The familiar keyword void in the header for the main means that main does not return a value. Methods that do not return a value simply do their job and return to the caller.

Notice that when calling a method of a class, the method call is prefaced with the object name followed by a dot (.) and then the method name and arguments. This is called ***dot notation*** and indicates which object the method should use. The syntax is:

```
//objectName.methodName( argument1, argument2, … ) // the number
of arguments varies
```

In the previous example, the nextInt method call is prefaced with input plus a dot. The name of the Scanner object is input. Similarly, the nextDouble method is called as:

```
double price = input.nextDouble( )
```

From this, the API of the nextDouble method can be deduced to be:

```
double nextDouble( )
```

Although the nextInt( ) and nextDouble( ) methods do not take any arguments, the empty parentheses after the method name still need to be included. Otherwise, the compiler will assume that nextInt or nextDouble is a variable name. The compiler will not be able to find the variable and will generate an error.

Remember to include parentheses after the method name when calling a method even if the method takes no arguments.

Language Arts and Java

Computer Terminology

A common expression is that you need to "walk the walk" and "talk the talk." That expression refers to people who say one thing, but do another. For example, a person may say taking care of the environment is important, but may not recycle. That person is talking the talk, but not walking the walk. With computer programming, it is often the opposite case. Programmers know how to walk the walk, but many do not talk the talk. That is, they can code programs, but do not know and use the correct computer terminology. Knowing the technical terms for various aspects of programming is essential. Professional programmers work in teams. You must be able to speak in a common language in order to communicate your design ideas to other people on the team.

REDPIXEL.PL/Shutterstock.com

Knowing and being able to use the correct terminology will become even more important as you progress through computing. Your applications will grow in size and become more complex. To deal with this complexity, a group of four computer scientists, Erich Gamma, Richard Helm, Ralph Johnson, and John Vlissides, wrote the now-classic book *Design Patterns*. In the book, the authors describe a set of patterns for combining classes in a program. The goal of using the design patterns is to create programs that are easy to write and to modify as the requirements change. Each pattern has a name. Teams designing programs can communicate design ideas by naming one or more patterns. For example, one team member might say, "The strategy pattern would be perfect for defining a variety of players in the game." The other team members would then understand the proposed design.

Assignment

In this chapter, you have been introduced to several technical terms related to the design of programs, specifically: class, object, and method. In the beginning of this chapter, a simile was presented that classes were like birdhouse designs and objects were like individual birdhouses built using the design. Each birdhouse has common design elements, but each birdhouse is also unique.

For this assignment, create another simile or metaphor for classes and objects. In a well-written paragraph, write definitions for the terms class, object, and method using your new metaphor. Compose the paragraph in complete sentences with proper grammar and punctuation. Proofread the paragraph and correct any spelling or other errors. Show your definitions to other students to be sure they are understandable and communicate your ideas effectively.

Using the Turtle

To illustrate using classes, you will work with the Turtle class. This class was written by the authors and the code is provided on the student companion website. It was inspired by Seymour Papert, inventor of Turtle Graphics. The Turtle class allows the programmer to create a turtle sprite, or image, that can move around a window to draw shapes. This is done by calling methods to move the turtle and to change the turtle's direction.

The turtle has a pen. When the pen is down and the turtle moves, it leaves a mark. This allows shapes to be drawn on the screen. When the pen is up, and the turtle moves, it does not leave a mark. This allows the turtle to move to another location without drawing.

Graphics Applications

The window for the turtle is created using a graphics application. The window is composed of pixels. A *pixel* is a picture element, which is the smallest dot that can be displayed by a computer monitor. Each pixel has a color, which is white by default. It also has an (x, y) location, which can be used to access the pixel.

The x coordinates start at 0 on the left side of the window and increase as the turtle moves from left to right. The y coordinates start at 0 at the top of the window and increase as the turtle moves from the *top* to the *bottom*. Thus, for a window that is 800 pixels wide and 500 pixels high, the top-left pixel in the window is (0, 0), the top-right pixel is (799, 0), the bottom-left pixel is (0, 499), and the bottom-right pixel is (799, 499). **Figure 6-2** shows the layout of the Java graphics window with some sample (x, y) locations.

FYI

Unlike a Cartesian coordinate system, *y* values in a Java graphics window increase from top to bottom.

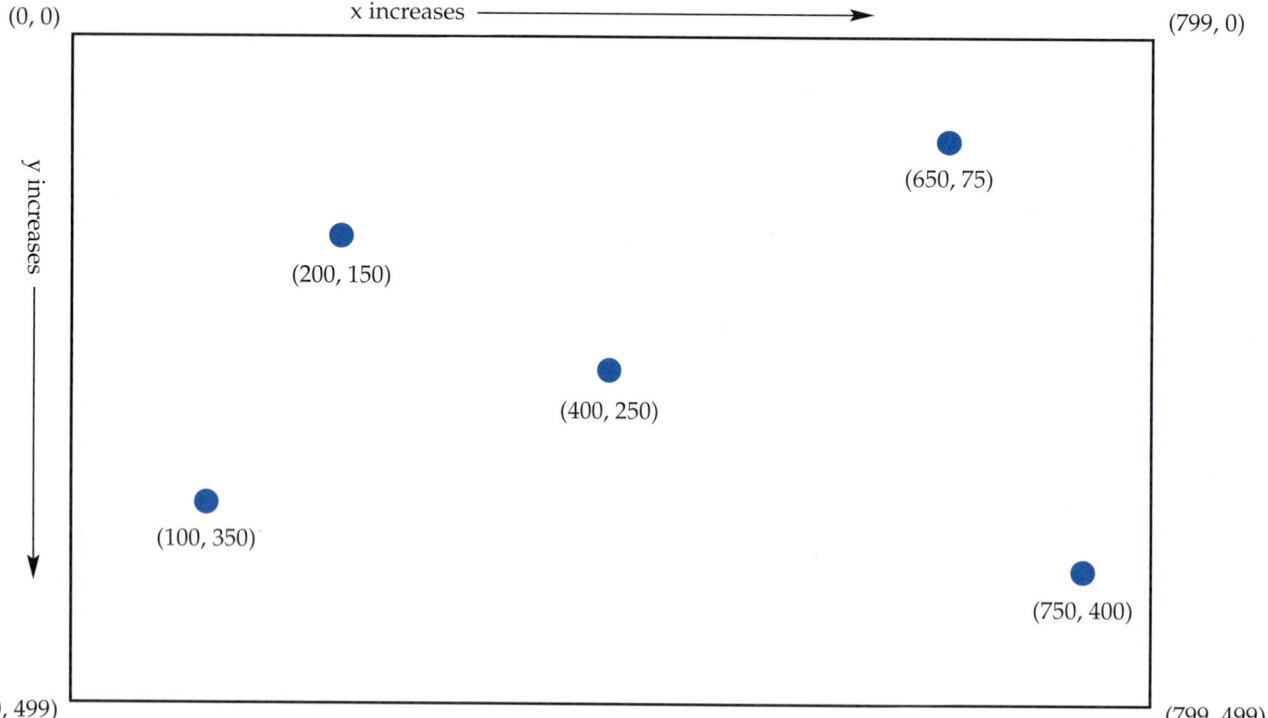

Figure 6-2. The positive *y* values extend from the top of the graphics window to the bottom. The positive *x* values extend from the left side of the graphics window to the right.

To use the Turtle class, there needs to be at least four .java files in the same folder: Turtle.java, Sprite.java, SpriteAnimator.java, and another Java file that is the application you write to direct the turtle's drawing. As a model for your application, the file SpriteApplication.java is provided. You will find these files on the student companion website. When you open the SpriteApplication.java file in jGRASP, change the name of the class to describe the drawing you want to make and save the file under that name. When all these files are collected in the same folder, compile them in this order:

1. SpriteAnimator.java
2. Sprite.java
3. Turtle.java
4. File you will use for drawing (SpriteApplication.java saved under a different name)

The code for the SpriteApplication.java file is shown in **Figure 6-3.** As you can see, a graphics application has a different format from the applications you have written so far. There is a lot here, but much of this code will not change from application to application. For now, ignore lines 15 through 29. What these

```
1   /** SpriteApplication
2   */
3   import javafx.application.*;
4   import javafx.scene.canvas.*;
5   import javafx.scene.*;
6   import javafx.scene.paint.*;
7   import javafx.stage.*;
8
9   public class SpriteApplication extends Application {
10
11      private String windowTitle = "Turtle Drawing";
12      private int windowWidth = 800;
13      private int windowHeight = 500;
14
15      private Group root;
16
17      @Override
18      public void start( Stage stage ) {
19         stage.setTitle( windowTitle );
20         Canvas canvas = new Canvas( windowWidth, windowHeight );
21         root = new Group( canvas );
22         stage.setScene( new Scene( root ));
23         stage.show( );
24         SpriteAnimator sa = new SpriteAnimator( );
25
26         buildScript( );
27         sa.playScript( );
28
29      }
30
31      public void buildScript( ) {
32
33         // student adds Turtle code here
34
35      }
36
37      public static void main( String [ ] args ) {
38         launch( args );
39      }
40   }
```

Figure 6-3. This is starter code for creating a graphics program. Notice the highlighted code, which is different from the applications you have written so far.

lines do will be explained later in this text. The highlighted parts of this application are the two areas you will change.

Look at lines 11 through 13. This is where you can set the title and the width and height of the window to your preferences. The given value for the title is Turtle Drawing. This is what will appear in the title bar of the window. To define a different title for the window, change the text inside the quotation marks on line 11. The given values for the window is a width of 800 pixels and a height of 500 pixels. For a different-size window, change the values for windowWidth and windowHeight on lines 12 and 13.

The other area you will change is the buildScript method that starts on line 31. In this method, you will add code to instantiate a Turtle object and call methods to move the Turtle around the window. Notice there is still a main method (lines 37 through 39) as required by the JVM, but its only job is to launch, or begin running, the application. The start method then runs, which opens a window and calls the buildScript method containing your code.

Turtle API

The first step in the buildScript method is to instantiate a Turtle object. The Turtle constructor has this API:

```
Turtle( drawingArea, double x, double y )
```

For the first argument, use the variable name root, which has been defined on line 15 and instantiated as a Group object on line 21. This is the drawing canvas on the window where the turtle's drawings will appear. The second and third arguments are the x and y locations in the window where the Turtle should first appear. These are double data types.

To create a Turtle object, choose a name (object reference), and call the constructor. For example, this Turtle object is named myrtle and will appear at the pixel located at (350, 250) in the drawing area of the window:

```
Turtle myrtle = new Turtle( root, 350, 250 );
```

Be sure to put this code inside the buildScript method. When a new turtle is created, it is facing the top of the screen and the pen is down. The pen is not a visible object.

The API specifies the x and y arguments are doubles. You may have noticed that 350 and 250 are integers. Remember, lower-precision values can be assigned to higher-precision variables. This allows int arguments to be sent to a method wherever doubles are specified in the API.

The Turtle class has methods that can be called to make drawings. **Figure 6-4** shows a few of these methods. The complete Turtle API is provided in the appendix. Most of the Turtle methods are void. They do not return a value. Only the getX and getY methods return a value. Those values are the *x* or *y* coordinate position of the turtle in the window. The values are returned as a double data type.

Notice also two methods have the name turnRight. The two versions differ in their arguments. One turnRight method does not take an argument. The other version takes one argument: the number of degrees to turn. The same can be said for the two turnLeft methods.

Methods that share a name but take different arguments are called *overloaded methods.* A class designer may find it useful to create overloaded methods so the programmer using the class does not need to learn multiple method names for similar functionality. The compiler can tell from the argument (or lack of an argument) which method is being called.

Method	Action
void forward(double pixels)	Moves the turtle forward by the number of pixels.
void backward(double pixels)	Moves the turtle backward by the number of pixels.
void turnRight( )	Turns the turtle 90 degrees to the right.
void turnRight(double degrees)	Turns the turtle right by the number of degrees.
void turnLeft( )	Turns the turtle 90 degrees to the left.
void turnLeft(double degrees)	Turns the turtle left by the number of degrees.
void moveTo(double x, double y)	Moves the turtle to the (x, y) location; the turtle's direction does not change.
void penDown( )	Sets the pen down; the turtle leaves a mark when the pen is down.
void penUp( )	Lifts up the pen; the turtle does not leave a mark when the pen is up.
double getX( )	Returns the x position of the turtle.
double getY( )	Returns the y position of the turtle.
void hide( )	Hides the turtle.
void show( )	Shows the turtle.

Goodheart-Willcox Publisher

Figure 6-4. These are some of the methods included in the custom Turtle class.

HANDS-ON EXAMPLE 6.1.1

Drawing a Rectangle

In this exercise, you will draw a rectangle using the Turtle class. The rectangle will be 150 pixels high and 75 pixels wide. Start with the algorithm. A possible way to draw a rectangle with the Turtle class is:

1. Instantiate a Turtle object.
2. Move forward the height of the rectangle (150 pixels).
3. Turn 90 degrees right.
4. Move forward the width of the rectangle (75 pixels).
5. Turn 90 degrees right.
6. Repeat steps 2–5 one time to complete the rectangle.

Looking at the Turtle methods, you can see that several methods could be of use. The forward method will move a Turtle object forward a certain number of pixels. The turnRight method will turn the Turtle object 90 degrees to the right.

Before beginning this exercise, download the chapter files from the student companion website. Locate the HOE06-01-01 folder. In this folder, you will find the files needed for this exercise.

1. Launch jGRASP, and open all the Java files in the HOE06-01-01 folder. The SpriteApplication.java file has been already renamed for you as DrawRectangle.java.
2. Compile each of the files in the order specified in the text.
3. Change windows so the DrawRectangle.java file is displayed in jGRASP. Follow the instructions in comment 1 to instantiate a Turtle object. This Turtle object is being named gertie. Put all the code in this exercise in the buildScript method beginning on line 32. Place the code under the comments. The first comment begins on line 37.

```
/*****1. Instantiate a Turtle named gertie at pixel (350, 250). */
Turtle gertie = new Turtle( root, 350, 250 );
```

4. Follow the instructions in comment 2 to compile and run DrawRectangle.java. You should see a window appear with a green turtle in the middle. To make the turtle window disappear so you can see the code, just click the close button (**X**) in the upper-right corner of the window. The window will reappear when you run the program again.

5. Follow the instructions in comment 3 to call the forward method to move gertie up 150 pixels. Remember to use the object name (gertie) and the dot notation (.) to call a method.

```
/***** 3. Move gertie forward 150 pixels. */
gertie.forward( 150 ); // move gertie up
```

6. Follow the instructions in comment 4 to turn gertie 90 degrees to the right.

```
/***** 4. Turn gertie right 90 degrees. */
gertie.turnRight( ); // face right
```

7. Follow the instructions in comment 5 to compile and run the application to see the turtle move up 150 pixels and turn right.

8. Follow the instructions in comment 6 to move gertie forward 75 pixels and turn right.

```
/***** 6. Move gertie forward 75 pixels and turn right. */
gertie.forward( 75 ); // move gertie right
gertie.turnRight( ); // face down
```

9. To finish the rectangle, follow the instructions in comment 7 to repeat the last four method calls. Although the code is repeated, the turtle's heading will be different, so adjust the comments to reflect this.

```
/***** 7. Repeat the last four instructions to complete the rectangle. */
gertie.forward( 150 ); // move gertie down
gertie.turnRight( );     // face left
gertie.forward( 75 );    // move gertie left
gertie.turnRight( );     // face up
```

10. Compile and run the application. The turtle should draw a complete rectangle, as shown.

11. Leave jGRASP and all files open.

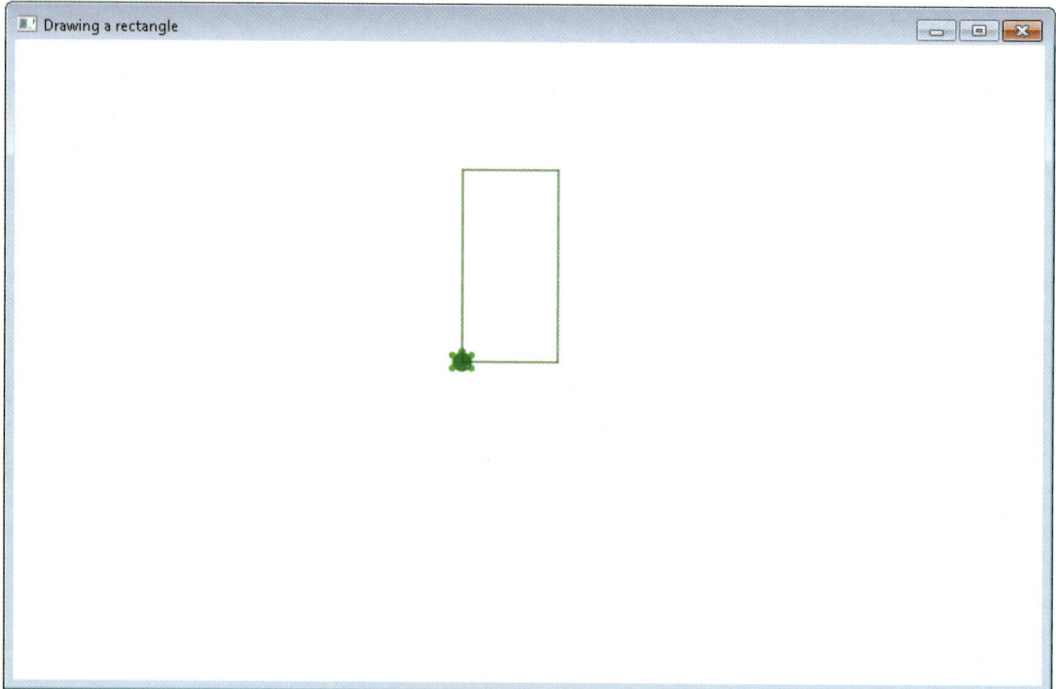

Goodheart-Willcox Publisher

Try It!

Other algorithms for drawing a rectangle will also work. For example, you could move the turtle backward and turn 90 degrees to the left. Your job is to create another Turtle object with a name of your choice. Add this new code at the bottom of the DrawRectangle.java file under the comment. Be sure, however, the code is still in the buildScript method. Start your new turtle at the location (250, 150). Then, draw another rectangle with a height of 250 and a width of 200 by moving the Turtle object backward four times with four left turns.

SECTION REVIEW 6.1

Check Your Understanding

1. The _____ tells programmers how to create objects and what methods are available for a class and how to call those methods.
2. Which special method creates objects?
3. What does an argument list consist of, separated by commas?
4. What do the (x, y) coordinates in the graphics window specify?
5. Creating an instance of a class requires a name for the object called its _____.

Build Your Vocabulary

As you progress through this course, develop a personal computer science glossary. This will help you build your vocabulary and prepare you for a career. Write a definition for each of the following terms and add it to your computer science glossary.

application programming
   interface (API)
constructor

dot notation
instance
null

object reference
overloaded method
pixel

Java Class Library

In the last section, you wrote code to draw a rectangle using a turtle. To do this, you instantiated a Turtle object and called methods of the Turtle class. You did not need to see the code in the Turtle class. In fact, if you did read the Sprite.java class, you would find some advanced code to respond to method calls and animate the turtle. The Turtle system was written by the authors of this text for your use.

The Java Class Library contains classes written by Java programmers for general-purpose use. These classes are designed to make it easy to create fun programs without having to write all the code yourself. You have already used the Scanner class for input. The Scanner class is just one of the more than 2,000 classes provided in the Java Class Library. This section explores a few more of the classes in the Java Class Library.

How does providing the Java Class Library on the Oracle website improve programming efficiency?

Learning Goals

- Describe the organization of the Java Class Library.
- Plan for user input Strings by calling Scanner methods.
- Generate random numbers.
- Compare and contrast static methods and static constants.

Terms

buffer	static method
immutable	token
package	tokenizing
pseudorandom number	

Library Overview

The Java Class Library has a rich set of predefined classes that can be used to create special features in your programs. The Random class allows us to generate random numbers for games. You have seen that String literals allow you to label output. The Scanner class makes input easy. The Math class provides methods for mathematical calculations, such as finding a square root, a number's absolute value, or raising a number to a power. These classes are automatically downloaded with the Java Development Kit (JDK). You may look up the API for any class in the Java Class Library on the Oracle website (www.oracle.com). Use the site's search function to locate the information for the version of Java you are using.

The classes are organized into packages according to their functionality. A *package* is a collection of related classes in one location. To use many of the classes in the Java Class Library, it is necessary to import the class. In previous chapters, you have imported the Scanner class from the java.util package. Some classes, however, such as String and Math, are in the java.lang package. The compiler automatically imports this package. You do not need to import these classes to use them.

As you have seen, the import statement has this syntax:

```
import packageName.Classname;
```

To import more than one class from the same package, use this syntax:

```
import packageName.*;
```

With this version of the import statement, the compiler will import all the classes in the package the program uses.

String Class

The char primitive data type can hold one character. Often, you need a data type that will hold many characters. For example, the name of a person, book, day, or month all contain more than one character. For these and similar types of data, the String class is used.

So far, you have used String literals. These are sequences of characters enclosed in double quotes. String literals are objects, but as you have used them they did not have names (object references). So, you were unable to reuse these String objects by name. In this section, you will explore creating named String objects and inputting Strings from the user.

The String class has many methods for searching Strings and for extracting characters from String objects. These methods are explored in depth in Chapter 10.

Creating Strings

A String can be created in multiple ways. One way is to call a constructor. A constructor of the String class is shown in **Figure 6-5.** This constructor creates a String object whose data consists of the characters in the String literal. If the String literal has no characters between the quotation marks, as in "", then the String object is created, but it is empty. For example:

```
String greeting = new String( "Hello" );
String empty = new String( "" );
```

Constructor	Explanation
String("String literal")	Creates a String object from the String literal.

Goodheart-Willcox Publisher

Figure 6-5. This is one constructor of the String class.

Another, simpler way to create a String is to assign a String literal to a String reference:

```
String greeting = "hello"; // greeting is the String name or reference
```

This statement works because a String literal is a String object. The statement assigns a String object to a String object reference.

The newly created String can then be assigned to another String:

```
String greeting2 = greeting;
```

After both statements have been executed, greeting and greeting2 refer to the same location: the String "hello" in memory. This sequence of statements is illustrated in **Figure 6-6A.**

Strings are immutable. *Immutable* means unable to be changed, so Strings are constants. After these two statements execute:

```
String food = "meat"; // food points to meat
food = "potatoes";     // now food points to potatoes
```

the String literal "meat" is still in memory, but the reference food now points to the location of the String "potatoes". In other words, the contents of the food

FYI

For many classes, java. is part of the package name, such as java.util. For JavaFX classes, the package name starts with javafx., such as javafx.scene.

FYI

The String class is in the java.lang package, so Strings can be used without an import statement.

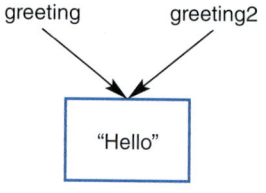

A Two references to the same object.

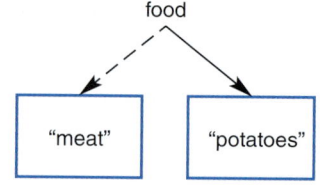

B Reference changes to another object.

Goodheart-Willcox Publisher

Figure 6-6. Two variables can refer to the same string (A), or the reference of a single variable can change to a different string (B).

reference changes, not the String itself. This sequence of statements is illustrated in **Figure 6-6B.**

Reading Strings with Scanner

You have already used the input methods for the Scanner class for reading int and double data types. The Scanner class also has two methods for reading Strings. These methods are shown in **Figure 6-7.** These methods differ in how white space is treated. Remember that white space is any:

- space;
- tab; or
- new line character.

Method	Action
String next( )	Returns the next input token as a String.
String nextLine( )	Returns the remainder of the line as a String.

Goodheart-Willcox Publisher

Figure 6-7. The next and nextLine methods can be used to input a String.

The next method will skip over any leading white space and start reading when it reaches a non-white-space character. The method continues collecting characters into a String until it reaches a white-space character. The white-space character is not put into the String and is not returned. This means any String that is read using the next method will not contain white space. Skipping white space and collecting non-white-space characters is called *tokenizing.* The String collected using this method is called a *token.* The nextInt and nextDouble methods also tokenize the input and return the next input token that fits the data type requested by the method.

In contrast, the nextLine method will collect and return all characters in the input, including white space, up to the new line character. The new line character is not returned, but it is skipped. Because of the difference in the way the nextLine method works, extra care must be taken when using the nextLine method in the same program with the next, nextInt, or nextDouble methods.

Inputting One Word

Here is an example where input is limited to the first word entered by the user. Assume that a Scanner object named input has been instantiated.

```
System.out.print( "Enter your name: " );
String firstName = input.next( );
System.out.println( "Hello, " + firstName );
```

If the user enters Joan Johnson at the prompt, then the output from line 13 will be:

```
Hello, Joan
```

because the next method reads only the first word entered. It stops when it reaches the space. What happens to the remaining characters (Johnson)? They are put into a buffer waiting for the next Scanner method to be called. A **buffer** is a storage area in memory; it is like a holding area. If no more Scanner methods are called before the program ends, the token is discarded.

HANDS-ON EXAMPLE 6.2.1

Inputting One Word

In this exercise, you will write a new program that creates a String object for someone's first name. The program will respond by saying "Welcome" and the first name.

1. Launch jGRASP. Close all open files, then begin a new Java file.
2. Create a new program called FirstName.java that imports the Scanner class and write the code described in the comments.

```java
/* Inputting a String
   your name here
*/
import java.util.Scanner;

public class FirstName {
    public static void main( String [ ] args ) {

        /* 2a. Create the Scanner object. */
        Scanner input = new Scanner( System.in );

        /* 2b. Ask for the user's first name
           and save the name as a String named firstName. */
        System.out.println( "Enter your first name and press enter" );
        String firstName = input.next( );

        /* 2c. Display "Welcome" and the user's first name. */
        System.out.println( "Welcome, " + firstName );
    } // end of main
} // end of program
```

3. Save the file in your working folder. Be sure to use the correct name.
4. Compile and run the program. Enter your name at the prompt.
5. Debug the program as needed.
6. Try entering your first and last name. The program should only return your first name.

Try It!

The genus name of the Giant Panda is *Ailuropoda* and the species is *melanoleuca*. Write a program that asks for the genus name and species. Make the genus name a String object. After the user enters the genus, display the sentence: "The giant panda is a member of the genus" and append the genus name entered by the user. Do the same operations for the species.

Inputting More Than One Word

It is possible to have the user enter two or more strings with a single prompt. The following shows an example of code that asks for the user's first and last names with one prompt. If the user enters Joan Johnson at the prompt, the input is divided into two String tokens because of the space between them. In this code, the first token is read with the first call to the next method and stored into the String firstName, and the last name is read with the second call to the next method and stored in the String lastName. Assume that a Scanner object named input has been instantiated.

```
System.out.print( "Enter your first and last name: " );
String firstName = input.next( );
String lastName = input.next( );
System.out.println( "Hello, " + firstName + " " + lastName );
```

The output from this code is:

```
Hello, Joan Johnson
```

However, there is a more efficient method for achieving this result. With this code (in another program, of course):

```
System.out.print( "Enter your full name: " );
String fullName = input.nextLine( );
System.out.println( "Hello, " + fullName );
```

if the user enters Joan Johnson at the prompt, the nextLine method reads the whole line, including spaces. The output is the same, but fewer lines of code are used.

HANDS-ON EXAMPLE 6.2.2

Inputting Multiple Words

In this exercise, you will create a program that asks the user to enter his or her full name. The program will display "Welcome" and the user's full name.

1. Launch jGRASP. Close all open files, then begin a new Java file.
2. Create a new program called FullName.java that imports the Scanner class and write the code described in the comments.

```
/* Inputting multiword Strings
   your name here
*/
import java.util.Scanner;

public class FullName {
    public static void main( String [ ] args ) {

        /* 2a. Create the Scanner object. */
        Scanner input = new Scanner( System.in );

        /* 2b. Ask the user for his or her first and last names,
           Use the nextLine method and place the input
           into one String object. */
        System.out.println( "Enter your full name and press enter" );
        String fullName = input.nextLine( );

        /* 2c. Display the full name after the word Welcome. */
        System.out.println( "Welcome, " + fullName );

    } // end of main
} // end of program
```

3. Save the file in your working folder. Be sure to use the correct name.
4. Compile and run the program. Enter your full name at the prompt.
5. Debug the program as needed.

Try It!

Modify the program created in Try It! 6.2.1 to accept both the genus and species of the Giant Panda as one entry. Place the user input into a single String object. Be sure to modify the prompt so the user knows what to enter. The program output should display the following information: "The genus and species of the Giant Panda are Ailuropoda melanoleuca."

Random Class

The Random class can generate seemingly random numbers. The numbers are "seemingly" random because with computers there is no such thing as truly random. Instead, the Random class uses a special algorithm to generate pseudo-random numbers. A *pseudorandom number* belongs to a set of a sequence of numbers long enough to appear to be random, but is not.

Like the Scanner class, the Random class is in the java.util package, so it will need to be imported. A second import statement can be added for the Random class:

```
import java.util.Scanner;
import java.util.Random;
```

or both classes can be imported with one statement:

```
import java.util.*;
```

The Random constructor decides where in that long sequence to start generating the random numbers. Refer to **Figure 6-8.** This constructor does not take any arguments. This statement will create a random-number generator named randGen:

```
Random randGen = new Random( );
```

Now that you have a random-number generator, you call the nextInt method to generate a random number:

```
int nextInt( int range );
```

The nextInt method takes one argument, which is the range of numbers from which to randomly choose. Refer to **Figure 6-8.**

For example, to simulate the roll of a six-sided die, a range of six numbers is needed: 1–6. Notice, however, the nextInt method returns a number from 0 up to, but not including, range. So, if you send 6 as the argument, the nextInt method will return values from 0 to 5. This is not the needed range of numbers. The needed range can be achieved by adding 1 to whatever number is returned from the method. The method call becomes:

```
int dieRoll = randGen.nextInt( 6 ) + 1;
```

The random-number generator still picks a number from the range 0–5, but the code then adds 1 to make the effective range of numbers 1–6.

Constructor or Method	Action
Random( )	Creates a random-number generator object.
int nextInt(int range)	Returns the next int in the sequence of pseudorandom numbers that is evenly distributed between 0 and (range – 1).

Goodheart-Willcox Publisher

Figure 6-8. The Random constructor generates random numbers. The nextInt method specifies the range from which the random numbers will be generated.

HANDS-ON EXAMPLE 6.2.3

Generating Random Numbers

In this exercise, you will create a random-number generator that makes numbers appear one at a time in the range from 1– 6. You will import the random-number generator class that allows the program to do this.

1. Launch jGRASP. Close all open files, then begin a new Java file.

2. Create a new program called RollOneDie.java that imports the Random class and write the code described in the comments.

```java
/* Using the Random Number Generator HOE 6.2.3
   your name here
*/
import java.util.Random;

public class RollOneDie {
    public static void main( String [ ] args ) {

        /* 2a. Create a Random number generator object */
        Random randGen = new Random( );

        /* 2b. To simulate the roll of a die, call the nextInt method
           with a range of 6. To generate numbers in the correct range,
           add 1 to the return value. Store the return value in an
           int variable named dieRoll. */
        int dieRoll = randGen.nextInt( 6 ) + 1;

        /* 2c. Display the random number by saying "The number is "
           and the number. */
        System.out.println( "The number is " + dieRoll );
    } // end of main
} // end of program
```

3. Save and compile the program. Then, run the program at least 10 times. Did you see all numbers from 1 to 6?

Try It!

Modify the RollOneDie.java program to simulate rolling two dice, and save it as RollTwoDice.java. You will not need to create another Random object. Just use the randGen object you already created. However, you will need a second random number. Output the numbers for each die and the total roll.

Note: although you might be tempted to generate a single number from 1 to 12, doing so would be incorrect. This is a logic error. The distribution of numbers from rolling two dice is different from generating one number between 1 and 12. With two dice, the number 7 is the most likely to occur, whereas with selecting one number from 1–12, the number 7 is as likely to occur as any of the other numbers.

Static Constants and Methods

Sometimes, it is inconvenient to create an object if you need just one calculation. For example, you want to calculate a square root or raise a number to a power. For these situations, some methods are designed to be used without creating an object. These methods are called static methods. A *static method* is called with a class name instead of an object reference. The methods are still called using the dot notation, but because there is no object, a static method is called using the class name instead of an object name.

Science and Java

Randomizing a Diabetes Study

Everyone needs the glucose carried by the blood as the main source of energy for all the cells in the body. Insulin helps glucose get from food into the cells. Insulin is produced by the pancreas, but for some people, it does not make enough of this hormone. In other cases, a person's body may not use insulin well. Both these situations result in glucose staying in the blood and not being used by cells. Diabetes is a disease that occurs when one's blood glucose is too high.

Over time, having too much glucose in your blood can cause major health problems. Some complications from diabetes include heart disease, stroke, kidney disease, eye problems, dental disease, nerve damage, and foot problems. Those at risk for developing diabetes generally have a family history of the disease, are over age 45, and are obese.

The occurrence of diabetes is increasing globally. According to the Centers for Disease Control and Prevention (CDC), the number of Americans with diabetes or prediabetes exceeds 100 million. More than 30 million Americans have fully developed the disease.

There is no known cure for diabetes. Making lifestyle and dietary changes can help a person with diabetes stay healthy. Many people with the disease take medications to help manage glucose levels. The use of new oral and injectable medications, insulin pens and pumps, and continuous glucose-monitoring meters have worked well in promoting the management of this disease. Patients are encouraged to take medication, develop a healthy diet, exercise regularly, and monitor their glucose levels.

A new tool for helping to manage diabetes has been studied. This method involves ingesting vinegar with meals. Preliminary studies have shown it is an effective and simple way to help manage

lola1960/Shutterstock.com

the disease along with medication and lifestyle changes. Always follow the advice of your health-care provider when it comes to treating disease and managing your health.

Assignment

Before beginning this exercise, download the chapter files from the student companion website. Launch jGRASP, open the StarterSnippet.java file, and save it as Diabetes.java in your working folder. You are part of a team who is managing a clinical trial with 24 diabetic patients. You are responsible for choosing which patients in your group will use the new vinegar method to help manage their diabetes. You are not allowed to know the patients you pick. You must use a random sampling. Each patient has been assigned a number from 1 to 24. Write a program that uses a random-number generator to select one patient at random. Repeat running the program until you have selected eight unique patient numbers. In other words, if the patient number has already been selected, run the program again.

Static Constants

In addition to defining static methods, some classes also define static constants. These static constants are provided as a convenience to programmers using the class. For example, the Color class provides predefined Color objects, such as Color.RED, for setting colors in graphics applications. Like static methods, static constants are referenced using the class name. The Color class is discussed in detail in the next chapter. Static constants are usually either predefined objects of the class or argument values for a method.

To illustrate using static constants for method arguments, take a look at another method of the Turtle class: setSpeed. This method sets the speed at which the turtle draws. The setSpeed method is not a static method, but three static constants are defined to be used as arguments. Refer to **Figure 6-9.**

By default, the turtle draws at medium speed, but the speed can be set to a slower or faster rate. The speed can also be set to no animation, which causes the whole drawing to appear at once. For example, this code will set the speed to the fastest rate available:

```
Turtle gertie = new Turtle( root, 350, 250 );
gertie.setSpeed( Turtle.FAST );
```

and this code will eliminate the animation:

```
gertie.setSpeed( Turtle.NO_ANIMATION );
```

You can see from the API the setSpeed method expects an argument that is an int. If you were to look at the Turtle code, you would see the static constants have been defined with these int values:

```
static final int FAST = 1;
static final int MEDIUM = 2;
static final int SLOW = 3;
static final int NO_ANIMATION = 0;
```

That means, for example, setting the speed to slow can be accomplished by either of these statements:

```
gertie.setSpeed( Turtle.SLOW );
gertie.setSpeed( 3 ); // also sets speed to slow
```

The value of the static constants becomes clear: if you want to slow down the turtle, it is much easier to remember "SLOW" than "3."

Method	Action
void setSpeed(int speed)	Sets the drawing speed. These static constants can be used to set the speed: Turtle.SLOW, Turtle.MEDIUM, Turtle.FAST, and Turtle.NO_ANIMATION.

Figure 6-9. The setSpeed method can be used with static constants to set the drawing speed of the turtle.

HANDS-ON EXAMPLE 6.2.4

Calling Methods from the Java Class Library ↱

In this exercise, you will create a program using the Turtle class to draw a rectangle with random width and height. You will also change the speed of the turtle. Before beginning this exercise, download the chapter files from the student companion website. Locate the HOE06-02-04 folder. In this folder, you will find the files needed for this exercise.

1. Launch jGRASP, and open the DrawRandomRectangle.java file.

2. The program requires random numbers, so you need to import the Random class. Follow the instructions in comment 1 below the other import statements.

   ```
   /***** 1. Import the Random class. */
   import java.util.Random;
   ```

3. Locate the buildScript method starting on line 37. Put all remaining code for this exercise in this method. Follow the instructions in comment 2 to instantiate a Turtle object named gertie.

   ```
   /***** 2. Instantiate a Turtle named gertie. */
   Turtle gertie = new Turtle( root, 350, 250 );
   ```

4. Follow the instructions in comment 3 to set the turtle's speed to fast.

   ```
   /***** 3. Set the Turtle's speed to fast. */
   gertie.setSpeed( Turtle.FAST );
   ```

5. Follow the instructions in comment 4 to instantiate a random number generator.

   ```
   /***** 4. Instantiate a random number generator. */
   Random randGen = new Random( );
   ```

6. Follow the instructions in comment 5 to generate a random value for the rectangle's height. The height should be between 10 and 100 pixels. Notice the number of values between 10 and 100 is actually 91 because both 10 and 100 are included in the range. Add the starting number of 10 to the return value of the nextInt method so the minimum value height is 10.

   ```
   /***** 5. Generate a random number for height between 10 and 100. */
   int height = randGen.nextInt( 91 ) + 10;
   ```

7. Follow the instructions in comment 6 to generate a random value for the rectangle's width. The width should be between 25 and 100 pixels. The number of values between 25 and 100 is 76 because both 25 and 100 are included. Add the starting number of 25 to the return value of the nextInt method.

   ```
   /***** 6. Generate a random number for height between 25 and 100. */
   int width = randGen.nextInt( 76 ) + 25;
   ```

8. Follow the instructions in comment 7 to output the height and width so you can check your work. When you run the program, this information will display in the **Run I/O** tab, rather than in the graphics window.

   ```
   /***** 7. Output the generated values for height and width. */
   System.out.println( "height: " + height + ", width: " + width );
   ```

9. Follow the instructions in comment 8 to draw the rectangle. The code used in Hands-On Exercise 6.1.1 is used here as well.

   ```
   /***** 8. Draw the rectangle using the height and width variables. */
   gertie.forward( height );  // move gertie up
   gertie.turnRight( );       // face right
   gertie.forward( width );   // move gertie right
   gertie.turnRight( );       // face down

   gertie.forward( height );  // move gertie down
   gertie.turnRight( );       // face left
   gertie.forward( width );   // move gertie left
   gertie.turnRight( );       // face up
   ```

10. Compile and run the program. Run the program several times to verify the rectangle's size is different and the turtle moves fast. Close the graphics window after each run before the next run.

11. Check the code for correct indentation. To do this, generate and then remove the code structure diagram (CSD).

Try It!

Create another Turtle object with a name of your choice, and begin the turtle at (250, 150). Add this new code at the bottom of the DrawRandomRectangle.java file under the comment. Set the turtle's speed to no animation so the drawing will appear all at once. Draw a rectangle with random ranges for the height and width you decide. Do not create another Random object. Just use the randGen object created in the code on the previous page. Similarly, reuse the height and width variables, but generate new random values for them. Choose a drawing direction for your turtle, either forward or backward. Run the program several times to verify the rectangle is drawn at different sizes.

FYI

The constant *e* is called Euler's Number after Leonhard Euler, the great 18th-century Swiss mathematician who discovered the number. It is also known as the Banker's Number because of its uses in financial calculations.

Constant	Value
PI	Approximation of the value of pi as a double.
E	Approximation of the value of the natural logarithm as a double.

Goodheart-Willcox Publisher

Figure 6-10. These are the two static constants in the Math class.

Math Class

The Math class has both static constants and static methods. The two static constants are shown in **Figure 6-10.** These constants are used by putting the name of the Math class and dot notation before the constant.

```
System.out.println( "PI is " + Math.PI ); // PI is 3.141592653589793
System.out.println( "E is " + Math.E );   // E is 2.718281828459045
```

Again, these constants are provided as a convenience to the programmer. They make it unnecessary to look up or memorize many digits of pi or to remember the value of *e*.

All methods in the Math class are static. Having all static methods makes sense because it would be wasteful of CPU time and memory to create an object, for example, just to calculate a square root. **Figure 6-11** shows a few of the static methods in the Math class.

The sqrt method is straightforward. It returns the positive square root of the argument.

```
double dResult = Math.sqrt( 9 ); // dResult gets 3.0
```

This method will always return a double. If you want the result as an int, you can type cast the return value:

```
int iResult = (int) Math.sqrt( 9 ); // iResult gets 3
```

The pow method raises the first argument to the power indicated by the second argument. To calculate 2^4, use this statement:

```
double powResult = Math.pow( 2, 4 ); // powResult gets 16.0
```

This method also returns only a double result. Like the sqrt method, the return value can be type cast to an int if desired. There is no direct method to square a number, so use the pow method with 2 as the second argument.

The round method rounds a floating-point number to the nearest integer. The method returns the rounded value as a long, but again you can type cast the return value to an int if desired.

```
long longRounded = Math.round( 5.673 ); // longRounded gets 6 as
//a long
int intRounded = (int) Math.round( 5.673 );  // intRounded gets 6
//as an int
```

Application of Method	Value
double abs(double n)	Returns the absolute value of *n*.
double pow(double base, double exponent)	Returns the base raised to the exponent power.
long round(double number)	Returns the closest integer to the floating-point number.
double sqrt(double n)	Returns the positive square root of *n* or an approximate value.

Goodheart-Willcox Publisher

Figure 6-11. These are some of the static methods in the Math class.

Coding Conundrum

1. A random number from 50 to 75 needs to be generated. The programmer has written this code:

```
Random randGen = new Random( );
int num50to75 = randGen.nextInt( 26 );
```

The code compiles and runs, but the numbers generated are between 0 and 25. How can you fix this conundrum to produce the correct results?

2. The circumference of a circle is calculated as π*d*, where *d* is the diameter of the circle. The programmer has written the following code, intending to use the static constant of the Math class for π.

```
double diameter = 10.5;
double circumference = PI * diameter;
```

However, the compiler returns this error:

```
error: cannot find symbol
double circumference = PI * diameter;
                       ^
symbol: variable PI
```

How can you fix this conundrum to produce the correct results?

3. The user is to input a complete sentence. The programmer has written this code to allow for spaces within the user input:

```
Scanner input = new Scanner( System.in );
System.out.println( "Enter a sentence" );
String sentence = input.nextString( );
```

However, the compiler returns this error:

```
error: cannot find symbol
String sentence = input.nextString( );
                  ^
symbol: method nextString( )
```

How can you fix this conundrum to produce the correct results?

4. The square root of 25 needs to be calculated. You know the correct value to be 5, so you will know if the output is correct. The programmer has written this code:

```
int five = Math.sqrt( 25 );
```

However, the compiler returns this error:

```
error: incompatible types: possible lossy conversion from double to int
int five = Math.sqrt( 25 );
           ^
```

How can you fix this conundrum to produce the correct results?

Math and Java

Rounding in Java

Often, the result of a floating-point expression is a number with many more decimal places of precision than are needed. Casting to an int data type will remove the decimal places, but at the risk of losing accuracy. Rounding the number is better than just lopping off the decimal places. The Math class has several methods to help return the answer that best represents the computed value.

For the following examples, first set this value for *a*:

```
double a = 4.67821;
```

Roman Samborskyi/Shutterstock.com

Apply what you learned in this chapter about the round( ) method. Its API is:

```
static long round( double a )
```

This returns the closest long to the argument.

```
long roundOff = Math.round( a ); // roundOff is set to 5
```

All the previous examples remove the decimal places altogether. What if some decimal places need to be kept? To solve this problem, some arithmetic can be done before rounding and after rounding. Use this algorithm:

1. Multiply the number by 10 for each decimal place you want to keep. If you need two decimal places, multiply by 100. For three decimal places, multiply by 1000, and so on.
2. Perform the rounding by calling the Math.round method.
3. Divide the result by the same power of 10 to put the decimal point back where it belongs.

Here is a step-by-step example to round the number 4.67821 to two decimal places:

```
double multiplyBy100 = 4.67821 * 100;                 // => 467.821
double rounded = (double) Math.round( multiplyBy100 );// => 468.0
double round2Places =  rounded / 100.0;               // => 4.68
```

The result is that round2Places is set to 4.68.

In another example, the number of places to round can be input by the user. This may be easier to use than multiplying by a power of 10. When multiplying by a power of 10, to keep six decimal places the value would be 1000000. It would be very easy to make an error by entering too many or too few zeros. Instead, use the Math.pow() method for multiplier and divider based on the number of decimal places input.

```
int power = 3; // Example for rounding to three decimal places
double places = Math.pow( 10, power );
double roundNPlaces = (double) Math.round( a * places ) / places;
```

Assignment

Before beginning this exercise, download the chapter files from the student companion website. Launch jGRASP, open the StarterSnippet.java file, and save it as RoundingSnippet.java in your working folder. Write a program that prompts the user for the number to round. Also, prompt the user for the number of decimal places. Run the program several times to verify the appropriate rounding is taking place.

The abs method returns the absolute value of its argument. If its argument is negative, the method returns the value as a positive. Otherwise, the method returns the value of its argument unchanged. Unlike the sqrt and pow methods, the abs method has several versions. The abs method will accept arguments of double, float, int, and long data types. It will return the result in the same data type as the argument. This is another example of an overloaded method. Here are sample results:

```
double dAbs = Math.abs( -5.7 );   // dAbs gets 5.7 as a double
float fAbs = Math.abs( -5.7f );   // fAbs gets 5.7 as a float

int iAbs = Math.abs( -5 );        // iAbs gets 5 as an int
long lAbs = Math.abs( -5L );      // lAbs gets 5 as a long
```

HANDS-ON EXAMPLE 6.2.5

Calling Math Methods and Inputting Strings

In this exercise, you will compute the size of the hypotenuse of a right triangle and then draw the triangle using the Turtle class. To make things simple, assume the right triangle is an isosceles triangle where sides *a* and *b* are the same length. In an isosceles right triangle, the length of the hypotenuse can be calculated using the Pythagorean Theorem, as shown. With an isosceles triangle, the two angles opposite the right angle (90 degrees) are both 45 degrees. To draw the triangle, tell the turtle to:

1. Go forward the length of a side (turtle moves up).
2. Turn right 90 degrees (face right).
3. Go forward the length of the side (turtle moves to the right).
4. Turn right 180 − 45 degrees. (The inner angle is 45 degrees, but the turtle needs to turn the outer angle, which is 180 − 45 degrees.)
5. Go forward the length of the hypotenuse.

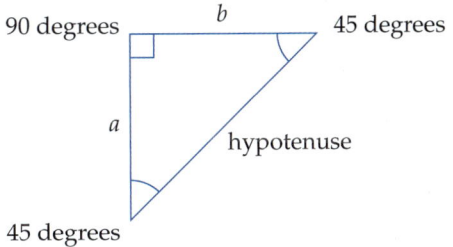

$$hypotenuse = \sqrt{a^2 + b^2}$$

where *a* and *b* represent the length of
the other two sides of the triangle

Goodheart-Willcox Publisher

Before beginning this exercise, download the chapter files from the student companion website. Locate the HOE06-02-05 folder. In this folder, you will find the files needed for this exercise.

1. Launch jGRASP, and open the DrawRightTriangle.java file. Put all code for this exercise in the buildScript method under the comments. Code to import Scanner and instantiate a Scanner object named input has already been included.
2. Follow the instructions in comment 1 to prompt the user for his or her first name.

```
/***** 1. Prompt the user for first name */
System.out.print( "What is your first name? " );
String firstName = input.next( );
```

3. Follow the instructions in comment 2 to prompt the user for the length of the sides. Include the user's name in the prompt.

```
/***** 2. Using the user's first name in the prompt, ask the user for the
        length of the two non-hypotenuse sides. */
System.out.print( firstName +
            ", what is the length of the two sides of the triangle? " );
double sides = input.nextDouble( );
```

4. Follow the instructions in comment 3 to calculate the length of the hypotenuse.

```
/***** 3. Calculate the hypotenuse length using the Pythagorean Theorem
        and output the value */
double hypotenuse = Math.sqrt( Math.pow( sides, 2 ) + Math.pow( sides, 2 ) );
System.out.println( "The length of the hypotenuse is " + hypotenuse );
```

5. Follow comment 4 to instantiate a **Turtle** object and draw the triangle.

```
/***** 4. Instantiate a Turtle and draw the triangle. */
Turtle triangle = new Turtle( root, 350, 250 );
triangle.forward( sides );
triangle.turnRight( );
triangle.forward( sides );
triangle.turnRight( 180 - 45 ); // turn outer angle
triangle.forward( hypotenuse );
```

6. Compile and run the program. Use a value for the sides between 50 and 100. Note the prompts for the person's name and length of the triangle sides will be displayed in the **Run I/O** tab, not in the graphics window. You might need to click on the jGRASP window to enter the input. If your calculation is correct, the turtle will draw a right triangle. An example is shown using a value of 100.

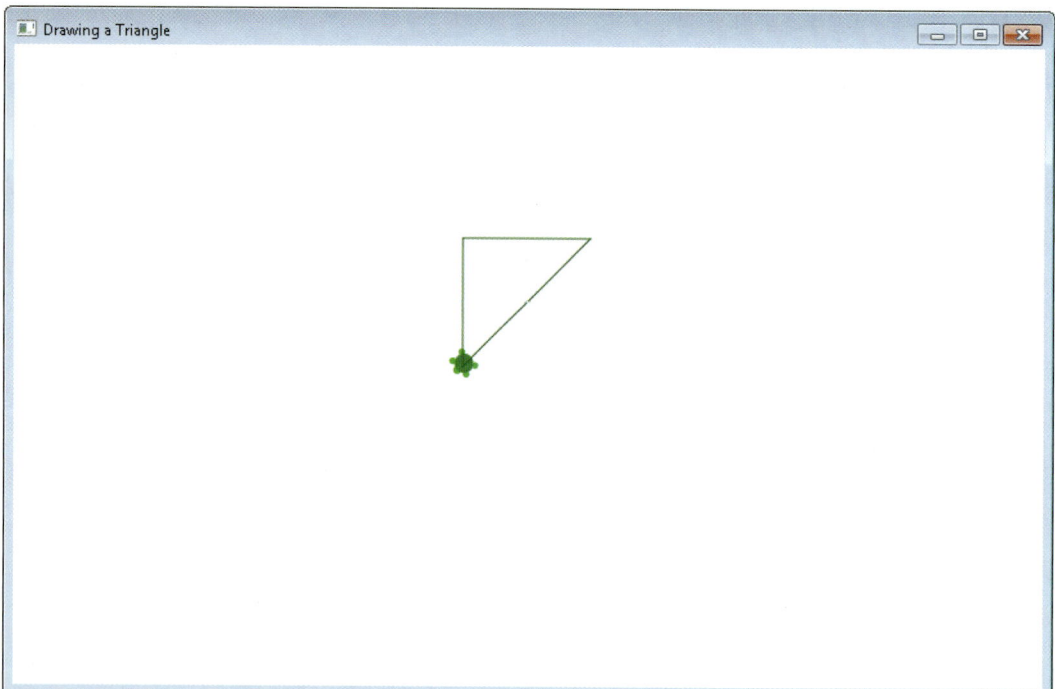

Try It!

Draw another triangle with a random size. Add this new code at the bottom of the DrawTriangle.java file under the comment. Instantiate a second Turtle object with a name of your choosing. Start the turtle at (400, 250). Generate a random number between 50 and 150. Use this random number for the length of the sides of the triangle. Output the length of the sides and the length of the hypotenuse to check your work, and draw the triangle.

SECTION REVIEW 6.2

Check Your Understanding

1. Why does the String class not need to be imported within a program?
2. Which method can be used to accept user input of multiple words?
3. When generating a random number, what number begins the range of possible numbers?
4. To call a static method, use the dot notation with the _____ name.
5. Which method in the Math class rounds a floating-point number to the nearest integer?

Build Your Vocabulary

As you progress through this course, develop a personal computer science glossary. This will help you build your vocabulary and prepare you for a career. Write a definition for each of the following terms and add it to your computer science glossary.

buffer	pseudorandom number	tokenizing
immutable	static method	
package	token	

Cooperative Coding

Making Letters

Capital letters can be made of lines or rectangles of different sizes. Investigate how you can make a word or your name or your initials using the Turtle class. Examine the examples shown. By squaring off the curves, almost any letter can be created. For example, a capital O is easy to illustrate by a rectangle.

PGMart/Shutterstock.com

Refer to **Figure 6-4** for some of the methods in the Turtle class. The moveTo, penUp, and penDown methods in the Turtle class can be used to write the letters. Call the hide method to make the turtle disappear when finished drawing a letter.

Assignment

Work with your team to identify a three-letter word, initials, or monogram for each member. Discuss how to create a single program to draw each member's word. How many methods will need to be used? Experiment with having each letter appear based on a random selection of the speed of the turtle. Write the code for the program, then run it. Debug as needed. Save the program as DrawLetters.java in your working folder.

1. What methods did your team decide are needed?
2. How did your team implement the random drawing speed?

Chapter Summary

Section 6.1 Introduction to Classes

- The application programming interface (API) tells programmers wishing to use the class how to create objects of that class, what methods are available, and how to call those methods; a constructor is used to create objects, and method calls are prefaced with dot notation.
- The Turtle class is an example of a custom class, and the API shows how it can be used to draw shapes in the graphics window.

Section 6.2 Java Class Library

- The Java Class Library contains classes organized into packages based on functionality, and many of the classes need to be imported from a package to be used in a program.
- The String class is used to store multiple characters, the Scanner next method inputs a single word, and the Scanner nextLine method inputs multiple words.
- Java can generate pseudorandom numbers, which are numbers that appear to be randomly selected from a large sequence of numbers, using the Random class; the Random class must be imported from the java.util package.
- A static method is a method called using the dot notation and the class name rather than with an object reference; a static constant contains a commonly used set value, such as the approximate value of pi or *e*, to ease the task of programming.

Chapter 6 Test

Multiple Choice

Select the best response.

1. What is the order of terms when calling a method found in a class?
   A. object.methodName(arguments)
   B. methodName,arguments(object)
   C. class.object.arguments
   D. (object) methodName.arguments

2. When no argument is required for a method, the parentheses should _____.
   A. contain a variable name
   B. contain zero
   C. be omitted
   D. be empty

3. Which of the following methods are in the Turtle class?
   A. goForward, penDown, moveTo
   B. goTo, forward, right
   C. penUp, left, backward
   D. moveTo, turnLeft, forward

4. Java classes downloaded with the JDK are found in the _____.
   A. Application Programming Interface
   B. Constructor bin directory
   C. Java Class Library
   D. Java Programming Catalog

5. Which of the following code calculates the input variable raised to the third power?
   A. Math.exp(3, input)
   B. Math.pow(input, 3)
   C. pow.Math(input) * 3
   D. Math.exp(input, exponent)

Completion

Complete the following sentences with the correct word(s).

6. When using the Turtle class, the origin (0, 0) of the graphics window is located in the upper-left corner of the window with the positive *y* direction extending _____.

7. The compiler automatically imports the _____ package, which includes the String and Math classes.

8. The Scanner class uses _____ to separate input tokens.

9. The correct range for the nextInt method of the Random class for a number between 0 and 9 is _____.

10. The two _____ constants in the Math class are PI and E.

Matching

Match the correct term with its definition.

   A. pow
   B. abs
   C. sqrt
   D. method call
   E. System.in

11. Method of the Math class that finds the square root.

12. Method of the Math class that finds the absolute value.

13. Default name for the keyboard.

14. Method of the Math class that raises a value to a specified power.

15. Examples include input.nextDouble( ) and gertie.moveTo(100, 50).

Application and Extension of Knowledge

1. You are in charge of name badges at a convention. Using the StarterSnippet.java file located on the student companion website, create a program that asks for the user's first name, middle initial, and last name. Then, the program should output "Hello, my name is" and the person's first name. Under this greeting, print the full name including the middle initial.

2. Different approaches to the same problem often exist. In Hands-On Example 6.1.1, you drew a rectangle using the forward method. Then in the Try It! exercise, you drew a rectangle using the backward method. Explore the Turtle methods shown in **Figure 6-4.** Using the Turtle files from Hands-On Example 6.1.1, draw another rectangle using the moveTo method. Hint: get the Turtle's starting *x* and *y* values using the getX and getY methods, then move the turtle to the four corners of the rectangle.

3. It is relatively simple to write the exponent of 10 for the numbers 10, 100, 1000, and so on. The exponents for these three numbers are 1, 2, and 3. Did you know that any number can be expressed as a power of 10? This is the meaning of a logarithm. That is, find the power of 10 that generates the number. The power of 10 that generates 2 is about 0.3, so the logarithm of 2 is about 0.3. Using the StarterSnippet.java file located on the student companion website, write a Java snippet that raises 10.0 to the power of 0.3. Use the Math.pow( ) method. Then, use the Math.log10( ) method to find the exponent of 10 for the number 8.0. The API for the log10 method is shown below. The log10 method returns the exponent of 10 that produces the argument num. Save your work as Log10Snippet.java.

```
static double log10( double num )
```

4. The Greek geographer Eratosthenes was the first person to estimate the Earth's circumference nearly accurately. With developments in technology, Earth's measurements, including the circumference, have been accurately measured. The circumference of the Earth is 40,075 kilometers (km). Using the StarterSnippet.java file located on the student companion website, write a program to find the radius of Earth. Use the formula r = C/2π. Display the radius and the diameter of Earth. The results have many decimals. Use what you know about rounding to eliminate the decimals.

5. You probably have noticed that Java performs calculations very quickly and never complains about doing the same operation over and over. A huge number of calculations is necessary for a weather forecast. Supercomputers operate at 33,860 trillion calculations per second. Meteorologists are happy with this. Rapid improvements in computer speed and expansion of observational capacity through deployment of weather satellites have improved forecasting. Write a one-page paper about how you think the speed and accuracy of computers used in weather forecasting affects society.

Online Activities

Complete the following activities, which will help you learn, practice, and expand your knowledge and skills.

Vocabulary. Practice vocabulary for this chapter using the e-flash cards, matching activity, and vocabulary game until you are able to recognize their meanings.

Communication Skills

Reading. *Reading for detail* involves reading all words and phrases, considering their meanings, and determining how they combine with other elements to convey ideas. Using this approach, read the first section in this chapter. Think through the way the author uses the words in each paragraph. After you have finished, decide if you have obtained a grasp of the content by reading for detail.

Writing. Everyone has a stake in protecting the environment. Taking steps as an individual to become more environmentally conscious is a behavior of responsible citizens. From a business standpoint, it may also help a company be more profitable. Write a list of actions a business can take to minimize risk to the environment.

Speaking. An *introduction* is making a person known to someone else by sharing the person's name and other relevant information. Create a script you might use to introduce yourself to another classmate whom you do not know well. What information will you tell this person about yourself? Then, use the script to introduce yourself to a classmate.

Listening. Listening with purpose will help you focus on important information a speaker is conveying. Rather than speaking during class, listen to the conversations around you. Make notes on which students contribute to discussions and the important points they make. What did you learn about your classmates by listening?

Portfolio Development

Digital File Formats. A portfolio will contain documents you created electronically as well as documents you have in hard-copy format that will be scanned. It will be necessary to decide file formats to use for both types of documents. Before you begin, consider

College and Career Readiness

the technology you might use for creating and scanning documents. You will need access to desktop publishing software, scanners, cameras, and other digital equipment or software.

For documents you create, consider using the default format to save the files. For example, you could save letters and essays created in Microsoft Word in DOCx format. You could save worksheets created in Microsoft Excel in XLSx format. If your presentation will include graphics or video, confirm the file formats necessary for each item. Use the appropriate formats as you create the documents.

Hard-copy items will need to be converted to digital format. Portable document format (PDF) is a good choice for scanned items, such as awards and certificates.

Another option is to save all documents as PDF files. Keep in mind that the person reviewing your digital portfolio will need programs that open these formats to view your files. Having all the files in the same format can make viewing them easier for others who need to review your portfolio.

1. Decide the strategy you will use for saving documents. Make note of where the technology is available for your use, such as your home computer or the school lab.

2. Document any special instructions needed to use the software or equipment. This will save time when you are ready to create or save files.

CTSOs

Ethics. Many competitive CTSO events include an ethics event or include an ethics compo-

nent as a part of another event. Ethics is a set of rules that define what is wrong and right. Business ethics is a set of rules that help define appropriate behavior in the workplace.

The ethics component of an event may be part of an objective test. However, ethics may also be a part of the competition in which teams participate to defend a given position on an ethical dilemma or topic. To prepare for an ethics event, complete the following activities.

1. Read the guidelines provided by your organization.

2. Make notes on index cards about important points to remember. Use these notes to study.

3. Ask someone to practice role-playing with you by asking questions or taking the other side of an argument.

4. Use the Internet to find more information about ethical issues. Find and review ethics cases that involve business situations.

Drawing

Sections

7.1 Java Graphics Components

7.2 Text and Color

7.3 JavaFX Shapes

You probably have heard the adages "Seeing is believing" and "A picture is worth a thousand words." Humans can be very visual creatures. Looking at a picture can help many people visualize what is written on the screen. Photographs are one medium for providing images. However, the human imagination can create images that cannot be photographed and shared. Sometimes it takes a computer-generated image (CGI) to visually share ideas.

Computer-generated images are used for applications on the web, personal computers, mobile devices, and many other situations. Developers who were creating CGI found they were often writing the exact same code as other programmers to produce a part of the graphics. This was very wasteful. In the last chapter, you learned that classes everyone can use are placed in the Java Class Library. Because graphics is a unique type of code, JavaFX was created as a sub-library to provide common graphics classes for everyone to use. This set of classes is ready-to-use code for producing special effects in a Java application. Whether you are creating graphics to help a user visualize a process or are producing characters for a video game, JavaFX helps create an infinite number of two-dimensional and three-dimensional graphic effects.

College and Career Readiness

Reading Prep

Before reading this chapter, review the learning goals for each section as well as the major headings. Compare the objectives to the headings. What did you discover?

While studying, look for the activity icon for:

- Vocabulary terms with e-flash cards and matching activities.
- Starter files for hands-on examples and other exercises.

These activities can be accessed at
www.g-wlearning.com/informationtechnology/1773

Chapter Glossary

alpha value: Amount of transparency.

arc: Section of an ellipse.

child class: New class that inherits methods and data from a parent class.

circle: Closed curve with each point on the curve the same distance from a center point; type of ellipse.

extends: Creates a child class while inheriting methods and data from a parent class.

font: Set of characters of a typeface in one specific style and size.

group: Virtual container within a scene that holds nodes in the JavaFX application.

hue: Mixture of the color contributions of the base colors.

line: Straight segment between two endpoints.

node: Object in a scene.

observable list: Master list of the nodes in the scene for updating the display automatically when the nodes change.

parent class: Class that can be extended to create new classes with the same methods and fields.

raster: Pixel-based graphic in which each individual pixel holds its own color value and is available for modification.

root: Always the first node.

scene: Controls the layout of all items in a group.

scene graph: Organizes all the effects in a scene.

shape: Any two-dimensional object defined in geometry.

stage: Window created by the start method where the graphical output is displayed.

typeface: Single design of text characters and punctuation.

vector: Graphic composed of shapes based on mathematical statements.

Java Graphics Components

Every graphics application creates a window that pops up, much like the Turtle application. In the three helper files for the Turtle, code was created by the authors for you to use. It is now time to explain what that code was doing for you. There are three key steps to creating a JavaFX Application: 1) create the stage, 2) set the scene, and 3) display the stage. The classes from the JavaFX library that facilitate this for a new application are Application, Scene, and Stage.

How can including graphics in a JavaFX application improve usability?

Learning Goals

- Distinguish the elements of the JavaFX Application class to create graphics.
- Plan a scene graph for creating JavaFX graphics.

Terms

child class

extends

group

node

observable list

parent class

scene

scene graph

stage

JavaFX Application Class

FYI

The FX in JavaFX stands for *effects*.

Basic graphics begin with simple shapes that are grouped together to create lovable characters and other objects in a scene. These primitive shapes are ovals, rectangles, lines, and more, as shown in **Figure 7-1.** Starting with these basic shapes, anyone can create an image tailored to the needs of a user or game player. JavaFX has a set of drawing classes that provide a variety of shape-building and grouping techniques that can be used to create output displayed in a window.

Every graphics application created in JavaFX derives from the Application class. The Application class provides methods that start the application and manage the graphics you create. In each new JavaFX class, start by creating a new class that extends the Application class. *Extends* means creating a new child class while inheriting all the methods and data from another class. When extending, you are making the new application a little different and keeping it separate from the original Application class.

Anytime you want to redefine a method in the parent class to do something different, the annotation @Override tells the compiler you are making your own version of the original method. The word *inherit* leads the original class to be called the parent class. A ***parent class*** is one that can be extended to create new classes with the same methods and fields. The new class is called the child class. A ***child class*** inherits methods and data from a parent class. Think of the child as an individual who inherits traits from the parent.

Alexandru Belpomo/Shutterstock.com

Figure 7-1. The final character (left) is composed of individual sections. Each section can be further separated into basic shapes such as circles and lines.

Import the Application class and use the word *extends* in the class definition. In this example, an application named JavaFXSampleApp is defined:

```
import javafx.application.Application;

public class JavaFXSampleApp extends Application {
   // insert code here
}
```

The architecture of the graphics window includes a group and a scene. Then graphic elements are added to the group and displayed by the start method.

JavaFX Group

The scene is the container for the application. A *group* is a virtual container within the scene that holds nodes in the JavaFX application. Refer to **Figure 7-2.** Nodes, which are objects such as shapes, text, buttons, and images, are added to a group. The objects are known as *group children*. The children are placed in a list. Every time a new object is added to the group, the name is added to the list of children. Think of a group as the cast of players in the application. Theatrical references are used here because an application can be thought of as a presentation.

Import the class that creates groups using this statement:

```
import javafx.scene.Group;
```

Notice that the Group class lives in the javafx.scene package.

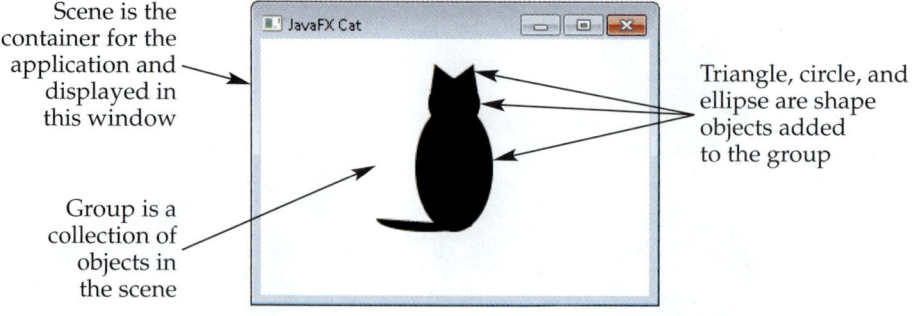

Scene is the container for the application and displayed in this window

Group is a collection of objects in the scene

Triangle, circle, and ellipse are shape objects added to the group

Goodheart-Willcox Publisher

Figure 7-2. These are the components of a JavaFX application.

To create a Group object, use the new keyword:

```
Group root = new Group( );
```

The first Group is always the root of the scene. The root is a node, and other nodes will be added to the group.

JavaFX Scene

A *scene* controls the layout of all items in a group. Import the class that creates scenes using this statement:

```
import javafx.scene.Scene;
```

Instantiate a scene on the stage using the new keyword. Provide the group name for the scene. Also provide the width and the height of the window.

```
Scene scene = new Scene( root, windowWidth, windowHeight );
```

JavaFX Stage

One of the methods that comes with the Application class is the start method. This method creates the window and sets its properties, such as its size and title. A *stage* is the window created by the start method where the graphical output is displayed.

Import the class that defines the stage using this statement:

```
import javafx.stage.Stage;
```

The main method is used to launch the application by calling the start method. The code to accomplish all this is shown below.

```
public class JavaFXSampleApp extends Application {
   @Override
   public void start( Stage mainStage ) {
      Group root = new Group( );            // create a Group container
      Scene scene = new Scene( root, 300, 200 );  /* create a Scene
                                              object that
                     contains the group and has width
                     of 300 pixels and height of 200 */
   }
   public static void main( String [ ] args ) {
      launch( args );  /* Application method that launches
              the JavaFX runtime and your JavaFX application */
   }
}
```

Language Arts and Java

Haiku

Haiku is a form of Japanese poetry. A haiku verse expresses a single feeling or impression. The word haiku is a combination of two Japanese words: *hai* meaning amusement and *ku* meaning verse.

simonidadj/Shutterstock.com

The structure of haiku is meant to be short. Haiku has no rhymes or punctuation. It consists of seventeen syllables on three lines. The first line should contain five syllables. The second should contain seven. The third line should have five syllables just as the first one does. A haiku should convey some sentiment. It is most interesting when a surprise is presented in the last five syllables.

Programmers, like other workers, occasionally experience a sense of frustration. They learn to take it lightly and find a way to defuse it. Humor is one way. Shown are some examples of haiku poetry written by computer users and posted on the Free Software Foundation's websites.

> *out of memory*
> *we wish to hold the whole sky*
> *but we never will*

> *three things are certain*
> *death taxes and lost data*
> *guess which has occurred*

> *yesterday it worked*
> *today it is not working*
> *coding is like that*

Look at the samples, count the syllables, and see if they give you a feeling or impression from the poet. In these lighthearted haiku, the frustration is expressed in the first 12 syllables. The humor is released in the last five syllables.

Assignment

You will create a haiku verse. Please use as many of the vocabulary words from this section as you can. Start with describing a situation in the first 12 syllables. Then, write a twist for the last five syllables. The unexpected shift is the part that produces the humor or prompts the emotional reaction. Two examples are shown using some of the vocabulary words.

> *observable child*
> *a node in the group and scene*
> *delights on the stage*

> *the compiler showed*
> *me errors I alone caused*
> *I fixed them all yay*

FYI

Remember, to make a JavaFX application, create a *stage* to display the *scene* that includes the *group*.

Stage properties can be modified by using the stage object name with the appropriate method. The setTitle method sets the title of the stage:

```
mainStage.setTitle( "Application 1" ); // change the title
```

The setScene method of the stage adds the scene to the stage:

```
mainStage.setScene( scene ); // add the scene to the stage
```

The show method displays the stage as set by the scene:

```
mainStage.show( ); // display what is on the stage
```

HANDS-ON EXAMPLE 7.1.1

Displaying the Window for a JavaFX Application

In this exercise, you will create an application to display the graphics window and set its title. It is a good practice to save the file each time after adding a section of code. Before beginning this exercise, download the chapter files from the student companion website. The JavaFXSampleApp.java file will be used as a starting point.

1. Launch jGRASP, and open the JavaFXSampleApp.java file.

2. Update the comments to include your name.

3. Beginning on line 6, notice all the classes for setting up this JavaFX application.

   ```
   // first import all the classes for setting up a JavaFX Application
   import javafx.application.Application;
   import javafx.scene.Group;
   import javafx.scene.Scene;
   import javafx.stage.Stage;
   ```

4. Note that you are using the Application class, but customizing it by adding the keyword extends to the class definition on line 11.

   ```
   public class JavaFXSampleApp extends Application {
   ```

5. Locate comment 1, and follow the instructions to add your own version of the start method and name the window mainStage. Add the @Override annotation.

   ```
   @Override // make your own version of the start method
   public void start( Stage mainStage ) {

   } // end start method
   ```

6. Locate comment 2, and follow the instructions to create a new Group and name it root.

   ```
   Group root = new Group( ); // create a Group container
   ```

7. Locate comment 3, and follow the instructions to create the scene object with the Group root, a width of 300, and a height of 200.

   ```
   Scene scene = new Scene( root, 300, 200 ); // create a Scene object
   ```

8. Locate comment 4, and follow the instructions to leave some white space (blank line), and use the setTitle method to set the title of the window to Hello World!

   ```
   mainStage.setTitle( "Hello World!" ); // change window title
   ```

9. Locate comment 5, and follow the instructions to add the scene to the stage using the setScene method.

   ```
   mainStage.setScene( scene ); // add the scene to the stage
   ```

10. Locate comment 6, and follow the instructions to display what is on the stage using the show method.

    ```
    mainStage.show( ); // display what is on the stage
    ```

11. Locate comment 7, and verify you still have a brace to end the start method. These can be inadvertently moved or accidentally deleted. Add a comment to indicate the purpose of the brace.

```
    } // end start method
```

12. If all went well, the main method should still be at the bottom of the file. If some mishap occurred, replace these lines in your file.

```
    public static void main( String [ ] args ) {

    } // end main
} // end Class
```

13. Locate comment 4, and follow the instructions to insert a call to the launch method, which calls the start method and opens the window.

```
    public static void main( String [ ] args ) {
        launch( args ); // start the JavaFX runtime and this JavaFX window
    } // end main
} // end Class
```

14. Compile the file.
15. Fix any compiler errors.
16. Run the program to see if the window appears on-screen. It should look similar to the one shown. You have created your first JavaFX application!

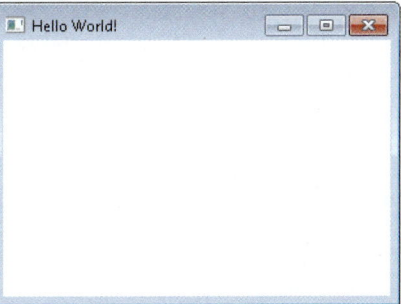

Goodheart-Willcox Publisher

Try It!

Save the JavaFXSampleApp.java file, rename the class to JavaFXFirstApp, and save the file under the new name. Modify the code by editing the scene to change the height and width of the window. Change the title of the window to your first and last name. Compile and run the program. Create a screenshot of your window to present to your instructor. You can create a screenshot by pressing the [Print Screen] key on the keyboard and then pasting it into image-editing software such as Paint or Photoshop.

Scene Graph

The *scene graph* organizes all the effects in a scene. It is an internal tree of nodes and groups. The programmer really does not interact with it. The scene graph uses a tree metaphor where a root, branches, and leaves are connected. A *node* is an object in a scene. Each branch or leaf is referred to as a node. **Figure 7-3** shows a generic scene graph and example nodes.

Each node holds the properties of the effect it contains. As you have seen, *root* is always the first node. An example of a leaf node is a text effect. An example of a branch node is a container for two or more effects.

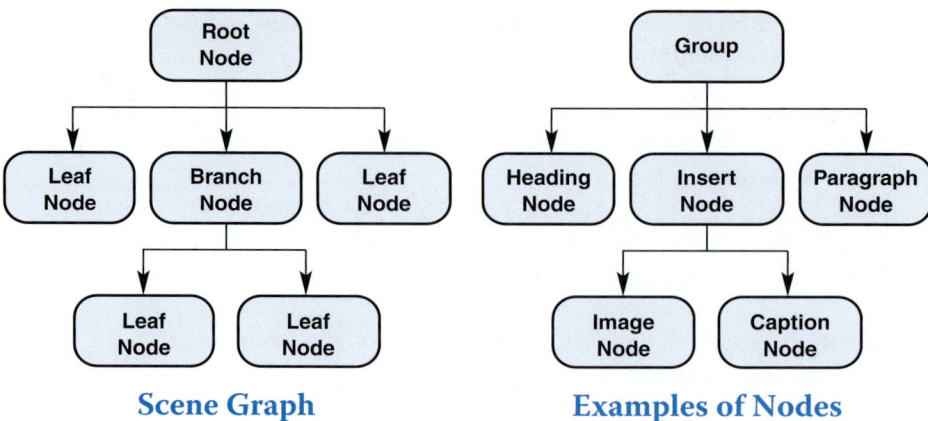

Scene Graph   **Examples of Nodes**

Goodheart-Willcox Publisher

Figure 7-3. A generic scene graph is shown on the left. On the right is a scene graph with example nodes.

Nodes

Nodes are referred to as children. The ***observable list*** is a master list of the nodes in the scene for updating the display automatically when the nodes change. This is kept in an **ObservableList** object. A *list* is a JavaFX structure that keeps the names of objects of the same type in a collection. You will need these import statements:

```
import javafx.collections.ObservableList;
import javafx.scene.Node;
```

Children

To add a node to the scene graph, add it to the list of children. First *get* the list, then *add* the new node.

```
// get the list of current nodes in the group
ObservableList<Node> nodesList = root.getChildren( );
// add a text object to it
nodesList.add( text );
```

When there is more than one child, use the **addAll** method and include all the nodes in the same method call:

```
// add several objects to the list
nodesList.addAll( text1, text2, text3 );
```

SECTION 7.1 REVIEW

Check Your Understanding

1. Which JavaFX class is the basis for every graphics application created in Java?
2. To modify a method from a parent class, the annotation _____ is used to make a different version of the original method.
3. A(n) _____ is a virtual container that holds all nodes in the JavaFX application.
4. In a JavaFX application, what is the window that displays the graphical output from the group?
5. What does the observable list contain?

Build Your Vocabulary

As you progress through this course, develop a personal computer science glossary. This will help you build your vocabulary and prepare you for a career. Write a definition for each of the following terms and add it to your computer science glossary.

child class	node	scene
extends	observable list	scene graph
group	parent class	stage

Text and Color

Two very important visual factors that create meaning for the viewer are color and the style of text. Both factors contribute to communication and the esthetic of an application. If used well, text styles and color enhance your message. If misused, they can hamper the message you are trying to communicate. Misuse includes random use of many colors and text styles.

Colors and text should never be used without regard to purpose. Good design invites the user to engage with the application. The style of text should be used to distinguish types of messages. Too many text styles make the screen look busy and can be distracting. Colors should be used to focus the user's attention, to set a mood, or convey an impression. The keyword to recall is screen *design.* Use of text styles and color should be thoughtful and purposeful in regard to the overall design of the application. JavaFX Text and Color classes are provided to work with text styles and color. The APIs for these classes are available on the Oracle JavaFX website.

Essential Question

What impact can a programmer's choice of color have on individuals with disabilities?

Learning Goals

- Create text in a JavaFX graphics window.
- Explain the application of color in JavaFX.

Terms

alpha value

font

hue

typeface

Text Component

A *typeface* is a single design of characters and punctuation, as shown in **Figure 7-4.** Common typefaces include Helvetica, Times, and Garamond. A *font* is a set of characters of a typeface in one specific style and size. For example, the typeface may be Palatino and the font may be bold, 11 point. *Point* is a unit of measurement commonly used in the graphic arts, printing, and publishing industries. In common usage, and especially in computer applications, *font* is used to mean *typeface,* but this is technically incorrect. In an application such as Microsoft Word, when you select the font, you are technically selecting the *typeface.* When you format a word as bold, you are selecting the *font.*

In Java, the JavaFX Text and Font classes are used to work with text, including selecting the typeface. The available text effects include what you would expect from using text in word processing. Typefaces (fonts), colors, and styles can be modified. Another feature is that text can be positioned anywhere in the application window. Apply what you learned about screen coordinates in Chapter 6 to use *x* and *y* locations to place the text on the screen. The sequence for placing text in the window is:

1. Import the Text class.
2. Create a Text object.
3. Set the properties.
4. Add the Text object to the stage.

FYI

There are 72 points per inch. So, 11-point type is 0.153 inches high on the printed page (11 ÷ 72 = 0.153). Sizes are relative on a computer display.

Typeface	Font	Sample Text
ITC Bauhaus™ Std	Light, 11 point	The Quick Brown Fox
	Medium, 11 point	The Quick Brown Fox
	Bold, 11 point	**The Quick Brown Fox**
	Heavy, 11 point	**The Quick Brown Fox**
ITC Century™ Std	Light, 11 point	The Quick Brown Fox
	Light, italic, 11 point	*The Quick Brown Fox*
	Bold, 11 point	**The Quick Brown Fox**
	Bold, italic, 11 point	***The Quick Brown Fox***
Garamond 3™ Std	Regular, 11 point	The Quick Brown Fox
	Italic, 11 point	*The Quick Brown Fox*
	Bold, 11 point	The Quick Brown Fox
	Bold, italic, 11 point	*The Quick Brown Fox*
Helvetica™ Std	Light, 11 point	The Quick Brown Fox
	Bold, 11 point	**The Quick Brown Fox**
	Roman, 11 point	The Quick Brown Fox
	Black, 11 point	**The Quick Brown Fox**

Goodheart-Willcox Publisher

Figure 7-4. The typeface is the collection of characters of one design, while the font is the style and size within a typeface. In computer applications, *font* is usually used to mean *typeface*.

Text Class

The Text class methods and properties are provided in the javafx.scene.text package. To use the Text class, add an import statement at the top of the program:

```
import javafx.scene.text.Text;
```

Create a text node in the scene by instantiating a Text object. There are three Text constructors. The number and type of arguments determine which constructor is used.

```
// 1. Create an empty Text object.
Text text = new Text( );

// 2. Specify coordinates on the screen and the text to be displayed.
Text text = new Text( double x, double y, String text );

// 3. Specify text to be displayed. Coordinates are by default. (0,0)
Text text = new Text( String text );
```

It is possible to change the text displayed, the location of the text in the scene, and the color. The API for the Text class includes these methods and properties along with many more. **Figure 7-5** shows the syntax and use of a few of the text effects. Search the Internet using the string Oracle Docs JavaFX Text to locate the full API.

Method	Action
setText(String value)	Sets the text to be displayed.
setX(double x)	Defines the *x* coordinate of left edge of text.
setY(double y)	Defines the *y* coordinate of bottom edge of text.
getText( )	Gets the string of the text.
setTextAlignment(TextAlignment value)	Sets the alignment of the text. Static constants include TextAlignment.LEFT, TextAlignment.CENTER, and TextAlignment.RIGHT.
setUnderline(boolean value)	Sets underline on for a true value and off for false.
setFill(Color object)	Sets the color of the text.
setStroke(Color object)	Sets the color of the outline of the text.

Goodheart-Willcox Publisher

Figure 7-5. The Text class includes these methods and many more for working with text.

Font Class

The Font class provides methods to display the text with a variety of properties. JavaFX can use any of the typefaces (fonts) on the computer running the JVM. Import the Font class from the javafx.scene text package. To add effects to the text, also import the FontPosture, FontWeight, and Color classes.

```
import javafx.scene.text.Font;        // Typeface
import javafx.scene.text.FontPosture; // Regular type or italic
import javafx.scene.text.FontWeight;  // Bold, Medium, Light,
                                      // Semi-Bold, etc.
import javafx.scene.paint.Color;      // Color for stroke
                                      // (outline) and fill (inside)
```

The three classes from the text package could be imported with one statement using the wild card (*):

```
import javafx.scene.text.*;
```

Some methods for managing the font are listed in **Figure 7-6.** These methods can be used with the Text object. All these methods create a font using the static font method of the Font class. This is a static method and does not require an object. Search the Internet using the string Oracle Docs JavaFX Font to locate the full API.

The Text class and Font class work hand-in-hand to render text on the stage. The Text class creates a node in the scene. The Font class is used to modify the properties of a Text object. There must be a Text object before the Font class can be used.

Method	Action
setFont(Font.font(String family, double size))	Sets the name and size of the font to be used. If the named font is not available, JavaFX replaces it with a font close to that family.
setFont(Font.font (String family, double fontWeight, double size))	The fontWeight argument can be any of the static constants FontWeight.NORMAL, FontWeight.LIGHT, FontWeight.EXTRA_LIGHT, FontWeight.MEDIUM, FontWeight.BOLD, FontWeight.EXTRA_BOLD, FontWeight.SEMI_BOLD, FontWeight.THIN, or FontWeight.BLACK.
setFont(Font.font(String family, double fontPosture, double size))	The fontPosture argument can be either of the static constants FontPosture.REGULAR or FontPosture.ITALIC.

Goodheart-Willcox Publisher

Figure 7-6. These are some methods for managing the font.

HANDS-ON EXAMPLE 7.2.1

Adding Text to a JavaFX Window ⤴

In this exercise, you will add text to display in the window. You will also select the typeface (font) and set the color of the text. Before beginning this exercise, download the chapter files from the student companion website. The JavaFXTextSnippet.java file will be used as a starting point.

1. Launch jGRASP, and open the JavaFXTextSnippet.java file. It will look very similar to the file you created in Hands-On Example 7.1.1.

2. Update the comment to include your name, and save the file.

3. Notice the new classes that are imported for this exercise. They are listed below with a comment.

```
import javafx.application.Application;
import javafx.collections.ObservableList; // added to keep track of nodes in the group
import javafx.scene.Group;
import javafx.scene.Node;          // added to specify observable list contains nodes
import javafx.scene.Scene;
import javafx.scene.text.*;        // added to display text, fonts, coordinates, etc.
import javafx.scene.paint.Color;      // added to set the color of the text
import javafx.stage.Stage;
```

4. Locate comment 1, and follow the instructions to instantiate a Text object named displayText.

```
Text displayText = new Text( ); // create a Text object
```

5. Locate comment 2, and follow the instructions to set the text properties. Set the String to be displayed to "Hello World!" the coordinates of the location in the window to (50, 30), and the font to Arial, bold, 30 points, and red.

```
/***** 2. Set the displayText properties. */
// Set String to display
displayText.setText( "Hello World!" );

// Set the location on the stage
displayText.setX( 50 );
displayText.setY( 30 );

// Set the font
displayText.setFont( Font.font( "arial", FontWeight.BOLD, 30 ) );

// Set the color
displayText.setFill( Color.RED );
```

6. Locate comment 3, and follow the instructions to add the text element to the scene graph. Get the list of nodes already in the scene graph, and then add the text to the list. At this point, there are no other children.

```
// Add the text to the group

// Define a list named effects and
// Assign it the list of current nodes in the group
ObservableList<Node> effects = root.getChildren( );

// Add the text object to the list
effects.add( displayText );
```

7. Compile the file. Fix any compiler errors.

8. Run the program to see the window appear on-screen with the text. It should look similar to the one shown.

Goodheart-Willcox Publisher

Try It!

Continue with the JavaFXTextSnippet.java file, rename the class to JavaFXTextTryIt, and save the file under the new name. Modify the code by adding a second text element to display your name. Use some of the text methods you have learned that have not been used in this Hands-On Example. Compile and run the program. Create a screenshot of your window to present to your instructor.

Goodheart-Willcox Publisher

Figure 7-7. RGB color is based on red, green, and blue. By combining these colors in various proportions, other colors can be created.

Color Component

The Color class was briefly introduced in the previous chapter. There is much more to learn about colors. Millions of colors can be generated by mixing a combination of red, blue, and green (RGB), as shown in **Figure 7-7.** Computer displays, televisions, and similar devices use RGB to generate colors. Color produced by light is different from color produced by pigment. Pigment is for paint and ink. Light is for computer displays. This discussion is exclusively on computer displays and RGB color.

Three bytes are used to define a color. The first byte is dedicated to the red contribution, the second byte to green, and the third byte to blue. Recall what you have learned about the byte data type. It can hold 256 different values. The value of 0 to 255 is called the color contribution. Zero is no color contribution, and 255 is saturated color. The mixture of the three color contributions of the base colors is known as the *hue.*

A mixture of (0, 0, 0) is no color contribution. The color displayed is black. The color (255, 255, 255) is all colors. The display shows the color white. It is white light passed through a prism that breaks down into all the colors of the rainbow. **Figure 7-8** shows the color contributions to create several colors.

Although millions of colors are available in the RGB model, agonizing over the exact mixture of colors is wasted effort. Unfortunately, each monitor has its own color production hardware. This means a color may appear almost the same, but different, on various monitors.

According to the American Optometric Association, about 10 percent of the population has some form of optical color deficiency. Most of these people experience reduced sensitivity to certain primary colors, such as red or green. Shades of red and shades of green look the same to them. Affected people have trouble distinguishing among blends of those colors. Browns and oranges look the same. Care must be taken to avoid color combinations that cause problems for people with color impairment. For example, if you want to use color to show

Red Value	Green Value	Blue Value	Color	Name
255	255	255		WHITE
0	0	0		BLACK
255	0	0		RED
0	255	0		GREEN
0	0	255		BLUE
0	255	255		CYAN
0	128	128		TEAL
255	255	0		YELLOW
128	0	128		PURPLE
255	0	255		FUCHSIA

Goodheart-Willcox Publisher

Figure 7-8. By changing the contribution of red, green, and blue, different colors can be created.

a contrast between two objects, coloring one red and the other green shows no contrast for someone with color impairment. Choose a color combination that is not in this category.

Color Class

JavaFX has a Color class that lives in the javafx.scene.paint package. This package was used earlier in this section. To use color in a JavaFX application, import the Color class:

```
import javafx.scene.paint.Color;
```

JavaFX colors are defined using four bytes. The first three bytes set the RGB values. The fourth byte sets the transparency of the color. The amount of transparency is called the *alpha value.* An alpha value of 1.0 is opaque, while 0.0 is fully transparent or completely see-through. If an alpha value is not explicitly given, the value 1.0 is used. To define a JavaFX color, use either the Color.rgb or Color.web method.

```
Color.rgb( int red, int green, int blue )        // implicit 1.0
                                                 // alpha
Color.rgb( int red, int green, int blue, double alpha )// explicit
                                                 // alpha
Color.web( String hexCode )
```

Hexadecimal notation, or hex, is discussed in Chapter 2. A hex web value for a color is #RRGGBB, where the pound sign (#) indicates a hexadecimal number. RR represents the two-digit contribution of red to the color, GG for green, and BB for blue. For example, the static constant Color.BROWN has a hex value of #A52A2A. The contributions are red: A5; green: 2A; and blue: 2A; as shown in **Figure 7-9.** Hex A5 equals 165 in decimal, and hex 2A equals 42 in decimal. Therefore, the two lines of code below produce the same color.

```
Color brown = Color.rgb( 165, 42, 42 );
Color brown = Color.web( "#A52A2A" );
```

Search the Internet using the string Oracle Docs JavaFX Color Class to locate the full API for using color in JavaFX. This page reports the color contribution in hexadecimal for each color, rather than decimal. So, the range is 0 to FF for each color contribution. Hex color is commonly used in web development.

FYI

A joke among graphic designers refers to the National Television Standards Committee (NTSC) standard for color encoding on monitors. They refer to NTSC color as *never the same color.*

Color Constant	Hex Value	RGB Decimal Values	Color
Color.BROWN	#A52A2A	(165, 42, 42)	
Color.RED	#FF0000	(255, 0, 0)	
Color.GREEN	#00FF00	(0, 255, 0)	
Color.BLUE	#0000FF	(0, 0, 255)	
Color.BLACK	#000000	(0, 0, 0)	
Color.WHITE	#FFFFFF	(255, 255, 255)	
Color.CADETBLUE	#5F9EA0	(95, 158, 160)	

Goodheart-Willcox Publisher

Figure 7-9. There are static constants in the Color class that provide named colors for easy use in your applications.

FYI

Apply what you have learned about calculators to use programmer mode to convert between RGB decimal and hex color codes.

It is possible to set the background color of the window. The window is controlled by the scene. So, adding an argument to the scene instantiation sets the background color for the window. An example is shown here.

```
Scene s = new Scene( root, 300, 300, Color.BURLYWOOD );
```

HANDS-ON EXAMPLE 7.2.2

Adding Color to a JavaFX Application

In the previous Hands-On Example, you added text to the graphics window, and this text was set in red. In this exercise, you will modify not only the color of the text, but the background color of the graphics window. Before beginning this exercise, download the chapter files from the student companion website. The JavaFXColorSnippet.java file will be used as a starting point.

1. Launch jGRASP, and open the JavaFXColorSnippet.java file. It will look like the file you created in Hands-On Example 7.2.1.

2. Locate comment 1, and follow the instructions to instantiate the scene and include a background color. Use the named color Color.GOLD.

```
Scene scene = new Scene( root, 300, 200, Color.GOLD ); // create a Scene
```

3. Compile and run the program. The graphics window should appear as shown.

Goodheart-Willcox Publisher

4. Locate comment 2, and follow the instructions to edit the displayText.setFill( ) method to generate an RGB red color instead of using a constant.

```
displayText.setFill( Color.rgb( 255, 0, 0 ) );
```

5. Compile and run the program. Compare the window contents with the prior window. How do they compare?

6. Edit the displayText.setFill( ) method again. This time, provide the hex value for red.

```
displayText.setFill( Color.web( "#FF0000" ) );
```

7. Compile and run the program. Compare the window content with the prior two windows. How do they compare?

8. Change the transparency on the text by adjusting the alpha to 0.2. Also, update the comment to reflect that the code now adjusts the alpha value.

```
// set text color to red, alpha 0.2
displayText.setFill( Color.web( "#FF0000", 0.2 ) );
```

9. Compile and run the program. The effect will be difficult to see because there are not any other components to see through the text other than the background. However, with the low alpha value, the text should be just visible over the background color, as shown.

Goodheart-Willcox Publisher

10. Locate comment 3, and follow the instructions to change the title of the window to JavaFX Text Color.

11. Compile and run the program.

Try It!

Search the Internet using the string Oracle Docs JavaFX Color Class to locate the full API for using color in JavaFX. Save the JavaFXColorSnippet.java file, rename the class to JavaFXColorTryIt, and save the file under the new name. Experiment with background colors and text colors as described in the API. Use at least one named color, one RGB color of your choice, and one web hex color. Add additional text elements as needed to show the different colors. Compile and run the program. Create a screenshot of your window to present to your instructor.

Text and Color Effects

JavaFX supplies many effects for text and color, including setting the fill and the stroke for text. The *fill* is the color on the inside of a character or shape. The *stroke* is the outline of a character or shape. Add the package javafx.scene.effect to use even more special effects. **Figure 7-10** shows some of the effects available for text in JavaFX. The pattern for using JavaFX effects is first to import the class, then create an object of the effect class, and finally use the setEffect methods with the effect object as the argument. For the complete API, search the Internet using the string Oracle Docs JavaFX Effects.

Effect Class	Effect on Text
DropShadow	Creates a shadow behind the text.
Glow	Makes the text appear to glow.
Reflection	Creates a reflected version of the text.

Goodheart-Willcox Publisher

Figure 7-10. These are some of the available effects that can be applied to text.

FYI

The effects shown in **Figure 7-10** can be applied to other objects as well as text.

The width of the stroke can be changed using the setStrokeWidth method. This method takes a double argument of the number of pixels wide the stroke should be. In this example, the fill is set to yellow, the stroke is set to red, and the width of the stroke is set at one pixel. Refer to **Figure 7-11.**

```
Scene scene = new Scene( root, 300, 200 ); // create a Scene object
scene.setFill( Color.LIGHTBLUE );

// set up text
// Text constructor with scene location and initial text
Text displayText = new Text( 25, 50, "Red Stroke" );
displayText.setFont( Font.font ( "Roman", 50 ) );
displayText.setFill( Color.YELLOW );
displayText.setStroke( Color.RED );
displayText.setStrokeWidth( 1.0 );
```

A *drop shadow* is a visual effect that makes an object look like it is raised above the background. Refer to **Figure 7-11.** A drop shadow can be added to text. The following example imports the javafx.scene.effect.DropShadow class, creates an instance of the DropShadow effect, and applies it to the displayText object.

```
import javafx.scene.effect.DropShadow;

// add drop shadow
DropShadow dropShadow = new DropShadow( );
displayText.setEffect( dropShadow );
```

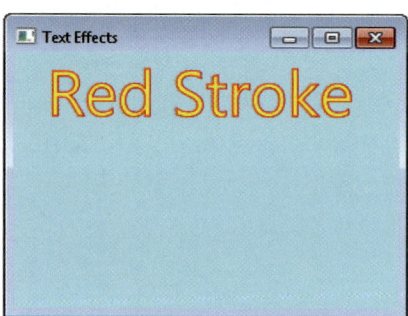

Stroke Added to Text

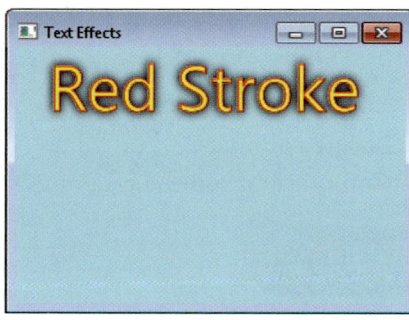

Drop Shadow Added to Text

Figure 7-11. A stroke is a line that appears around the outside of a shape. A drop shadow makes it appear as if the shape is floating above the surface.

HANDS-ON EXAMPLE 7.2.3

Applying Text Effects

In the previous Hands-On Example, you changed the background color of the graphics window and adjusted the transparency of the text color. In this exercise, you will make the text color fully opaque, edit the text, and then apply a reflection effect to the text. Before beginning this exercise, download the chapter files from the student companion website. The JavaFXTextEffectsSnippet.java file will be used as a starting point.

1. Launch jGRASP, and open the JavaFXTextEffectsSnippet.java file.
2. Rename the class to TextApp_HOE07_02_03, update the comment to reflect the new function of the application, and save the file under the new name.
3. Remove the alpha argument from the displayText.setFill statement (or set it to 1.0), and modify the displayText.setText statement so the text is See Me?
4. Add an import statement for the Reflection class. Place this with the other import statements at the beginning of the code.

```
import javafx.scene.effect.Reflection;
```

5. Create an instance of the Reflection class. Place this after the code that sets the color of the text for organization.

```
// add reflection effect
Reflection reflection = new Reflection( );
```

6. On the next line, set the effect.

```
displayText.setEffect( reflection );
```

7. Compile and run the program. Your graphics window should appear as shown. Notice how a reflection of the text is displayed below the text. You have applied a special effect to the text.

Goodheart-Willcox Publisher

Try It!

Save the JavaFXTextEffectsSnippet.java file, rename the class to JavaFXTextEffectsTryIt, and save the file under the new name. Comment out the reflection effect code by adding two forward slashes (//) at the beginning of the two lines. This changes the code to inline comments but makes it easy to return to code later if needed by simply removing the forward slashes. Change the text that will be displayed to any String of your choosing. Add the code needed to apply the glow effect to the text. Refer to **Figure 7-10.** Do not forget to import the class. Compile and run the program. Create a screenshot of your window to present to your instructor.

SECTION 7.2 REVIEW

Check Your Understanding

1. Which class is used to display characters in a JavaFX window?
2. The mixture of the three color contributions in RGB color is known as the _____.
3. What describes the transparency of a color?
4. Which type of numbers are used as arguments for the Color.web( ) method?
5. To apply an effect to text in a JavaFX application, import the appropriate effect class, create an object of the class, then use the setEffect method with the _____ as the argument.

Build Your Vocabulary

As you progress through this course, develop a personal computer science glossary. This will help you build your vocabulary and prepare you for a career. Write a definition for each of the following terms and add it to your computer science glossary.

alpha value   font   typeface
color contribution   hue

JavaFX Shapes

The building blocks of JavaFX graphics are shapes. The classes for generating the shapes of circle, line, rectangle, and other polygons are members of the javafx.scene.shape package. These shapes become nodes in the scene graph. Once created, the fill and stroke of each shape can be colored. Shapes can be combined to create characters or icons for the application. You will learn in a later chapter how to make these shapes animated and clickable to create a graphical user interface. Because the screen is composed of distinct pixels arranged in a rectangular grid, producing corners and round shapes is problematic. However, the individual shape classes provide methods to handle this problem.

Learning Goals

- Explain the process for creating primitive shapes in JavaFX.
- Compare and contrast creating primitive shapes with using the Path class in JavaFX.

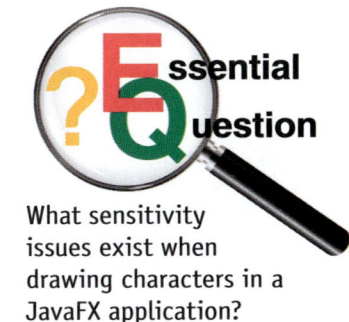

What sensitivity issues exist when drawing characters in a JavaFX application?

Terms

arc	raster graphic
circle	shape
line	vector

Two-Dimensional Shapes

Two-dimensional graphics are flat in appearance. Their location is in the plane of the display screen. In Java, a *shape* is any two-dimensional object defined in geometry. Examples are lines, circles, rectangles, and arcs.

Vector graphics are composed of shapes based on mathematical statements. These shapes can be joined together to create a complex shape. Each individual shape retains its properties, but contributes to the look of the final shape. Vector graphics are easily scaled and moved while retaining a smooth appearance. JavaFX creates vector shapes.

Vector graphics contrast to raster graphics. A *raster graphic* is a pixel-based graphic in which each individual pixel holds its own color value and is available for modification. Paint programs generally produce raster graphics. Raster graphics can become blurry or jagged when scaled or moved. **Figure 7-12** demonstrates the difference in appearance of raster and vector graphics.

The Shape class is the parent class for all primitive two-dimensional shapes in JavaFX. There are four main steps to create a 2D shape.

1. Import the class for the shape desired.
2. Create an instance of the class for the shape desired.

Venimo/Shutterstock.com

Figure 7-12. Notice how the raster image becomes jagged when resized.

3. Set the properties of the shape.

4. Add the shape object to the scene graph.

All classes in a package can be imported with one import statement using the wild card character (*):

```
import javafx.scene.shape.*;
```

When adding multiple shapes to the scene, the order in which they are added to the list is the order in which they are drawn. If shapes overlap, the latter one drawn will hide the overlapping part of the earlier shape. If that is not what you want, change the order in which the shapes are added.

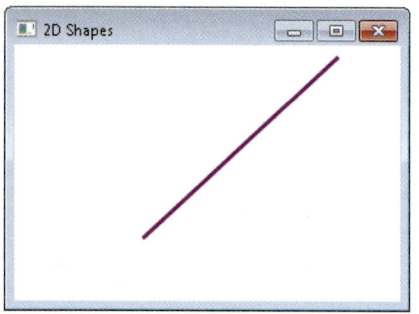

Goodheart-Willcox Publisher

Figure 7-13. A line is a straight segment between two endpoints.

Line

A *line* is a straight segment between two endpoints. **Figure 7-13** shows a line with endpoint coordinates of (100, 150) and (250, 10) and stroke color of purple. The Line class is used to draw lines. It is found in the javafx.scene.shape package. **Figure 7-14** shows the API.

To create an instance, use the new keyword:

```
Line line = new Line( );
```

Next, set the coordinates of the two endpoints of the line using the set x and y methods. The argument for these methods is a double data type. However, you can use the int data type because the compiler will promote it to a double type. For example:

```
line.setStartX( 20 );
line.setStartY( 40 );
line.setEndX( 400 );
line.setEndY( 160 );
```

Once the endpoints are specified, add the line to the group:

```
// get the list of current nodes in the group
ObservableList<Node> effects = root.getChildren( );
// add the shapes to it
effects.add( line );
```

Alternatively, one of the constructors can directly accept the coordinates for the line. This is another method for specifying the endpoints. For example,

Constructor	Arguments
Line( )	Creates an empty instance of Line.
Line(double startX, double startY, double endX, double endY)	Creates a new instance of Line with endpoints of (startX, startY) and (endX, endY).

Method	Action
setStartX(double x1)	Sets the x coordinate of first endpoint.
setStartY(double y1)	Sets the y coordinate of first endpoint.
setEndX(double x2)	Sets the x coordinate of second endpoint.
setEndY(double y2)	Sets the y coordinate of second endpoint.

Goodheart-Willcox Publisher

Figure 7-14. This is the API for the Line class.

to draw the line between (20, 40) and (400, 160), use this statement:

```
Line line = new Line( 20, 40, 400, 160 );
```

Circle

A *circle* is a closed curve with each point on the curve the same distance from a center point. It is a type of ellipse. **Figure 7-15** shows a circle with its center at (150, 100) having a radius 50, fill color blue, stroke color orange, and stroke width of 2. The figure also shows that the turtle in the custom Turtle used in earlier chapters is a drawing composed of circles. The Circle class is used to draw circles. Like the Line class, it is found in the javafx.scene.shape package. **Figure 7-16** shows the API.

To create a circle instance, use the new keyword:

```
Circle circle = new Circle( );
```

Next, set the coordinates of the center of the circle and the radius using the set *x*, *y*, and radius methods. The argument for these methods is a double data type.

```
circle.setCenterX( 150 );
circle.setCenterY( 100 );
circle.setRadius( 50 );
```

Lastly, add the circle to the group:

```
// get the list of current nodes in the group
ObservableList<Node> effects = root.getChildren( );
// add the shapes to it
effects.add( circle );
```

Arc

An *arc* is a section of an ellipse. The length of the arc is correlated to the number of degrees on a base ellipse, which may be a circle. **Figure 7-17** shows the numbering system for the degrees in a circle on which an arc is based.

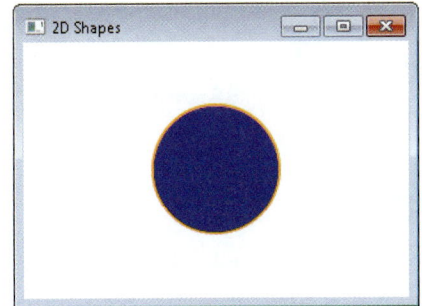

Circle Drawn in the Application Window

Turtle Graphic Composed of Circles

Goodheart-Willcox Publisher

Figure 7-15. A circle is a type of ellipse. Notice how circle primitives can be combined to create a turtle.

Constructor	Action
Circle( )	Creates an empty instance of Circle.
Circle(double radius)	Creates a new instance of Circle with given radius.
Circle(double xCenter, double yCenter, double radius)	Creates a new instance of Circle at a given position and radius.
Circle(double radius, Color fillColor)	Creates a new instance of Circle of a given color and radius.
Method	**Action**
setCenterX(double x1)	Sets the *x* coordinate of the center.
setCenterY(double y1)	Sets the *y* coordinate of the center.
setRadius(double x2)	Sets the radius of the circle.
setFill(Color.color)	Sets the fill color.
setStroke(Color.color)	Sets the stroke color.
setStrokeWidth(double w)	Sets the width of the stroke.

Goodheart-Willcox Publisher

Figure 7-16. This is the API for the Circle class.

Math and Java

Graphing a Linear Equation

Drawing a graph of a linear equation using a JavaFX line shape is pretty straightforward. Algebra has several forms for expressing the equation of a line. For ease of calculation, use the slope intercept form:

$$y = mx + b$$

where m is the slope of the line and b is the y-intercept.

You learned in algebra that two points determine a line. Calculate the two points using an arbitrary value for x in each point. The challenge is that the Cartesian coordinate system is different from the Java graphics window. Recall that the y values in the Java window count up from the top of the window to the bottom. There are no negative x or y values in the window. The y values can be correctly found simply by modifying the equation to include multiplying by -1. This will provide a reflection of the line in the window so it appears in the correct orientation for the Cartesian coordinate system.

Two lines can be drawn for the x and y axes to display a coordinate system. The coordinates of that origin can be added to the x and y values to translate the line into position on the axes. The rest is simple algebra and drawing of shapes and text.

Assignment

Download the chapter files from the student companion website. The GraphOfLineSnippet.java file will be used for this activity. Open this file in jGRASP. Update the comments to add your name, and save the file. Compile and run the program. Code to draw axes and a sloped line has been created for you. Close the graphics window, and study the code to see how the elements have been constructed.

Under the third comment, add code needed to label the two axes and to display the equation using the JavaFX Text component. Test the program. Then, change the values for slope (m) and y-intercept (b) under the second comment to verify the correct graph is being drawn.

For an extra challenge, apply what you have learned about the Scanner class to ask the user to input the values for slope and y-intercept. Note that the prompt will not appear in the JavaFX window, rather in the **Run I/O** tab at the bottom of the jGRASP window. After modifying the code, compile and run the application. Enter various values for the slope and y-intercept. Verify the output is correct. Create a screenshot of your stage, and submit it along with your program for grading.

Cartesian Coordinate System

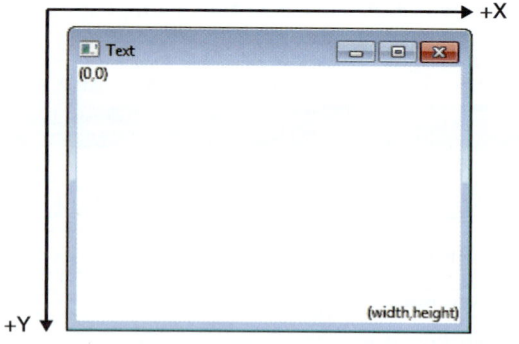

Java Graphics Window

Goodheart-Willcox Publisher; BaMic_illustrations/Shutterstock.com

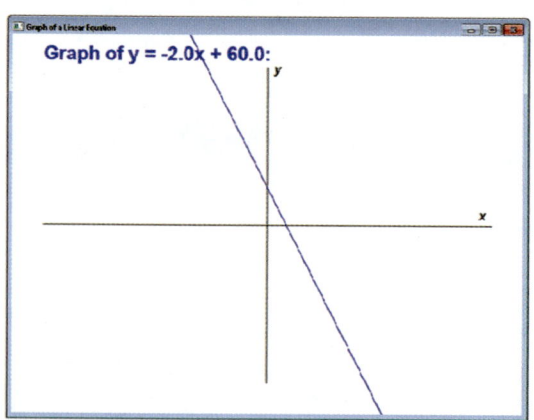

Goodheart-Willcox Publisher

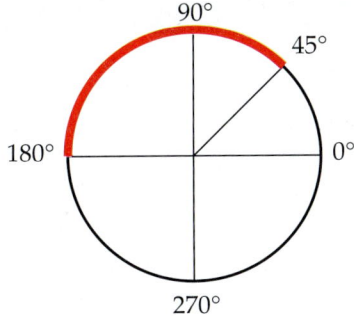

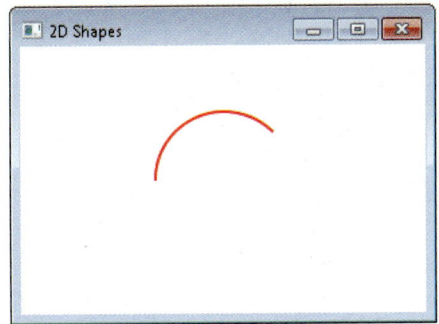

Degree Measurements
for an Arc

Arc Drawn in the
Graphics Window

Goodheart-Willcox Publisher

Figure 7-17. The numbering system for an arc is based on the degrees in a circle or ellipse.

Zero degrees is the far-right point of the circle. Ninety degrees is straight up, and so forth. The figure also shows how the arc would appear in the graphics window.

The Arc class is used to draw arcs. It is found in the javafx.scene.shape package. To create an instance, use the new keyword:

```
Arc arc = new Arc( );
```

An arc is a shape, so it inherits all setter methods for a shape. The properties setCenterX and setCenterY are the same as for a circle. The shape of the underlying ellipse is determined by the setRadiusX and setRadiusY properties. If these two properties are equal, an arc of a circle is drawn. The length of the arc is determined by the setStartAngle and the setLength methods. To specify all these properties in the object creation, use the following constructor.

```
Arc arc = new Arc( double centerX, double centerY, double radiusX,
    double radiusY, double startAngle, double length );
```

Unique to the arc is the setType method. There are three static constants for this method: ArcType.OPEN, ArcType.CHORD, or ArcType.ROUND. **Figure 7-18** shows the effect of each constant.

The following code will draw an open arc based on a 100-pixel-radius circle with its center at (200, 200), starting at 45 degrees, and ending at 180 degrees with a red stroke and transparent fill.

```
Arc arc = new Arc( 200, 200, 100, 100, 45, 135 );
arc.setStrokeWidth( 5.0 );
arc.setStroke( Color.RED );
arc.setFill( null );
arc.setType( ArcType.OPEN );
```

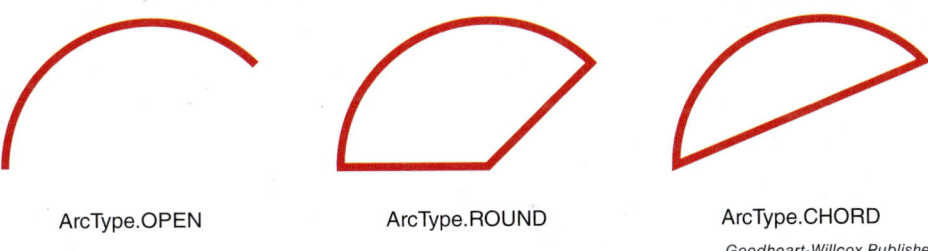

ArcType.OPEN   ArcType.ROUND   ArcType.CHORD

Goodheart-Willcox Publisher

Figure 7-18. There are three types of arcs that can be created with JavaFX.

Science and Java

JavaFX Rainbow

Sir Isaac Newton is to thank for the explanation of the nature of light as demonstrated in prisms and rainbows. He studied the bending and spreading of light. The description of these experiments and his discoveries detailing light phenomena was published in 1704 in the book, *Opticks.*

Newton discovered that light is made up of wavelengths of color. He demonstrated this by using a prism. The prism bent a beam of sunlight into wavelengths of different hues. The order in which the colors appear is based on the wavelength of the color. Red has the longest wavelength and violet the shortest. The order never changes. Up until Newton published this theory, the accepted explanation was that light was colorless.

mikeshinmaksim/Shutterstock.com

When you see a rainbow in the sky, water vapor is close by. The water acts like a prism. Rainbows often appear after a rainstorm where you are between the sun and the water in the air. The order of the colors in a rainbow is always the same because of the wavelengths of color. A mnemonic device that physics students use to remember the order is *Roy G. Biv*, which stands for red, orange, yellow, green, blue, indigo, and violet.

Assignment

Your assignment is to write a program that draws a rainbow composed of open arcs of colors in the correct order. Download the chapter files from the student companion website. The JavaFXScienceAndJava.java file will be used for this activity. Open the file in jGRASP, compile, and run the program. The red arc has been completed for you. Close the graphics window, and examine the code.

Add code for the remaining open arcs. Set the stroke of each arc to 5.0. Reduce the x and y radii by 5 for each color. Create a constant as a color for a clear fill. Refer to Chapter 4 if you need a refresher on how to create a constant. By setting the alpha value for this color to 0.0 and specifying the constant as the fill color for each arc, the arc will appear as a single line as the fill color will be transparent.

Goodheart-Willcox Publisher

Compile and run the program. Check that the arcs are properly aligned to create the rainbow. Debug the program as needed. Create a screenshot of your stage, and submit it along with your program for grading.

Notice that the length is specified in the number of degrees covered by the arc. In this case, 135 is entered so the arc will end at the 180-degree point on the base circle (45 + 135 = 180).

Rectangle

The rectangle shape is determined by its location in the scene and its width and height. There are four constructors for rectangles. All four create an instance of a Rectangle. One just creates the object. One additionally specifies its location and dimensions. One also supplies only its dimensions. One supplies its dimensions and its fill color as well.

```
Rectangle rectangle = new Rectangle( );
Rectangle rectangle = new Rectangle( double x, double y, double
    width, double height );
Rectangle rectangle = new Rectangle( double width, double height );
Rectangle rectangle = new Rectangle( double width, double height,
    Color fill );
```

It is possible to specify a rounded rectangle by setting two additional properties as double data types: arcHeight and arcWidth. The following code will draw a blue rounded rectangle at (100, 150) with width 200 and height 75. There is no constructor to specify the arc height and arc width for the rounded corners. These require extra setters. To make the inside of the rectangle transparent, use the value null for the setFill color.

```
Rectangle rectangle = new Rectangle( 100, 150, 200, 75 );
rectangle.setArcHeight( 50 );
rectangle.setArcWidth( 50 );
rectangle.setStroke( Color.BLUE );
rectangle.setFill( null );
```

Other Primitive Shapes

There are many other primitive shapes that can be drawn using JavaFX. All shapes are created in the same manner as lines, circles, arcs, and rectangles. **Figure 7-19** shows the API for a few of the other JavaFX primitive shapes. For the complete API for each, search the Internet using the string Oracle Docs JavaFX Shapes. Note: drawing polygons is not discussed here because it uses a data structure called a list, which has not been covered.

Constructor	Property Methods
Ellipse ellipse = new Ellipse( );	setCenterX, setCenterY, setRadiusX, setRadiusY
Polygon polygon = new Polygon( );	polygon.getPoints( ).addAll(list of vertices)

Goodheart-Willcox Publisher

Figure 7-19. These are some of the primitive shapes that can be created with JavaFX.

Coding Conundrum

The task is to display a rectangle in the graphics window for a JavaFX application. The programmer wrote this code. It compiles cleanly, but the graphics window is blank. A conundrum! What is the problem with this code?

```
1  // first import all the classes for setting up a JavaFX Application
2  import javafx.application.Application;
3  import javafx.collections.ObservableList;
4  import javafx.scene.Group;
5  import javafx.scene.Scene;
6  import javafx.scene.shape.*; // new package to enable drawing shapes
7  import javafx.scene.paint.Color;
8  import javafx.stage.Stage;
9
10 public class Conundrum extends Application {
11
12   @Override /* make our own version of the start method that gets used
13          instead of the one that comes with Application class */
14   public void start( Stage mainStage ) {
15
16      Group root = new Group( );              // create a Group container
17      Scene scene = new Scene( root, 100, 150 );// create a Scene
18
19      /***** Draw a rectangle at (30, 30),
20          width=60, height=80
21          Fill with GREEN */
22      Rectangle rect = new Rectangle( 30, 30, 60, 80 );
23      // set rectangle properties
24      rect.setArcWidth( 30 );
25      rect.setArcHeight( 40 );
26      rect.setFill( Color.GREEN );
27
28      mainStage.setTitle( "Green Rectangle" );   // change the title
29      mainStage.setScene( scene );         // add the scene to the stage
30      mainStage.show( );                   // display what is on the stage
31   }
32
33   public static void main( String [ ] args ) {
34      launch( args ); /* Application method that launches
35             the JavaFX runtime and your JavaFX application */
36   }
37 }
```

HANDS-ON EXAMPLE 7.3.1

Drawing JavaFX Shapes

Drawing basic shapes using JavaFX is a simple process. Basic shapes can be combined to create more complex graphics. In this exercise, you will use JavaFX to create a cartoon face. Before beginning this exercise, download the chapter files from the student companion website. The starter file JavaFXShapesSnippetApp.java is used in this exercise.

1. Launch jGRASP, and open the JavaFXShapesSnippetApp.java file. Notice that the import statement is already included for the javafx.scene.shape package.

2. Locate comment 1, and follow the instructions to draw two lines. Name the objects line1 and line2.

```
/***** 1. Draw a purple line whose endpoints are (54, 76) and (50, 70).
      Do not include the coordinates in the constructor. Use the setters to do that for you.
      Set the stroke to 3. */
Line line1 = new Line( );
line1.setStartX( 54 );
line1.setStartY( 76 );
line1.setEndX( 50 );
line1.setEndY( 70 );
line1.setStroke( Color.PURPLE );
line1.setStrokeWidth( 3 );
/* Draw a second purple line whose endpoints are (66, 76) and (50, 70)
      Include the coordinates in the constructor. Set the stroke to 3 */
Line line2 = new Line( 66, 76, 50, 70 );
line2.setStroke( Color.PURPLE );
line2.setStrokeWidth( 3 );
```

3. At the bottom of the program, add the lines to the group.

```
/***** Add the shapes to the group */
// get the list of current nodes in the group
ObservableList<Node> effects = root.getChildren( );
// add the shapes to list
effects.addAll( line1, line2 );
```

4. Compile and run the program. Your application should appear as shown with two short purple lines at an angle to each other.

Goodheart-Willcox Publisher

5. Locate comment 2, and follow the instructions to draw two circles. Name the objects circle1 and circle2.

```
/***** 2. Draw a circle whose center is at (48, 55) with radius 8
      and stroke orange, stroke width 2, fill with blue. */
Circle circle1 = new Circle( 48, 55, 8 );
// set circle1 properties
circle1.setFill( Color.BLUE );
circle1.setStroke( Color.ORANGE );
circle1.setStrokeWidth( 2 );

/* Draw a second circle whose center is at (72, 55) with radius 8
   and stroke orange, stroke width 2, fill with blue */
Circle circle2 = new Circle( 72, 55, 8 );
// set circle2 properties
circle2.setFill( Color.BLUE );
circle2.setStroke( Color.ORANGE );
circle2.setStrokeWidth( 2 );
```

6. Add the circles to the group. Add these objects to the existing statement.

```
effects.addAll( line1, line2, circle1, circle2 );
```

7. Compile and run the program. Your application should appear as shown with the two circles above the two purple lines.

Goodheart-Willcox Publisher

8. Locate comment 3, and follow the instructions to draw an arc. Name the object arc.

```
/***** 3. Draw a red chord arc with center (60,85)
        with x-radius and y-radius 15 starting at 200 degrees,
        and length 140 degrees. Add a black stroke, width 3. */
Arc arc = new Arc( 60, 85, 15, 15, 200, 140 );
// set arc properties
arc.setFill( Color.RED );
arc.setStroke( Color.BLACK );
arc.setStrokeWidth( 3 );
arc.setType( ArcType.CHORD );
```

9. Add the arc to the group. Add the object to the existing statement.

```
effects.addAll( line1, line2, circle1, circle2, arc );
```

10. Compile and run the program. Your application should appear as shown with the arc below the two purple lines. You have created a JavaFX application using shapes!

Goodheart-Willcox Publisher

Try It!

Continue using the JavaFXShapesSnippetAnswer.java file. Following the instructions in the Try It! comment, add a rounded rectangle around the other shapes in the scene to suggest a face. Name the object rect. Use the setArcWidth and setArcHeight methods to make the corners of the rectangle round. Add an RGB color of your choice for the fill with an alpha value that allows the other shapes to show through the rectangle. Compile and run the program. Create a screenshot of your window to present to your instructor.

Java Path

Irregular shapes can be drawn using the JavaFX Path class. This class provides much more than will be covered in this text. For the complete API, search the Internet using the string Oracle Docs JavaFX Path. Two of the classes will be discussed here: MoveTo and LineTo. These classes can be used, for example, to draw a triangle.

Drawing irregular shapes takes planning. Using a sheet of graph paper, draw a rectangle the size of the stage. Then, draw the irregular shape you want to create. In this case, the shape is a triangle. Use the squares on the graph paper to determine the coordinate locations of the endpoints. **Figure 7-20** shows a sketch of a triangle on graph paper and estimate of vertices. The stage has dimensions (300, 200). Using the Path class, this triangle can be easily drawn. First, move to one of the vertices. Then, draw a line to a second vertex, then another line to the third vertex, and back to the first. Fill with a color, and *voilà*, a triangle!

The Path class is found in the javafx.scene.shape package. Import this package to use the class.

```
import javafx.scene.shape.*; // package to enable drawing shapes
```

To create a path, the pattern is the same as for other shapes: import the package, create the shape object, set the properties, and add the object to the scene. The following code draws the triangle sketched in **Figure 7-20.** Notice how the coordinates in the drawing are used in the code.

```
Path triangle = new Path( ); // Create the shape object
// Set the properties
triangle.getElements( ).add( new MoveTo( 150, 20 ) );
triangle.getElements( ).add( new LineTo( 210, 170 ) );
triangle.getElements( ).add( new LineTo( 20, 140 ) );
triangle.getElements( ).add( new LineTo( 150, 20 ) );
triangle.setFill( Color.RED );

/***** Add the shapes to the group
       get the list of current nodes in the group */
ObservableList<Node> effects = root.getChildren( );
// add the shapes to list
effects.add( triangle );
```

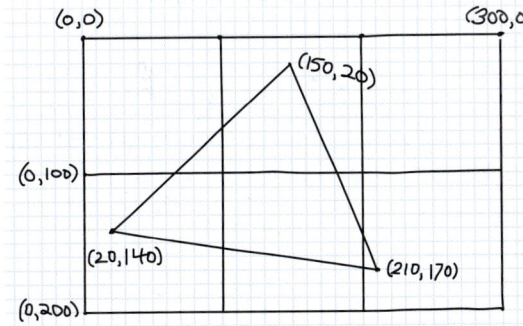
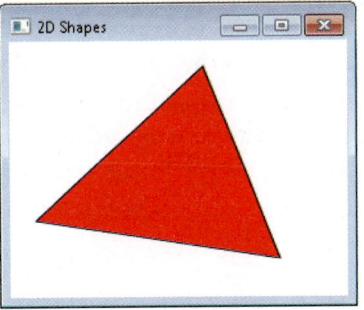

Figure 7-20. When drawing an irregular shape, planning it on graph paper helps identify the coordinate locations needed to draw it.

HANDS-ON EXAMPLE 7.3.2

Drawing a Graphic Character with Several Shapes ➦

Basic shapes can be overlapped and used together to form more complex shapes. In this exercise, you will create the silhouette of a cat. Start by planning the final image on graph paper. Before beginning this exercise, download the chapter files from the student companion website. The starter file JavaFXCatSnippetApp.java is used in this exercise.

1. On a sheet of graph paper, draw a rectangle the size of the stage (300 by 200). Number the coordinates.

2. Sketch the character. In this example, the character is a silhouette of a cat. Determine which shapes best represent the various elements of the character.

3. Identify key coordinates and dimensions, as shown.

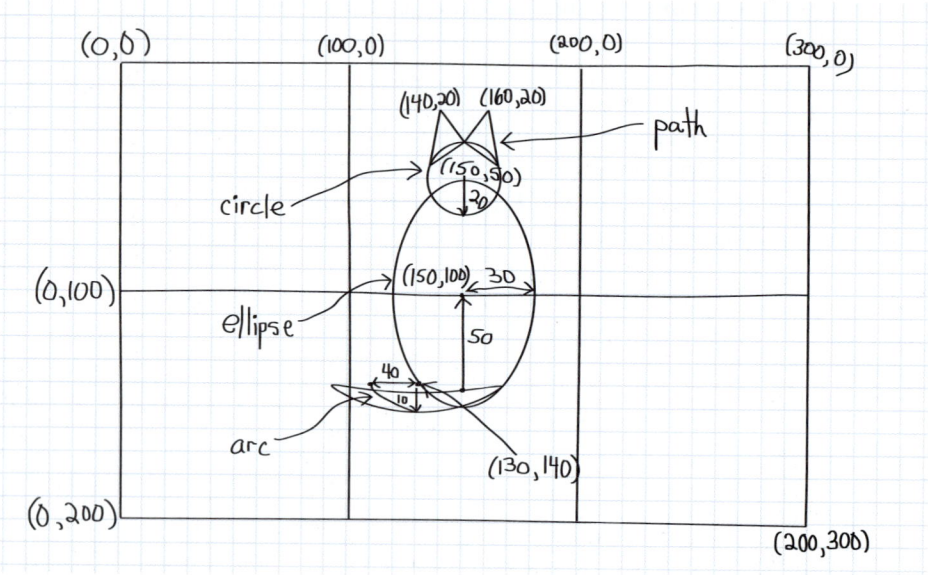

Goodheart-Willcox Publisher

4. Launch jGRASP, and open the JavaFXCatSnippet.java file. Add your name to the comment. Code for drawing the head, body, and tail has been provided for you.

5. Compile and run the program. To finish the cat, two ears are needed. Close the graphics window.

6. Follow the instructions in the fourth comment to draw the ears as paths. Name the objects ear1 and ear2. Begin with ear1. Notice the path will create a closed area and fill it with color without drawing the third side of the triangle. It is not required to draw the line going back to the **MoveTo** location.

```
/***** 4. Draw two black ears as triangle paths with vertices
      (135, 20), (148, 30), (132, 40) and
      (165, 20), (152, 30), (168, 40). */
Path ear1 = new Path( );
ear1.getElements( ).add( new MoveTo( 135, 20 ) );
ear1.getElements( ).add( new LineTo( 148, 30 ) );
ear1.getElements( ).add( new LineTo( 132, 40 ) );
ear1.setFill( Color.BLACK );
```

7. Add the new shape to the list.

```
// add the shapes to list
effects.addAll( head, body, tail, ear1 );
```

8. Compile and run the program. Check that the first ear is drawn, then close the graphics window.

9. On your own, add the code to draw the path for ear2. The code is very similar to that used for the first ear except the object name is different and the coordinates are different. Do not forget to add the new shape to the list.

10. Compile, and run. You have used JavaFX shapes to draw a graphic character!

Goodheart-Willcox Publisher

Try It!

On a sheet of graph paper, sketch a stage. Label the coordinates of the stage. Then, sketch a five-point star. Identify the coordinates of the star. Launch jGRASP, and open the JavaFXStarSnippet.java file. Rename the class JavaFXStar, and save the file under that name in your working folder. Add all the code needed to draw the star in the graphics window. Use the JavaFX Path class to draw it. Compile and run the program. Create a screenshot of your window. Submit your sketch, the program file, and the screenshot to your instructor.

SECTION 7.3 REVIEW

Check Your Understanding

1. Which type of graphics is composed of individual pixels?
2. The general pattern for creating shapes in a JavaFX application is _____ the class for the shape, create a(n) _____ of the class for the shape, set the _____ for the shape, and add the shape object to the _____.
3. What determines which shape appears on top when shapes overlap?
4. Which class allows drawing of irregular shapes?
5. When designing an irregular shape to be created in a JavaFX application, make a sketch on graph paper that identifies the _____ locations of all the shapes to be added.

Build Your Vocabulary

As you progress through this course, develop a personal computer science glossary. This will help you build your vocabulary and prepare you for a career. Write a definition for each of the following terms and add it to your computer science glossary.

arc	line	shape
circle	raster	vector

Cooperative Coding

Drawing a Train

Nothing brings giggles to a child's day like the sight of the circus train! Working with your group, you will create a JavaFX application that displays a circus train made of JavaFX shapes. The design will be up to you.

bilha Golan/Shutterstock.com

Assignment

As a group, decide how many cars the train will have. Each group member will write the code to draw one car or the engine. Determine who will code each car in the train. You must also make decisions on specifications before beginning to code so every car will look like it belongs to the train. Make a preliminary sketch on graph paper. Determine the height and width for each car. Determine the names of the cars. Decide on the location of each car. Keep it simple. The more complex the shapes, the longer it will take to complete. Write a document that lists all the specifications. If changes are made during development, update the document.

Create a master class Java file named Circus-Train. Save this in a common working folder where all team members can access it. Add all import statements you will need for color, shapes, text, scene, group, observable list, nodes, application, and stage. Each team member will use this file as a starting point. Once the individual cars are coded in separate files, the code will be copied and pasted into the master file to complete the project.

Individually, open the CircusTrain.java file, rename the class to reflect the car that will be drawn, and save the file under the new name in the common working folder. For example, the person coding the caboose might name the class CircusTrainCaboose. Each shape in this file might be named starting with the word caboose. This will avoid duplicate names for the shapes when the code is combined. Duplicate names will cause problems with the compiler.

Once all team members have completed their individual cars, meet as a group to elect a member to combine the code into a single master file. Assign another team member to test the final file. The remaining team members should be responsible for debugging the program. When the project is clean, discuss the project as a group. Submit the specification document, individual sketches, individual class files, combined final class file, and a screenshot of the train for grading.

1. Were there any obstacles the group had to overcome in working on a common coding project?

2. You likely found this activity a tad messy. As a group, speculate how custom classes could be used to help streamline this project and make for a more elegant solution.

Chapter Summary

Section 7.1: Java Graphics Components

- The JavaFX Application class is the basis for every graphics application created in Java, and graphics exist in a scene that contains groups of graphic elements to be displayed on a stage.
- The scene graph organizes all the effects in a scene where each element is a node, and all nodes are contained in the observable list.

Section 7.2: Text and Color

- The Text class is used to create text in a JavaFX graphics window, and the Font class is used to control various properties of the text.
- Color for graphics created in JavaFX is based on the RGB model, and the color contribution of each color results in a hue, but color can also be specified with a hexadecimal number; the javafx.scene.effect package allows control over special effects such as stroke color and drop shadow.

Section 7.3: JavaFX Shapes

- Primitive shapes are created using JavaFX by first importing the class for the shape desired, then creating an instance of the class for the shape desired, setting the properties of the shape, and finally adding the shape object to the scene graph.
- The Path class is used to create irregular shapes, and the process is basically the same as for creating primitive shapes except the coordinate location of each vertex on the shape must be specified, which requires planning to be successful.

Chapter 7 Test

Multiple Choice

Select the best response.

1. What is the window in which graphical output is displayed?
   A. group
   B. stage
   C. scene
   D. observable list

2. Which of the following statements is correct to import the Text class to create text?
   A. import javafx.scene.text.Text;
   B. import scene.text.TEXT;
   C. import javafx.group.Text;
   D. import scene.text.Text;

3. The hue created with the RGB combination of (0, 0, 0) is:
   A. black
   B. white
   C. transparent
   D. This is an invalid combination.

4. Which of the following methods are valid text special effects achievable with JavaFX?
   A. DropShadow, Glow, Reflection
   B. DropShadow, Lipstick, Mirror
   C. setFill, Display, Glow
   D. setStroke, Reflection, setStrokeHeight

5. Which class is used to create a closed curve with each point on the curve equidistant from a center point?
   A. Arc
   B. Bounded
   C. Circle
   D. Line

Completion

Complete the following sentences with the correct word(s).

6. The _____ library contains code used to create graphics in Java.

7. Use the methods in the _____ class to add a text string to the stage.

8. The hexadecimal number for displaying the color white is _____.

9. The function of the alpha value is to display the degree of _____.

10. To create a rectangle requires the x and y coordinate locations of the starting point and the _____ and _____ of the rectangle.

Matching

Match the correct term with its definition.
   A. ObservableList<Node> effects = root.getChildren ( );
   B. setStroke(Color.PURPLE);
   C. group
   D. extends
   E. overriding

11. Virtual container of shapes, buttons, text, and images.

12. Creating a new class while inheriting methods and data from another class.

13. Compiling a new version of the original method.

14. Current nodes in the group.

15. Sets the color of a line.

Application and Extension of Knowledge

1. The Oracle website contains a tutorial you can use to extend your knowledge on text effects. Conduct an Internet search using the string Oracle Docs Using Text and Text Effects in JavaFX. In the search results, locate the link for the Using Text and Text Effects in JavaFX tutorial on the Oracle website. Complete the tutorial. Then, write a program that demonstrates what you learned. Create a screenshot of your stage, and submit it along with your program for grading.

2. Search various reliable sources for information about colors and typefaces. Identify characteristics that are considered easy or difficult for humans to see. Write a two-page summary of your findings. Conclude the summary with your recommendations for three colors to avoid and two typefaces to use in a Java application.

3. RGB color can generate 256 × 256 × 256 different colors. Calculate that number. Then, write a JavaFX program that shows three circles each with an RGB fill color codes that differs by 1 in one of the color contributions. For example, you may decide to use the values (128, 128, 128), (128, 128, 129), and (128, 128, 130) to vary the blue contribution. Include text that states the number you calculated for the total of colors RGB can generate. Were you able to see the difference among the three colors? Create a screenshot of your stage, and submit it along with your program for grading.

4. On a sheet of graph paper, sketch a mouse character. Recall that mouse ears are round, and their tails are thinner than a cat's. You may want to draw the mouse in a crouched position rather than upright like the cat created in this chapter. On the sketch, identify the needed coordinate locations. Then, apply what you learned about drawing the cat to write a program to draw the mouse. Create a screenshot of your stage, and submit it along with your program for grading.

5. Water consists of two hydrogen atoms and one oxygen atom. Locate an image representing a water molecule. Notice that the hydrogen atoms are not equidistant from each other. Then, write a program to draw a 2D image of a water molecule. Use any colors to represent the atoms. Include text to identify the atoms in the molecule. Create a screenshot of your stage, and submit it along with your program for grading.

Online Activities

Complete the following activities, which will help you learn, practice, and expand your knowledge and skills.

Vocabulary. Practice vocabulary for this chapter using the e-flash cards, matching activity, and vocabulary game until you are able to recognize their meanings.

Communication Skills

Reading. *Environmental print* refers to the words you come across in your daily life, such as signage, logos, and advertisements. Locate an example of environmental print in your classroom or school, and read the words. What is the purpose or meaning of the environmental print?

Writing. To become career ready, it is necessary to utilize critical-thinking skills in order to solve problems. Identify an example of a problem you needed to solve that was important to your success at school or work . Write an explanation of how you applied critical-thinking skills to arrive at a solution.

Speaking. Most people in the United States act as responsible and contributing citizens. How can a person demonstrate social and ethical responsibility in times when disaster relief is needed in the community? Participate in a group discussion about how citizens can go beyond the minimum expectations of helping others in the community.

Listening. Your purpose for listening will change depending on the situation. You will become a better listener if you know your purpose for listening and change your behavior accordingly. What was your purpose for listening to your instructor lecture to the class today?

Portfolio Development

College and Career Readiness

File Structure. After you have selected a file format for your documents, determine a strategy for storing and organizing the materials. The file structure for storing digital documents is similar to storing hard-copy documents. First, you need a place to store each item. Ask your instructor where to save your documents. This could be on the school's network or a flash drive of your own. Next, decide how to organize related files into categories. For example, Certificates might be the name of a folder with a subfolder Community Service Certificates and another subfolder named School Certificates. Appropriate certificates would be saved in each subfolder. The names for folders and files should be descriptive, but not too long.

1. Decide on the file structure for your documents.
2. Create folders and subfolders on the school's network drive or flash drive on which you will save your files.

CTSOs

Parliamentary Procedures. *Parliamentary procedure* is a process for holding meetings so they are orderly and democratic. In the parliamentary procedure competitive event, the participants must demonstrate understanding of these procedures, such as *Robert's Rules of Order.* Applying the procedures dictated by *Robert's Rules of Order* is an effective way to conduct a formal meeting.

This is a team event in which the group will demonstrate how to conduct an effective meeting. In addition, an objective test may be administered to each person on the team that will be evaluated and included in the overall team score. To prepare for the parliamentary procedure, complete the following activities.

1. Read the guidelines provided by your organization.
2. Study parliamentary procedure principles by reviewing *Robert's Rules of Order.*
3. Practice proper procedures for conducting a meeting. Assign each team member a role for the presentation.
4. Visit the organization's website and look for the evaluation criteria or rubric for the event. This will help you determine what the judges will be looking for in your presentation.

Selection

Sections

The Java programs you have created so far in this text have been linear. Statements were lined up and executed in the order listed. That means every execution of the program was the same. The same statements were executed in the same order. Even though user input may have provided different data to alter the value of the results, the statements were always executed in the same order.

This chapter introduces selection, or decision-making. Selection allows one of two or more different branches of code to be executed based on the current truth-value of a condition. For example, some privileges are age-based. Say Glenn wanted to get his driver's license. In Glenn's state, the driving age is 16. The condition would be: is Glenn's age greater than or equal to 16? If he is 16 or older, the truth-value is true. If Glenn is 15 or younger, the truth-value of that condition is false. If Glenn is 16 or older, he can get a driver's license. If Glenn is less than 16, he is too young to get a driver's license. In this chapter you will create conditions, evaluate their truth-value, and branch to execute different code segments. You will also use a selection structure called *switch* that allows choosing one of a set of options based on a value of a variable.

With the added versatility of selection comes an added complexity. Algorithm development must provide for all possible results of conditions. The need for diagramming the flow of the program and careful testing escalates. This chapter will help to develop skills for flowcharting and testing techniques.

Reading Prep

College and Career Readiness

Prior knowledge is experience and information a person already possesses. Your prior knowledge of a topic allows you to make sense of new information quickly. Before reading this chapter, read the chapter title. What does it tell you about what you will be learning? How does this chapter relate to information you already know?

While studying, look for the activity icon 📲 for:

- Vocabulary terms with e-flash cards and matching activities.
- Starter files for hands-on examples and other exercises.

These activities can be accessed at
www.g-wlearning.com/informationtechnology/1773

Chapter Glossary ↗

case: Each choice for a switch statement.

condition: Boolean expression.

control flow statement: Statement that causes different statements to be executed based on certain conditions.

equality operator: Tests if two variables or expressions evaluate to the same value.

equalsIgnoreCase method: Compares the characters in Strings considering uppercase and lowercase letters to be equal.

equals method: Method of the String class provided to perform an equivalent function to determine if the characters of the Strings are the same.

if-then statement: Selection statement.

logical operator: Allows the combination of Boolean expressions into more complex expressions.

nested selection: One selection placed inside another selection statement.

relational operator: Compares two numeric expressions to see if one is greater, lesser, or both are equal.

scope: Block of code in which a variable is defined; a variable cannot be used outside its scope.

selection statement: Allows the program to choose a set of statements to execute based on a certain condition at the time the selection statement is processed.

switch statement: Type of selection statement that tests the value of a variable and selects statements based on any number of choices.

true block: Set of statements to execute when the condition of an *if-then* statement is true.

truth table: Shows the truth-value of a compound selection statement for each component in the statement.

truth-value: Whether a Boolean statement is true or false; used to determine branching to execute different code segments.

unary operator: Operator in which only one expression is used.

Conditions

By default, the flow of a program is sequential, that is, statements are executed in the order in which they are coded. This provides for a very linear, straightforward execution of one statement after another. If a method call is used, such as objectNode.setFill(Color.RED), the program control leaves the linear path, performs the statements of the method, and returns to the exact place from where it was called. Control of the program falls to the very next statement in linear sequence after the method call.

Additionally, a programmer may change the flow based on certain conditions. This chapter and the next will provide many statements to control the flow during execution of a Java program. The conditions that determine the flow are Java expressions that simplify to a true or false value rather than a numeric value. These are also referred to as logical or test conditions.

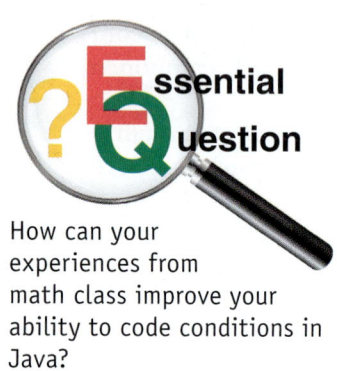

Essential Question

How can your experiences from math class improve your ability to code conditions in Java?

Learning Goals

- Construct condition expressions using equality, relational, and logical operators.
- Diagram operator precedence.

Terms

condition	relational operator
control flow statement	truth-value
equality operator	unary operator
logical operator	

Operators and Conditions

A statement that causes different statements to be executed based on certain conditions is called a ***control flow statement.*** These types of statements are divided into two main categories, selection statements and repetition statements. Repetition statements are discussed in the next chapter. A selection statement allows the program to choose a set of statements to execute based on a certain condition at the time the selection statement is processed. Many programming languages call selection statements conditionals.

A ***condition*** is a Boolean expression. That means a condition can be judged as true or false. Conditions often compare objects or variables. Often the expressions are complex and must be simplified to determine the truth-value. The ***truth-value*** is whether a Boolean expression is true or false. The boolean data type has two reserved words that hold truth-values: true and false.

Recall what you learned about operators in Chapter 5. Operators are generally binary. This means that only two variables can be operated on at a time. Occasionally, only one expression is used. A ***unary operator*** is one in which only one expression is used. An example you have used before is the increment operator x++.

FYI

When doing your own research into other programming languages, always remember *selection statements* are the same thing as *conditionals.*

Selection statements are composed of distinct parts:
- Boolean condition to test
- block of code to execute if the expression is true
- optional block of code to execute if the expression is false

Key to the process of a selection statement is the condition to test. Java provides operators to build expressions to test. The operators help to build equality or inequality, relational, and logical expressions. The requirement is that these expressions simplify to a true or false result.

Equality Operators

An *equality operator* tests if two variables or expressions evaluate to the same value. Java supplies two equality operators, as shown in **Figure 8-1.** The equal to operator tests if two elements are equal. The symbol in Java for this operator is two equals signs (==). The not equal to operator tests if they are not equal. The Java symbol for this operator is an exclamation point followed by an equals sign (!=).

For example, suppose the variable housesOnBlock contains the int 12. The expression housesOnBlock == 12 evaluates to true. However, the expression housesOnBlock != 12 evaluates to false.

When testing equality, take care to ensure two equal signs are used. One equal sign indicates assignment. The expression housesOnBlock = 12 will assign the value of 12 to the variable, rather than test the truth-value.

Operator	Java Symbol	Description
Equal to	==	Tests if both elements are equal.
Not equal to	!=	Tests if the two elements are unequal.

Goodheart-Willcox Publisher

Figure 8-1. The two equality operators in Java are equal to and not equal to.

Relational Operators

Java also provides relational operators. A *relational operator* compares two elements to see if one is greater, lesser, or both are equal. These are the same as used in algebra. The main difference between relational expressions in Java versus algebra is that in algebra, they are statements of truth. In Java, a relational expression is more like a question: is this relationship true? Java has four relational operators, as shown in **Figure 8-2.** Note that the Java symbols are similar to those used in algebra.

Continue with the example in which the variable housesOnBlock contains 12. The expression housesOnBlock > 12 evaluates to false. The expression housesOnBlock < 12 evaluates to false. The expression housesOnBlock >= 12 evaluates to true. The expression housesOnBlock <= 12 evaluates to true.

Operator	Java Symbol	Description
Greater than	>	Tests if the first element is greater than the second.
Less than	<	Tests if the first element is less than the second.
Greater than or equal to	>=	Tests if the first element is greater than *or* equal to the second.
Less than or equal to	<=	Tests if the first element is less than *or* equal to the second.

Goodheart-Willcox Publisher

Figure 8-2. These are the four relational operators in Java.

Logical Operators

Java has a third set of operators. A *logical operator* allows the combination of Boolean expressions into more complex expressions. These operators are **NOT, AND,** and **OR,** as shown in **Figure 8-3.** The Java symbol for the **NOT** operator is the exclamation point (!). The Java symbol for the **AND** operator is two ampersands (&&). The Java symbol for the **OR** operator is two piping symbols (||). The piping symbol is found on most keyboards above the [Enter] key. It is usually entered by holding down the [Shift] key and pressing the backslash key.

Recall what you learned in math class about Venn diagrams. **AND, OR,** and **NOT** are the same terms used in Venn diagrams to show set relationships, as illustrated in **Figure 8-4. AND** is the conjunction or intersection of the two sets. **OR** is the disjunction or union of the two sets. **NOT** is the negation or complement of the set.

NOT

The **NOT** operator is a unary operator that changes the truth-value of an expression from true to false or vice versa. Consider this example. Suppose a >= b is a true statement. Using the **NOT** operator, !(a >= b) evaluates to false. Use parentheses to show the comparison is being made first and then the result is negated.

AND

Both operands of an **AND** expression must be true for the entire expression to be true. If either of the two operands are false, the **AND** expression is false. Suppose a >= b is a true statement and c < d is a true statement. Using the **AND** operator, a >= b && c < d evaluates to true.

OR

If either of the two operands is true, the **OR** expression is true. Both operands of an **OR** expression must be false for the entire expression to be false. For example, suppose a >= b is a true statement and c < d is a false statement. Using the **OR** operator, a >= b || c < d evaluates to true.

Operator	Java Symbol	Description	Function
NOT	!	Changes the truth value.	Negation
AND	&&	Tests if both expressions are true.	Conjunction
OR	||	Tests if at least one expression is true.	Disjunction

Goodheart-Willcox Publisher

Figure 8-3. Java contains three logical operators.

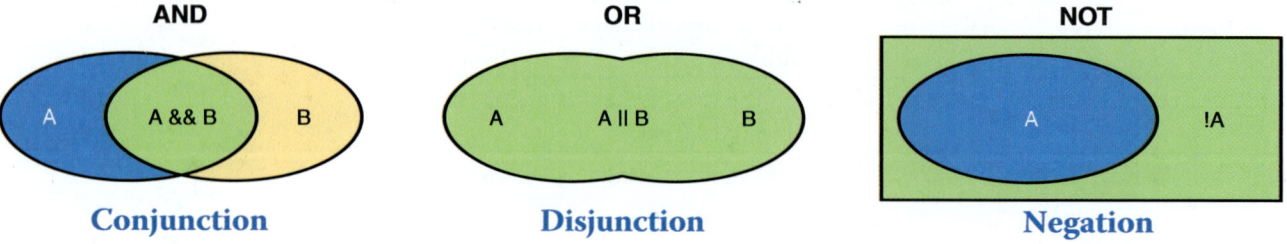

Goodheart-Willcox Publisher

Figure 8-4. The logical operators can be compared to Venn diagrams.

Operator Precedence

As you learned in Chapter 5, expressions may combine many values and operators to result in a single value. The order of precedence in Java expressions is the same as the order of operations in algebra. **Figure 8-5** shows Java's order of precedence with the logical and relational operators introduced in this chapter. The new operators are highlighted to show where they fall in the order.

The **NOT** operator has the highest precedence of operators. It is performed immediately after the parentheses are cleared.

The logical **AND** operator has a higher precedence than the logical **OR** operator. Suppose three boolean variables are defined as true or false.

```
boolean a = true;
boolean b = false;
boolean c = false;
```

Following the order of precedence, the expression:

```
a || b && c
```

is simplified by applying the && first. Then, apply the ||. Thus, the statement is evaluated as:

1. b && c is false
2. a || false is true

The truth value of the expression is true.

Step	Operation	Action
1	( )	Perform all the operations inside a set of parentheses first. With nested parentheses, start with the innermost set of parentheses. Within each pair of parentheses, follow the next rules.
2	!	Scan the expression from left to right. Perform the logical unary operation **NOT**.
3	* / %	Scan the expression from left to right. Perform each of these three operations as they are encountered.
4	+ −	Scan the expression from left to right. Perform each of these two operations as they are encountered.
5	> >= < <=	Scan the expression from left to right. Perform each of these four relational operations as they are encountered.
6	== !=	Scan the expression from left to right. Perform each of these two operations as they are encountered.
7	&&	Scan the expression from left to right. Perform the **AND** logical operation as it is encountered.
8	||	Scan the expression from left to right. Perform the **OR** logical operation as it is encountered.
9	= += −= *= /= %=	After the expression on the right of the assignment is evaluated, store the value in the variable on the left of the equals sign.

Figure 8-5. The order of precedence for Java including operators.

If **OR** should be evaluated first, use parentheses to raise the order of precedence. The expression:

```
( a || b ) && c
```

is simplified by evaluating the expression within the parentheses first. Then, apply the &&. Thus, the statement is evaluated as:

1. a || b is true
2. true && c is false

The truth value of the expression is false.

HANDS-ON EXAMPLE 8.1.1

Evaluating Conditions ↪

Evaluating conditions is a key to computer programming. It is important to understand how to construct proper statements for evaluating conditions. Before beginning this exercise, download the chapter files from the student companion website. The EvaluatingConditionsSnippet.java file will be used as a starting point.

1. Launch jGRASP, and open the EvaluatingConditionsSnippet.java file. Add your name to the comments, and save the file. The first two examples of conditions are coded for you. They show examples of the equality operators.

```
int a = 5;
int b = 3;
boolean test;

System.out.println( "Evaluating Conditions" );
System.out.println( "Let a = " + a + ", b = " + b );

/***** 1. is equal to operator */
test = a == b; // Apply operator precedence! The == operator is evaluated first, then the
    truth-value is assigned
System.out.println( " 1. " + a + " == " + b + " is " +  test );

/***** 2. is not equal to operator */
test - a != b;
System.out.println( " 2. " + a + " != " + b + " is " + test );
```

2. Compile and run the program. Verify the results.
3. Follow this pattern, and complete the instructions for comments 3 through 6. Create conditions using the operators given in the comments.
4. Compile and run the program. Verify the results.
5. Locate the close of the block comment for comment 7. Delete this, and close the comment at the end of the first line in the comment.

```
/***** 7. && AND operator */
// Define two more variables
int c = 4;
int d = 7;
System.out.println( "Let c = " + c + ", d = " + d );

test = a >= b  &&  c < d; // Apply operator precedence!
System.out.println( "7. a >= b  &&  c < d  is " + test );
```

6. The code at comment 7 uses a compound expression. Apply operator precedence. Compile and run the program. Verify the results.
7. Create a compound expression for the condition for comment 8. Use the variables a, b, c, and d in the condition.

```
/***** 8. || OR */
```

8. You will now write expressions using the keywords true and false. Locate the close of the block comment for comment 9. Delete this, and close the comment at the end of the first line in the comment.

```
/***** 9. The reserved word true */
test = true;
System.out.println( "9. The reserved word true is " + test );
```

9. Compile and run. Verify the result.

10. Create a condition for comment 10. You can copy and paste the code from comment 9 and modify it as needed.

11. Compile and run the program. Verify the result.

12. Create the expressions indicated in comments 11 through 14. Use the proper Java symbols for the operators indicated in the comments. Use the reserved words true and false in the conditions required.

13. Compile and run the program. Verify the result.

14. Select the output in the **Run I/O** tab by clicking and scrolling over it. Then, press the [Ctrl][C] key combination to copy it. Open a word-processing document, and paste the content. Save this document as EvaluatingConditionsOutput in your working folder.

15. Change the values of the variables in the program and run it again. Compare the output to the original output. Copy the new output into the document.

16. Write a paragraph in the document to describe what you learned in this activity. Save the document, and submit it for grading.

Try It!

On a piece of paper, write the Java condition for this situation:

A person is old enough to drive (16), but not old enough to vote in federal elections (18).

By hand, check the boundary values of 15, 16, 17, 18, and 19. Then, continue working with the EvaluatingConditionsSnippet.java file. At the bottom of the program, add the code, and run the program to check your answer.

SECTION REVIEW 8.1

Check Your Understanding

1. Conditions are _____ expressions, which are judged as true or false.
2. What is the Java symbol for testing equality?
3. Which logical operator uses the exclamation point as the negation operator?
4. The operator for the logical **AND** is _____.
5. In terms of operator precedence, what is the correct order for **AND**, **OR**, and **NOT**?

Build Your Vocabulary

As you progress through this course, develop a personal computer science glossary. This will help you build your vocabulary and prepare you for a career. Write a definition for each of the following terms and add it to your computer science glossary.

condition
control flow statement
equality operator

logical operator
relational operator

truth-value
unary operator

Selection Statements

Selection statements provide the branching capability in computer programming. Some languages refer to these as conditional statements. The capability is the same, but the syntax and name selection are specific to Java.

You use selection statements every day in your conversations. "If you want to purchase this item, please step over to this register." "If you finish your homework early, we will have time to play a game together; otherwise it will be bedtime." "If you want to see a movie, we can go at 8:00 pm; otherwise we could go at 10:00 pm; or else if you would like to go bowling, we could do that." "If we have five people going, I can drive my car. If six people go, we will need two cars. For 11 people, we will need three cars." These same patterns are used in coding. The structures are the same. First, you evaluate the condition, which often follows an "if," then you decide based on the truth-value of the conditions which option you want to select.

A selection statement can always be rewritten to use "if." Suppose the selection statement is, "Whoever is going to study for the science test should meet in the library study room at 3:30 today." It can be rewritten to "*If* you are going to study for the science test, *then* you should meet in the library study room at 3:30 pm today." Translating from words to code is a key skill to develop for computer programming.

Essential Question

Why should Java applications contain branching capabilities?

Learning Goals

- Select the appropriate form of the if-then statement for an algorithm.
- Explain the function of the switch statement.
- Assess the scope and visibility of a given variable.
- Inspect code for common errors with selection statements.

Terms

case	selection statement
if-then statement	switch statement
scope	true block

Selection Statement Construction

A **selection statement** allows the program to choose a set of statements to execute based on a certain condition at the time the selection statement is processed. A basic selection statement begins with the keyword if. Refer to **Figure 8-6.** The keyword if alerts the compiler that a decision must be made. This is followed by a condition enclosed in parentheses. If the condition is true, a selection of code is executed. The code that is selected to execute is called the **true block** of code. If the condition is not true, the selection is skipped and control flows to the next statement after the if statement.

The Oracle documentation for Java refers to a selection statement as an **if-then statement.** However, the word *then* is never used in an if statement. The only keywords for if-then statements are if and else.

There are several forms of the if-then statement. These include if-then, if-then-else, and if-then-else-if. Each of these is discussed here.

If-Then Statement

The if-then statement tests a condition and provides a single selection or branch. It is one that is used in English. It means the same thing in Java. Meet the condition, have the consequence!

If you behave, then you get a treat.

In this if-then statement, the condition is *you behave.* The consequence is *you get a treat.*

The syntax for an if-then statement is:

```
if ( condition ) {
   // statements to select
}
```

The statements to select are executed *only* if the condition is true. If the condition is false, control jumps to the end of the if-then statement. For example:

```
if ( math test is tomorrow ) {
   // study for math test
}
```

You would study for the math test if the test is tomorrow. If the math test is not tomorrow, you would continue on to the next activity.

Figure 8-6 illustrates the control flow of an if-then statement. If the condition is true, the selected statements are executed. Then, the next sequential statements are executed. If the condition is false, the selected statements are skipped, and the next sequential statements are executed.

Recall that Java is punctuation driven. If the selection code consists of more than one statement, Java requires the true block of code be put inside braces. If a block consists of only one statement, the braces are optional. However, it is a good programming practice to always include the braces. A semicolon is not coded after the if condition; a semicolon is used to terminate statements within the true block of code.

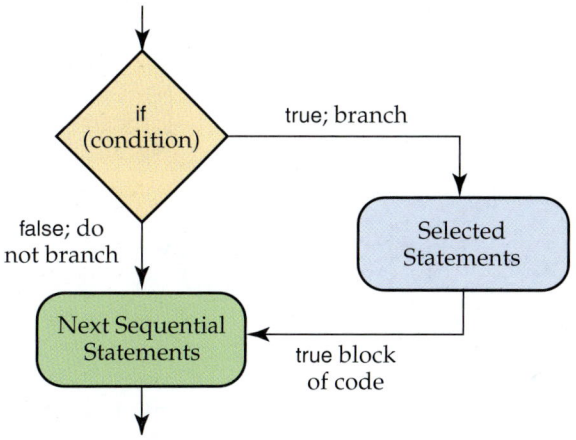

Goodheart-Willcox Publisher

Figure 8-6. The most basic selection statement is the if-then statement.

If-Then-Else Statement

The if-then-else statement provides two possible selections to execute. Which selection is executed depends on the truth-value of the condition. If the condition is true, one selection—the true block—will be processed. If the condition is false, the other selection—the false block—will be processed. In either case, control returns to the next sequential statement in the program.

If you behave, then you get a treat, else you will have a time-out.

The syntax for an if-then-else statement is:

```
if ( condition ) {
   // statements to select if condition is true
} else {
   // statements to select if condition is false
}
```

The first statements to select are executed *only* if the condition is true. If the condition is false, the second statements to select are executed. After the statements to select are executed, whichever ones those are, control returns to the next statement in the program. For example:

```
if ( math test is tomorrow ) {
   // study for math
```

(Continued)

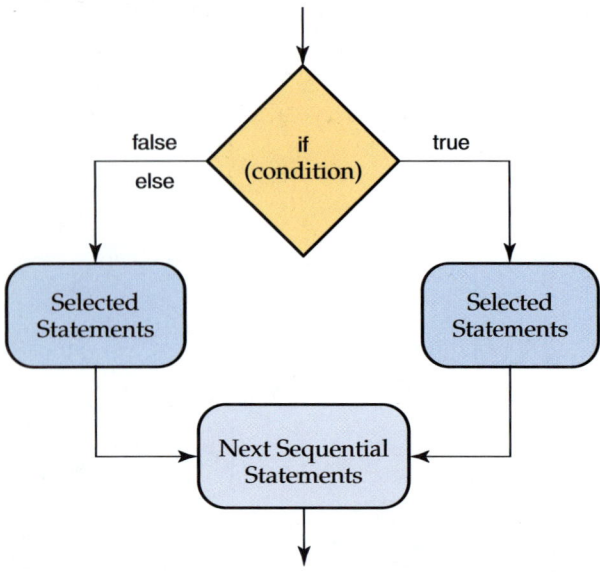

```
} else {
  // relax
}
// perform next activity
```

You would study for the math test if the test is tomorrow. If the math test is not tomorrow, you would relax. After relaxing, you would continue on to the next activity.

Figure 8-7 illustrates the control flow of an if-then-else statement. If the condition is true, one set of selected statements is executed. Then, the next sequential statements are executed. If the condition is false, the other set of selected statements is executed, and then the next sequential statements are executed.

Goodheart-Willcox Publisher

Figure 8-7. The if-then-else statement allows one of two sets of selected statements to be executed.

If-Then-Else-If Statement

An if-then-else-if statement is used in the case where a false result requires another decision. **Figure 8-8** illustrates the control flow of an if-then-else-if statement. It is possible to have many else-if clauses in the statement, as shown. For example:

```
if ( math test is tomorrow ) {
   // study for math
}
else if ( science test is tomorrow ) {
    // study for science
}
else {
    // relax
}
// perform next activity
```

If the math test is tomorrow, you would study for it. If the math test is not tomorrow, you would check to see if the science test is tomorrow. If the science test is tomorrow, you would study for it. If neither the math test nor the science test is tomorrow, you would relax. After relaxing, you would continue on to the next activity.

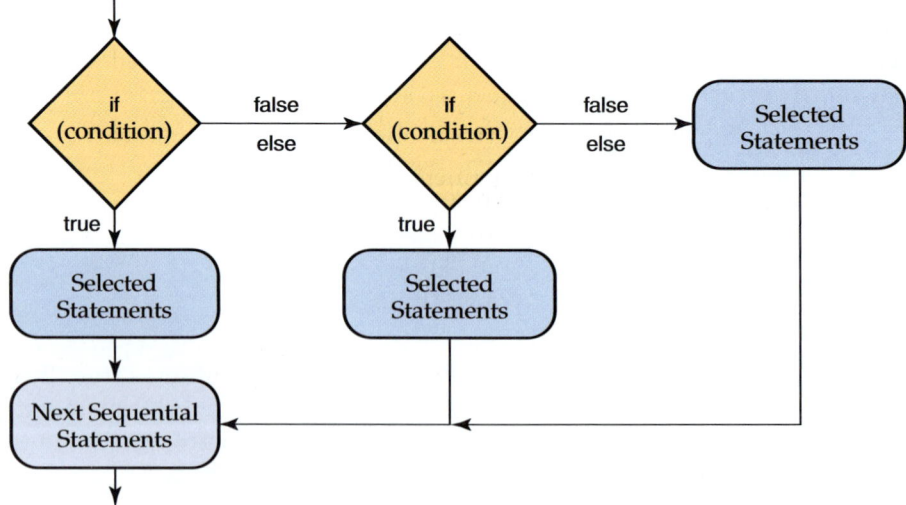

Goodheart-Willcox Publisher

Figure 8-8. An if-then-else-if statement allows another decision to be made when the original condition is false.

Science and Java

Converting Temperature

Humans are very good at seeing specific wavelengths of visible light. Slight variations in wavelength immediately let us see millions of colors. However, humans are not good at assessing temperature. The perception of temperature tends to be relative. For example, you place three glasses of water in front of you. One contains hot water, one room-temperature water, and one cold water. Place a finger from each hand in the cold and hot water at the same time. After a minute, place those same fingers in the middle glass. The water will feel hot and cold at the same time! Thus, instruments are used to measure temperature. The most common instrument for measuring temperature is the thermometer.

Scientists have many scales to measure temperature. The earliest scale was developed by Daniel Fahrenheit in 1724. The Fahrenheit scale (F) is still used as the standard in the United States, the Bahamas, Belize, the Cayman Islands, and Palau. The rest of the world uses the Celsius scale (C), also known as the centigrade scale. Celsius is the temperature scale in the International System of Units (SI). Celsius is named after a Swedish astronomer Anders Celsius who developed the Celsius scale in 1744. As scientists became more knowledgeable about thermodynamics, they needed a temperature scale that measured when all thermal motion among atoms ceases, absolute zero. To accommodate this need, the Kelvin scale (K) was invented by Lord Kelvin in 1848.

The Kelvin scale is based on the Celsius scale. In the Celsius scale, water freezes at 0 degrees C and boils at 100 degrees C. In the Kelvin scale, water freezes at 273.15 degrees K and boils at 373.15 degrees K. Absolute zero is 0 degrees Kelvin. It is easy to convert from Celsius to Kelvin by adding 273.15. The conversion from Celsius to Fahrenheit and back again is more complicated. The conversion formulas are:

- $°C = 9/5°F - 32$
- $°F = 1.8°C + 32$
- $°K = °C + 273.15$

NASA

Assignment

Write out the pseudocode to complete a program that converts temperatures from one scale to the two others. First, ask the user which scale to use for the original temperature. Then, ask what the original temperature is. Use an if-then-else-if control structure to calculate the results.

Your project is to write a program from your pseudocode. A test plan is provided:

- The surface of the Sun is 5800°K. What is that in degrees C and degrees F?
- Human body temperature is 37°C. What is that in degrees F and degrees K?
- The average surface temperature of Earth is 57°F. What is that in degrees K and degrees C?
- Use values that you know, such as the freezing and boiling point of water.

HANDS-ON EXAMPLE 8.2.1

Using Selection Statements 🔗

A selection statement allows the program to make decisions and react based on conditions. This is the foundation of most computer programs. In this activity, you will create selection statements. Before beginning this exercise, download the chapter files from the student companion website. The SelectionSnippet.java file will be used as a starting point.

1. Launch jGRASP, and open the SelectionSnippet.java file. Add your name to the first comment, and save the file. A first selection statement is coded for you.

2. Compile and run the program. The output will appear in the **Run I/O** tab.

   ```
   Selection Practice
   The score is now 1
   ```

3. Locate comment 1, and following the instructions, write the code to define two integers and test the condition if a and b are equal. Print the result.

   ```java
   int a = 6;
   int b = 8;
   if ( a == b ) {
       System.out.println( "a and b are equal." );
   } else {
       System.out.println( "a and b are not equal." );
   } // end if
   ```

4. Compile and run the program. The result should be:

   ```
   a and b are not equal.
   ```

5. Following the instructions in comment 2, write an if-then-else-if statement that tests the relationship between two numbers. Use the existing a and b variables. If a equals b, say so. If not, test if a is greater than b. If yes, say so. If not, say a is less than b. Think about this: why is testing for if a less than b not needed?

   ```java
   if ( a == b) {
       System.out.println( "a and b are equal." );
   }
   else if ( a > b ) {
       System.out.println( "a is greater than b." );
   }
   else {
       System.out.println( "a is less than b." );
   } // end if
   ```

6. Compile and run the program. The message should read:

   ```
   a is less than b.
   ```

7. Modify the code so the value of a is 16. Compile and run the program. The message should read:

   ```
   a is greater than b.
   ```

8. Modify the code so the value of a is 8. Compile and run the program. The message should read:

   ```
   a is equal to b.
   ```

   You have created selection statements!

Try It!

Continue working with the SelectionSnippet.java file. At the bottom of the file, add code to ask the user for ingredients for a salad. Define a String named ingredients. Set it to "Your salad will have". Prompt the user for a 1 or 0 on three ingredients of your choice. For example:

```
Do you want shredded carrots? (1=yes, 0=no)
```

- If the input is 1, add "carrots" to the ingredients. Repeat for two other ingredients. At the end, print the String
- ingredients. Hint: remember to import the Scanner class and create a Scanner object. Also remember to insert a
- space before each added ingredient.

• • • • • • • • • • • •

Switch Statement

On occasion, there may be several options to take based on the value of a single int, char, byte, short, or String variable. The *switch statement* is a type of selection statement that tests the value of a variable and selects statements based on any number of choices. Note, though, that the switch statement does not apply when there is a range of numbers. Each choice for a switch statement is a constant value called a *case.* The flow of the switch statement is continued within the cases. To interrupt this path, use the break keyword.

FYI

Other programming languages may call a switch statement a case statement.

```
switch ( variable ) {
   case value1:   statements;
      break;
   case value2:
   case value3:   statements;
      break;
   case value4:   statements;
      break;
   default:       statements;
}
```

The default statement, which is optional, is executed when a variable value does not match any value in the case statements. In addition, after a match has been found between the switch variable and a case value, processing will continue to the next statement until the end of the switch statement or until a break statement is reached. If the processing of the selection is finished, use the break keyword to leave the case statement and execute the next sequential statement after the switch statement. Note that between case value2 and case value3 there is no break. That means the same statements will be selected for either of these values.

Say you know the number of the president's term, but want to print the name of the president. An if-then-else-if structure could be used for this, but it would be very messy. However, switch statements can organize the code in an easy-to-read format, even for 45 or more cases. An example of the code for four of the presidents follows.

```
int term = 16;
String president;
switch ( term ) {
   case 1:  president = "George Washington";
      break;
   case 16: president = "Abraham Lincoln";
      break;
   case 44: president = "Barack Obama";
      break;
   case 45: president = "Donald Trump";
      break;
   default: president = "Not available for that term";
      break;
}
System.out.println( president );
```

The JVM will process this code looking for a match to the value of the variable term, which in this example is 16. Since the break statement is used after each case, once a match is found control skips the remaining cases as well as the default and jumps to the code to print the name contained in the president variable. The output from this code would be:

Abraham Lincoln

Math and Java

Calculating the Day of the Week

One of the more enjoyable pastimes for fans of number theory is to create a new formula to calculate the day of the week on which a certain date falls. There are many such formulas. Christian Zeller, a 19th century German mathematician, devised one of these formulas. It is known as Zeller's Congruence.

Dragon Images/Shutterstock.com

What complicates the calculation for finding the day of the week is that February has either 28 or 29 days. Because of its variability, February must be handled last. Zeller changed January and February dates to the year before, counting them as months 13 and 14.

Another consideration for this calculation is the amount of calendar reform that has taken place over the years. Zeller's Congruence works for any date on the Gregorian calendar. The Gregorian calendar is the one used in the United States and most of the world.

Zeller's Congruence makes use of integer variables and modulus:

```
day = ( month + 13(month+1)/5 + year %
100 + (year % 100)/4 + year/100/4 –
   2(year/100) ) % 7
```

In the case the result is negative, use the Math.abs() method to find the absolute value of the day. The result is a number from 0 to 6, with 0 representing Saturday, 1 Sunday, and so on.

Assignment

Write a Java program to calculate the day of the week based on Zeller's Congruence. Use the following if-then statements to adjust any January or February dates. Use a switch statement to output the name of the day of the week.

```
if ( month == 1 ) {
  month = 13;
  year--;
}
if (month == 2 ) {
  month = 14;
  year--;
}
```

Test your program with today's date and then your birthday for this year. Compare the results to a calendar to verify correct output. Then, enter your birthdate to find out on which day of the week you were born.

HANDS-ON EXAMPLE 8.2.2

Rolling Dice with a Switch Statement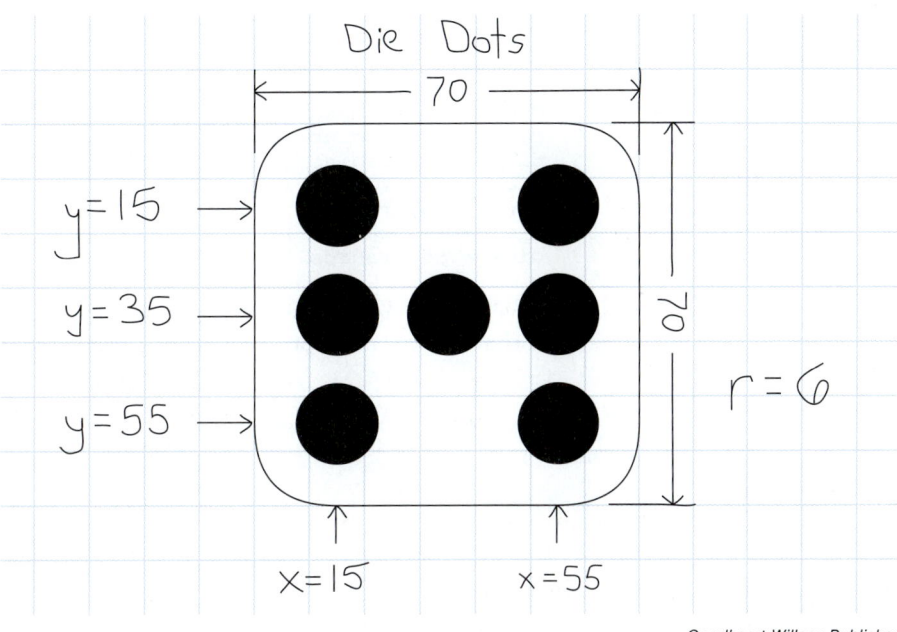

To draw all faces on a six-sided die, seven dots (circles) are required. The algorithm is: get a random number between 1 and 6, and then draw the die face and the required number of circles with JavaFX. Before beginning this exercise, download the chapter files from the student companion website. The DrawDieSnippet.java file will be used as a starting point.

1. On graph paper, sketch a rounded rectangle with the seven circles in place, as shown. Note the centers and radius of each circle.

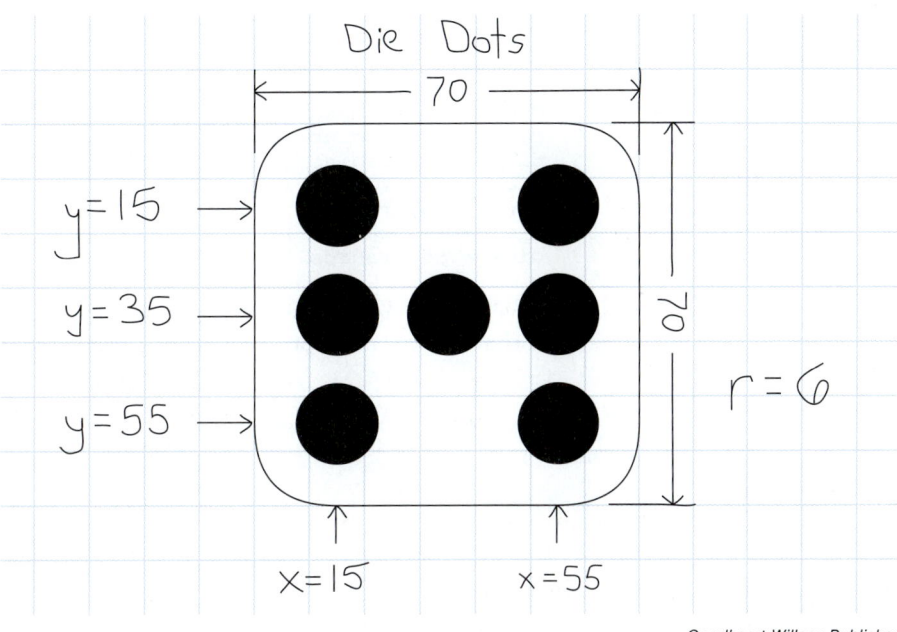

Goodheart-Willcox Publisher

2. Launch jGRASP, and open the DrawDieSnippet.java file. Add your name to the first comment, and save the file. Code to create a stage has been provided for you.

3. Compile and run the program. The graphics window is displayed with the title **Roll of One Die**. Close the window.

4. Locate comment 1, and follow the instructions to apply what you have learned about random numbers to code a random-number generator and to print the output. Note that the Random class is already imported. The number generated will display in the **Run I/O** tab. Use this to verify the correct die is displayed in the JavaFX window.

    ```
    /***** 1. Get a random number between 1 and 6 and output it */
    Random randGen = new Random( );
    int dieRoll = randGen.nextInt( 6 ) + 1;
    System.out.println( "The roll is " + dieRoll );
    ```

5. Locate comment 2, and follow the instructions to apply what you have learned about JavaFX shapes to draw a rounded rectangle for the outline of a die. Locate the rectangle at (115, 40). These values have been assigned to variables for you as dieX and dieY. Use the width and height from your sketch. Set the arc width and height to 10 to round the corners. Set the fill to white and the stroke to black.

    ```
    /***** 2. Draw a rounded rectangle for the die.
         Set arc width and arc height to 10,
         Set the fill color to white and the stroke color to black */
    Rectangle die = new Rectangle( dieX, dieY, 70, 70 );
    ```

(Continued)

```
die.setArcWidth( 10 );
die.setArcHeight( 10 );
die.setFill( Color.WHITE );
die.setStroke( Color.BLACK );
```

6. Locate comment 3, and follow the instructions to draw the die by adding it to the list of effects.

```
/***** 3. Draw the die by adding die to effects list */
effects.add( die );
```

7. Locate comment 4, and follow the instructions to define the seven dots needed for the face of the die. First, set a common radius for the dots. Then, define each of the seven dots needed. The location is relative to the corner of the die. Add the coordinate of the rectangle to the x and y coordinates of the circle. Refer to your sketch for the centers of the circles. The first one is completed for you.

```
/***** 4. Define the radius and seven dots for the face of the die
          1    5
        2  4  6
          3    7
*/
int radius = 8;
Circle dot1 = new Circle( dieX + 15, dieY + 15, radius );
Circle dot2 = new Circle( dieX + 15, dieY + 35, radius );
Circle dot3 = new Circle( dieX + 15, dieY + 55, radius );
Circle dot4 = new Circle( dieX + 35, dieY + 35, radius );
Circle dot5 = new Circle( dieX + 55, dieY + 15, radius );
Circle dot6 = new Circle( dieX + 55, dieY + 35, radius );
Circle dot7 = new Circle( dieX + 55, dieY + 55, radius );
```

8. Compile and run the program. Verify the die face (rectangle) is drawn. The dots will not be drawn yet.

9. Follow the prompt 5 to add a **switch** statement that draws the appropriate dots for each value of the **dieRoll**.

```
switch ( dieRoll ) {
    case 1: effects.addAll( dot4 );
        break;
    case 2: effects.addAll( dot1, dot7 );
        break;
    case 3: effects.addAll( dot1, dot4, dot7 );
        break;
    case 4: effects.addAll( dot1, dot3, dot5, dot7 );
        break;
    case 5: effects.addAll( dot1, dot3, dot4, dot5, dot7 );
        break;
    case 6: effects.addAll( dot1, dot2, dot3, dot5, dot6, dot7 );
        break;
    default: break;
}
```

10. Compile and run the program. You have used a **switch** statement to draw faces of a die! Verify the die face is drawn correctly and the number of dots matches the value output in the **Run I/O** tab. Run the program several times until all six die faces have been displayed. Another way to verify that all die rolls are correctly displayed is to comment out the statement that generates the random number for the die roll, and instead hard-code each die number in turn until you have tested all six numbers. Then uncomment the code to randomly generate the number again.

Try It!

Save the DrawDiceSnippet.java file, rename the class to DrawDiceSnippetTryIt, and save the file under the new name. Modify the code to draw a pair of dice. Think about what will be replicated and what will stay the same. Move the first die left on the stage by editing the dieX variable. Create new variables for the x and y values of the second die. Reuse the random-number object to generate the roll for the second die. Add code to draw the second die with the correct number of dots to match the random number. Create a screenshot of your window to present to your instructor along with your program.

Scope and Visibility of Variables

The *scope* of a variable is the block of code where it is defined. If a variable is defined within a selection, the variable disappears after the selection is processed. If a variable is defined within a method, it is visible from the point where it was defined until the end of the method.

In the example below, area is defined in the first selection. Its scope is the first selection only. To be used outside of this selection, it must be defined again, as it is on line 21. The length, width, radius, and figure variables are defined for the main method. Therefore, they can be referenced anywhere in the program.

```
 5 public class ScopeExampleSnippet {
 6
 7   public static void main( String [ ] args ) {
 8
 9      double length = 10;
10      double width = 30;
11      double radius = 15;
12
13      char figure = 'r'; // r for rectangle, c for circle
14      if ( figure == 'r' )  {
15         double area = length * width;
16         System.out.println( "The area of the rectangle is " + area );
17      }
18      else if ( figure == 'c' ) {
19         double area = Math.PI * Math.pow( radius, 2 );
20         System.out.println( "The area of the circle is " + area );
21      } // end if
22
23      System.out.println( area );
24   } // end main
25 } // end class
```

Line 23 tries to reference area after the if-then-else-if statements are complete. At that point, area is not available, and the compiler sends an error message:

```
ScopeExampleSnippet.java:24: error: cannot find symbol
     System.out.println( area );
                         ^
   symbol:   variable area
   location: class ScopeExampleSnippet
 1 error
```

To write the code so that area is available for printing after the if-then statement, define area before the statement begins and do not redefine area within the true and false blocks.

Common Errors with Conditions

Obviously, syntax errors are common issues when writing conditions. Errors related to scope are also common. Beyond these, logic errors may produce incorrect results in a selection statement.

A common logic error occurs when a programmer inadvertently uses a semicolon at the end of the condition in an if statement. This is fine with the compiler because the semicolon simply terminates the if statement. The logic error is the selected statements are no longer conditional and will always be executed. This is the type of error that is very difficult to uncover except through very extensive testing of the code.

```
if ( condition );  // logic error to have a semicolon here!
{
   // doSelectionStatements
}
```

Remember that there is no semicolon after the if condition in a selection statement. The braces are the punctuation that open and close the true and false blocks. In Java Style, the brace should be at the end of the condition. If you follow Java Style, this error likely will not occur.

Errors that occur in the condition itself may involve incorrect operator precedence. Always check your conditions on paper. For example, suppose the intended test for the boolean variable a is **NOT** equal to the conjunction of a **AND** b. You write this code:

```
a == ! a && b
```

The above expression will first negate a and then test the **AND** part because of operator precedence. If you want to first combine a and b before applying the test, use parentheses:

```
a == ! ( a && b )
```

Another problem occurs when the boundary values are not properly handled. If you are testing a range of numbers, always check whether the numbers at the ends of the range should be used or not. For example, in the condition below, verify that 10 is *not* to be considered a value and that 12 *is* to be considered a value.

```
( ( age  >  10 ) && ( grade <= 12 ) )
```

SECTION REVIEW 8.2

Check Your Understanding

1. What is a selection statement called in Java?
2. The punctuation that encloses the condition in a selection statement in Java is _____.
3. What is the constant value in each choice for a switch statement called?
4. What determines the scope of a variable?
5. Identify the common error in this selection statement.

```
if ( condition );
{
    // doSelectionStatements
}
```

Build Your Vocabulary

As you progress through this course, develop a personal computer science glossary. This will help you build your vocabulary and prepare you for a career. Write a definition for each of the following terms and add it to your computer science glossary.

case	scope	switch statement
if-then statement	selection statement	true block

Helpful Conditions

Conditions can become quite complex. They can be developed with if-then statements inside if-then statements. This complexity is often necessary to develop algorithms to solve some problems. You might think of a test grade you received. You could write an if-then statement that gives a letter grade, with a score of 90 percent or above being an *A*. However, if the score were 98 or above, you would earn an *A+*. To achieve that result requires another if-then statement to be inserted, or nested, within the "90 and above" statement. In this section, you will learn to nest selection statements and to apply that to write a familiar game.

Learning Goals

- Apply nested conditionals.
- Explain De Morgan's laws.
- Construct truth tables.

Terms

nested selection

truth table

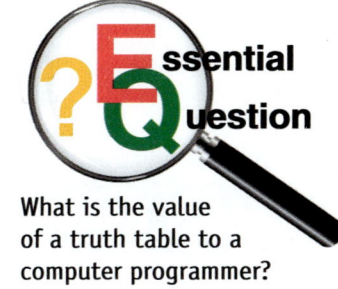

Essential Question

What is the value of a truth table to a computer programmer?

Nested Selection Statements

A *nested selection* is one selection statement placed inside of another selection statement. You can use one if-then or if-then-else statement inside another if-then or if-then-else statement. When this is done, it is called *nested if* statements in Java. The key to nesting is the placement of the braces. To ensure the selection statements are nested, verify the braces of the first selection statement are *outside* of the nested selection statement.

For example, suppose you are working in a carry-out kitchen. Someone places an order. Your responsibility is to add toppings to all sandwiches. If the person did not order a sandwich, then toppings are irrelevant. You only want to add toppings if they said yes to a sandwich. A sample nested if snippet for this is:

```java
int order = 1;
int sandwich = 1;
int salad = 2;
boolean mushrooms = true;
boolean cheese = false;
String ingredients = "Toppings for sandwich are ";

if ( order == sandwich ) {
  if ( mushrooms ) {
    ingredients += " mushrooms";
  }
  if ( cheese ) {
    ingredients += " cheese";
  }
} // end if sandwich
```

The two "toppings" conditions are nested within the "if order is sandwich" condition.

In Java programming, there are many options and many tradeoffs in selecting which selection statement to use. Some coders resist using nesting. They think it is too hard to read. They prefer to use a single compound condition that eliminates the need to nest. However, if that single condition becomes too complex, the tradeoff is to nest.

In the choice of conditions, the first concern is always for correctness. Second is efficiency of execution and readability by future programmers. Take care not to code conditions that do not need to be evaluated.

For example, in a case of comparing two numbers, there are only three possible situations: the two numbers are equal, the first is greater than the second, or the first is less than the second. Once you have checked for equal and greater than, you do not need to test for less than.

Additionally, always use self-documenting code and ample comments. Comments should always reflect the task being accomplished, not what the code is doing. For example, a comment that says // for loop is not as useful as // generate 10 random numbers.

HANDS-ON EXAMPLE 8.3.1

Writing Nested Selection Statements ⤤

In this exercise, you will explore nesting selection statements. Before beginning this exercise, download the chapter files from the student companion website. The NestedSelectionSnippet.java file will be used as a starting point.

1. Launch jGRASP, and open the NestedSelectionSnippet.java file. Notice two booleans have been defined. First you will investigate a simple if statement nested within another if statement.

2. Locate comment 1, and remove the comment code (//) on the statement.

```
System.out.println( "1. simple nest" );
```

3. Add code to demonstrate an if statement nested within another if statement.

```
if ( levelUp ) {
    System.out.println( "Level Up is " + levelUp );

    if ( gameOver ) {
        System.out.println( "Game Over is " + gameOver );
    } // end if gameOver

} // end if levelUp
System.out.println( "end of nested ifs" );
```

4. Predict what the output will be. Then, compile and run the program to verify your prediction.

5. Change the value of gameOver to false. Again, predict what the output will be, and then compile and run the program to verify your prediction.

6. Change the value of levelUp to false and gameOver to true. Predict what the output will be, and then compile and run the program to verify your prediction.

7. Change the value of both variables to false. Predict what the output will be, and then compile and run the program to verify your prediction. What you have observed is the action of nested if statements. Now you will investigate a switch statement nested within an if statement.

8. Locate comment 2, and remove the comment codes (//) from the four lines of code after the comment. The code sets up a random number you will use to test whether a gamer levels up. Level up only occurs on multiples of 4. Check to see there is a multiple of 4, and then determine which level the player has reached. Notice that the switch statement does not have a default case. In this example, the only possible values for

numbers are 4, 8, and 12. Add the following lines of code to demonstrate the switch statement nested within an if statement.

```java
if ( number % 4 == 0 ) { // check for a multiple of 4
    System.out.print( "Level Up! " );

    switch ( number ) {
        case 4:
            System.out.println( "You are now on the Ruby Level." );
            break;
        case 8:
            System.out.println( "You have reached Emerald Level." );
            break;
        case 12:
            System.out.println( "You are on the Diamond Level." );
            break;
    } // end switch

} // end if number is multiple of 4
```

9. Compile and run the program. If you are having difficulty getting this to work, check the location of the braces. The key to nesting is the placement of the end braces. To ensure the selection statements are nested, verify the braces of the first selection statement are outside the nested selection statement.

Try It!

Continue using the NestedSelectionSnippet.java file. Add a pair of nested if statements where one of the statements is an if-then-else statement. Try this: If it is Friday night, would you like to go to a movie or go bowling?

De Morgan's Laws

The work of British mathematician Augustus De Morgan, **Figure 8-9,** has made setting up conditions a little easier to understand. De Morgan is famous for his study of Boolean algebra and set theory. He developed a set of rules to help write expressions that are equivalent. De Morgan was striving for the simplest translation of the condition to test. Because of his work and that of another mathematician, Emil Post, the concept exists of a truth table that demonstrates all possibilities of two variables being either true or false.

At times, you may write a complex, nested condition that can be simplified. Logicians, such as De Morgan, have developed equivalent logical expressions over the years. De Morgan's contributions are known as De Morgan's laws.

- **NOT** (A **AND** B) is equivalent to (**NOT** A) **OR** (**NOT** B)
- **NOT** (A **OR** B) is equivalent to (**NOT** A) **AND** (**NOT** B)

How can these laws help to simplify logical expressions? Take a look at an example. Suppose A represents the boolean variable haveCake. Then let B represent the boolean variable eatCake. The adage, "You can't have your cake and eat it too," can be expressed as **NOT** (A **AND** B). In Java, this is:

```java
! (haveCake && eatCake)
```

De Morgan's laws say an equivalent expression is (**NOT** A) **OR** (**NOT** B). Using the cake example, this says "You can't have your cake, or you can't eat your cake":

```java
( ! haveCake ) || ( ! eatCake )
```

Public Domain

Figure 8-9. British mathematician Augustus De Morgan developed a set of rules to help write equivalent statements.

Consider another example. You are rolling dice in a game called *Eighty-Four*. The object is to keep rolling until you roll an 8 or a 4 or reach a score of 10. You score a point every time you do not roll an 8 or a 4. When you roll an 8 or a 4, you stop rolling. If you keep rolling long enough to get to 10, you win. The conditional test for earning a point is: roll is **NOT** 8 **AND** roll is **NOT** 4. This condition in Java is:

```
!( roll == 8 ) && !( roll == 4 )
```

This is an example of the right-hand side of the second law: (**NOT** A) **AND** (**NOT** B). De Morgan says this is the same test as: **NOT** (roll is 8 **OR** roll is 4). The equivalent Java condition is:

```
!( ( roll == 8 ) || ( roll == 4 ) )
```

These laws will come in very handy in the next chapter, which covers repetition. In that chapter, you learn to write the Java statements to play this game.

Truth-values for a	Truth-values for NOT a
true	false
false	true

Goodheart-Willcox Publisher

Figure 8-10. This truth table shows the truth-values for all possible combinations of the single variable a (a and !a).

Truth Tables

A good way to visualize how Boolean values are related is by writing the relationships in a truth table. A ***truth table*** shows the truth-value of a compound selection statement for each component in the statement. For variable a, there are two truth-values: a could be true or a could be false. What is the result for **NOT** a? **Figure 8-10** is a truth table showing the truth-values for a and !a. Read across each line. When a is true, read to the right to find that !a is false. When a is false, !a is true.

Suppose there are two boolean variables, a and b. The truth-values for each could be true or false. That makes a combination of four cases when using the && and || operators. **Figure 8-11** is a truth table for the four cases for two variables. Use the two left-hand columns to locate the states of a and b. Then, read across to find the truth-value for each operator.

Truth-value for a	Truth-value for b	a && b	a || b
true	true	true	true
true	false	false	true
false	true	false	true
false	false	false	false

Goodheart-Willcox Publisher

Figure 8-11. This truth table shows the truth-values for all possible combinations of the two variables a and b.

Language Arts and Java

Applying AND and OR

There are many uses for the logical **AND** and **OR** in computing and communication. De Morgan's laws are likely the most famous. In the study of philosophy, there are laws about existence that employ && and ||. For example, nothing can exist and not exist at the same time. This is called the law of noncontradiction. The statement A && !A is always false. The law of the excluded middle states that either something exists or it does not exist. The statement A || !A is always true. The famous Hamlet soliloquy applies the excluded middle where it begins "To be, or not to be…"

Another example of the use of && is the law of double negation that states A && !!A is always true. An example of the double negative in improper English is, "I am not doing nothing." While some people who use such an expression think of it as having double emphasis, it is actually a contradiction of what they are trying to say, which is "I am not doing anything."

fizkes/Shutterstock.com

Another example of using && and || is the law of absorption: A && (A || B) = A || (A && B) = A. In this law, you can see that it does not matter

what the truth-value of B is. B is absorbed into the expression and only A remains.

While imprecision in the spoken word is often forgiven and reinterpreted by the listener to what the speaker intended to convey, precision in computing is paramount. Practice in precision will help your coding ability.

Assignment

Write meaningful sentences that demonstrate each of the logical expressions. They are summarized below for your convenience. Deliver the sentences to your teacher for review.

De Morgan's law #1: NOT (A **AND** B) is equivalent to (**NOT** A) **OR** (**NOT** B)
- Write a sentence using **NOT** (A **AND** B)
- Write a sentence using (**NOT** A) **OR** (**NOT** B)

De Morgan's law #2: NOT (A **OR** B) is equivalent to (**NOT** A) **AND** (**NOT** B)
- Write a sentence using **NOT** (A **OR** B)
- Write a sentence using (**NOT** A) **AND** (**NOT** B)

Noncontradiction: A **AND NOT** A is always false
- Write a sentence using A **AND NOT** A

Excluded Middle: A **OR NOT** A is always true
- Write a sentence using A **OR NOT** A

Double Negation: A **AND NOT NOT** A is always true
- Write a sentence using A **AND NOT NOT** A

Absorption #1: A **AND** (A **OR** B) is equivalent to A
- Write a sentence using A **AND** (A **OR** B)

Absorption #2: A **OR** (A **AND** B) is equivalent to A
- Write a sentence using A **OR** (A **AND** B)

HANDS-ON EXAMPLE 8.3.2

Writing Truth Tables in Java ↗

With some basic character graphics, you can make truth tables for De Morgan's laws using Java statements. Before beginning this exercise, download the chapter files from the student companion website. The DrawTruthTableSnippet.java file will be used as a starting point.

1. Launch jGRASP, and open the DrawTruthTableSnippet.java file.
2. Compile and run the program. The output in the **Run I/O** tab is a truth table, as shown.

```
Truth Table for De Morgan's Law #1:
    ----------------------------------------------------
    |  A    |   B    | NOT(A AND B) | (NOT A) OR (NOT B) |
    ----------------------------------------------------
    | true  |  true  |    false     |       false        |
    ----------------------------------------------------
```

Goodheart-Willcox Publisher

3. Examine the statements provided in the snippet. The horizontal dashed lines are produced by printing out the String horizontalLine in line 17. The titles of the truth table columns are from line 18, and the String horizontalLine is printed again by line 19. Lines 22 and 23 define two boolean variables, A and B, and they are given true values. The statement on lines 24 and 25 is the output that uses De Morgan's first law with A and B. The tabs (\t) are used for uniform spacing because true and false have a different number of letters. The vertical piping symbols help to make the boxes for the truth table. This statement will not change, only the values for A and B will change.
4. Locate comment 1, and follow the instructions to change the value of B to false.

   ```
   B = false;
   ```

5. Copy lines 24 through 27, and paste them after the statement B = false.
6. Compile and run the program. Examine the output. Notice the printout statements were not changed. Java used the new value for B and output the new result.

```
Truth Table for De Morgan's Law #1:
    ----------------------------------------------------
    |  A    |   B    | NOT(A AND B) | (NOT A) OR (NOT B) |
    ----------------------------------------------------
    | true  |  true  |    false     |       false        |
    ----------------------------------------------------
    | true  | false  |    true      |       true         |
    ----------------------------------------------------
```

Goodheart-Willcox Publisher

7. Locate comment 2, and follow the instructions to add code to change the value of A to false and B to true. Copy the output statements, and paste them below this code. Compile and run the program. Examine the output. You should have a third line in the truth table: false, true, true, true.
8. Locate comment 3, and follow the instructions to add code to produce the output for A is false and B is false.
9. Compile and run the program. Examine the output. Note that the truth-values in the two right-hand columns are always equal. That is De Morgan's first law: the two conditions are equivalent.
10. Locate comment 4, and follow the instructions to create a truth table for De Morgan's second law. Verify that the two right-hand columns are equivalent. You have made truth tables for De Morgan's laws!

Try It!

Continue working with the DrawTruthTableSnippet.java file. At the bottom, add statements to produce the truth table for double negation. Show that for A is true or A is false, **NOT** (**NOT** A) is the same as A. This truth table should have three rows: title, A is true, and A is false. Adjust the String horizontalLine to accommodate just one variable and one expression. Compile and run the program. Create a screenshot of the output or copy and paste it into a word-processing document. Submit this along with the program for grading.

Coding Conundrum

The programmer has coded the following to compare the value of two variables and determine if they are equal. The test values of 5 and −10 are clearly unequal. However, the program does not compile. It is a conundrum! Find the two errors in this code.

```
11    int a = 5;
12    int b = -10;
13    if ( a = b )  {
14        System.out.println( "a equals b" );
15    else {
16        System.out.println( "a does not equal b" );
17    } // end if/else
```

SECTION REVIEW 8.3

Check Your Understanding

1. What is it called when one selection statement is placed within another selection statement?
2. The relationship **NOT** (A **AND** B) = (**NOT** A) **OR** (**NOT** B) is one of two laws attributed to whom?
3. What does ! (A || B) equate to?
4. What does a truth table demonstrate?
5. How many truth conditions are there for two boolean variables?

Build Your Vocabulary

As you progress through this course, develop a personal computer science glossary. This will help you build your vocabulary and prepare you for a career. Write a definition for each of the following terms and add it to your computer science glossary.

nested selection   truth table

Comparing Objects

A comparison of objects can be used in the condition of the selection statement. Objects are different from primitive data types. For example, an int has exactly one value. The identifier of the int points to the memory location that contains that one number. However, objects may have many variables within them. All that will not fit into one memory location. When you create a new object from a class, the identifier of the object (object reference) is used to refer to the location of the actual data in the new object. So, the object reference contains a memory location, not data.

How are branching decisions applied in video games?

Learning Goals

- Explain the need for a dedicated method for String comparisons.
- Apply the equals( ) method.
- Compare and contrast the equals( ) method and the equalsIgnoreCase( ) method for Strings.
- Program a game using selection.

Terms

equals method   equalsIgnoreCase method

Overview of Comparisons

The relational operators (>, >=, <=, <) operate only on primitive numeric operands. The logical operators (!, &&, ||) operate only on Boolean operands. The equality operators (==, !=) operate on any primitive numeric or reference type. Remember that object references point to the location of the object. When object references are compared using the equality operators, the location of the objects are compared. To compare the data in any two objects, it is necessary to use the equals method. Thus, each class must contain an equals method if their object data are to be compared. In fact, each class does have an equals method inherited from the Object class.

In this code:

```
String student1 = "Jerry";
String student2 = "Maria";
```

The reference student1 points to where the letters Jerry are stored in memory. For example, student1 might contain the memory address: 1ABCD458. The reference student2 points to where the letters Maria are stored in memory. For example, student2 might contain the memory address: 1F25CD64. If these two references are compared using the equality operator:

```
if ( student1 == student2 ) // compares 1ABCD458 and 1F25CD64
```

then the locations of each String in memory are compared, as shown in **Figure 8-12**. To compare the letters in the String, the equals method must be used. The equals method compares the data in two objects. For Strings, the equals method compares whether two Strings contain exactly the same characters. While the equals method works on any objects, this section will discuss the application to String objects.

Equals Method

Strings are a special case because they are objects of a class. Using the equality operator on two Strings merely compares the location of the Strings in memory. It is not possible, therefore, to compare the characters within the Strings with an equality operator (==). The *equals method* is a method of the String class provided to perform an equivalent function to determine if the characters of two Strings are the same. Because it is a method, equals must be used with an object of the String class. The API for the equals method is shown in **Figure 8-13**.

In this code, two Strings are instantiated and then compared using the equals method:

```
String figure1 = new String( "rectangle" );
String figure2 = new String( "circle" );
if ( figure1.equals( figure2 ) ) {
  // execute selected statements
}
```

The condition evaluates to false, and the selected statements are skipped.

Using the equality operator in the condition with Strings leads to an erroneous result. With these Strings:

```
String figure1 = new String( "rectangle" );
String figure2 = new String( "rectangle" );
```

using the equality operator:

```
System.out.println( figure1 == figure2 ); // outputs false
```

produces an output of false because the memory locations are different. However, using the equals method:

```
System.out.println( figure1.equals( figure2 ) ); // outputs true
```

produces an output of true because the characters are being compared and they are exactly the same.

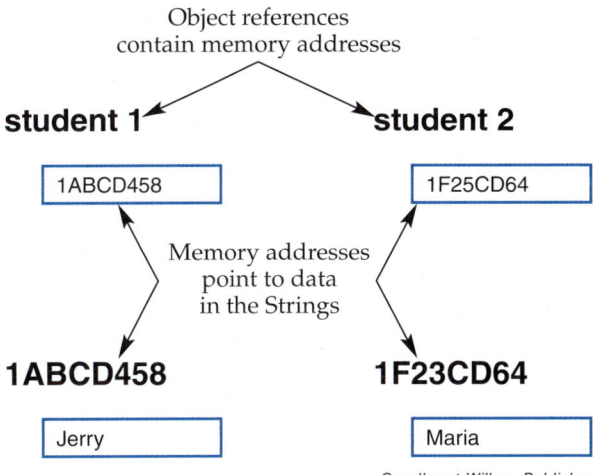

Figure 8-12. Comparing object references compares the locations of Strings, not of the characters.

EqualsIgnoreCase Method

Getting user input for Strings is problematic. The user may not enter the character case for an answer that the program is expecting. For example, Yes and YES are different Strings. Comparing these Strings using the equals method would return false because the characters are different. Fortunately, the String class provides a method to compare Strings while ignoring the case.

Method	Action
boolean equals(String anotherString)	Compare the first String to the second String, considering case.
boolean equalsIgnoreCase(String anotherString)	Compare the first String to the second String, ignoring case considerations.

Figure 8-13. The API for the equals and equalsIgnoreCase methods.

The *equalsIgnoreCase method* compares the characters in a String considering uppercase and lowercase letters to be equal. The API for the equalsIgnoreCase method is given in **Figure 8-13.**

In this example, differences in case are handled by the equalsIgnoreCase( ) method:

```java
String figure1 = new String( "Circle" );
String figure2 = new String( "circle" );
System.out.println( figure1.equals( figure2 ) );          // false
System.out.println( figure1.equalsIgnoreCase( figure2 ) ); // true
```

Rock-Paper-Scissors

The classic game Rock-Paper-Scissors uses branching control statements, selections, and String comparisons. The rules are two people choose either rock, paper, or scissors. Rock beats scissors. Scissors beats paper. Paper beats rock. The same choice is a draw. For this example, one of the players will be the computer. The algorithm is:

1. Select rock, paper, or scissors at random for the computer's choice.
2. Prompt the user to enter rock, paper, or scissors as a choice.
3. If the input is not valid (ensure the entry is one of those three words), then say input is not valid.
4. Else check for a user win or loss.

Think about how to check for a user win or loss. At this point in the algorithm, the computer and the user both have an object selected.

5. Check if the objects are the same. If so, say it is a draw.
6. Suppose the player chooses rock. There are two cases left to consider: computer has paper (loss) or computer has scissors (win).
7. Suppose the player chooses paper. There are two cases left to consider: computer has rock (win) or computer has scissors (loss).
8. Suppose the player chooses scissors. There are two cases left to consider: computer has rock (loss) or computer has paper (win).

Once an outcome is determined, there is no need to keep checking. Use a combination of the if-then-else with nested if-then.

```
if both have the same, it is a draw
else if player has rock            // computer has paper or scissors
   if computer has paper, then player loses     // paper beats rock
      else player wins                    // rock beats scissors

   else if player has paper       // computer has rock or scissors
      if computer has rock, then player wins     // paper beats rock
      else player loses                  // scissors beats paper

else // player must have scissors, computer has rock or paper
   if computer has rock, then player loses // rocks beats scissors
      else player wins                 // scissors beats paper
```

HANDS-ON EXAMPLE 8.4.1

Playing Rock-Paper-Scissors ↗

Review the algorithm for Rock-Paper-Scissors. Practically every topic of this chapter is used in this game: conditions, switch, if-then, if-then-else, if-then-else-if, and De Morgan's laws. You will apply these concepts to code a Rock-Paper-Scissors game. Before beginning this exercise, download the chapter files from the student companion website. The RockPaperScissorsSnippet.java file will be used as a starting point.

1. Launch jGRASP, and open the RockPaperScissorsSnippet.java file. Some code has already been added for you. Add your name to the comment at the top, and save the file.

2. Locate comment 1. To select a choice for the computer, use a random number from 1 to 3 and a switch statement.

```
/***** 1. Select rock, paper, or scissors at random for computer choice */
String computer = "";
Random rand = new Random( );
int choice = rand.nextInt( 3 ) + 1;
switch ( choice ) {
    case 1: computer = "Rock";
        break;
    case 2: computer = "Paper";
        break;
    case 3: computer = "Scissors";
        break;
}
```

3. Locate comment 2. Apply what you have learned to prompt the user to enter a choice.

```
/***** 2. Prompt the player to enter rock, paper, or scissors */
Scanner input = new Scanner( System.in );
System.out.println( "Play Rock, Paper, Scissors" );
System.out.print( "Enter Rock, Paper, or Scissors to start. " );
String player = input.next( )
```

4. Locate comment 3. Add code to display the computer choice.

```
/***** 3. Display computer choice */
System.out.print( "Computer has " + computer + "." );
```

5. Locate comment 4. Add code to validate the input. First, there needs to be a test for a bad input. You need to check if the player did **NOT** enter rock **OR** paper **OR** scissors. Use De Morgan's laws to test the **NOT AND** combination. If it is bad input, then the game is over.

```
/***** 4. If the input is not valid (check that the entry is not good)
        then say the input is not good, GAME OVER */
if ( !player.equalsIgnoreCase( "Rock" ) &&
    !player.equalsIgnoreCase( "Paper" ) &&
    !player.equalsIgnoreCase( "Scissors" ) ) {

    System.out.println( player + " is not a valid choice. Game Over." );
}
```

6. Locate comment 5. Now that good input has been verified, check for a draw.

```
/***** 5. Else check for player wins or loses
        if both have the same, it is a draw */
else if ( player.equalsIgnoreCase( computer ) ) {

    System.out.println( "It's a draw!" );
}
```

7. Locate comment 6. Since it has been determined not to be a draw, if the player has rock, the computer can only have paper or scissors. The user's selection should be allowed to be case-insensitive. So, when checking

the user's selection, use the **equalsIgnoreCase** method. However, a specific **String** has been assigned to the computer's selection. Therefore, when checking the computer's selection, just use the **equals** method.

```
/***** 6. else if player has rock, check if computer has paper or scissors */
else if ( player.equalsIgnoreCase( "Rock" ) ) {

    // if computer has paper, then player loses - paper beats rock
    if ( computer.equals( "Paper" ) ) {

        System.out.println( "Paper covers Rock. You lose." );
    }
    // else computer has scissors - rock beats scissors
    else {

        System.out.println( "Rock breaks Scissors. You win!" );
    }
}
```

8. In the same pattern, add the code for checking if the player chooses paper; if the computer has rock it is a win, else if the computer has scissors it is a loss.

```
// else if  player has Paper, check computer has rock or scissors
else if ( player.equalsIgnoreCase( "Paper" ) ) {

    // if computer has rock, then player wins - paper beats rock
    if ( computer.equals( "Rock" ) ) {

        System.out.println( "Paper covers Rock. You win!" );
    }
    // else player loses - scissors beats paper
    else { // computer has scissors

        System.out.println( "Scissors cuts Paper. You lose!" );
    }
}
```

9. After checking for player having rock or paper, the only thing the player can have at comment 7 is scissors. Check for the computer having rock or paper.

```
// player has scissors, computer has rock or paper
// if computer has rock, then player loses - rock beats scissors
else if ( computer.equals( "Rock" ) ) {

    System.out.println( "Rock breaks Scissors. You lose!" );
}
else { // else computer has paper, scissors beats paper

    System.out.println( "Scissors cuts Paper. You win!" );
}
```

10. You can test the game by removing the random selection for the computer. Hard code the computer's selection to rock, then run the program three times with the player choosing rock, then paper, then scissors. Continue to test by hard coding the computer's selection to scissors and then paper. Run the program with the three possible user selections for each computer selection. Make sure the results are correct. If not, then review the statements to be certain the code is correct.

11. Restore the random number generator, and test all paths. Make a chart for the possible outcomes. Check them off as they occur.

Try It! 🔗

Before beginning this exercise, download the chapter files from the student companion website. The GuessMyNumberSnippet.java file will be used as a starting point. On paper, write an algorithm to play one round of Guess My Number. Assign a random number from 1 to 5, and ask the player to guess it. If the player guesses the number, the player wins. If the player's guess is less than the number, say "Too Low!" Otherwise, say "Too High!" Test the algorithm with a friend. Then, when you are sure the algorithm works, write the code for the algorithm in the GuessMyNumberSnippet.java file. Use the algorithm steps as comments in the file. Submit the file for grading.

SECTION REVIEW 8.4

Check Your Understanding

1. The logical operators (!, &&, ||) operate only on what data types?
2. When an object is created, the identifier (object reference) holds the _____ of the object's data.
3. Why is the expression (string1 == string2) invalid to check if two Strings are equal?
4. When using the equals method to compare the Strings Yes and YES, what is the result?
5. Which String method compares strings without regard to uppercase or lowercase characters?

Build Your Vocabulary

As you progress through this course, develop a personal computer science glossary. This will help you build your vocabulary and prepare you for a career. Write a definition for each of the following terms and add it to your computer science glossary.

equals method   equalsIgnoreCase method

Cooperative Coding

Divisibility Rules

A popular activity in number theory is divisibility. A number is divisible by another if the remainder is zero. Finding all of the divisors of a number is a recreational activity people engaged in before there were computers. Today, you could use a computer to perform the division or use the mod operator in Java to determine divisibility. However, to make it more game-like, people wrote rules for divisibility so that you could determine divisibility without actually dividing. You already know several of these rules. Examples are:

goodluz/Shutterstock.com

- A number is divisible by 5 if the last digit is 0 or 5.
- A number is divisible by 2 if it is even.
- A number is divisible by 100 if the last two digits are zero.

The following table shows many more rules.

Divisor	Rule
1	Every number is divisible by 1.
2	A number is divisible by 2 if the last digit is 0, 2, 4, 6, or 8.
3	A number is divisible by 3 if the sum of the digits is divisible by 3.
4	A number is divisible by 4 if the last two digits are divisible by 4.
5	A number is divisible by 5 if the last digit is 0 or 5.
6	A number is divisible by 6 if it is divisible by 2 and 3.
8	A number is divisible by 8 if the last 3 digits are divisible by 8.
9	A number is divisible by 9 if the sum of the digits is divisible by 9.
10	A number is divisible by 10 if the last digit is 0.
12	A number is divisible by 12 if it is divisible by 3 and 4.

Assignment

Working with your team, write a program using selection statements to determine divisibility of a four-digit number the user provides one digit at a time. You will echo the result to the user. Each member of the team will write the selection statement for one or more of the divisors. Set a boolean to true if the user's number is divisible by your divisor or false if it is not. Other team members may rely on your result; see the rule for a divisor of 6. In the rule for 8, you need to build a number out of the last three digits. Put all the code together in one program named Divisibility.java. Use the Boolean results to write out one statement that lists all the divisors of the number. Test for the divisibility rules in the table.

1. How did you build a number out of the last three digits for rule 8?
2. How effective was dividing the work among team members? Could coding efficiency be improved by dividing the work in a different way?

Chapter Summary

Section 8.1: Conditions

- A selection statement is a type of control flow statement used to allow the program to choose a set of statements to execute based on a certain condition that evaluates to either true or false using equality, relational, and logical operators.
- All operators, including equality, relational, and logical, must be applied following an order of precedence, which is similar to the order of operations in arithmetic.

Section 8.2: Selection Statements

- The basic selection statement in Java is an if-then statement, but there are variations, including if-then-else and if-then-else-if.
- The switch statement tests the value of a variable and selects statements based on any number of choices, and the break statement is used to interrupt the processing of cases in a switch statement.
- The scope of a variable is the block of code where it is defined, which may be as broad as at the method level or as limited as at the selection-statement level.
- There are several common errors made by programmers when coding conditional statements, including using a semicolon after the condition in an if-then statement, not maintaining proper operator precedence, and improper handling of boundary values for a range.

Section 8.3: Helpful Conditions

- A nested selection statement is a selection statement placed inside another selection statement.
- British mathematician Augustus De Morgan developed a set of rules to help write expressions that are equivalent, which became known as De Morgan's laws; truth tables help simplify writing compound conditions.
- A truth table shows the truth-value of a compound selection statement for each component in a statement demonstrating all possibilities of the two variables being either true or false.

Section 8.4: Comparing Objects

- The equality operators cannot be used to compare the characters in Strings because they will compare the memory addresses of where the Strings are stored, not the characters in the Strings.
- The equals method is a method of the String class provided to perform an equivalent function to determine if the characters of two Strings are the same, and it must be used with an object of the String class.
- The equalsIgnoreCase method compares the characters in two Strings considering uppercase and lowercase letters to be equal.
- The classic game Rock-Paper-Scissors uses branching control statements, selections, and String comparisons.

Chapter 8 Test

Multiple Choice

Select the best response.

1. Which of the following are logical operators?
   A. **AND**, **NOT**, **OR**
   B. greater than, less than
   C. greater than or equal to, less than or equal to
   D. switch statements

2. In the program below, what term is missing?

```
if ( x == y) {
  System.out.println( "x and y are equal."
    );

} else if ( x > y ) {

  System.out.println( "x is greater"
  + "than y." );
}

{
  System.out.println( "x is less than y."
      );
} // end if
```

   A. else
   B. else if
   C. elseif
   D. No term is missing.

3. Which of the following is used to test the value of a variable and select a statement based on cases?
   A. break
   B. switch
   C. **OR**
   D. equals

4. Which of these conditions will evaluate to true if string1 is "Circle" and string2 is "circle"?
   A. if (string1.equalsIgnoreCase(string2))
   B. if (string1.equals(string2))
   C. if (string1 == string2))
   D. if (string1 > string2))

5. Which of these expressions is the valid way to compare the characters in string1 and string2?

A. (string1 == string2)

B. (string1 != string2)

C. (string1 => string2)

D. (string1.equals(string2))

Completion

Complete the following sentences with the correct word(s).

6. For one boolean variable, there are _____ truth-values.

7. A(n) _____ statement can be used instead of some if-then-else-if statements.

8. The _____ of a variable is the block of code in which it is defined.

9. _____ laws can be applied to simplify logical expressions.

10. The _____ method is a method of the String class that can compare characters of two Strings without regard to case.

Completion

Match the correct term with its definition.

A. scope

B. condition

C. equals method

D. if-then

E. if-then-else

11. Compares the characters in two strings.

12. Statement that offers one choice depending on the condition.

13. Statement that offers several choices depending on the condition.

14. If a variable is defined at the beginning of the main method, then it is visible for the entire execution of the main method.

15. Determines how a program branches.

Application and Extension of Knowledge

1. In mathematics, the Law of Trichotomy states there are only three choices for the relationship of any two numbers: the numbers are equal, the first is greater than the second, or the first is less than the second. Write the algorithm for this law. Then, write a program that prompts the user for two numbers, compares the numbers, and then tells the user (prints) which of the three relationships is true.

2. The manager of Jackson Supply wants a program that displays the domestic shipping price of an order. If the order goes to Alaska, the shipping is $12; to Hawaii, the shipping is $23; and to the rest of the United States is $9. Develop the algorithm using branching control statements. Write this program to ask the user for the destination state, and then display the destination and cost of shipping.

3. The periodic table of elements displays the abbreviation, full name, and atomic number of the element. Look in a chemistry textbook or check the Internet for a copy. Create a program that requires the user to enter the abbreviation of one of the Noble Gases and then displays the full name and atomic number for the element. Develop the algorithm first using branching control statements, and then write the program.

4. Create a program that provides feedback to the user based on the letter grade he or she enters. For example, if the user enters "A," the feedback may be "Excellent!" Develop the algorithm to include feedback for the grades A through F. Then, write the program. Ask the user for the letter grade, and then print an assessment for the grade. Be sure to account for an invalid entry. Use a switch statement to select which case is a match.

5. The table below determines a letter grade for a test score. Analyze this data. Determine the best selection statement to use to apply the rules in the table. Then, develop an algorithm using branching control statements to assign a letter grade based on the score the user enters. Write the program using the selection statement you selected.

Letter Grade	Score
A	90 or above
B	79.9 or above
C	69.9 or above
D	59.9 or above
F	Below 59.9

Online Activities

Complete the following activities, which will help you learn, practice, and expand your knowledge and skills.

Vocabulary. Practice vocabulary for this chapter using the e-flash cards, matching activity, and vocabulary game until you are able to recognize their meanings.

Communication Skills

Reading. *Decoding* is applying knowledge of the relationships between letters and sounds, including letter patterns, to correctly pronounce written words. This is achieved by recognizing individual letters and combinations of letters and matching them to their respective sounds. When you *sound out* a word, you are decoding it. Look at the list of key terms, and select one that is unfamiliar to you. Look at the letters that create the word and think about the sounds each one makes. Decode the term by sounding it out.

Writing. *Tone* is the way a writer expresses an attitude toward a topic in his or her writing. Just as you can tell a person's tone of voice when you listen, you can often tell a writer's tone when reading his or her writing. The tone might be happy, solemn, or humorous. Write several paragraphs that describe your usual tone when completing a typical writing assignment that you submit for a grade.

Speaking. Select three of your classmates to join a formal discussion panel. Assign each person a specific role, such as leader, timekeeper, or recorder. Hold a panel discussion on the pros and cons of teens being allowed to drive at age 16. The leader should keep the panel on task and encourage fair discussion. The recorder should make notes of the information discussed. Afterward, hold an informal discussion to decide the most important information to include in the notes. Create a final document of the discussion notes to distribute to the class.

Listening. A *barrier* is anything preventing clear, effective communication. *Internal barriers* are within an individual, such as fatigue, hunger, or wandering thoughts. During class, attempt to recognize any internal barriers to listening. How can you fight internal barriers?

Portfolio Development

College and Career Readiness

Certificates. Exhibiting certificates you have received in your portfolio reflects your accomplishments. For example, a certificate might show you have completed training in a particular programming language. Another one might show you are certified in a technical area.

Include any certificates that show tasks completed or your skills or talents. Remember that this is an ongoing project. Plan to update when you have new certificates to add.

1. Scan the certificates that will be in your portfolio.

2. Give each document an appropriate name and save in a folder or subfolder.

3. Place the hard-copy certificates in a container for future reference.

4. Record these documents on your master spreadsheet that you started earlier to record hard-copy items. You may list each document alphabetically, by category, date, or other convention that helps you keep track of each document you are including.

CTSOs

Extemporaneous speaking. Extemporaneous speaking is an event that demonstrates the ability to speak effectively on a topic without advance preparation. This event allows you to display your communication skills, specifically your ability to organize and deliver an oral presentation.

At the competition, several topics will be presented from which to choose. Time limits will be in place for creating the speech and for delivering it. The evaluation will be based on your content, organization, coherence, and structure of the speech. In addition, verbal and nonverbal skills, as well as the tone and projection of voice, will be evaluated. To prepare for an extemporaneous speaking event, complete the following activities.

1. Ask your instructor for several topics so you can practice making impromptu speeches.

2. After you have a practice topic, jot down the ideas and points to cover. An important part of making this type of presentation is that you will have only a few minutes to prepare. Being able to write down your main ideas quickly will enable you to focus on what you will actually say in the presentation.

3. Practice the presentation. You should introduce yourself, review the topic that you will present, defend the topic, and conclude with a summary.

4. Ask your instructor to play the role of competition judge as you deliver the presentation. Afterward, ask for feedback from your instructor. You may also consider having a student audience listen and provide feedback.

5. For the event, bring paper and pencils to record notes. Supplies may or may not be provided.

Repetition

Sections

9.1 Java Loops

9.2 Applying Loops

Repetition is something you do in many ways every day. For example, walking is a repetition of steps. One way scientists define walking is the controlled state of falling forward. To start walking, you lean forward, then swing a foot out to stop from falling. Then you swing the other foot out. To keep walking, you repeat the same two-footed movements. You continue this repetition for a certain number of steps while you are exercising or until you arrive at your destination. A daily routine is also a repetition of tasks. Repetition is at the very core of human nature.

All procedures are coded using one or more of three basic control structures. The three basic structures are sequence (executing commands in the order they are written), selection (making decisions), and finally repetition. In the last chapter, you learned about selection as a means of controlling programs. Computers are very good at repetition, which is another way of directing code. Not only can computers repeat steps, they also can perform this repetition very rapidly. Computers are faster than humans, and they can repeat each task exactly the same way. Also, computers never get tired of repeating.

Reading Prep

College and Career Readiness

Before reading this chapter, preview the chapter content by examining its headings, key terms, and illustrations. Write three to four sentences on what you think this chapter will cover.

While studying, look for the activity icon ⤴ for:

• Vocabulary terms with e-flash cards and matching activities.
• Starter files for hands-on examples and other exercises.

These activities can be accessed at
www.g-wlearning.com/informationtechnology/1773

Chapter Glossary

condition-controlled loop: Loop that repeats until a boolean expression (condition) tests false; while loop.

counter-controlled loop: Loop regulated by a counter; for loop.

do/while loop: Loop that iterates a set of statements while a boolean expression is true.

file pointer: Mechanism that keeps track of the next data to read in the external file.

for loop: Loop that iterates a block of code a predetermined number of times.

infinite loop: Loop that executes forever (never ends); results from poor programming.

initialization: Code that gives the counter in the loop its beginning value.

iteration: Each repetition of the statements in a loop.

loop: Block of code, or statements, that is repeated as many times as required.

loop body: Consists of zero, one, or more valid Java statements that are repeated within a loop.

loop counter: Variable defined in the initialization; used to keep track of the number of iterations of the loop.

looping condition: Requirement for repeating statements in a loop.

nested loop: One loop placed inside another loop.

opening a file: Coding the File and Scanner constructors to read an external file.

paper check: Manually writing out a description of the sequence of instructions to be executed and then analyzing it for proper logic.

while loop: Loop that iterates a set of statements while a boolean expression is true; statements may not be executed at all if the tested condition is false when the loop begins.

Java Loops

Most of the applications you have seen to this point have been written without any parts repeated. A key exception is using the Turtle class to draw a square. That application made use of duplicated blocks of code. The application stepped through the block of code sequentially, performing the same steps four times to draw a square and two times to draw a rectangle. Writing the code once and asking the computer to repeat those statements allows the program to be more efficient and results in a shorter, more elegant solution.

How does using loops allow a coder to be more efficient?

Learning Goals

- Describe the function of loops in a computer program.
- Apply counter-controlled loops in Java.
- Design condition-controlled loops in Java.
- Differentiate while and do/while loops in Java.

Terms

condition-controlled loop	loop
counter-controlled loop	loop body
do/while loop	loop counter
for loop	looping condition
infinite loop	nested loop
initialization	while loop
iteration	

Overview of Loops

A *loop* is a block of code, or statements, that is repeated as many times as required. These statements are called loops because some statements are executed and then the program circles back and executes the statements again, making a loop. Each repetition of the statements in a loop is called an *iteration.* The *looping condition* is the requirement for repeating the statements in a loop. Loops can improve the efficiency of the code.

In Java it is possible to write a loop inside another loop. This is similar to placing a selection statement within another selection statement, which you learned about in Chapter 8. A loop placed inside of another loop is called a *nested loop.*

In Chapter 6, you learned how to use the custom Turtle class to draw shapes. One of the example applications draws a square. For this application, the same two statements were repeated four times. **Figure 9-1** shows a snippet from the Square.java drawing application and the square drawn by the snippet. Notice which statements are repeated.

A regular hexagon has six equal sides and six equal angles. To draw this, the turtle would move forward six times and turn right six times. Apply what you have learned in math class about exterior angles to calculate the number of

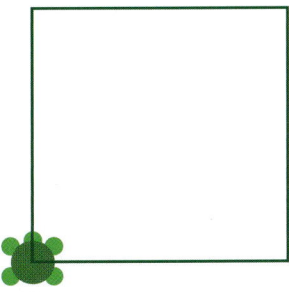

```
Turtle turtle = new Turtle( root, 250, 250 );
// drawing a square, pre-loops
turtle.forward( 100 );
turtle.turnRight( );

turtle.forward( 100 );
turtle.turnRight( );

turtle.forward( 100 );
turtle.turnRight( );

turtle.forward( 100 );
turtle.turnRight( );
```

Figure 9-1. This square is drawn with four identical statements.

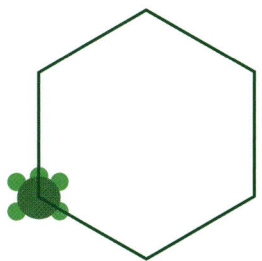

```
// drawing a hexagon, pre-loops
turtle.forward( 75 ); // side 1
turtle.turnRight( 60 );

turtle.forward( 75 ); // side 2
turtle.turnRight( 60 );

turtle.forward( 75 ); // side 3
turtle.turnRight( 60 );

turtle.forward( 75 ); // side 4
turtle.turnRight( 60 );

turtle.forward( 75 ); // side 5
turtle.turnRight( 60 );

turtle.forward( 75 ); // side 6
turtle.turnRight( 60 );
```

Figure 9-2. This hexagon is drawn in the same way the square shown in Figure 9-1 was drawn. Notice there are six sets of identical statements.

degrees the turtle needs to turn for a hexagon. Divide the total degrees in the exterior angles of a polygon by the number of sides:

$$360° \div 6 = 60°$$

Figure 9-2 shows a code snippet and output from the Hexagon.java drawing application file. This application is available for download on the student companion website. Notice how many lines of code are needed compared to drawing a rectangle. Now, imagine drawing a 20- or 40-sided figure. The code would be very long, very long indeed. Even though copying and pasting code could help you create the program, if the code ever needed to be modified, it would take a while.

Fortunately, the repeating of statements can be eliminated. You can tell Java to do the repeating. There are several Java statements you can use to repeat other statements. Each has its own way to do the repeating.

- for loop
- while loop
- do/while loop

The next few sections show how to use each type of loop to draw a square. Keep a look out for which of the loops is the most efficient to use for this application.

For Loop

The *for loop* iterates a block of code a predetermined number of times. Use this loop when you know the exact number of times the statements need to be repeated. The general form of the for statement can be expressed as:

```
for ( initialization; test; loop update ) {   // loop header
    statement(s)                               // loop body
}                                              // loop end
```

FYI

Although advanced coders sometimes write for loops with empty bodies, empty bodies are not useful for the applications in this textbook. They also can cause compiler errors or unexpected results when the for loop runs.

FYI

In the early days of computing, the programming language FORTRAN set aside the letters *I* through *N* for the first letter of an integer variable name. This automatically reserved memory for an integer variable. *I* and *N* are the first two letters in the word *integer*.

Figure 9-3 explains the parts of a for statement. Take note of the punctuation. Proper punctuation is critical. Semicolons follow the initialization and test. Note that there is no semicolon at the end of the for loop header. Placing a semicolon here would indicate the for loop body is empty. **Figure 9-4** shows an example of a for loop.

The *initialization* is the code that gives the counter in the loop its beginning value. The initialization statement is executed only once when the for loop begins. Most often, the initialization is a variable declaration and assignment statement. If the variable is an integer, *i* is classically used as its name because it is the first letter of *integer*. For example, int i = 1 may be used. However, any legal variable name can be used. Sometimes the variable defined in the initialization and used to keep track of the number of iterations of the loop is called the *loop counter*. A for loop is a *counter-controlled loop* because it is regulated by a counter.

The scope of a variable defined within the initialization is the loop only. So, the loop counter cannot be used after the loop. This is useful because if there is more than one for loop in the program, the name of the loop counter can be reused in each loop.

After the initialization statement is executed, the test expression is evaluated. If the value of the test expression is true, the loop body is executed. If the value of the test expression is false, the loop stops. Any test expression that never becomes false will cause the for loop to execute forever. Coders call a loop that executes forever an *infinite loop.*

The *loop body* consists of zero, one, or more valid Java statements that are repeated. The statements in the loop body are indented for easy reading. Java style dictates that the opening brace is written on the same line as the for statement declaration. The ending brace is written on its own line directly under the *f* in *for.*

If the for loop body has more than one statement, braces enclosing all statements are required. If the loop body has only one statement, braces are optional. However, braces should be included for clarity. It is Java style to always use these braces even when they are not needed. If you omit the braces at first and then later add a statement to the loop body, it is easy to forget to add the newly required braces at that time.

After the loop body is executed, the loop update is executed. The loop update determines how the loop counter is changed. This is how the iterations of the loop are advanced.

Statement Part	Location	Explanation
for	Loop header	Statement name.
( )	Loop header	Parentheses are used to enclose the loop control statements.
initialization	Loop header	Code that should be executed once as the for loop begins. Notice that the initialization ends with a semicolon (;).
test	Loop header	Boolean expression that determines whether the loop continues. Notice that the test expression ends with a semicolon (;).
loop update	Loop header	Determines how the loop counter is changed. Notice there is no semicolon at the end of the loop update.
{}	Loop body	Braces enclose all statements that are to be repeated.
statements(s)	Loop body	Valid Java statements that are repeated.

Goodheart-Willcox Publisher

Figure 9-3. These are the parts of a for statement and where each is located.

```
for ( int i = 1; i <= 5; i++ ) {
   System.out.println( i ); // statement to repeat
} // end for
```

Goodheart-Willcox Publisher

Figure 9-4. This is an example of a for loop.

The action in a for statement is:

1. Initialize the counter.
2. Evaluate the test expression.
3. If the test expression is false, exit the loop.
4. If the test expression is true, execute the statements in the loop body.
5. Perform the loop update.
6. Go to step 2.

The flowchart in **Figure 9-5** shows a for loop with the sequence and path of the control.

Writing the application to draw a square with a for loop is shown below. The loop is in lines 39–43. The output from this application is the same as shown in **Figure 9-1.**

```
34 public void buildScript( ) {
35    // student adds Turtle code here
36    Turtle turtle = new Turtle( root, 250, 250 );
37
38    // drawing a square using a for loop
39    for ( int i = 1; i <= 4; i++ ) {
40
41       turtle.forward( 100 );
42       turtle.turnRight( );
43    } // end for
44
45 } // end buildScript
```

Goodheart-Willcox Publisher

Figure 9-5. This flowchart shows the sequence and control path that exists in a for statement.

Coders can be very creative using for loops. In a for loop, the initial value can be any data type. Also, the loop update can be any expression that changes the value of the loop counter. To count by fives, increment by five. To count down, use a subtraction and change the test. For example, the for loop in **Figure 9-6** starts at 14 and counts down to zero by two. Think about the number of times this loop iterates. The statements inside this loop will be executed eight times. In the example shown in **Figure 9-7,** the loop generates the positive odd integers less than ten, each on a separate line.

```
for ( int counter = 14; counter >= 0; counter -=2 ) {
   // statements to repeat
} // end for
```

Goodheart-Willcox Publisher

Figure 9-6. This for loop has an initial value for counter of 14 and updates by subtracting two from counter until the value of counter is less than zero.

```
// This for loop will output the positive odd integers less than 10
for ( int odd = 1; odd < 10; odd += 2 ) {
   System.out.println( odd );
} // end for
```

Goodheart-Willcox Publisher

Figure 9-7. This for loop outputs all positive odd integers less than 10.

There is rarely only one way to solve a problem. Different algorithms may produce the same result. Examine the code snippets shown in **Figure 9-8.** Verify the output for each is identical to what is generated by the snippet in **Figure 9-7.**

Algorithm change: declare a separate integer variable for the odd numbers

```java
int odd = 1;

for ( int i = 1; i <= 5; i++ ) { // coder knows there are 5 odds
    System.out.println ( odd );
    odd += 2;
} // end for
```

Algorithm change: check for odd numbers in the loop body

```java
for ( int number = 1; number < 10; number++ ) {
    if ( number % 2 == 1 ) { // remainder 1 is odd number
        System.out.println( number );
    } // end if
} // end for
```

Goodheart-Willcox Publisher

Figure 9-8. These code snippets will produce the same output as the snippet shown in Figure 9-7. Notice how the algorithm is different for each.

HANDS-ON EXAMPLE 9.1.1

Evaluating for Loops ⤴

Before beginning this exercise, download the chapter files from the student companion website. Launch jGRASP, and open the StarterSnippet.java file. Rename the class to ForLoopExercises, and save the file under the new name. Evaluate each of the following for loop examples. Determine the number of times each for loop executes in the following examples. Once you have predicted the number of iterations for each, enter the first snippet into the ForLoopExercises program and run the program. Replace the first snippet with the second snippet, and so on. If you do something that results in an infinite loop, click the **End** button on the **Run I/O** tab in jGRASP to terminate the program.

1. Predict how many iterations this for loop will perform. Then, run the code to test your prediction.

```java
for ( int i = 1; i <= 12; i++ ) {
    System.out.print ( i  + " " );
}
System.out.println( );
```

2. Predict how many iterations this for loop will perform. Then, run the code to test your prediction.

```java
for ( int i = 1; i < 12; i++ ) {
    System.out.print ( i  + " " );
}
System.out.println( );
```

3. Predict how many iterations this for loop will perform. Then, run the code to test your prediction.

```java
for ( int i = 1; i <= 12; i += 2 ) {
    System.out.print ( i  + " " );
}
System.out.println( );
```

4. Predict how many iterations this for loop will perform. Then, run the code to test your prediction.

```java
for ( int i = 10; i > 0; i-- ) {
    System.out.print ( i + " "   );
}
System.out.println( );
```

5. Predict how many iterations this for loop will perform. Then, run the code to test your prediction.

```java
for ( int i = 10; i >= 2; i-- ) {
    System.out.print( i   + " " );
}
System.out.println( );
```

6. Predict how many iterations this for loop will perform. Then, run the code to test your prediction.

```java
for ( int i = 1; i <= 12; i += 3 ) {
    System.out.print( i   + " " );
}
System.out.println( );
```

7. Predict how many iterations this for loop will perform. Then, run the code to test your prediction.

```java
int loopCounter;
int loopEnd = 25;
for ( loopCounter = 0; loopCounter < loopEnd;  loopCounter++ ) {
    System.out.print( loopCounter + " " );
}
System.out.println( );
```

8. Predict how many iterations this for loop will perform. Then, run the code to test your prediction.

```java
for ( int k = 10; k >= 10; k-- ) {
    System.out.print( k   + " " );
}
System.out.println( );
```

9. Predict how many iterations this for loop will perform. Then, run the code to test your prediction.

```java
for ( int k = 10; k <= 0; k++ ) {
    System.out.print( k   + " " );
}
System.out.println( );
```

10. Predict how many iterations this for loop will perform. Then, run the code to test your prediction.

```java
for ( int k = 10; k >= 0; k++ ) {
    System.out.print( k + " "   );
}
System.out.println( );
```

Try It!

Ensure the SpriteAnimator.java, Sprite.java, Turtle.java, and SquareUsingForLoop.java files are in the same folder. Launch jGRASP, and open the SquareUsingForLoop.java file. Rename the class Spiral, and save the file under the new name. Compile and run the program. The program currently draws a square. Modify the application to use a turnRight(89) statement and increase the length of the side by 1 each iteration. Iterate 200 times. Use the Turtle setSpeed(Turtle.FAST) method to speed things up. Experiment with the angle setting, the increments to the side, and the number of sides. In a word-processing document, write a statement that describes what was changed and how that affected the drawing. Write a second statement that describes a surprise you encountered. Submit the Java application and the word-processing document for grading.

While Loop

The *while loop* iterates a set of statements while a boolean expression is true. Use this loop when you want to repeat the same statements until something changes. A while loop is a *condition-controlled loop* because it repeats until a boolean expression (condition) tests false. The general form of the while statement can be expressed as:

```java
while ( boolean test ) {  // loop header
    statement(s)          // loop body
}                         // loop end
```

Language Arts and Java

Poetic Pattern

Some poets write poetry that does not rhyme. It is called free verse or blank verse. It conveys meaning by using images. Gertrude Stein was a poet who wrote free verse. Stein was an American novelist, playwright, lecturer, and art collector as well as a poet. She was born in Pennsylvania in 1874 and raised in California. In 1903, Stein moved to Paris and made France her home for the remainder of her life. She lived until 1946.

Stein hosted many gatherings in Paris where the leading figures of modernism in literature and art would get together. These famous people included Henri Matisse, Ernest Hemingway, F. Scott Fitzgerald, and Pablo Picasso. The first illustration shows a stamp based on a painting Picasso created that depicts spring.

neftali/Shutterstock.com

The second illustration is a stamp showing a painting of Gertrude Stein done by Picasso. Stein wrote a poem in 1913 called "Sacred Emily."

It was published in 1922. Part of that poem is shown below.

> Rose is a rose is a rose is a rose.
> Loveliness extreme.
> Extra gaiters.
> Loveliness extreme.
> Sweetest ice-cream.
> Page ages page ages page ages.

Moroz Yurii/Shutterstock.com

Assignment

Examine the excerpt of the poem. It contains some repeated phrases. Identify these phrases. Then, write the pseudocode to print the poem using loops. Finally, create a Java application that prints this excerpt. Name the application SacredEmily, and save the file in your working folder.

Statement Part	Location	Explanation
while	Loop header	Statement name.
( )	Loop header	Parentheses are used to enclose the Boolean test expression.
{ }	Loop body	Braces enclose all statements that are to be repeated.
statement(s)	Loop body	Valid Java statements that are repeated.

Goodheart-Willcox Publisher

Figure 9-9. These are the parts of a while statement and where each is located.

Figure 9-9 explains the parts of a while statement. Take note of the punctuation. Proper punctuation is critical.

The opening brace is written on the same line as the while statement declaration. The statements are zero, one, or more valid Java statements that are repeated. The ending brace is written on its own line directly under the *w* in *while.* Like the for loop, the braces are optional if the loop body has zero or one statements. However, they should be included for clarity. It is Java style always to use these braces even when they are not needed. The statements in the loop body are indented for easy reading.

The action in a while statement is:

1. Evaluate the boolean test expression.
2. If the boolean test expression is false, exit the loop.
3. If the boolean test expression is true, execute the statements in the loop body.
4. Go to step 1.

The flowchart in **Figure 9-10** shows a while loop with the sequence and path of the control. It is the responsibility of the coder to guarantee the value of the boolean test expression is changed by the loop statements. If the boolean test expression always remains true, an infinite loop results.

The while loop is useful when you are waiting for something to happen. Think about the Guess My Number game. This game was introduced in Chapter 8. In that chapter, the game was coded to play only one round. However, in the real game, players keep guessing until they guess the number. Correctly guessing the number is the condition the players are waiting to have happen. Below is the code for playing the full version of Guess My Number.

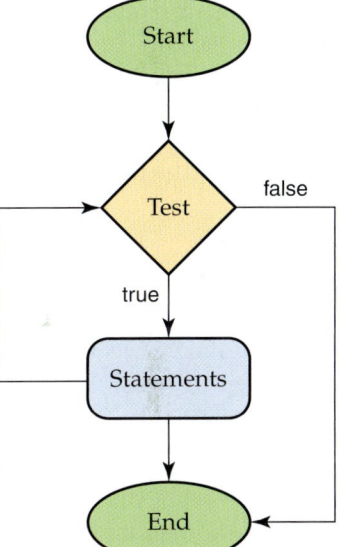

Goodheart-Willcox Publisher

Figure 9-10. This flowchart shows the sequence and control path that exists in a while statement.

```java
public static void main( String [ ] args ) {
    Scanner input = new Scanner( System.in );
    Random rand = new Random( );

    int secretNumber = rand.nextInt( 100 ) + 1;

    System.out.println( "I'm thinking of a number from 1 to 100" );
    System.out.print( "What is your guess? " );
    int guess = input.nextInt( );

    while ( guess != secretNumber ) { //begin while loop

        if ( guess > secretNumber ) {
            System.out.print( "Too high! " );
        } else {
        System.out.print( "Too low! " );
        }

        System.out.print( "Guess again " );
```

(Continued)

```
        guess = input.nextInt( );
    } // end while

    System.out.println( "Congratulations! You guessed correctly." );

} // end main
```

The action of a while loop is to test first, and then execute. In the above code, if the player guesses the value of the secretNumber variable on the first try, the boolean expression would be false at the beginning of the loop. No statements inside the loop braces would be executed. The control jumps to the "congratulations" message.

HANDS-ON EXAMPLE 9.1.2

Creating a while Loop

There are many applications for the while loop. In this exercise, you will create an application that asks the user for an integer. The application will determine if the integer is odd, even, or zero. The application will stop when the user enters a zero.

1. Launch jGRASP, and open the StarterSnippet.java file.

2. Add your name to the comments at the beginning, rename the class to EvenOrOdd, and save the file under the new name.

3. Create an instance of the Scanner class.
   ```
   Scanner input = new Scanner( System.in );
   ```

4. Prompt the user to enter an integer or enter zero to stop the program.
   ```
   System.out.print( "Enter an integer. ( 0 to stop ):  " );
   ```

5. Declare the variable to hold the integer and get a value from the keyboard.
   ```
   int entry = input.nextInt( );
   ```

6. Write the while loop with the condition that the integer is not zero. Enter both braces now so you do not forget the closing brace later. Include a comment to document what the closing brace represents.
   ```
   while ( entry != 0 ) {
   } // end while
   ```

7. Apply what you have learned about conditionals and binary operators to test the input for an even or odd number and to report the result to the user. Put this code between the braces of the while loop. It is not necessary to use braces in an if statement when there is only one statement in the block, but they are included for clarity.
   ```
   if ( entry % 2 == 0 ) {
       System.out.println( "You entered an even number." );
   } else {
       System.out.println( "You entered an odd number." );
   }
   ```

8. Prompt for another integer and scan the entry. This action provides an opportunity to change the condition of the while loop so it is not an infinite loop. This code is also part of the while loop. Put this code after the code above and before the closing brace of the loop.
   ```
   System.out.print( "Enter another number ( 0 to stop ): " );
   entry = input.nextInt( );

} // end while
```

9. After the closing brace for the loop, tell the user the program is over.

```
System.out.println( "Zero input received. Program terminated." );
```

10. Compile and run the program. Test with two odd numbers, two even numbers, and then zero. Debug as needed.

Try It!

When rolling two six-sided dice, the most likely outcome is seven. Apply what you have learned about the Random class to create an application that simulates the rolling of two dice. Display the values for the dice at each roll. The application should keep rolling while the roll is not seven and quit when the roll is seven. Save the program as RollSevens.java in your working folder.

Do/While Loop

The *do/while loop* iterates a set of statements while a boolean expression is true. It is a type of condition-controlled loop. Use this loop to be certain the set of statements is executed at least once. The general form of the do/while statement can be expressed as:

```
do {
   statement(s)                         // loop body
} while ( boolean test expression ); // loop end
```

Figure 9-11 explains the parts of a do/while statement. Take note of the punctuation. Proper punctuation is critical. Any variable that is used in the while condition must be defined *before* the do/while statement. A variable defined within the loop body of the do/while statement is out of scope for the while condition.

Braces enclose all statements that are to be repeated. The opening brace is written on the same line as the do statement declaration. The ending brace is written on its own line directly under the *d* in *do*. Zero, one, or more valid Java statements that are to be repeated make up the loop body. The statements in the loop body are indented for ease in reading. Unlike the other two types of loops, the braces *are* required even if there are no statements in the loop body. Immediately following the ending brace, the while keyword is placed, followed by the boolean test expression inside parentheses. For this type of loop *only*, a semicolon is used to terminate the loop statement.

Statement Part	Location	Explanation
do	Loop header	Statement name.
{}	Loop body	Braces enclose all statements that are to be repeated.
statement(s)	Loop body	Valid Java statements that are repeated.
while ( )	Loop ender	Immediately following the ending brace, the Boolean expression is inside the parentheses.
;	Statement terminator	For this type of loop *only*, a semicolon terminates the loop statement.

Goodheart-Willcox Publisher

Figure 9-11. These are the parts of a do/while statement and where each is located.

Math and Java

Going Viral: Populations and Chain Letters

A popular Internet meme is the chain letter post. A note arrives in your inbox, via Twitter, or in another social-media app. It promises great rewards to the recipient if the post is sent to 10 new people in the next 20 minutes, and ultimate dangers if the post is not sent on. Assume the post is first sent to 10 people. Then, each of those 10 people sends it along to 10 new people (10 × 10 = 100 people). Then in 20 more minutes, each of those 100 people send it along to 10 new people (100 × 10 = 1000 people), and so on.

Rawpixel.com/Shutterstock.com

If everybody plays along and forwards the post as instructed, and no one receives the post a second time, how much time elapses before the post is received by every person in the world? Apply what you have learned to develop pseudocode by means of the USAP problem-solving model:

- **Understand the problem (U):** Count the number of times you multiply by 10 until the number reaches the earth's total population, then compute the elapsed time.

- **Break the problem into small doable parts (S):** 1) Find Earth's population (about 7.6 billion), 2) program a loop to perform the multiplication, 3) count the iterations, and 4) calculate the elapsed time.

- **Develop the algorithm (A):** Set a counter to 0 because the first ten messages will be sent at the beginning before any time elapses. Set a variable for people to 10. Loop to multiply people by 10 and increment the counter. The loop will stop when people reaches or exceeds the world population. Examine the counter to find the number of iterations. Multiply the counter by 20 minutes to find how many minutes it takes to max out.

- **Write the pseudocode (P):**

```
count = 0
people = 10
repeat
    people *= 10
    count++
until people >= 7.6 billion
```

Assignment

Write the Java application to calculate this result. Research the current world population. Name the application ViralLetter, and save it in your working folder. Pay close attention to the proper application of the mathematical formulas. For an extra challenge, write the code to calculate the elapsed time converted from only minutes to minutes, hours, days, months, and years. Use logic to deduce the mathematical formulas to use for the conversion. After you have tested and debugged your program, write a paragraph in a word-processing document to discuss your findings. Save it in your working folder, and submit your program and the document for grading.

The action in a do/while statement is:

1. Execute the statements in the braces.
2. Evaluate the boolean test expression.
3. If the boolean expression is false, exit the loop.
4. If boolean expression is true, go to step 1.

The flowchart in **Figure 9-12** shows a do/while loop with the sequence and path of the control. The do/while loop always executes the set of statements *before* the test. This guarantees the statements are executed at least once. By contrast, the statements in for and while loops are executed *only* if the boolean expression is true at the beginning of the loop.

It is possible to use a do/while loop to draw a square with the Turtle class. The code using a do/while loop is shown below. Compare this to the snippet shown in **Figure 9-1.** Although the algorithms are different, both snippets produce the same drawing.

```
public void buildScript( ) {

    Turtle turtle = new Turtle( root, 250, 250 );

    double startX = turtle.getX( );
    double startY = turtle.getY( );

    do {
       turtle.forward( 100 );
       turtle.turnRight( );
    } while ( turtle.getX( ) != startX || turtle.getY( ) != startY );

} // end buildScript
```

The action of a do/while loop is to execute first and then test. In the above snippet, the first side is drawn, and then a test is performed on the turtle's location to see if it has returned to the starting coordinates. If the turtle is at the starting point, the loop stops.

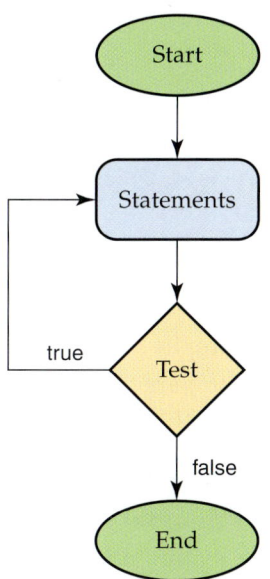

Goodheart-Willcox Publisher

Figure 9-12. This flowchart shows the sequence and control path that exists in a do/while statement.

HANDS-ON EXAMPLE 9.1.3

Creating a do/while Loop

In computer programming, there are often many ways to solve a problem. In Hands-On Example 9.1.2, you created an application using a while loop. The same program output can be achieved with a do/while loop.

1. Launch jGRASP, and open the EvenOrOdd.java file created in Hands-On Example 9.1.2.
2. Change the class to EvenOrOddDo.java, and save the file under the new name.
3. Delete the prompt asking the user to enter an integer. Scanning will be done inside the loop.
4. Remove the initialization of entry in the declaration. Change that line to:

```
int entry;
```

5. Replace the while loop with a do/while loop. Include the input statement at the beginning of the loop.

```
do {

    System.out.print( "Enter an integer. ( 0 to stop ):   " );
    entry = input.nextInt( );

    if ( entry % 2 == 0 ) {

        System.out.println(  "You entered an even number." );
    } else {

        System.out.println( "You entered an odd number." );
    } // end if/else

} while ( entry != 0 ); // end do/while

System.out.println( "Zero input received. Program terminated." );
```

6. Compile and run the program. Test the output with odd and even numbers and zero. Debug as needed.

7. Compare the structures of EvenOrOdd and EvenOrOddDo programs. What happens when the user enters a zero at the beginning of the loop?

Try It!

A common function in video games is to ask the player if he or she wants to play again. This can be achieved by enclosing the entire game code within a large do/while loop. Why would a do/while loop be used instead of a while loop? Write a do/while loop snippet that would enclose the game code. Use a comment to represent the game code, such as: // game code goes here. At the end of the game, ask the player if he or she wants to play the game again. Ask for input of *Y* or *N*. Accept input of *Y* or *y* as a positive response and anything else exits the loop. Save the application as PlayAgainSnippet.java in your working folder.

SECTION 9.1 REVIEW

Check Your Understanding

1. What is each repetition of the statements in a loop called?
2. Which type of loop should be used if the programmer knows the exact number of times to repeat the statements?
3. _____ is a variable declaration and assignment statement at the beginning of a for loop.
4. What type of loop is a while or do/while loop?
5. Which type of loop always executes the set of statements before the test?

Build Your Vocabulary

As you progress through this course, develop a personal computer science glossary. This will help you build your vocabulary and prepare you for a career. Write a definition for each of the following terms and add it to your computer science glossary.

condition-controlled loop	initialization	loop counter
counter-controlled loop	iteration	looping condition
do/while loop	loop	nested loop
for loop	loop body	while loop
infinite loop		

Applying Loops

There is a wide range of opportunities to apply loops in Java code. Reading items from a text file is a good example. Sometimes, the quantity of data coders want to read and process can be large. Perhaps the problem is to list the first names and four-digit telephone extensions of everyone in a 50-person company. In that case, prompting the user to enter the first name and extension for everyone can be time-consuming. The user can easily make mistakes. If, however, all those names and extensions are in a text file, the program can have a loop to read each line and process each name and extension in turn. In so doing, the program easily reads and processes the data without the user's help. This is just one example of how loops can be applied.

Learning Goals

- Diagram the process of reading from an external file in Java.
- Identify common looping techniques.

Terms

file pointer

opening a file

paper check

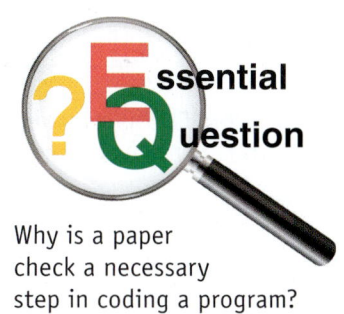

Essential Question

Why is a paper check a necessary step in coding a program?

Reading from External Files

There are many times a program may need to read the data in an external file. To read from a file, four things need to be done:

1. Import the java.io and java.util packages.
2. Add throws FileNotFoundException to main.
3. Open the file.
4. Read the file using a while loop controlled by the Scanner class hasNext method.

You already know how to read data using methods from the Scanner class. Fortunately, those methods can also be used to read from a file. You just need to tell Scanner that you want the input to come from a file rather than from the keyboard.

Coding to Read from a File

To read from a file, you need a constructor from the File class. Opening a file is done by first instantiating a File object using the File constructor, which takes the name of the file as an argument. The API for the File constructor is shown in **Figure 9-13.** Then, use a different Scanner constructor that takes the File object as an argument. The API for this constructor is also shown in **Figure 9-13.** In Java, coding the File and Scanner constructors to read an external file is called *opening a file.*

While data is being read from a file, a file pointer moves through the file. The *file pointer* is a mechanism that keeps track of the next data to read in the external file. When all data in the file has been read, the program should stop reading and move on to the other statements in the program. The Scanner

Constructor	Explanation
File(String filename)	A constructor of the File class; the name of the file to read is sent as an argument.
Scanner(File inFile)	A constructor of the Scanner class; the constructor opens the File object inFile for reading.

Goodheart-Willcox Publisher

Figure 9-13. These constructors are needed to read content in an external file.

Method	Explanation
boolean hasNext( )	Scanner method that returns true if there is more data in the file to read and false if the end of the file has been reached.

Goodheart-Willcox Publisher

Figure 9-14. The hasNext method tests if the end of an external file has been reached.

class contains the hasNext method that tests if the end of the file has been reached. Its API is shown in **Figure 9-14.** The hasNext method returns true if there is more data to read in the file. It returns false when the end of the file has been reached.

The hasNext method is perfect to use in the condition of a while loop:

```
while ( file.hasNext( ) )
```

Why does the while loop condition not need to be coded as the following?

```
while ( file.hasNext( ) == true )
```

This statement is redundant. The return value of the hasNext method is already a boolean. If the method returns true, the condition will be true. So, this second statement asks if true == true. If the method returns false, the second statement asks if true == false, which is a false condition.

Something to consider in reading from a file is that the file may not be where the Scanner constructor looks for it. If that is the case, the Scanner constructor "throws an exception." It means that processing will stop if the exception is not handled by the coder. To alert Java a missing file might occur, add a phrase to the end of main: throws FileNotFoundException. If the Scanner constructor cannot find the file or if the file name is misspelled, the program will stop with a FileNotFoundException. Chapter 11 discusses how to prevent the program from stopping and how to handle the error.

It is important to add the throws FileNotFoundException clause to the main when you want to read from a file. Without this clause, the compiler will generate an "unreported exception" error.

Application of Reading a File

Consider the text file shown in **Figure 9-15.** This is a list of runners who participated in a 5K race. Each runner's time appears on a line followed by the name of the runner. Depending on the number of racers, the file could have more data. If 50 racers ran the 5K, then the file would have 50 lines, each with a running time and a racer name.

racerTimes.txt - Notepad

```
15.5 Garcia
16.25 Chang
14.32 Smith
13.25 Runningbear
25.77 Umeki
15.75 Gupta
```

Goodheart-Willcox Publisher

Figure 9-15. This text file can be read by a Java program. Notice how the data is formatted, each line containing a numerical value followed by text.

Writing the Code

This program simply reads the data from the file racerTimes.txt and outputs each value:

```
import java.io.*;  // import File and FileNotFoundException
import java.util.*; // import Scanner
```

(Continued)

```
public class RaceAnalysis1 {

  public static void main( String [ ] args ) throws
    FileNotFoundException {

    // open the file
    File inFile = new File( "racerTimes.txt" );
    Scanner file = new Scanner( inFile );

    // read the data in the file
    while ( file.hasNext( ) ) {
      double racerMinutes = file.nextDouble( );
      String racerName = file.next( );

      System.out.println( racerMinutes + " " + racerName );
    } // end while

  } // end main
} // end class
```

In the body of the while loop, one line is read at a time. In this example, each line has two pieces of data: the racer's time and the racer's name. The time of the racer is a double, so the program uses the nextDouble method to read the time. The racer's name is a single-word String, so the program uses the next method to read the name. After reading each line, the program outputs the data it has just read, as shown in **Figure 9-16.** This program can be run with any text file in this format even if you do not know how many lines are in the file. The program simply reads the data line-by-line until the hasNext method returns false to indicate there is no more data to read.

Determining a Minimum or Maximum Value

If we want to determine which racer won the race, we need to find the shortest (or minimum) running time. To find the minimum of a set of values, initialize a minimum variable to a value that is *larger* than any possible value. This ensures the first value read will be lower than the initial value, and it can be saved as the first minimum. As each value is read, compare the value to the current minimum. If the value is less than the current minimum, save the value as the new minimum.

To find the maximum of a set of values, initialize a maximum variable to a value that is *smaller* than any possible value. This ensures the first value read

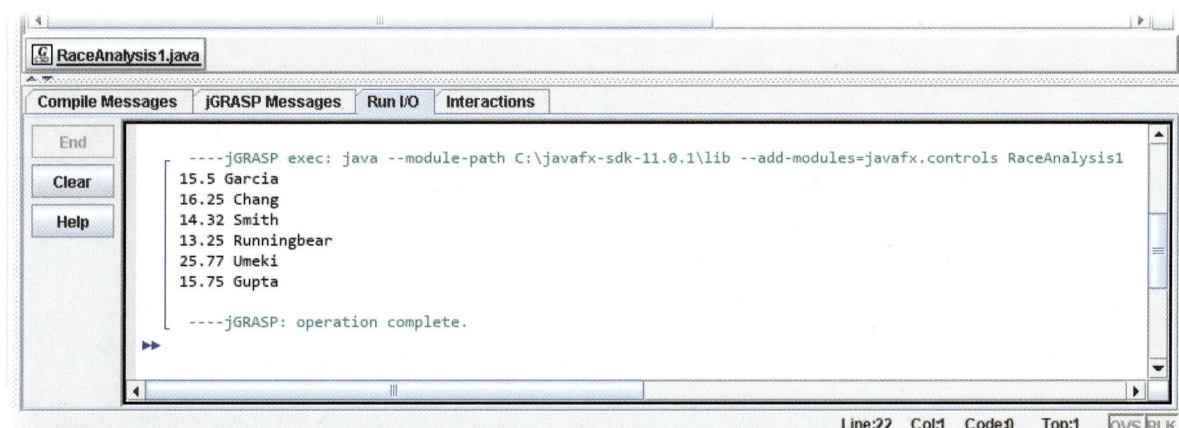

Figure 9-16. The Java program has read the external text file shown in Figure 9-15 and output its contents.

will be higher than the initial value, and it can be saved as the first maximum. As each value is read, compare the value to the current maximum. If the value is greater than the current maximum, save the value as the new maximum.

The Java Class Library provides static constants to use for initializing maximum and minimum values, as shown in **Figure 9-17.** In the next Hands-On Example, Double.MAX_VALUE is used to initialize the minimum value.

To find the winner of the 5K race, determine the shortest time. The shortest time is the lowest, or minimum, number of minutes. Use this algorithm to find the minimum of a set of values:

1. Initialize the minimum variable to a value larger than any value expected in the file.
2. Define a String to hold the name of the winner.
3. Open the file.
4. While there is more data in the file to read:

   A. Read the racer's time.

   B. Read the racer's name.

   C. If the racer's time is less than the current value of minimum:

      i.   Save the racer's time to the minimum variable.
      ii. Save the racer's name to the winner's name String.

5. Output the minimum variable and the winner's name String.

This code uses the above algorithm to find the winning time and winner's name:

```java
// initialize the winning time and winner's name
double winningTime = Double.MAX_VALUE;
String winnerName = "";

// open the file
File inFile = new File( "racerTimes.txt" );
Scanner file = new Scanner( inFile );

// read the data in the file
while ( file.hasNext( ) ) {
   double racerMinutes = file.nextDouble( );
   String racerName = file.next( );

   // if this racer's time is less than the minimum,
   //   save the new minimum and the racer's name
   if ( racerMinutes < winningTime ) {
     winningTime = racerMinutes;
     winnerName = racerName;
   }

} // end while

// output the winner and minimum time
```

(Continued)

Constant	Explanation
Integer.MIN_VALUE	Minimum value of an int: −2147483648
Integer.MAX_VALUE	Maximum value of an int: 2147483647
Double.MIN_VALUE	Minimum positive value for a double: 4.9E-324
Double.MAX_VALUE	Maximum value for a double: 1.7976931348623157E308

Figure 9-17. The Java Class Library provides these static constants that can be used to initialize minimum and maximum values. The values are the limits of the data type.

```
String s = String.format( "The winner is %s with a time of %.2f "
+ "minutes.", winnerName, winningTime );
System.out.println( s );
```

The output of this program is shown in **Figure 9-18.** Verify this is correct by looking at the values in the file. Code to find the slowest runner or maximum time would be similar.

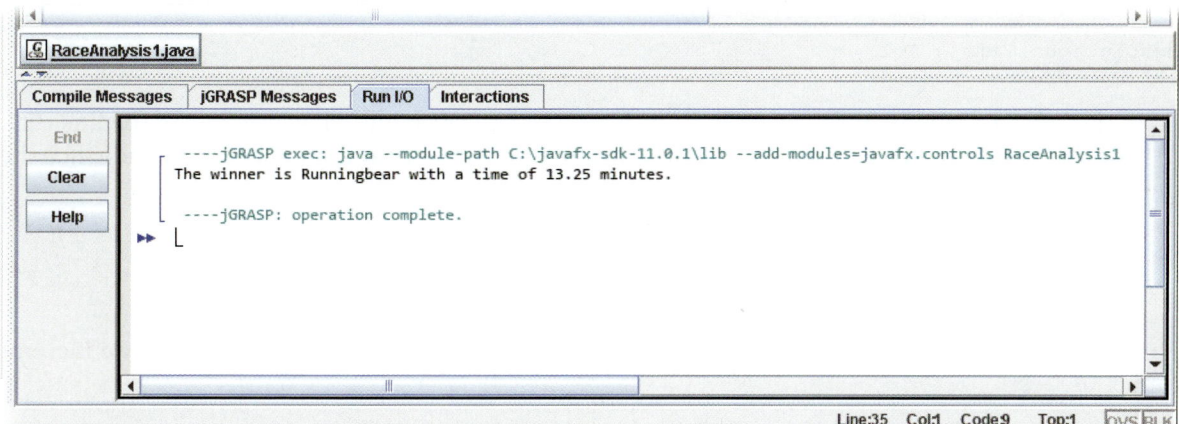

Figure 9-18. The Java program has read the external text file and determined which runner had the lowest time.

HANDS-ON EXAMPLE 9.2.1

Applying Loops to Find an Average

In this exercise, you will create a program to read data in an external file. The program will calculate the average of all the racers' times in the file. A total and a counter are used for this. To calculate the average, find the total of all the times and count the number of racers. Then, calculate the average as:

```
Total time / number of racers
```

This algorithm will be used:

1. Open the file.
2. Set total to 0.
3. Set counter to 0.
4. While there is more data in the file to read:
   A. Read the minutes.
   B. Read the racer's name.
   C. Add the minutes to the total.
   D. Add 1 to the counter.
5. Calculate the average as the total divided by the counter.
6. Output the average.

Before beginning this exercise, download the chapter files from the student companion website. The RaceAnalysis1.java file is used as a starting point.

1. Launch jGRASP, and open the RaceAnalysis1.java file.
2. Update the header comment to your name.

3. Rename the class to RaceAnalysis2, and save the file under the new name. Be sure the racerTimes.txt file is located in the same folder.

4. Examine the program. There is already a while loop coded. Compile and run the program. Currently, this program simply outputs the times and names of the runners. To find an average, you need to add up the individual times and count the number of racers.

5. Declare and initialize two variables for keeping track of the total time and the number of racers. Ensure this code is *before* the while loop starts so the scope of these variables is at the main level. They will need to be used after the while loop has terminated. Place this code before the code for opening the file.

```
// initialize the total and counter
double totalTime = 0.0;
int numberOfRacers = 0;
```

6. Retain the code for reading the time for each racer and the racer's name. However, delete the line that prints the time and name. The racer's name is not used in this application, but it must be read to advance the file pointer to the next racer time.

```
// read the data in the file
while ( file.hasNext( ) ) {
    double racerMinutes = file.nextDouble( );
    String racerName = file.next( );
```

7. Before the end of the while loop, add two statements that build the sum of the racer times and increment the number of racers.

```
    totalTime += racerMinutes; // add racer's time to the total
    numberOfRacers++;          // increment the count of racers

} // end while
```

8. After the end of the while loop and before the end of the main, calculate and output the average time.

```
// calculate and output the average
double averageTime = totalTime / numberOfRacers;
String s = String.format( "The average time for the racers is %.2f "
            + "minutes.", averageTime );
System.out.println( s );
```

9. Compile and run the program. Verify the output matches what is shown. Debug as needed.

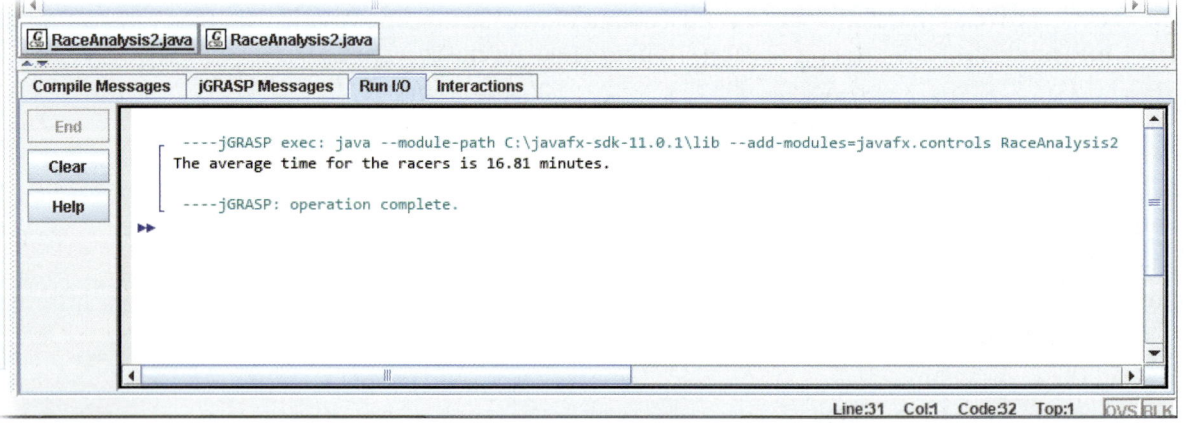

Goodheart-Willcox Publisher

Try It!

Open the RaceAnalysis3.java file. Rename the class RaceAnalysis4, and save the file under the new name. Add code to find the longest time for the racers, or the maximum time. Refer to the example in the text for finding the winner (lowest time), and adapt it as needed. For the output, print the name of the racer with the highest time and what that time is in a complete sentence. For an extra challenge, find the range of the times. Modify the loop statements to calculate the difference between the best time and the worst time.

Troubleshooting Loops

The programmer determines the problem's solution. He or she studies the problem specification and writes an algorithm. If the solution requires repetition, the programmer will decide which type of loop structure to use. Loops are a well-used concept in programming. The ability to write them is a requirement for all high-level programming languages. Loops are used so often that coders have collected good techniques for applying loops in programs.

Forming Loop Conditions

Care must be taken in forming loop conditions. If you know what event will stop the loop, then the loop condition will be that the event did *not* occur. For example, in the Guess My Number application, guessing the correct number will stop the loop. Therefore, the looping condition asks if the guess is correct, so as long as the guess in incorrect, the loop continues.

When writing compound Boolean expressions, see if they can be simplified. Look to De Morgan's laws to see if they will be helpful to simplify expressions.

Testing Loop Conditions

Two common errors are that the loop starts at an incorrect data value or ends before all cases are considered. It is often helpful to paper check the loop conditions. A *paper check* is manually writing out a description of the sequence of instructions to be executed and then analyzing it for proper logic. It is very difficult to catch a logic error while the program is running. A paper check allows the coder to experience what the computer will do when the program is running.

Verify the loop starts at the right place and ends at the right place to handle every case in the algorithm. It is easy to miss the ends of the range being tested. The solution may be as simple as using less than or equal to (<=) instead of just less than (<) in the test. However, the problem may be more complicated to solve. Paper checking will reveal omissions.

Verify any data required is available and there is enough input to complete the execution of the loop. For example, if names and numbers are required to be read in an external file, be sure all lines of data are complete in the file. Also check that the file is in the proper location and the file name is spelled correctly in the program code.

Check the truth-value of the test. Is it what is needed when the statements are executed? For example, in a video game, the truth-value for the lives variable is not equal to zero during gameplay. When the lives variable becomes zero, the game is terminated. Verify that the truth-value changes under the correct conditions.

Ensure the loop terminates at some point. In a for loop, check that the loop update gets you closer to the termination condition. Check that the while statements update the condition so that it will eventually evaluate to false to exit the loop. Locate the statements that change the loop condition and paper check them.

Choosing a Control Structure

Java has three loop statements. The coder must decide which type of loop is the best to apply in a given situation. The needs of the algorithm help the coder to make this determination.

Science and Java

Worldwide Malaria

Malaria occurs mostly in underdeveloped tropical and subtropical areas of the world. Africa is most affected due to a combination of factors. A very efficient mosquito (*Anopheles gambiae*) native to that area is responsible for high transmission of the disease. The female mosquito carries with it a parasite called *Plasmodium falciparum.* Local warm-weather conditions often allow the mosquito to transmit malaria year-round. There is no winter to kill the insects. Scarce resources and wars have hindered control of the disease.

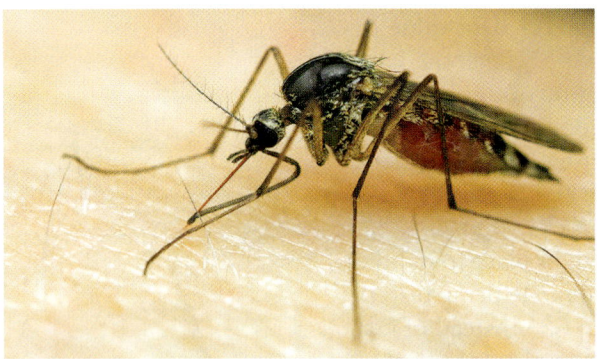

corlaffra/Shutterstock.com

According to the World Health Organization (WHO), 91 countries reported a total of 216 million cases of malaria in 2016, an increase of 5 million cases over the previous year. The global tally of malaria deaths for that year reached 445,000 deaths, about the same number reported in 2015. The WHO African Region continues to account for about 90 percent of malaria cases and deaths worldwide. Fifteen countries, all but one in sub-Saharan Africa, experience 80 percent of the global malaria incidents.

The WHO has many activities to promote disease eradication. For example, it recommends the use of mechanical barriers, such as insecticide-treated mosquito nets (ITNs). Between 2014 and 2016, a total of 582 million ITNs were delivered throughout the world.

Assignment

As a member of the WHO African Region team, you have been informed of the following data.

In sub-Saharan Africa, 16 countries accounted for more than 80 percent of ITN deliveries in the 2014–2016 period. The table shows the delivery statistics. Outside sub-Saharan Africa, most deliveries of ITNs were accounted for by eight countries, which is also reflected in the table.

Sub-Saharan Africa	ITN*	Outside Sub-Saharan Africa	ITN*
Nigeria	78.0	India	15.5
Democratic Republic of the Congo	61.2	Myanmar	11.0
Uganda	35.6	Indonesia	7.7
Ethiopia	33.0	Pakistan	5.1
United Republic of Tanzania	29.2	Cambodia	5.0
Ghana	19.6	Afghanistan	4.4
Mozambique	17.6	Bangladesh	4.4
Côte d'Ivoire	16.9	Yemen	2.9
Kenya	16.9		
Senegal	15.1		
Burkina Faso	14.6		
Mali	14.1		
Sudan	13.6		
Madagascar	12.7		
Malawi	12.4		

*millions

Write a Java program that calculates a generic average. Name the application Average, and save the file in your working folder. Accept input from the keyboard, and compute the average for each set of numbers input. Include a String input for the user to enter a title for the data. In this case, the user will enter either "Sub-Saharan Africa ITNs" or "Outside Sub-Saharan Africa ITNs". Count the number of entries. Set a flag of input == −1 to terminate the input. Calculate the average. Output the title and the average. Run the program twice, once for each group. Once your program is error-free, submit it for grading.

Use a for loop for counter-controlled applications. Any time the number of iterations is known in advance, a for loop can be used most efficiently. Any algorithm that requires counting up or counting down a fixed number of times is ideal for a for loop.

Use a while loop for condition-controlled applications. Any time the number of iterations is not known in advance, a while loop can be used most efficiently. The while loop iterates until a signal is received to terminate the loop. In a while loop, the condition is tested before any statements are performed, so the statements may or may not be executed based on the result of the test. Any algorithm that does not require the statements within the loop to be executed at least once is ideal for a while loop.

Use a do/while loop for any condition-controlled application in which the number of iterations is not known, and it is important for the statements in the loop to be executed at least once. In a do/while loop, the condition is not tested until the statements have been executed at least once.

It is possible to combine two types of looping nested in the same program. For example, if the program called for finding prime numbers, the developer would find numbers that have nonzero remainders when using integer division by any positive integer other than 1. Begin by having the user enter an integer number. Then, use a do/while loop to test for nonzero remainders of integer division. The number is divided first by 2 and increased by 1 each time. The factor is the divisor of the mod operation. When the remainder of the number divided by the factor is *not* 0 for all divisions, the number is prime. The program ends when a remainder is 0 or the factor is half of the number. An algorithm for this is:

```
user enters first number
set prime to true
set factor to 2
do
{
   if number % factor is not 0
      increment factor
   else
      set prime to false
} while prime is true and factor <= number / 2 //end of loop

if prime is true,
   print the number is prime
else
   print the number is not prime
```

HANDS-ON EXAMPLE 9.2.2

Exploring Loops and Data Integrity

The best way to explore loops and data integrity is to practice by doing a paper check. In this exercise, you will test the data that you want to read from a file. Before beginning this exercise, download the chapter files from the student companion website. The file census.txt contains the increase in population and city names for the eight fastest growing cities in the United States from 2016 to 2017 according to the US Census Bureau. It needs some correction before it can be read by Java. In this example, you will read the data and look for formatting errors, missing data, and duplicate entries that might interfere with the loop.

1. Open the census.txt file in Notepad or another plain-text editor.

2. The second line includes the state. Delete the state since the other entries do not include a state.

3. The third line has a formatting error. It includes a comma. Delete the comma.

4. Lines 5 and 6 are duplicates. Delete one of them or the loop will read both lines. This will give the wrong answer if you average or total the populations.

5. The data for Seattle is entered in reverse order. Make the necessary exchange of data. The loop will not work if the numerical data is reversed with String data.

6. The population data for Charlotte is missing. It is 15551. Add this data to the file.

7. The last line contains a decimal. Delete the decimal point and the number after it since there cannot be half of a person.

8. Save the file with the name censusRevised.txt.

9. Decide which kind of loop is best to read the file. With paper and pencil, write what is read after each loop iteration. Is there a way to exit the loop? What exception would occur if you did not correct the file?

Try It!

Open the USPopulationByState.txt file. Perform a paper check on the data to see if it is in the correct format for being read by a loop. Save the file as USPopulationByStateRevised.txt. What is the best loop structure to use? How can you write a program that finds the state with the maximum population increase? Write this program and test it to verify the data.

Coding Conundrum

A programmer has coded the following loops. However, the output is not as intended. It is a conundrum! Identify the errors in these code snippets. Explain the problem, describe the consequence, and propose a fix.

1. The loop body is executed just once, not five times as expected.

```
for ( int i = 0; i < 5; i++ ); {
    System.out.println( "Clap along!" );
} // end for loop
```

2. The loop repeats without ending.

```
boolean happy = true;
while ( happy ) {
    System.out.println( "Clap along if you feel"
            + " like happiness"
            + " is the truth." );
}
```

3. The program will not compile.

```
Scanner getInput = new Scanner( System.in );
String happiness;
do }
    System.out.print( "R U happy? (Y/N) " );
    happiness = getInput.next( );
{ while ( happiness.equalsIgnoreCase( "y" ) );
```

SECTION 9.2 REVIEW

Check Your Understanding

1. Which Java packages must be imported in order to read an external file?
2. What mechanism keeps track of the next data to be read in an external file?
3. Reading the file using a while loop is controlled by which Scanner method?
4. A(n) _____ is a method to manually describe and analyze a sequence of instructions being followed.
5. Any algorithm that requires counting up or counting down a fixed number of times is ideal for a(n) _____ loop.

Build Your Vocabulary

As you progress through this course, develop a personal computer science glossary. This will help you build your vocabulary and prepare you for a career. Write a definition for each of the following terms and add it to your computer science glossary.

file pointer   opening a file   paper check

Cooperative Coding

Reading a Statistics File

Sports management is a broad field that involves all aspects of business as it relates to sports. In both professional and amateur sports, a large group of sports-management professionals is working to facilitate the league, a particular event, or even an individual athlete's progress. The specific responsibilities of a sports-management professional are generally related to his or her chosen sport. Some examples focus on team management, event management, facility management, marketing, economics, or finance.

The number of teams and organizations seeking qualified candidates for sports-management jobs is very high. Many colleges and universities offer sports-management programs for both bachelor and master degrees. Professionals in the field require strong abilities when it comes to oral expression, oral comprehension, speech clarity, written expression, and written comprehension. Sports-management professionals also require a diverse array of knowledge, which often includes education and training, psychology, counseling, and customer and personal relations. Some candidates will also have knowledge of coding and statistics as well as experience with the sport itself. This is an example of a hybrid job.

Approximately 130,000 people currently work in the spectator-sports industry. This sector is where many of the most coveted sports-management positions are. With a team of sports professionals, managers are required to keep track of scores and ratings of athletes in various venues.

Assignment

Your task is to acquire skills at manipulating statistics. Work with the group of classmates as assigned by your instructor. As a group, decide which of your school's teams you want to investigate. Then, collect information about the members of the team. Identify each team member's last name and the points he or she scored in the past season. For individual sports such as tennis or wrestling, record the number of wins the person had. The athletic department should have a record of these statistics.

Create a text file to contain the information you collected. When entering the data in the text file, place each member on a separate line, last name first and then score. Create the file using a plain-text editor, not a word processor, because there should be nothing in the file except the text.

Using what you have learned about looping, code a Java application that determines and displays the highest score, the average, and the total number of team members. Discuss as a group which looping technique will be most efficient. Identify the conditions to use. Then, paper check the logic before coding the program. Select one or two members of your group to code the project. The remaining members should test and debug the program.

1. Which looping technique did you choose, and why?
2. What is the condition(s) being tested in your loop?

Aspen Photo/Shutterstock.com

Chapter Summary

Section 9.1: Java Loops
- A loop is a block of code that is repeated as many times as required by the looping condition; each time the block is repeated is called an iteration; loops can improve the efficiency of the code.
- The for loop iterates a block of code a predetermined number of times as determined by the loop counter, which is why this type of loop is called a counter-controlled loop; take care to avoid coding an infinite loop that repeats forever.
- The while loop iterates a set of statements while a boolean expression is true, which is why it is called a condition-controlled loop; take care to update the loop condition in the loop body to avoid coding an infinite loop.
- The do/while loop iterates a set of statements while a boolean expression is true, and the statements will always be executed at least once, unlike a while loop in which the statements will only be executed if the expression is true. As with the other loops, take care to update the loop condition in the loop body to avoid coding an infinite loop.

Section 9.2: Applying Loops
- To read the data in an external file, first import the java.io and java.util packages, then add throws FileNotFoundException to main, open the file, and read the file using a while loop controlled by the Scanner class hasNext method.
- Paper check the logic in a loop by manually writing out a description of the sequence of instructions being followed and then analyzing it; always choose the best type of loop to meet the needs of the problem.

Chapter 9 Test

Multiple Choice
Select the best response.
1. What is true about all types of loops?
   A. They repeat 10 times.
   B. They may contain statements to be repeated.
   C. They repeat forever.
   D. They will not repeat.
2. When a variable is defined within the initialization of a loop, what is the scope of the variable?
   A. The main method.
   B. The loop and all subsequent loops.
   C. The loop only.
   D. There is no impact on the scope.
3. Which loop repeats a fixed number of times?
   A. while
   B. repeat until
   C. infinite
   D. for
4. Which loop is a condition-controlled loop?
   A. for
   B. do/while
   C. System.out.println ( );
   D. break
5. Which method is used to determine if any more data exists in the external file?
   A. filePointer
   B. hasNext
   C. FileNotFoundException
   D. java.io

Completion
Complete the following sentences with the correct word(s).
6. Setting a variable to a beginning value before the main logic begins is called _____.
7. No instructions following a(n) _____ will ever be executed because the loop never ends.
8. A do/while loop is a(n) _____-controlled loop.
9. The _____ keeps track of the next data to be read in the external file.
10. Coding the File constructor and Scanner constructor to read from an external file is called _____.

Matching
Match the correct term with its definition.
   A. file pointer
   B. paper check
   C. iteration
   D. counter-controlled
   E. condition-control
11. Loop that repeats until a certain event or value is encountered.
12. Keeps track of the location in an external file.
13. One execution of the instructions within a loop.
14. Loop that repeats until a variable is incremented to a certain value.
15. Manual check for proper logic.

Application and Extension of Knowledge

1. Write a Java application that nests two for loops that include loop counters. Write the second loop between the braces of the first. Be careful to use different variables for the loop counters. In the inside loop, print the values of the two counters each time an iteration occurs. Use small numbers for the iterations. Just a few will provide ample data for you to draw conclusions. What do you observe about how the loops are processed?

2. Apply what you have learned about regular polygons and turtle graphics to draw a polygon that has so many sides it begins to look like a circle. State the number of sides used and the angle you used in the turnRight( ) statement. Which type of loop will you select for this task?

3. Find a copy of *Mindstorms: Children, Computers, and Powerful Ideas* by Seymour Papert in the school or local library. Read the sections about turtle graphics. Write a paragraph that summarizes what you learned.

4. Write a program that applies the algorithm in Section 9.2 to check for a prime number. Accept input from the user of an integer and test whether it is prime. Output a meaningful result.

5. Life expectancy for the United States population has been statistically derived by the Centers for Disease Control. For example, 2.3 percent of white females, 2.9 percent of black females, and 2.8 percent of Hispanic females alive in 2014 will live to the age of 100. Write a program that lets the user enter these three percentages. Then, have the program compute the average of the percentages using a loop and print out the following statement: "The average percentage of white, black, and Hispanic females alive in 2014 who will likely live to age 100 is _____."

Online Activities

Complete the following activities, which will help you learn, practice, and expand your knowledge and skills.

Vocabulary. Practice vocabulary for this chapter using the e-flash cards, matching activity, and vocabulary game until you are able to recognize their meanings.

Communication Skills

Reading. A *prefix* is added to the beginning of a word to create a new meaning. A *suffix* is added to the end of an existing word to create a new meaning. *Affix* means to add a prefix or suffix to a root word. Locate three words in this or another chapter that have a prefix or suffix. Determine the meaning of the affixes, and then the meaning of the whole word. Look for clues in the surrounding text to help you understand what the word means.

Writing. *Figurative language* is a word or phrase used in a way that is different from its normal or literal meaning. Figurative language makes the idea more interesting to the reader. An example is to say snow is a "clean, white blanket" to describe "snow on the ground." Select a section of this chapter and rewrite selected paragraphs in your own words. Highlight the sentences in which you used figurative language.

Speaking. Being able to verbally *retell* or *summarize* what you read can help you confirm your understanding of the material and demonstrate you comprehend the subject. Select a visual from this chapter such as a table, photo, or other illustration. Explain to the class what the visual means and how it relates to the chapter topic.

Listening. Begin a conversation with a classmate you do not know very well. Ask the person about the clubs, sports teams, or other groups to which he or she belongs. Actively listen to what this person says. Build on his or her ideas by sharing your own thoughts. Next, summarize and retell what the person said in the conversation. Did you really hear what was being said? How is having a conversation with someone with whom you do not normally speak different from having a conversation with a friend or family member?

Portfolio Development

College and Career Readiness

Community Service. Community service is an important quality to show in a portfolio. Serving the community shows a candidate is well rounded and socially aware. Many opportunities are available for young people to serve the community. You might volunteer for a park clean-up project. Perhaps you might enjoy reading to residents in a senior-living facility. Maybe raising money for a pet shelter appeals to you. Whatever your interests, there is sure to be a related service project.

1. Create a Microsoft Word document that lists service projects or volunteer activities in which you have taken part. Use the heading Community Service on the document along with your name. List the name of the organization or person you helped, the dates of service, and the activities that you performed. If you received an award related to this service, mention it here.

2. Save the document in an appropriate folder.

3. Update your spreadsheet to reflect the inclusion of this Community Service document.

CTSOs

Written Events. Many competitive events for CTSOs require students to write a paper and submit it either before the competition or when the student arrives at the event. Written events can be lengthy and take a lot of time to prepare, so it is important to start early. To prepare for a written event, complete the following activities.

1. Read the guidelines provided by the organization. The topic to be researched will be specified in detail. Also, all final formatting guidelines will be given, including how to organize and submit the paper. Ask questions about any points you do not understand.

2. Begin researching early. Research may take days or weeks, and you do not want to rush the process.

3. Set a deadline for yourself so you can write at a comfortable pace.

4. After you write the first draft, ask an instructor to review it for you and provide feedback.

5. Once you have the final version, go through the checklist for the event to make sure you have covered all the details. Your score will be penalized if you do not follow instructions.

6. To practice, visit your organization's website and select a written event in which you might be interested. Research the topic and then complete an outline. Create a checklist of guidelines that you must follow for this event. After you have completed these steps, decide if this is the event or topic that interests you. If you are still interested, move forward and start the writing process.

String Processing

Sections

String processing is widespread in computing. For example, commerce websites provide forms in which customers enter names, addresses, telephone numbers, and credit card information. Those customers may enter this information in any format. It is then a program's job to massage the data into a standard format to be saved in the retailer's database.

As another example, search engines such as Google allow users to search the web for terms of interest by entering text into a search form. The search engines have automated programs called bots that continually seek and read web pages. The bots extract keywords and store those finds in a large database. Because of these databases, the search engine can respond quickly to user searches by displaying relevant web pages that match the user's search terms. All this requires string processing. In this chapter, you will learn how to process strings to perform a variety of tasks.

Reading Prep

Before reading this chapter, write the main headings for each section, leaving space under each. As you read the chapter, write three points you learned that relate to each heading.

College and Career Readiness

While studying, look for the activity icon ➦ for:

- Vocabulary terms with e-flash cards and matching activities.
- Starter files for hands-on examples and other exercises.

These activities can be accessed at
www.g-wlearning.com/informationtechnology/1773

Chapter Glossary

empty String: String object that has been created, but contains no characters.

index: Number representing the relative position of a character in a String; first index is always 0.

shortcut concatenation operator: Appends a String or other data type to an existing String; represented in Java with the symbol +=.

substring: String composed of a sequence of zero to all the characters in a String in the original order.

traverse: Visit each character in a String in order one at a time.

wrapper class: Encloses, or "wraps," a primitive type into an object, which can then be used to call methods.

Creating Strings

A Java String is a sequence of characters. Strings are objects, rather than a primitive type such as ints or doubles. The String class is in the java.lang package so no import is needed to use Strings. Strings can be created in several ways. As with any object, a String can be created using one of the many String constructors. However, as explained in Chapter 6, programmers typically do not use the String constructors because other ways of creating Strings are easier to use.

One of those ways to create a String is by assigning a String literal to a String reference as you have done for input prompts and output labels. Another way to create a String is to receive a String as a return value from a method. For example, the next and nextLine methods of the Scanner class return the user input or file contents as a String. Finally, characters can be added to existing Strings by concatenating data of any type. This includes other Strings. As discussed in Chapter 4, to *concatenate* means to join or link together.

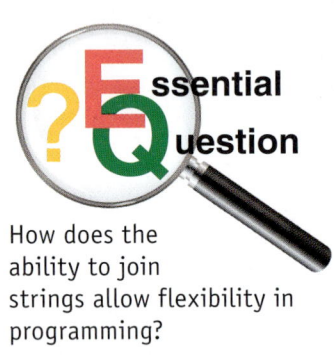

How does the ability to join strings allow flexibility in programming?

Learning Goals

- Discuss an overview of String literals.
- Identify the methods used to input Strings into programs.
- Create concatenated Strings.

Terms

empty String

shortcut concatenation operator

Defining String Literals

Strings can be created by assigning a String literal to a String object reference. As you have previously learned, a String literal is any sequence of characters enclosed within quotation marks (""). Creating Strings in this way works because Java implements String literals as String objects. The String reference is the name of the String object.

Programs you have created to this point used String literals mainly for input prompts and output labels. For example, these are String literals:

```
"Enter the code letter >"
"1234567890"
"The answer is "
```

Java implements String literals as String objects. Creating a named String object is as simple as assigning a String literal to a String reference. Some examples are:

```
String prompt = "Enter the code letter >";
String digits = "1234567890";
String label = "The answer is ";
String empty = ""; // an empty String
```

An *empty String* is a String object that has been created, but contains no characters.

String Input Review

As discussed in Chapter 6, the Scanner class in the java.util package provides two methods for inputting a String into a program. As a review, these methods are shown in **Figure 10-1.** Both methods return a String, which typically is assigned to a String reference. Examples of the use of these methods are:

```
Scanner input = new Scanner( System.in );

System.out.print( "Enter a word > " );
String word = input.next( );

System.out.println( "Enter a sentence > " );
String sentence = input.nextLine( );
```

Method	Action
String next( )	Returns the next input token as a String.
String nextLine( )	Returns the remainder of the line as a String.

Goodheart-Willcox Publisher

Figure 10-1. These methods are used to input a String.

Both methods work on user input, as well as data in a text file. The methods differ in their handling of white-space characters. Recall, white space is spaces, tabs, and new line characters.

The next method skips any leading white space, then builds a String with successive characters until another white-space character is found. The String is then returned from the method without the terminating white-space character, which is left in the input.

In contrast, the nextLine method builds a String composed of every character, including white-space characters, until a new line character is found. The new line character is not returned as part of the String, but the character is removed from the input.

Building Strings Using Concatenation Operators

Java provides two concatenation operators for adding characters to an existing String. These operators are shown in **Figure 10-2.** These concatenation operators work similarly to their cousin operators: the addition operator (+) and shortcut addition operator (+=). The difference is that both String concatenation operators append characters to a String.

As you have previously learned, the concatenation operator (+) builds a String by combining characters from Strings or a variable of another data type. One of the operands must be a String, but the other operand can be any data type. The JVM converts the variable of the other data type to a String, then combines the two Strings in order left to right.

The *shortcut concatenation operator* (+=) appends a String or other data type to an existing String. This operator is useful for incrementally building Strings as the program runs.

FYI

Concatenate seems like a pretty big word to use for a plus sign. It actually comes straight from Latin. *Con* means together, and *caten* means chain. So, *concatenate* means to chain together.

Operator	Example	Action
Concatenation (+)	operand1 + operand2	Results in a String composed of operand1 followed by operand2. One of the operands must be a String; the other operand can be any data type.
Shortcut concatenation (+=)	string += operand	Appends operand, which can be any data type, to an existing String.

Goodheart-Willcox Publisher

Figure 10-2. These are the concatenation operators for working with Strings.

A useful method in the String class for tidying up a String is the trim method. This method is shown in **Figure 10-3.** The trim method removes any leading or trailing white-space characters from a String. This can be especially useful when concatenating Strings.

Method	Explanation
String trim( )	Returns a copy of the String with leading and trailing white-space characters removed.

Goodheart-Willcox Publisher

Figure 10-3. The trim method of the String class can be used to clean up a String.

HANDS-ON EXAMPLE 10.1.1

Building a Sentence

In this exercise, you will build a sentence. The user will enter one word at a time and will signal the end of the sentence by entering a period. The String concatenation operators will help in building the sentence one word at a time. First, think of the algorithm. Given that the number of words in the sentence will not be known and can vary from one run of the program to another, it makes sense to use a while loop that ends when the user enters a period. For the sentence, we can start with an empty String and build the sentence incrementally as the user enters words. An algorithm for the whole program is:

1. Define an empty String for the sentence.
2. Prompt the user for a word.
3. While the word is not a period:
   A. Add the word plus a space to the sentence.
   B. Prompt the user for the next word.
4. Trim the sentence to remove the final space.
5. Append the period to the sentence.
6. Output the sentence.

Before beginning this exercise, download the chapter files from the student companion website. The SentenceBuilderSnippet.java file will be used as a starting point.

1. Launch jGRASP, and open the SentenceBuilderSnippet.java file.
2. Examine the code and the comments. Notice the steps of the algorithm are reflected in the comments.
3. Locate comment 1, and follow the instructions to define the period as a constant.

```
/***** 1. Define the period as a constant. */
final String PERIOD = ".";
```

4. Locate comment 2, and follow the instructions to create an empty String named sentence. You will build the sentence word by word. Notice that no space is between the quotation marks. If a space were included, the sentence would contain one space rather than being empty.

```
/***** 2. Create an empty sentence. */
String sentence = "";
```

5. It is time to write the while loop. Locate comment 3, and follow the instructions to perform the priming read. That is, prompt the user for the first word.

```
/***** 3. Prompt the user for the first word. */
System.out.println( "Enter a word, period to stop" );
String word = input.next( );
```

6. Locate comment 4, and follow the instructions to write the while loop header. The condition will be that the user has not entered a period. Remember, use the equals method to compare Strings. In this case, the condition is that the word is not a period, so the not operator (!) goes before the comparison.

```
/***** 4. Write the while loop header. */
while ( !word.equals( PERIOD ) ) {
```

7. Locate comment 5, and follow the instructions to write the loop body. Append the word plus a space to the growing sentence, then update the loop condition by reading the next word. In this step, a space is needed between the quotation marks.

```
/***** 5. Write the loop body:
       append the word and a space to the sentence
       and prompt for the next word. */
sentence += word + " ";
System.out.println( "Enter another word or period to stop" );
word = input.next( );
```

8. Locate comment 6, and follow the instructions to add the closing brace for the while loop.

```
/***** 6. End the while loop. */
} // end while loop
```

9. At this point, the while loop is complete. The user has entered a period to indicate the sentence has no more words. After each word, however, the program has added a space, so the String in the sentence variable has a space in the last position. Adding the period after the space is improper. Instead, call the trim method to remove the final space, and then add the period to the sentence. Locate comment 7, and follow the instructions.

```
/***** 7. Tidy up the sentence by removing the space after the last word,
       then append the period to the sentence. */
sentence = sentence.trim( );
sentence += PERIOD;
```

10. Locate comment 8, and follow the instructions to output the complete sentence.

```
/***** 8. Output the sentence. */
System.out.println( "Your sentence is:" );
System.out.println( sentence );
```

11. Compile and run the application. Test the program multiple times. Enter a different number of words each time. Debug as needed.

Try It!

Write a similar program that allows a user to add multiple adjectives for an item. First, output the item. You can choose what the item is: a box, a game, a day, or something else. Then, ask the user to enter adjectives or the word "done." The program should build a phrase with each adjective followed by a comma and a space. For example, if the item is a "day," and the user successively enters "A," "long," "hot," "sunny," "done", then the program would output "A long, hot, sunny, day." The trim method is not needed for this program. It would be nice to remove the last comma, however. The next section of this chapter introduces the substring method, which could be useful to achieve that end.

Language Arts and Java

Parts of Speech Game

A fun way to practice parts of speech is to play a game where a story is built around words the player enters. The player is prompted for random words from parts of speech, and the words are weaved into a story. For example, suppose this clever sentence is used as the original story:

> "The quick brown fox jumped over the lazy dog."

The player can be asked to enter an adjective, such as hairy. That new word is used as a replacement for the adjective quick. The story now becomes:

> "The hairy brown fox jumped over the lazy dog."

Undrey/Shutterstock.com

Assignment

Write a Java program to play this game. First, write a relatively plain story. Then, identify nouns, adjectives, and verbs you want replaced by words the player enters. Choose words you think will make your story much funnier when they are replaced by random words. When the program runs, prompt the player to enter a word for a particular part of speech. Continue requesting words from the player for all of the words you identified in your story. Build the story as a String using the concatenation operator as the player supplies each word. When this process is completed, output the story string for the player to see. This is an example of how the game might progress:

```
Enter an adjective: hairy
Enter a noun: refrigerator
Enter a verb in past tense: wandered
Enter a noun: banana
Your story is: The hairy brown
    refrigerator wandered over the lazy
    banana.
```

Name the class PartsOfSpeechGame, and save the file under this name in your working folder. Test and debug the program as needed.

SECTION REVIEW 10.1

Check Your Understanding

1. A String literal is any sequence of characters enclosed between _____.
2. What are the two Scanner methods for inputting a String?
3. What is a String that contains no characters called?
4. To use the String concatenation operators, at least one of the operands must be a(n) _____.
5. Which method can be used to remove leading and trailing white-space characters from a String?

Build Your Vocabulary

As you progress through this course, develop a personal computer science glossary. This will help you build your vocabulary and prepare you for a career. Write a definition for each of the following terms and add it to your computer science glossary.

empty String   shortcut concatenation operator

String Methods

The String class has a number of methods that can be used for powerful searches and for manipulating Strings. One method, length, returns the number of characters in the String. Another method, charAt, extracts a character using its position in the String. New Strings can be created from parts of an existing String using the substring method. Letters in a String can be converted to uppercase using the toUpperCase method or to lowercase using the toLowerCase method. A String can be searched using the indexOf method, which locates a sequence of characters within a String. The replace method works like the search-and-replace function of a word processor. It substitutes all occurrences of one sequence for another sequence of characters. These methods are discussed in this section.

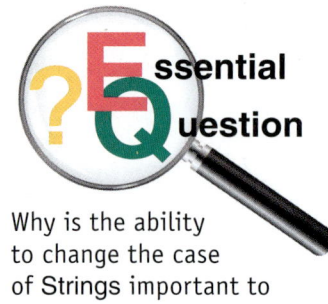

Essential Question

Why is the ability to change the case of Strings important to proper coding of programs?

Learning Goals

- Find the length of a String.
- Determine the index of a character in a String.
- Describe how to extract a substring from a String.
- Create code to change the case of characters in a String.
- Explain how to search a String for characters.
- Replace selected characters in a String.

Terms

index   substring

Length of a String

A Java String can contain from zero to an almost unlimited number of characters. To process a String, it is important to know how many characters it has. The length method of the String class counts the number of characters in a String. This includes letters, digits, special characters, and spaces as well as any other Unicode character. The API for the length method is shown in **Figure 10-4.** Like any method, the length method is called using an object reference, specifically a String object reference, and the dot notation. The length method does not take any arguments. Remember, however, to code the empty parentheses after the method name. The length method returns the number of characters as an int.

Method	Explanation
int length( )	Returns the number of characters in the String.

Goodheart-Willcox Publisher

Figure 10-4. The length method of the String class counts the number of characters in a String.

For example, these statements create String literals and call the length method:

```
String message = "Hello World!";
int lengthMsg = message.length( ); // lengthMsg gets 12

String three = "3";
int lengthThree = three.length( );    // lengthThree gets 1

String empty = "";
int lengthEmpty = empty.length( ); // lengthEmpty gets 0
```

Notice that the space within the String is counted as a character. Notice also that a number within quotes is a String, not a numeric value.

String Indexes

Often, you will need to refer to individual characters in a String. To allow for this, each character in a String is assigned an index. An *index* is a number representing the relative position of a character in a String. The first character in a String always has the index 0. Indexes are incremented by 1 for each succeeding character. Many String methods take indexes as arguments.

Figure 10-5 shows each character's index for this String:

```
String message = "Hello World!";
```

The index of the first character (H) is 0. The index of the second character (e) is 1, and so on. Even the space has an index, which is 5. The index of the last character in the String is always one less than the length of the String. So, with the length of message being 12, the last character's index is 11. The exclamation point can be referred to, therefore, as having index 11. This can be found with the length method: message.length( ) − 1.

FYI

The last character in a String is *always* located at length( ) − 1.

Extracting Characters from a String

A *substring* is a String composed of a sequence of zero to all the characters in a String in the original order. For example, for the String "Hi!", a substring could be any of these:

```
"H"
"Hi"
"Hi!"
"i"
"i!"
"!"
"" // empty
```

However, "iH" is *not* a substring because the order of the letters is not the same as in the original String.

The String class provides two versions of the substring method for extracting characters from a String, as shown in **Figure 10-6.** Both methods take indexes as arguments and return a String. The first version of the substring method takes just one argument, which is the index of the first character to begin the substring. The substring is created starting from the character at the argument index and continuing to the end of the String. Some example method calls for the first method include:

```
String message = "Hello World!";
String sub1 = message.substring( 6 ); // sub1 is World!
String sub2 = message.substring( 3 ); // sub2 is lo World!
String sub3 = message.substring( 11 );  // sub3 is !
```

Character	H	e	l	l	o		W	o	r	l	d	!
Index	0	1	2	3	4	5	6	7	8	9	10	11

Goodheart-Willcox Publisher

Figure 10-5. Each character in a String has an index that can be used to reference it.

Method	Explanation
String substring(startIndex)	Returns the characters in the String beginning at startIndex and continuing to the end of the String.
String substring(startIndex, endIndex)	Returns the characters in the String beginning at startIndex and continuing up to, but not including, endIndex.

Goodheart-Willcox Publisher

Figure 10-6. The substring method of the String class has two versions for extracting characters.

The second version of the substring method takes a second argument to specify the end point of the substring. This substring method then collects characters from startIndex up to, but *not* including, the endIndex. Some example method calls for the second method include:

```
String message = "Hello World!";
String sub4 = message.substring( 0, 5 );   // sub4 is Hello
String sub5 = message.substring( 1, 7 );   // sub5 is ello W
String sub6 = message.substring( 0, 1 );   // sub6 is H
```

If any of the indexes are invalid, the substring method will generate a StringIndexOutOfBoundsException error. A start index would be invalid if it is less than zero. An end index would be invalid if it is less than zero or greater than the length of the String.

Hands-On Example 10.2.1

Formatting String Input

In this exercise, you will develop a program that simulates processing a web form. Specifically, you will code the function to format a phone number. The user will enter a phone number as ten digits. The program will then format the phone number as the area code in parentheses followed by a space, the next three digits of the phone number, a hyphen, and then the last four digits of the number. For example, if the user enters 1234567890, your program will display the number as (123) 456-7890. Here is an algorithm that starts with an empty String and builds the formatted phone number a little at a time:

1. Start the formatted String as an empty String.
2. Add a left parenthesis.
3. Extract the area code, and add it to the formatted String.
4. Add a right parenthesis and a space.
5. Extract the next three digits, and add them to the formatted String.
6. Add a hyphen.
7. Extract the last four digits, and add them to the formatted String.

The substring methods should be helpful for this problem. Before beginning this exercise, download the chapter files from the student companion website. The FormatPhoneSnippet.java file will be used as a starting point.

1. Launch jGRASP, and open the FormatPhoneSnippet.java file.
2. Examine the code and the comments. Notice the steps of the algorithm are reflected in the comments.

3. Locate comment 1, and follow the instructions to define a constant for the required length. Also, define the inputPhone variable as a String. This must be done now so the scope is outside the do/while loop because the variable will be used in the while condition.

```
/***** 1. Define NUMBER_OF_DIGITS as a constant (10)
       and the input phone number as a String */
final int NUMBER_OF_DIGITS = 10;
String inputPhone; // define before do/while loop
```

4. Locate comment 2, and follow the instructions to write a do/while loop to ensure the user enters exactly ten characters for the phone number. It is important to verify the user has indeed entered ten digits because the substring method will generate a StringIndexOutOfBoundsException error if an index does not exist.

```
/***** 2. Write the do/while loop; the input phone number must be 10
       characters. Use the length method to count the number of
       characters */
do {
    System.out.println( "Enter your phone number; 10 digits only" );
    inputPhone = input.next( );
    if ( inputPhone.length( ) != NUMBER_OF_DIGITS ) {
        System.out.println( "Please enter " + NUMBER_OF_DIGITS + " digits!" );
    }
} while ( inputPhone.length( ) != NUMBER_OF_DIGITS );
```

5. Locate comment 3, and follow the instructions to define an empty String for the formatted phone.

```
/***** 3. Define an empty String for the formatted phone number */
String formattedPhone = "";
```

6. Locate comment 4, and follow the instructions to build the formatted phone number using the substring method and concatenation.

```
/***** 4. Build the formatted String by calling substring and using
       concatenation */
formattedPhone += "(" + inputPhone.substring( 0, 3 ) + ") ";   // area code
formattedPhone += inputPhone.substring( 3, 6 ) + "-";     // exchange
formattedPhone += inputPhone.substring( 6 );            // last 4 digits
```

7. Locate comment 5, and follow the instructions to output the formatted phone.

```
/***** 5. Output the formatted phone number */
System.out.println( "Your phone number is " + formattedPhone );
```

8. Compile and run the program. Enter a phone number with nine digits. Then, enter a phone number with 11 digits. Notice if you enter too few or too many digits, the program continues to prompt for a phone number. Finally, enter a ten-digit phone number. Verify the phone number is correctly formatted.

Try It!

Another item that users enter into a web form is a credit card number. Write a program that accepts a 16-digit credit card number and formats the number so a space separates every four digits. For example, if the user enters 1234567890123456, the program should output: 1234 5678 9012 3456. Be sure to ensure the user has entered exactly 16 digits. Name the application FormatCreditCard, and save it in your working folder.

Changing Case

The String class has two methods for changing the case of letters: toUpperCase and toLowerCase. These methods are described in **Figure 10-7.** These methods return a new String with all letters converted to uppercase or lowercase. The original String is unaffected. These methods affect only letters. Other characters in the String are returned unchanged.

In the following statements, the first two print the return value from the method without assigning the converted String to a variable. The third statement

Method	Explanation
String toUpperCase( )	Returns a copy of the String with all letters in uppercase.
String toLowerCase( )	Returns a copy of the String with all letters in lowercase.

Figure 10-7. These methods can be used to change the case of all letters in a String.

demonstrates that the original String, which is stored in message, has not been changed.

```java
String message = "Hello World!";
System.out.println( message.toUpperCase( ) );  // HELLO WORLD!
System.out.println( message.toLowerCase( ) );  // hello world!
System.out.println( message );                 // outputs Hello World!
```

Always remember, the toUpperCase and toLowerCase methods return a *copy* of the String. To modify the original String, assign the return value to the original String.

```java
String message = "Hello World!";
message = message.toUpperCase( );
System.out.println( message ); // message is now HELLO WORLD!
message = message.toLowerCase( );
System.out.println( message ); // message is now hello world!
```

HANDS-ON EXAMPLE 10.2.2

Formatting a String Input Name

In this exercise, you will develop another program that simulates processing a web form. Specifically, you will code the function to format a name. The user will enter his or her name. The program will then format the name to start with an uppercase letter followed by all other letters in lowercase. An algorithm could be to start with an empty String and build the name:

1. Start the formatted name as an empty String.
2. Extract the first character in the name, convert to uppercase, and add it to the formatted name.
3. Extract the remaining characters in the name, convert to lowercase, and add them to the formatted name.

The toUpperCase, toLowerCase, and substring methods will be helpful for this problem. Before beginning this exercise, download the chapter files from the student companion website. The FormatNameSnippet.java file will be used as a starting point.

1. Launch jGRASP, and open the FormatNameSnippet.java file.
2. Examine the code and the comments. Notice the steps of the algorithm are reflected in the comments.
3. Locate comment 1, and follow the instructions to prompt the user for a first name. Read this name using the next method of the Scanner class.

```java
/***** 1. Prompt for the user's first name. */
System.out.println( "Enter your first name:" );
String firstName = input.next( );
```

4. Locate comment 2, and follow the instructions to define an empty String for the formatted name.

```java
/***** 2. Define the formatted name as an empty String.  */
String formattedName = "";
```

5. Locate comment 3, and follow the instructions to build the String.

```
/***** 3. Build the formatted string by calling toUpperCase,
      toLowerCase, and substring and using concatenation. */
String firstLetter = firstName.substring( 0, 1 );  // extract first letter
firstLetter = firstLetter.toUpperCase( );      // convert to uppercase
formattedName += firstLetter;           // add to formatted name

String restOfName = firstName.substring( 1 ); // extract second letter to the end
restOfName = restOfName.toLowerCase( );      // convert to lowercase
formattedName += restOfName;           // add to formatted name
```

6. Locate comment 4, and follow the instructions to output the formatted name.

```
/***** 4. Output the formatted name. */
System.out.println( "Your name is " + formattedName );
```

7. Compile and run the program. Enter your own name with a random mix of uppercase and lowercase letters. Check that the program outputs your name with only the first letter in uppercase.

Try It!

Add on to this program the ability to format first and last names. Read the last name as well as the first name each as a separate String using the next method twice. Remember to add a space to the formatted name between the first and last names. Fully test and debug your program.

Searching for Characters

Knowing if a String contains specific sequences of characters, and knowing where those characters are located, can be helpful. Perhaps the program knows what characters to extract from a String, but does not know where those characters begin. The String class has an indexOf method that locates the index of the first character in a search string. The argument for the method is the search string. The search is case-sensitive. If the search string is not found, the method returns −1.

Actually, the String class provides multiple variations of the indexOf method. The simplest is highlighted here in **Figure 10-8.** Here are some examples of the indexOf method in action using the String shown in **Figure 10-5.**

Remember to check the return value from the indexOf method. Attempting to use the return value as an index when the search string is not found (−1) will result in a StringIndexOutOf-BoundsException error.

```
String message = "Hello World!";
int indexE = message.indexOf( "e" );       // indexE gets 1
int indexOfSpace = message.indexOf( " " ); // indexOfSpace gets 5
int indexO = message.indexOf( "o" );       // indexO gets 4 ->
                                           // first occurrence
int indexL = message.indexOf( "l" );       // indexL gets 2 ->
                                           // first occurrence
int indexW = message.indexOf( "w" );       // indexW gets -1 -> not
                          // found because of case-sensitivity
int indexEnd = message.indexOf( "orld!" ); // indexEnd gets 7 ->
                                           // first character
```

Method	Explanation
int indexOf(String searchString)	Returns the index of the first character of searchString in the String or −1 if searchString is not found.

Goodheart-Willcox Publisher

Figure 10-8. This is the simplest variation of the indexOf method in the String class, which is used to locate the index of the first character in a search string.

Math and Java

Mathematics Quotations Quiz

Many mathematicians throughout history were philosophers as well. They have contributed many great quotes to our culture. René Descartes is famous for saying, "I think, therefore I am." Isaac Newton credited his ability to invent with the knowledge of those who came before him. He said, "If I have seen further, it is by standing on the shoulders of giants."

NASA, Bob Nye

Georgios Kollidas/Shutterstock.com

Albert Einstein once said, "Do not worry too much about your difficulties in mathematics, I can assure you that mine are still greater." NASA mathematician Katherine Johnson stated, "We will always have STEM with us. Some things will drop out of the public eye and will go away, but there will always be science, engineering, and technology. And there will always, always be mathematics."

Assignment

Apply what you have learned about string processing to create a quiz in which you provide these four quotations, one at a time. Prompt players to enter the name of the person credited with the quotations.

Take care with input validation. Players may enter the last name or the entire name. You cannot anticipate what the user will try to enter. Apply what you have learned about substrings to isolate the last name in the input string. Make your determination on the correct spelling of just the last name, and allow the answer to be in any case. If the answer is correct, add 1 point to the score and congratulate the player. If incorrect, display the correct name. Provide a percentage of correct answers at the end.

HANDS-ON EXAMPLE 10.2.3

Searching for Strings 🔗

In this exercise, you will develop a program that searches a file for a sequence of characters. The user will enter the sequence of characters to find. The program will read the file and output each line that contains that sequence of characters, including the line number. To make it simple, your program will search itself (the .java file). Here are the steps:

1. Prompt the user for the search string.
2. Open the file to read.
3. Use a boolean flag variable to indicate whether the searchString has been found. Set the flag stringFound to false at the beginning and to true when the searchString is found.
4. While there is data in the file, read a line, increment the line number, and search for the search string. If the search string is found, output the line number and the line with the search string surrounded by two asterisks (**) to make it easier to locate the sequence of characters in the output; set stringFound to true.
5. If the whole file has been read and stringFound is still false, output a message that the search string was not in the file.

The indexOf, length, and substring methods will be helpful for this problem. Before beginning this exercise, download the chapter files from the student companion website. The SearchStringsSnippet.java file will be used as a starting point.

1. Launch jGRASP, and open the SearchStringsSnippet.java file.
2. Examine the code and the comments. Notice the steps of the algorithm are reflected in the comments.
3. Locate comment 1, and follow the instructions to prompt the user for the search string. Read this using the nextLine method of the Scanner class to allow for spaces in the search string.

```
/***** 1. Prompt the user for the search string.
    Use nextLine to allow for multiple words. */
System.out.println( "Enter a word or phrase to find:" );
String searchString = input.nextLine( );
```

4. Locate comment 2, and follow the instructions to open the program file.

```
/***** 2. Open this file */
File inputFile = new File( "SearchStringsSnippet.java" );
Scanner file = new Scanner( inputFile );
```

5. Locate comment 3, and follow the instructions to set up some variables. The boolean flag stringFound starts as false and is set to true if the search string is found. The line number starts at 0 and is incremented each time a line is read.

```
/***** 3. Define a boolean flag, stringFound as false,
    and initialize the line number to 0 */
boolean stringFound = false;
int lineNumber = 0;
```

6. Locate comment 4, and follow the instructions to begin the while loop to read the file.

```
/***** 4. Begin the while loop to read the file line by line */
while ( file.hasNext( ) ) {
```

7. Locate comment 5, and follow the instructions to increment the line number. This should be inside the loop.

```
/***** 5. Read a line, increment the line number,
    and search for the searchstring */
String line = file.nextLine( );
lineNumber++;
int indexOfFound = line.indexOf( searchString );
```

8. Locate comment 6, and follow the instructions to output the line if the search string is found. This should also be inside the loop. If the search string is found, create a new String with the formatted line number. Start a System.out.println where the output is built using concatenation. Add the formatted line number. Use substring to collect the characters before the search string, then concatenate those characters as well as the

two asterisks into the output String. To extract the search string from the line, use substring with the start index found by indexOf, and calculate the end index by adding the length of the search string. Follow those characters by two more asterisks and the remainder of the line.

```
/***** 6. If the search string is found,
     output the line number and the line, enclosing
     the searchstring in **, and set stringFound to true. */
if ( indexOfFound != -1 ) {
   String lineNo = String.format( "%2d ", lineNumber );
   System.out.println( lineNo
      + line.substring( 0, indexOfFound )
      + "**"
      + line.substring( indexOfFound, indexOfFound + searchString.length( ) )
      + "**"
      + line.substring( indexOfFound + searchString.length( ) ) );
   stringFound = true;
} // end if
```

9. Locate comment 7, and follow the instructions to finish the while loop.

```
/***** 7. Finish the while loop. */
} // end while loop
```

10. Locate comment 8, and follow the instructions to check whether the search string was found in the file. If not, output a message to the user.

```
/***** 8. Check if the search string was found in any line. */
if ( !stringFound ) {
   System.out.println( searchString + " was not found." );
} // end if
```

11. Compile and run the program. Enter a search string that will be found on multiple lines, (perhaps indexOfFound). Notice how the output is formatted with line numbers for where the search string is found and the search string is surrounded by asterisks. Run the program again with a search string that will not be found.

Try It!

Modify the SearchStringsSnippet application, rename the class to SearchMarsInNASA, and save the file under the new name in your working folder. Edit the code to open the file NASA.txt. Run the program, and search the file for "Mars". Modify the program to count how many times the search string occurs in the news feed. At the end of the program, output that number. Try the program again and search for a different String.

Replacing Characters

There may be times where you need to selectively replace characters in a string. The String class provides two versions of the replace method for doing this. These methods can be thought of as "search and replace." The methods take a target character or String and return a copy of the String with each occurrence of that target replaced by the specified character or String. The methods are described in **Figure 10-9.** One version accepts the target and replacement as

Method	Explanation
String replace(char oldChar, char newChar)	Returns a copy of the String with all occurrences of the single character oldChar replaced with the single character newChar.
String replace(String oldString, String newString)	Returns a copy of the String with all occurrences of oldString replaced with newString.

Goodheart-Willcox Publisher

Figure 10-9. Two versions of the replace method in the String class can be used to selectively replace characters in a String.

Coding Conundrum

1. You intend to define a String named pet using this statement:

```
String pet = 'dog';
```

but the compiler generates three errors:

```
...error: unclosed character literal
String pet = 'dog';
           ^

...error: unclosed character literal
String pet = 'dog';
               ^

...error: not a statement
String pet = 'dog';
           ^
```

It is a conundrum! How can you fix this code?

2. You want to extract the first letter from this String:

```
String name = "Mary Lee";
System.out.println( name.substring( 0, 0 ) );
```

but nothing is printed. It is a conundrum! How can you fix this code?

3. You intend to convert a String to uppercase using this code:

```
String city = "Rome";
city.toUpperCase( );
System.out.println( city );
```

but the output is still "Rome". It is a conundrum! How can you fix this code?

---

a char data type. The other version accepts the target and replacement as a String data type. Both versions are case-sensitive.

For example, assume your instructor originally posted a message that the course will have two exams. Then, he or she decides the word "test" would be less intimidating than "exam." Using these statements, the instructor can easily make this change:

```
String policy = "This course will have two exams. Each exam is "
+ "worth 50 points.";
policy = policy.replace( "exam", "test" );
System.out.println( policy );
// output is: This course will have two tests. Each test is worth
// 50 points.
```

Notice the first occurrence of exam is plural in the original String (exams) and the replacement (test) is singular. However, the s is retained because it is not part of the "search and replace." Like the toUpperCase and toLowerCase methods, the replace method returns a *copy* of the original String. To change the original String, assign the return value to the original String, as in the example above.

Since the method is case-sensitive, any characters in the original String that vary in case from the replacement String will not be changed. In the following example, only the first occurrence of "exam" is replaced. The second occurrence does not match the replacement String because "Exams" starts with an upper-case letter.

```java
String policy2 = "This course will have two exams. Exams are worth "
+ "50 points.";
policy2 = policy2.replace( "exam", "test" );
System.out.println( policy2 );
// output is: This course will have two tests. Exams are worth
   50 points.
```

HANDS-ON EXAMPLE 10.2.4

Decoding a Cryptogram 🔗

A cryptogram is an encoded message. Each letter in the cryptogram represents a letter in the decrypted message. For example, if the letter *C* represents the letter *X*, then all letters *C* in the cryptogram would be *X* in the message. Each cryptogram has a unique set of letter mappings. Here is a simple example: bffgfg. For this cryptogram, the letter mappings are:

    b = w
    f = e
    g = d

that is, all *b*s in the cryptogram are *w*s in the solution, all *f*s in the cryptogram are *e*s in the solution, and all *g*s in the cryptogram are *d*s in the solution. The user solves the puzzle by guessing a cryptogram letter and a replacement letter until the solution is revealed. For bffgfg in the above example, the solution is *weeded*.

In this exercise, you will write code that displays a cryptogram. The user will then input target letters and replacement letters until the cryptogram is solved. The user may give up by entering a "?" for the target letter. In that case, the game will be over and the program will display the solution. Before beginning this exercise, download the chapter files from the student companion website. The DecoderSnippet.java file will be used as a starting point.

1. Open the adages.txt file in Notepad or other text editor, and examine the file. It contains 12 cryptograms and solutions that are adages. Each cryptogram occupies two lines in the file: the first line is the cryptogram and the second line is the cryptogram's solution. Your program should randomly select one of the adages to present to the user. In some cases, the coded letter and the solution letter are the same. Close the text file.

2. Launch jGRASP, and open the DecoderSnippet.java file.

3. Examine the code and the comments.

4. Locate comment 1, and follow the instructions to generate a random number between 1 and 12. This will be used to select the cryptogram and adage from the text file.

```java
/*****1. Generate a random number for selecting the adage. */
final int NUMBER_ADAGES = 12; // 12 adages in the file
Random randGen = new Random( );
int randAdage = randGen.nextInt( NUMBER_ADAGES ) + 1;
```

5. Locate comment 2, and follow the instructions to read the selected adage and cryptogram.

```
/***** 2. Open the file adages.txt
          and read the selected adage and its cryptogram. */
Scanner file = new Scanner( new File( "adages.txt" ) );
String adage = "", codedAdage = "";

for ( int i = 1; i <= randAdage; i++ ) {
   adage = file.nextLine( );
   codedAdage = file.nextLine( );
}

String cryptogram = codedAdage;
String solution = adage;
```

6. Locate comment 3, and follow the instructions to display the cryptogram and the first word in the adage as a hint.

```
/***** 3. Display the cryptogram and the first word in the solution
          as a hint. */
System.out.println( "The cryptogram is: \n" + cryptogram );

// Find the space after the first word
int indexOfSpace = solution.indexOf( " " );

// First word is from index 0 up to space - 1
String firstWord = solution.substring( 0, indexOfSpace );

System.out.println( "Hint! The first word is " + firstWord );
```

7. Locate comment 4, and follow the instructions to create a do/while loop and read the user's guess.

```
/***** 4. In a do/while loop read the user's code letter.
          Check if the code letter is ? indicating the user gives up.
          If not, read the substitution letter and make the replacement.
          The while condition is that the user has not guessed the solution
          and has not given up. */
String codeLetter; // define before do/while
do {
   System.out.println( "Enter the code letter or ? to give up" );
   codeLetter = input.next( );

   if ( ! codeLetter.equals( "?" ) ) { // not giving up
      System.out.println( "Enter the letter to substitute for " + codeLetter );
      String substituteLetter = input.next( );

      //  add more code here from comment 5
   }
} while ( !cryptogram.equalsIgnoreCase( solution ) && !codeLetter.equals( "?" ) );
```

8. Locate comment 5, and follow the instructions to make the substitution requested by the user. Note that this code must go in the body of the do/while loop where indicated by the // Add more code here from comment 5 comment added in the previous step. First, convert the code letter to lowercase to ensure it will match letters in the cryptogram. Also, convert the substitute letter to uppercase to clearly indicate which letters have been solved.

```
/***** 5. Convert code letter to lowercase and
          substitute letter to uppercase, then replace
          code letter with substitute letter and display cryptogram. */

codeLetter = codeLetter.toLowerCase( );
substituteLetter = substituteLetter.toUpperCase( );

cryptogram = cryptogram.replace( codeLetter, substituteLetter );

System.out.println( "The cryptogram is now\n" + cryptogram );
```

9. Locate comment 6, and follow the instructions to determine if the loop ended because the user gave up or if the user solved the cryptogram. If the user gave up, display the solution. If the user solved the cryptogram, congratulate the user. Users appreciate recognition.

```
/***** 6. Check whether user gave up or solved the cryptogram. */
if ( codeLetter.equals( "?" ) ) { // did user give up?
   System.out.println( "The adage is: " + solution );
} else {
   System.out.println( "Congratulations! You decoded the cryptogram!" );
}
```

10. Compile and run the program. Test to ensure the program functions as intended. You might want to "cheat" by looking in the text file for the answer.

Try It!

This program could be improved by allowing the user to undo the last guess. Change the name of the class to DecoderWithUndo, and save the file under the new name. Then, modify the program so the user can enter a caret (^) when prompted for the code letter to undo the last substitution. You will need two variables that store the last values passed to the replace method. Initialize these variables to an empty String before the do/while loop, and update the variables each time the user makes a substitution. If the user tries to undo the last guess, but the last values are empty, then output a message, "Nothing to undo." If you are able to undo the substitution, call the replace method with the variables reversed. Then, reset the last values to empty Strings to indicate that there is nothing more to undo.

SECTION REVIEW 10.2

Check Your Understanding

1. Which method in the String class is used to find the number of characters in a String?
2. Which index is given to the first character in a String?
3. If an index sent as an argument to the substring method is invalid, what exception is generated?
4. Which two methods change the case of all letters in a String?
5. What value is returned by the indexOf method if the search string cannot be found?

Build Your Vocabulary

As you progress through this course, develop a personal computer science glossary. This will help you build your vocabulary and prepare you for a career. Write a definition for each of the following terms and add it to your computer science glossary.

index   substring

Processing Strings Character by Character

Some applications need to process every character in a String one by one. Perhaps the processing is counting the number of words by counting the spaces in the String. Another application may verify that a password contains the required number of digits or capital letters. In these applications, a for loop can be used to extract each character in the String and process that character individually. The String class provides the charAt method for extracting individual characters as the char data type rather than as substrings.

Primitive data types are not objects. Methods cannot be called on primitive data types. To add the capability of calling methods for a primitive data type such as char, the Java Class Library provides wrapper classes. Specifically, the Character wrapper offers methods for examining the char data type to see if the character is uppercase or lowercase and if the character is a digit or a letter.

Learning Goals

- Explain extracting characters by traversing a String.
- Use a wrapper class.

Terms

traverse   wrapper class

Essential Question

Why is the ability to extract characters from a string important to programming?

Extracting Characters

The charAt method returns the character at a specified index in the String. **Figure 10-10** shows the API for the charAt method. Notice that the method returns a char, not a String. Here are some examples of using the charAt method:

```
String message = "Hello World!";
char firstCharacter = message.charAt( 0 );
  // firstCharacter gets H
char lastCharacter = message.charAt( message.length( ) - 1 );
  // lastCharacter gets !
```

A for loop used with the charAt method is handy for traversing a String character by character. To *traverse* means to visit each character in a String in order

FYI

Remember, the last letter in any String is located at the index length( ) − 1.

Method	Explanation
char charAt(int index)	Returns the character in the String at the specified index as a char data type.

Goodheart-Willcox Publisher

Figure 10-10. The charAt method can be used to extract the character at a specified index in a String.

one at a time. To traverse a String with a for loop, use the index of the character as the loop-control variable.

Forward Traversing

A String can be traversed forward from the first character to the last. This process is as follows.

1. Initialize the loop-control variable (i) to the index of the first character (0).
2. Use the condition i < string.length( ) to test if the loop has processed each character.
3. Inside the loop, extract the current character as charAt(i).
4. For the loop update, add one to the loop-control variable (i++) to move to the next index.

The variable i is commonly used for a loop counter, but this can be any variable name. The code below traverses a String forward from the first character. This creates a String that consists of each character in the original String separated by a space:

```
String message = "Hello World!";
String output = ""; // start with an empty String

for ( int i = 0; i < message.length( ); i++ ) {
   output += message.charAt( i ) + " ";  // append character and
                                         // space to output
}

System.out.println( output ); // output: H e l l o   W o r l d !
```

Backward Traversing

There may be times when you need to start at the end of the string and traverse to the beginning. A String can be traversed backward from the last character to the first. This process is as follows.

1. Initialize the loop-control variable (i) to the index of the last character: length( ) − 1.
2. Use the condition i >= 0 to test if i has taken on the value of each index.
3. Inside the loop, the code is the same as for forward traversing: extract the current character as charAt(i).
4. For the loop update, subtract one from the loop-control variable (i--) to move backward to the next lower index.

The following code will output the characters in the String in reverse order.

```
String message = "Hello World!";
String output = "";

for ( int i = message.length( ) - 1; i >= 0; i-- ) {
   output += message.charAt( i ); // append character to output
}

System.out.println( output ); // output is: !dlroW olleH
```

FYI

Be careful *not* to use the condition i <= string.length( ) because string.length( ) is an invalid index.

Science and Java

Gene Sequencing

One of the tasks of genetic research is gene sequencing. Strings of genetic information are compared and analyzed for the percentages of difference and similarity. The sequences are divided into smaller segments for analysis and then spliced together for the whole genome. For the human genome, this comparison could take days of high-speed computing.

Dmytro Zinkevych/Shutterstock.com

In the case of nucleic acid, the components are combinations of adenine (A), cytosine (C), guanine (G), and thymine (T). The four components are called DNA bases. Arrangements of these bases define subsequences.

Assignment

Before beginning this exercise, download the chapter files from the student companion website. The GeneSequencingSnippet.java file will be used as a starting point. Suppose two sample 20-nucleotide sequences have these arrangements:

Sequence 1:   ATACGCCTGACCCTATAAAG
Sequence 2:   ATCAGCTTGAACCTAGATAG

Apply what you have learned about traversing strings to identify the differences in the two sequences. Write a program to count the differences and calculate the percentage of the 20 nucleotides that are different. Subtract that percentage from 100 percent to compute the percentage of similarity between the two nucleotide sequences. Output the two sequences, the number of differences, the percentage of difference, and the percentage of similarity. Hand-check your results by comparing the two given sequences.

HANDS-ON EXAMPLE 10.3.1

Checking for Palindromes

In this exercise, you will write code that checks whether a word, phrase, or sentence is a palindrome. A palindrome is text that is spelled the same way backward and forward. For example, *radar*, *otto*, and *racecar* are all palindromes. These famous sentences are also palindromes: *Able was I ere I saw Elba* and *Madam I'm Adam*. Here is an algorithm for determining if a palindrome exists:

1. Prompt the user for the text.
2. Use a for loop with two variables, one as the loop-control variable at the beginning of the String and a second variable representing the index at the end of the String.
3. Traverse the String comparing the characters at corresponding indexes. Compare the first character to the last character, and the second character to the second-to-the-last character, and so on.
4. If any of the two characters being compared do not match, the String is not a palindrome.
5. If the loop finishes without finding characters that do not match, the String is a palindrome.

The charAt method and a for loop will be helpful for this problem. The figure shows how this comparison will work for the String RACECAR. What becomes obvious is that the second half of the String is being checked at the same time as the first half. This leads to the conclusion that only half the String needs to be traversed. Also obvious is the middle character of a String with an odd number of characters is not checked. That is fine because the middle character does not have to match another character. The middle character will be the same middle character both forward and backward. Before beginning this exercise, download the chapter files from the student companion website. The PalindromeSnippet.java file will be used as a starting point.

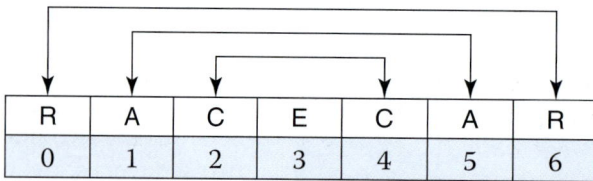

Goodheart-Willcox Publisher

1. Launch jGRASP, and open the PalindromeSnippet.java file.
2. Examine the code and the comments.
3. Locate comment 1, and follow the instructions to prompt the user for a String to check. Make a copy of the String in uppercase to allow comparison without regard to case. In this way, *Radar* and *radar*, for example, will both be found to be palindromes and the original String is unchanged.

```
/***** 1. Prompt the user for a word, phrase, or sentence. */
System.out.println( "Enter a word, phrase, or sentence:" );
String sentence = input.nextLine( );
sentence = sentence.trim( ); // remove leading and trailing spaces
String checkString = sentence.toUpperCase( ); // to disregard case
```

4. Locate comment 2, and follow the instructions to initialize the flag variable, which will be used to track whether or not any pairs of characters do not match.

```
/***** 2. Set up the isMatch flag variable.
      isMatch is initialized to true
boolean isMatch = true;
```

5. Locate comment 3, and follow the instructions to set up the for loop header and the second index. The condition sends the forwardIndex only halfway through the String by dividing the length by 2. The state of the flag variable isMatch is also part of the condition. If isMatch is set to false, the condition will no longer be true and the loop will end.

```
/***** 3. Set up the for loop header. The forwardIndex will start at 0 and
       move forward halfway through the String.
       The backIndex will start at the end of the String and move backward
       halfway through the String. In the for loop condition, check whether
       isMatch is still true. If not, the for loop will end. */
int backIndex = checkString.length( ) - 1;
for ( int forwardIndex = 0;
    forwardIndex < checkString.length( ) / 2 && isMatch;
    forwardIndex++ ) {
```

6. Locate comment 4, and follow the instructions to extract the characters at each index. If the characters are not equal, then the String is not a palindrome. Setting the isMatch flag to false will cause the for loop to end.

```
/***** 4. Extract the characters at forwardIndex and backIndex
       and compare. If not equal, set isMatch to false. */

char fCh = checkString.charAt( forwardIndex );
char bCh = checkString.charAt( backIndex );

if ( fCh != bCh ) {
    isMatch = false;
}
```

7. Locate comment 5, and follow the instructions to decrement the backIndex variable and finish the for loop. Note: the forwardIndex variable is incremented by the for loop automatically.

```
/***** 5. Decrement backIndex. */
    backIndex--;

} // end of for loop
```

8. Locate comment 6, and follow the instructions to check whether the sentence is a palindrome. The for loop ends either because two corresponding characters were found not to be equal (isMatch is false) or all the characters in the sentence were found to be equal (isMatch is still true). Checking isMatch will tell you which event occurred.

```
/***** 6. Check isMatch to see if sentence is a palindrome. */
if ( isMatch ) {
    System.out.println( sentence + " is a palindrome." );
} else {
    System.out.println( sentence + " is not a palindrome." );
}
```

9. Compile and run the program. Enter a sentence or word that is a palindrome to test. Then, enter at least one word or sentence that is not a palindrome. Debug the program as needed.

Try It!

A simpler approach to this program is to create a reversed version of the String and compare it to the original. Using this method, a flag variable is not needed and only one index is used. Write this version of the program. Remember to use the equalsIgnoreCase method to compare the reversed version with the original String.

Inputting a Character

Scanner provides methods for inputting every primitive type except char. Inputting a char *is* possible. It just takes a few steps and involves extracting a character from a String.

1. Read the value as a String.
2. Extract the first character from the String.

For example, you may have a program that asks the user to enter his or her middle initial. For the purposes of the program, you need this as a char, not a String. This code accepts the input and then converts it to a char:

```
System.out.println( "Enter your middle initial:" );
String middleString = input.next( ); // read as a String
char middleInitial = middleString.charAt( 0 ); //extract the first
                                                //character
```

If the user enters B, then the char variable middleInitial will get the value B. However, note that if the user enters Belinda, middleInitial will still get the value B.

Wrapper Classes

One challenge with working with the char data type is that primitive types do not have methods. For example, there is not a toUpperCase method for a char. Further, sometimes a program needs to check whether a character is a digit or a letter or whether the character is uppercase or lowercase. For these situations, the Java Class Library provides wrapper classes.

A *wrapper class* encloses, or "wraps," a primitive type into an object, which can then be used to call methods. This is done transparently without any action on the programmer's part. Wrapper classes are provided for all primitive data types, as shown in **Figure 10-11.** Most wrapper classes have the same name as the primitive type with a capital letter. The two exceptions are the Integer class for the int data type and the Character class for char data type. All wrapper classes are in the java.lang package, so an import statement is not needed to use a wrapper class.

The Character class provides some static methods for determining information about a char, such as whether the char is a digit or a letter or whether the char is uppercase or lowercase. These methods are shown in **Figure 10-12.** They take the char variable as an argument and return true or false. Remember that static methods need to be called with the name of the class, in this case Character. For example, examine this code:

Primitive Data Type	Wrapper Class
byte	Byte
short	Short
int	Integer
long	Long
float	Float
double	Double
boolean	Boolean
char	Character

Goodheart-Willcox Publisher

Figure 10-11. All primitive data types have a wrapper class that can be used to apply methods.

```
char mystery = '1';
System.out.println( "Is " + mystery + " a letter? "
        + Character.isLetter( mystery ) );    // false
System.out.println( "Is " + mystery + " a digit? "
        + Character.isDigit( mystery ) );      // true

mystery = 'm';
System.out.println( "Is " + mystery + " a letter? "
        + Character.isLetter( mystery ) );    // true
System.out.println( "Is " + mystery + " a digit? "
        + Character.isDigit( mystery ) );      // false
System.out.println( "Is " + mystery + " uppercase? "
        + Character.isUpperCase( mystery ) ); // false
System.out.println( "Is " + mystery + " lowercase? "
        + Character.isLowerCase( mystery ) ); // true
```

Method	Explanation
boolean isDigit(char ch)	Returns true if the character ch is a digit and false if not.
boolean isLetter(char ch)	Returns true if the character ch is a letter and false if not.
boolean isUpperCase(char ch)	Returns true if the character ch is an uppercase letter and false if not.
boolean isLowerCase(char ch)	Returns true if the character ch is a lowercase letter and false if not.

Figure 10-12. These static methods of the Character wrapper class can be used to determine certain information about a char.

Two other static methods of the Character wrapper class convert a char to uppercase or lowercase. These methods are shown in **Figure 10-13.** They take the char as an argument and return a converted copy of the char. Examine this code:

```
char letterM = 'm';
System.out.println( letterM + " in uppercase is "
            + Character.toUpperCase( letterM ) ); // outputs M
System.out.println( letterM + " in lowercase is "
            + Character.toLowerCase( letterM ) ); // outputs m
```

Like the toUpperCase and toLowerCase methods in the String class, these methods do not directly change the char variable. To change the case of the char variable, assign the return value of the method to the char itself:

```
letterM = Character.toUpperCase( letterM ); // letterM gets M
```

Method	Explanation
char toUpperCase(char ch)	Returns an uppercase version of ch.
char toLowerCase(char ch)	Returns a lowercase version of ch.

Figure 10-13. These static methods of the Character wrapper class are used to change the case of an individual char.

HANDS-ON EXAMPLE 10.3.2

Checking for a Valid Password

In this exercise, you will develop a program that checks if a user's password meets the criteria for an acceptable password. In your company, a password must be at least eight characters long, contain at least one uppercase letter, and contain at least two digits. The algorithm is:

1. Begin a do/while loop.
2. Read the password.
3. Traverse the password.

   A. If the current character is a digit, count it.

   B. If the current character is an uppercase letter, set a flag.

4. Repeat while the password does not meet the requirements.

The for loop that will be nested inside the do/while loop does not need to count the characters in the password. The length method can be used in the do/while condition to check the length. Before beginning this exercise, download the chapter files from the student companion website. The PasswordCheckerSnippet.java file will be used as a starting point.

1. Launch jGRASP, and open the PasswordCheckerSnippet.java file.
2. Examine the code and the comments.
3. Locate comment 1, and follow the instructions to define variables used in the do/while loop.

```
/***** 1. Define variables used in do/while loop. */
final int MIN_LENGTH = 8;
final int MIN_DIGITS = 2;

int countDigits = 0; // counter for number of digits
boolean foundCap = false; // flag variable for uppercase letter
String password = "";
```

4. Locate comment 2, and follow the instructions to read a new password and reset the state of the checker variables. Forgetting to reset these variables for each new password is a common error. If the variables are not reset, the result could be the accumulation of the digit count from one password to the next or the acceptance of a password if any of the preceding passwords contained an uppercase letter.

```
/***** 2. Begin the do/while loop, read the password
       and reset checker variables. */
do {

    System.out.println( "Enter a password" );
    password = input.next( );

    // reset checker variables
    countDigits = 0;
    foundCap = false;
```

5. Locate comment 3, and follow the instructions to traverse the password, counting the digits and setting the flag if a character is uppercase.

```
/***** 3. Traverse the password, checking for digits and uppercase. */
for ( int i = 0; i < password.length( ); i++ ) {

    char ch = password.charAt( i );   // extract the character at i

    if ( Character.isDigit( ch ) ) {    // if digit, count it
        countDigits++;
    }

    if ( Character.isUpperCase( ch ) ) {   // if uppercase, set flag
        foundCap = true;
    }

} // end for loop
```

6. Locate comment 4, and follow the instructions to write the do/while condition. This condition can be derived using De Morgan's laws. An acceptable password is:

   countDigits >= MIN_DIGITS && password.length( ) >= MIN_LENGTH && foundCap

so the while condition is that the password is not acceptable:

   !(countDigits >= MIN_DIGITS && password.length( ) >= MIN_LENGTH && foundCap).

Applying De Morgan's laws, the condition becomes:

   !(countDigits >= MIN_DIGITS) || !(password.length( ) >= MIN_LENGTH) || !(foundCap).

Simplifying this, the condition is:

```
/***** 4. Write the loop condition. */
} while ( countDigits < MIN_DIGITS || password.length( ) < MIN_LENGTH || !foundCap );
```

7. Locate comment 5, and follow the instructions to inform the user the password is acceptable when the do/while loop has ended.

```
/***** 5. Inform the user */
System.out.println( "Your password is accepted." );
```

8. Compile and run the program. Check for proper operation by entering a password eight characters in length and containing one uppercase character and two digits. Then, enter invalid passwords. Notice the program repeats until a valid password is entered.

Try It!

The rules for passwords vary from one company or organization to another. Write a password checker that verifies the following rules for a password.
- Must be at least 10 characters.
- Must contain at least one uppercase letter.
- Must contain at least one lowercase letter.
- Must contain at least one digit.

SECTION REVIEW 10.3

Check Your Understanding

1. Which method is used to return the character at a specified index in a String?
2. What does it mean to traverse a String?
3. To input a single character as a char, first use the next method of the Scanner class to read the character as a String data type. What must be done next?
4. To call methods on primitive types, the Java Class Library provides _____ classes.
5. Which method in the Character class determines if a char variable is a digit?

Build Your Vocabulary

As you progress through this course, develop a personal computer science glossary. This will help you build your vocabulary and prepare you for a career. Write a definition for each of the following terms and add it to your computer science glossary.

traverse   wrapper class

Cooperative Coding

Coding a Form Letter

The biggest event of the year is coming up for your school. Every student should be involved. Today, most communication is handled by e-mail, text, or social media. Receiving a physical letter in the mail is so unusual that it will make an impression. Every student must be contacted and encouraged to attend. A form letter can be used for this. A form letter is sent to many people, but contains the same message body. Only the recipient's name changes on each letter.

SpeedKingz/Shutterstock.com

Assignment

Work with your team to write a form letter to tell students what is going to happen, encourage them to participate, and give them tips on how to prepare. The student's name will appear in two places in the letter:

- in the greeting
- at the point where you ask for participation

You will also use the student's year at school. The student name and year will be contained in a text file that is read by your program.

All students should receive the letter, but for this activity, include only four or five students. Create a plain-text file that contains the first name, last name, and year in school for each student, such as "Tamira Jackson freshman." Each student's data should appear on separate lines. Create another text file that contains the segments of the letter you want to personalize.

Apply what you have learned about reading from a file, String concatenation, and escape sequences to develop a program that formats and outputs the letter. The program should output one letter for each student in your text file. After printing the four letters, copy the output from the **Run I/O** tab, and paste it in a word-processing document. Each student letter should be its own page in the document.

1. How are escape sequences implemented in your team's program?
2. Brainstorm with your team other applications of a form letter. What are some examples?

Chapter Summary

Section 10.1: Creating Strings

- A Java String is a sequence of characters and can be created with constructors, assigning a literal to a String reference, and as a return value from a method.
- The Scanner class in the java.util package contains two methods for inputting a String: next method and nextLine method.
- Strings can be concatenated to combine them.

Section 10.2: String Methods

- The length of a String is found using the length method of the String class, which counts the number of characters in the String.
- Characters in a String are identified by an index, which always begins at 0 and ends at one less than the length of the string.
- A substring is part of a String, and the substring method is used to extract characters from a String; an invalid index argument to this method will generate a StringIndexOutOfBoundsException error.
- The case of all letter characters in a String can be changed using the toUpperCase and toLowerCase methods.
- The indexOf method will find the index of the first character in a String that matches a specified case-sensitive search string; if the search string is not found, the method returns –1.
- The replace method is used to find characters in a String and replace them with different characters; the search string is case-sensitive.

Section 10.3: Processing Strings Character by Character

- The charAt method can be used when traversing a String either forward or backward; the charAt method returns the character at a specified index, and this method can be used in conjunction with the next method to get input as a char data type.
- Primitive data types do not have methods, but a wrapper class encloses a primitive data type into an object that can then be used to call methods.

Chapter 10 Test

Multiple Choice

Select the best response.

1. What is the act of combining two Strings into one String called?
   - A. addition
   - B. concatenation
   - C. substring
   - D. wrapper class
2. What does the trim method do?
   - A. Shortens a String by a specified number of characters.
   - B. Removes leading and trailing white-space characters from a String.
   - C. Breaks apart a String created by concatenation.
   - D. Removes all characters in a String that follow a white-space character.
3. Which method is used to determine how many characters are in a String?
   - A. numChars
   - B. chars
   - C. length
   - D. sequence
4. What does it mean to traverse a String?
   - A. To send the String across the program.
   - B. To break apart the String into multiple Strings.
   - C. To join one String to another String.
   - D. To visit each character in order one at a time.
5. Which Character class method determines if a char variable is a digit?
   - A. isDigit
   - B. toUpper
   - C. isNumber
   - D. makeDigit

Completion

Complete the following sentences with the correct word(s).

6. A String reference is the _____ of the String.
7. Creating a String literal effectively creates a String _____.
8. A(n) _____ is a new String created from zero or more characters of another String.
9. The replace method is _____-sensitive.
10. The _____ method returns the character at a specified index in the String.

Matching

Match the correct term with its definition.

A. toUpperCase

B. empty String

C. wrapper class

D. index

E. isLetter

11. Method to change the case of all characters in a String.

12. Number representing the relative position of a character in a String.

13. Test for an alphabetic character.

14. Has no characters.

15. Elevates primitive types to objects.

Application and Extension of Knowledge

1. In Hands-On Example 10.2.4, you created a program that asks the user to decode a cryptogram. Develop a program to encrypt a phrase using a key entered by the user.

2. Before beginning this exercise, download the chapter files from the student companion website. The BirthdayInPiSnippet.java file will be used as a starting point. This file contains a String named pi that consists of the first 10,000 digits of the mathematical constant pi (π). Add a prompt for users to enter his or her birthdate in the *mmdd* format. Search pi for that substring, and let the user know at what index the birthdate is found or that it was not found at all. This will be a fun activity for users on the next π Day (March 14).

3. Do you know the rally cheer? Someone yells, "Gimme a T!" Then, the crowd yells back, "T!" This continues while the word *team* is spelled. Then, the person yells, "Whaddaya got?" The crowd yells back, "TEAM!" Develop a program to replicate the rally cheer. Prompt the user to enter a word for the cheer. This can be any word. The program then separates, or parses, the word into letters and prints "Gimme a" and each letter. Be sure to have the program check for letters that should be preceded by "an" instead of "a." When each letter is printed, echo the letter in uppercase. Finish with the "Whaddya got?" line and the original word in uppercase letters.

4. Research the five states of matter. Identify one structure that exists in each state. Develop a Science Quiz program that prompts the student to enter the name of the state of matter for each structure described. Include a notification that spelling counts. If the student responds correctly, give congratulations and add 1 to the student's score. If not, supply the correct response. At the end, report the percentage of correct responses.

5. Before beginning this exercise, download the chapter files from the student companion website and use the starter file AnagramsSnippet.java. An anagram is a rearrangement of the letters in one word to form another word. For example, ITEM, MITE, TIME, and EMIT are all arrangements of the same four letters. Determine an algorithm to create all of the arrangements of the letters in a given word composed of four letters. There are 24 possible arrangements of four letters, but not all are words. Manually scan the output to find actual words. Then, develop a program to create the arrangements of those that are words. Test with the words LOVE, LIVE, and STOP.

Online Activities

Complete the following activities, which will help you learn, practice, and expand your knowledge and skills.

Vocabulary. Practice vocabulary for this chapter using the e-flash cards, matching activity, and vocabulary game until you are able to recognize their meanings.

Communication Skills

Reading. *Visual support* is used to communicate an idea using images or objects. It can be used instead of words or in addition to them. Illustrations that accompany written text are visual supports that help you understand the content. Select an illustration from this chapter that serves as a visual support. Analyze and evaluate if the information is easy to understand. How does it relate to the written text?

Writing. *Writing style* is the way in which a writer uses language to convey an idea. Select a page or pages of notes you have taken during a class. Evaluate your writing style and the relevance, quality, and depth of the information. What did you do well? What do you need to improve?

Speaking. *Self-confidence* is being certain and secure about one's own abilities and judgment. Practice your speaking skills and self-confidence by volunteering to read a special feature in this chapter aloud to your classmates. Stand in the front of the room, control your voice, and read loudly and clearly. How would you rate your speaking skills?

Listening. *Implicit* means hinted at without being expressed. When information is implicit, there may be hints or related ideas, but it is not said directly. For example, suppose you ask your instructor what will be on an exam. Your instructor responds that he or she cannot tell you what will be on the test, but suggests studying specific topics. The implicit idea is the list of suggested topics will be on the test. Have a conversation with a classmate about tonight's homework. Identify any implicit information from him or her. How were you able to understand the implicit messages?

Portfolio Development

College and Career Readiness

Schoolwork. Academic information is important to include in a portfolio in order to show your accomplishments in school. Include items related to your schoolwork that support your portfolio objective. These items might be report cards, transcripts, or honor roll reports. Diplomas or certificates that show courses or programs you completed should also be included. Other information can be included as a list, such as relevant classes you have taken.

1. Create a Microsoft Word document that lists notable classes you have taken and activities you have completed. Use the heading Schoolwork on the document along with your name.

2. Scan hard-copy documents related to your schoolwork, such as report cards, to serve as samples. Place each document in an appropriate folder.

3. Place the hard-copy documents in the container for future reference.

4. Update your spreadsheet.

CTSOs

EVENT PREP

Public Speaking. Public speaking is delivering a prepared speech to a large-group audience. This event enables participants to demonstrate communication skills of speaking, organizing, and making an oral presentation. The topic for this event will be posted on the website of the CTSO. You will be given research, prepare, and practice before going to the competition. Review the specific guidelines and rules for this event for direction as to topics and props that are permitted for the presentation. To prepare for a public speaking event, complete the following activities.

1. Read the guidelines provided by your organization. Review the topics from which you may choose to make a speech.

2. Locate a rubric or scoring sheet for the event on your organization's website.

3. Confirm if visual aids may be used in the presentation and the amount of setup time permitted.

4. Review the rules to confirm if questions will be asked or if you will need to defend a case or situation.

5. Make notes on index cards about important points to remember. Use these notes to study. You might also be able to use these notes during the event.

6. Practice the presentation. You should introduce yourself, review the topic that you are presenting, defend the topic, and conclude with a summary.

7. After the presentation is complete, ask for feedback from your instructor. You may consider also having a student audience listen and provide feedback.

Managing Input and Output

Sections

You have already used several strategies for entering data into a computer program. Input can be received from the keyboard and from an external file using the Scanner class. Until now, the output of your programs has appeared in the **Run I/O** tab of jGRASP or the JavaFX graphics window. This output has not been permanent. It has disappeared when the program ends. A more permanent solution is to create the output as a file.

Another topic that must be considered is the location of files. Any time a piece of code has to deal with an element not contained in its program, like an external text file, errors can happen. The file may have been moved and may not be where the program expects it to be or the file name may be different than expected. Attempting to create a file where the program does not have permission to write is also a problem. Java lets the program know when an exceptional error condition arises that may suddenly terminate the program. Exceptions handlers are tools the coder can use to prevent system crashes or data loss. Previous chapters have discussed some situations that can cause exceptions.

That is not all! Once the data has been successfully entered, it may not be the correct information. It is the job of the coder to anticipate any mishaps and prepare for successful data entry. In this chapter, you will learn how to handle writing files as well as dealing with exceptions and input errors in a swift and elegant way.

College and Career Readiness

Reading Prep

Before reading this chapter, preview the illustrations. Translate the information in the illustrations into words. In what ways do you think the illustrations support the content?

While studying, look for the activity icon for:

- Vocabulary terms with e-flash cards and matching activities.
- Starter files for hands-on examples and other exercises.

These activities can be accessed at
www.g-wlearning.com/informationtechnology/1773

Chapter Glossary

append: Keep the data in an existing file and add new data to the end of the file.

catching an exception: Exception handler receives the exception object from the runtime system.

checked exception: One that must be handled or acknowledged by the application.

exception handler: Code that tells the runtime system what to do when a specific exception occurs.

input validation: Process of ensuring information entered into a program is acceptable.

stack trace: History of the methods called before the error occurred.

throwing an exception: Sending an exception object to the runtime system.

unchecked exception: One that is reported by the RunTimeException class and does not need to be handled by the programmer.

Handling Exceptions

I n the previous chapter, you learned how to input data from a file using the Scanner class. The applications created in that chapter were based on the assumption that the file could be opened and read. If the file was not there, the program stopped. The runtime system reported the problem, but the message could be confusing to a user. Java has an error-catching scheme called exceptions. *Exception* is a shortcut for *exceptional event.* In this section, you will learn about exceptions and how to handle them when they occur.

How important is the programmer's handling of exceptions to the end user of the application?

Learning Goals

- Explain the concept of throwing and catching an exception.
- Use a try/catch/finally block to handle exceptions.

Terms

catching an exception
checked exception
exception handler

stack trace
throwing an exception
unchecked exception

Overview of Exceptions

When the normal flow of execution is interrupted by an exceptional error that makes continuation of the program impossible, one of two things happens. The method that is active at the time sends an exception object to the runtime system. Or, the runtime system creates an exception object. This exception object contains information about the error. The exception must be handled or the program will terminate. Examples of exceptions are shown in **Figure 11-1.** Search the Internet using the string Oracle Docs Java Exception to locate the full API for the Exception class.

In keeping with the urgent nature of the error, the metaphor is that the program "tosses" this exception object to the runtime system. Hence, sending an exception object to the runtime system is called ***throwing an exception.***

The runtime system gives the programmer the ability to intervene and accommodate for exceptions. The programmer can code an exception handler. An ***exception handler*** is code that tells the runtime system what to do when a specific exception occurs. ***Catching an exception*** means the exception handler receives the exception object from the runtime system. The exception handler then takes the action the programmer coded for the event. For example, if the program cannot find a file, the programmer could code an exception handler to notify the user the file was not found, and prompt the user for a revised file name.

In addition, it is possible to create your own exceptions. However, the best path is to plan in advance to make sure conditions do not arise that could cause exceptions.

Example	Programming Response
FileNotFoundException	Open file with error-checking.
InputMismatchException	Validate input.
NoSuchElementException	Check for end of file.
StringIndexOutOfBounds	Monitor end of string.

Goodheart-Willcox Publisher

Figure 11-1. These are some of the exceptions that may occur.

This section focuses on the mechanisms of an exception and how to generically recover.

Error-Catching

When an exception occurs, the JVM reports the error and any nongraphical program is terminated. Java exceptions fall into two groups: checked and unchecked. A *checked exception* must be handled or acknowledged by the application. In the previous chapter, you added throws FileNotFoundException to the main method header. This was to acknowledge that a FileNotFoundException could occur. An *unchecked exception* is one that is reported by the RunTimeException class. Unchecked exceptions do not need to be handled. The ArithmeticException for division by zero and InputMismatchException are unchecked exceptions.

When the program throws an exception, either checked or unchecked, the program can try to recover from the error if the programmer has coded a try/catch/finally combination. For example, try to open a file. If it does not work, catch the exception and deal with it. Whether an exception occurs or not, perform the code in the optional finally block.

```
try {
    // code that might cause an exception
} catch ( ExceptionType name ) {
    // code to execute if this exception occurs
} catch ( ExceptionType name ) {
    // code to execute if this exception occurs
}
finally {  // optional
    // code to execute whether exceptions occurred or not
}
```

Example of Error-Catching

Refer to **Figure 11-2.** This is the SearchStringsSnippet program you edited in Hands-On Example 10.2.3. The program opens an external file and searches for a string. In this case, the program file itself is opened. When creating this program, you may have thought there were mysterious statements in the snippet. Examine the code again. Line 6 contains an import statement that brings in all the exception code, among other classes. Line 9 adds a throws FileNotFoundException that alerts the runtime system that there may be a problem. Lines 18–20 are where the potential for the input-output (I/O) problem exists.

In order for this program to find the file, the assumption is the SearchStringsSnippet.java file is in the same folder as the .class file. That makes sense because the compiler automatically stores the .class file in the same folder as the .java file. However, assume the file was not there or it had been renamed. At line 20, when the Scanner class does not find the file object, the exception processing swings into high gear:

1. The Scanner constructor creates a FileNotFoundException object.
2. The Scanner constructor tosses the object to the runtime system.
3. The runtime system looks for a try/catch block...but there is none. Crash!

The output for the I/O problem is:

```
Exception in thread "main" java.io.FileNotFoundException:
    SearchStringsSnippet.java
```

More output than this is usually printed that shows a stack trace. A *stack trace* is the history of the methods called before the error occurred. What is important here, however, is that the file was not found.

```java
1 /** search for strings
2    <your name here>
3 */
4
5 import java.util.Scanner;
6 import java.io.*;
7
8 public class SearchStringsSnippet {
9    public static void main( String [ ] args ) throws IOException {
10
11       Scanner input = new Scanner( System.in );
12
13       /***** 1. Prompt the user for the search string.
14            Use nextLine to allow for multiple words. */
15       System.out.println( "Enter a word or phrase to find:" );
16       String searchString = input.nextLine( );
17
18       /***** 2. Open this file. */
19       File inputFile = new File( "SearchStringsSnippet.java" );
20       Scanner file = new Scanner( inputFile );
21
22       /***** 3. Define a boolean flag, stringFound as false,
23            and initialize the line number to 0. */
24       boolean stringFound = false; // flag indicates whether string is found
25       int lineNumber = 0;
26
27       /***** 4. Begin the while loop to read the file line by line. */
28       while ( file.hasNext( ) ) {
29
30          /***** 5. Read a line, increment the line number,
31               and search for the searchstring. */
32          String line = file.nextLine( );
33          lineNumber++;
34          int indexOfFound = line.indexOf( searchString );
35
36          /***** 6. If the search string is found,
37               output the line number and the line, enclosing
38               the searchstring in **, and set stringFound to true. */
39          if ( indexOfFound != -1 ) {
40             String lineNo = String.format( "%2d ", lineNumber );
41             System.out.println(  lineNo
42                + line.substring( 0, indexOfFound )
43                + "**"
44                + line.substring( indexOfFound, indexOfFound + searchString.length( ) )
45                + "**"
46                + line.substring( indexOfFound + searchString.length( ) ) );
47          stringFound = true;
48          } // end if
49       /***** 7. Finish the while loop. */
50       } // end while loop
51
52       /***** 8. Check if the search string was found in any line. */
53       if ( !stringFound ) {
54          System.out.println( searchString + " was not found." );
55       }
56    } // end main
57 } // end program
```

Figure 11-2. This is the SearchStringsSnippet program created in Chapter 10.

Try/Catch Block

The following example shows how to use the try/catch block to avoid the disaster of a file not being found. This is a modification of the SearchStringsSnippet.java file. On line 24, the exception object is named e. When the exception object is created, error messages are incorporated. Use the getMessage method with the exception object e to find out what the error message is. This is shown on line 25. Print the message using the System.err.println( ) statement, which finds and outputs the error message for you. You can also add your own message to the user, as shown in lines 27–29.

```
18 /***** 2. Try to open this file. */
19 try {
20
21   File inputFile = new File( "SearchStringsSnippet.java" );
22   Scanner file = new Scanner( inputFile );
23
24 } catch ( FileNotFoundException e ) {
25   System.err.println( "Caught FileNotFoundException: " +
        e.getMessage( ) );
26
27     System.out.println( "The file SearchStringsSnippet.java "
          "cannot be "
28   + "found.\nPlease place it in the folder with the "
29   + "class file and try again." );
30 }
```

If the SearchStringsSnippet.java file is not in the same folder as the .class file, the output is:

```
Enter a word or phrase to find: print
Caught FileNotFoundException: SearchStringsSnippet.java (The
system cannot find the file specified)
The file SearchStringsSnippet.java cannot be found.
Please place it in the folder with the class file and try again.
```

The catch block and exception error reporting are designed for situations in which the processing cannot continue without the error being resolved. Often, the program can handle the exception by providing the users with options to correct the problem on the spot.

Another type of exception is the NoSuchElementException. If a program attempts to read beyond the end of the file, the Scanner method will throw the exception. This exception is unchecked, so a try/catch block is not required. If desired, however, you could catch a NoSuchElementException and output a useful message to the user.

HANDS-ON EXAMPLE 11.1.1

Using try/catch

In this exercise, you will get input from the user. You will use a try/catch block to handle an IllegalArgumentException when a nonpositive number is used as an argument for the nextInt method of a Random object. Before beginning this exercise, download the chapter files from the student companion website. The IllegalArgumentSnippet.java file will be used as a starting point.

1. Launch jGRASP, and open the IllegalArgumentSnippet.java file.

2. Examine the code. Because an IllegalArgumentException is not a checked exception, we do not need to add a throws clause to the main method.

3. Locate comment 1, and follow the instructions to write a prompt for the user to enter the upper bound for a set of random numbers. Then, get the input using the Scanner object.

```
/***** 1. Prompt for an integer for argument for Random number object. */
System.out.println( "Enter the upper bound for the random numbers: " );
bound = input.nextInt( );
```

4. Locate comment 2, and follow the instructions to create an instance of the Random class.

```
/***** 2. Instantiate Random object. */
Random rand = new Random( );
```

5. Locate comment 3, and follow the instructions to set up the try block for getting a random number.

```
/***** 3. Set up try/catch for sending an illegal argument to the Random nextInt method. */
try {
    int randNumber = rand.nextInt( bound );
    String s = String.format( "The upper bound for random numbers is %d. " +
                "The number generated is %d.", bound, randNumber );
    System.out.println( s );
```

6. Set up the catch block to display the exception error, if it occurs.

```
} catch ( IllegalArgumentException e )  { // exit gracefully
    System.err.println( "The random numbers may not be negative or zero.\n" +
                "Caught exception: " + e.getMessage( ) );
} // end try/catch
```

7. Compile and run the program. Use the test data set of 5, 0, and −2. Verify the results. You have used an exception handler to provide exceptional error results!

Try It!

Continue working in the IllegalArgumentSnippet.java file. Change the catch block to set up a loop to capture a positive integer. Rename the class as IllegalArgumentTryIt, and save the file under the new name.

SECTION REVIEW 11.1

Check Your Understanding

1. What is an event generated as a result of an error that prevents the program from continuing?
2. Which object contains information about an error thrown by the runtime system?
3. Distinguish between checked and unchecked exceptions.
4. What is the name of the block that can be used to attempt to recover from an exception gracefully?
5. Which method would be used to report the error to the user in a prompt?

Build Your Vocabulary

As you progress through this course, develop a personal computer science glossary. This will help you build your vocabulary and prepare you for a career. Write a definition for each of the following terms and add it to your computer science glossary.

catching an exception
checked exception

exception handler
stack trace

throwing an exception
unchecked exception

Language Arts and Java

Translations

The process of providing the appropriate language for a user is called localization. Translations, idioms, and more are determined with the user in mind. At the current state of technology, automatic translations by code are not perfect. Sometimes, the idioms of a language are misconstrued as a literal translation is attempted.

CHM3N/Shutterstock.com

In the following example, an English idiom is translated to French and Spanish in literal, word for word fashion.

English	French	Spanish
Tell it like it is	*Le dire comme il est*	*Decirlo como es*

However, in French, the equivalent phrasing is *to call a cat a cat*:

> *Appeler un chat un chat*

In Spanish, the idiom is *not to have hairs on your tongue*:

> *No tener pelos en la lengua*

A way to ensure the translation of text in your program is as pleasing to the user as possible is to use external files that contain confirmed translations. The best route, of course, is to avoid including idioms in your programs. However, there are phrasings other than idioms that are awkward when automatically translated.

Assignment

Before beginning this exercise, download the chapter files from the student companion website. The LocalizationSnippet.java file will be used as a starting point along with the LocalSpanish.txt, LocalEnglish.txt, and LocalFrench.txt files. The text files contain a paragraph that will be read by the program. You have contracted a team of linguists with preparing the localization for a program. Apply what you learned about Strings and reading from a file to create a Java application. The algorithm is to prompt a user for the desired language and open the appropriate language file. Output the three Strings from the selected file. If you speak another language, create a local text file for that language and incorporate it into the program.

Data Validation and Output

One of the biggest issues facing all programmers is getting the user to enter the correct information into the program. Sometimes the directions provided in the program are poor or the user simply does not understand what is supposed to be entered. Sometimes the user just hits [Enter] before he or she has had a chance to enter the information. Getting information from a file can also be a problem. Even if reading from the file was successful, trouble arises if the data is formatted differently from how the programmer is expecting it to be arranged. It is not just the programmer who can be frustrated. A confused user may likely leave the program and not supply the input needed. It is a significant problem.

Data output is also a concern for programmers. In previous chapters, you learned how to format output to the Run I/O tab of jGRASP. Using the escape sequences \n and \t and the String.format method, clean and clear output was achieved. You learned how to apply rich text output to a JavaFX application using a text object. However, in both cases, the output was lost once the program ended. To make the output permanent, an output file can be created.

Where does data validation rank in terms of importance to the programmer?

Learning Goals

- Apply techniques to validate input from the user.
- Explain how to write a String to a file.

Terms

append   input validation

Input Validation

Input validation is the process of ensuring the information entered into a program is acceptable. Valid information has the correct data type and is the correct response to the prompt. The best way to avoid problems with user input is to provide meaningful prompts. However, mishaps will happen. To allow for human error, programmers must predict any way in which a user may input invalid data.

Java is heavily dependent on data types. Because of this, extra care needs to be taken to validate what the user has entered before trying to use that data in the algorithm. Creating a loop for data entry allows the programmer to accept the input, validate it, and then prompt the user to try again if the input is not correct.

Numeric Input Validation

Consider the phone number format program from Chapter 10. The program requested ten digits be entered, but it did not check that they were really numbers. Trust was placed on the user to enter digits from 0 to 9. If you rely on the try/catch method, the program might terminate. Using a do/while provides a second chance for the user to supply good input.

```
18 /***** 2. Write the do/while loop; the input phone number must
      be 10
19    characters. Use the length method to count the number of
```

(Continued)

```
20   characters. */
21 do {
22   System.out.println( "Enter your phone number; 10 digits only" );
23   inputPhone = input.nextLine( );
24
25   if ( inputPhone.length( ) != NUMBER_OF_DIGITS ) {
26      System.out.println( "Please enter " + NUMBER_OF_DIGITS
27                 + " digits!" );
28   }
29 } while ( inputPhone.length( ) != NUMBER_OF_DIGITS );
```

On a sample run of this program, nonnumeric characters were entered. Since there were ten characters, the input was accepted, as shown in **Figure 11-3.**

This input error can be prevented by adding a loop enclosing lines 18–29 to validate the digits are numeric. An algorithm to validate the numeric digits uses these steps:

1. Define a boolean named numeric to govern the loop cycles.
2. Then loop, checking for nonnumeric characters using the isDigit method of the Character wrapper class until all the digits are checked, or the end of the String is reached.

Write the following lines of code within the loop that checks for ten digits.

```
32 boolean numeric;   // define before do/while loop
33 // check for numeric characters
34 int i = 0;
35 numeric = true;
36 while ( i < inputPhone.length( ) && numeric ) {
37
38   char digit = inputPhone.charAt( i );
39   if ( !Character.isDigit( digit ) ) {
40
41      numeric = false;
42   }
43   i++; // increment the index of the String
44 } // end while checking for numeric
```

Coding Conundrum

Amira is attempting to append to her journal file on drive G:, but the program cannot find her file. The exception handler reports there is an IO exception, but does not report the error message. It is a conundrum! What is the problem with her code?

```
/* Open a file
   Amira
*/
import java.util.*;
import java.io.*;
public class OpenFile {
   public static void main( String [ ] args ) {
      try {

         File inputFile = new File( "g:\\AmiraJournal.java" );
         Scanner file = new Scanner( inputFile );

      } catch ( IOException e ) {
         System.err.println( "Caught IOException: "  );

      } // end try/catch
   } // end main
} // end class
```

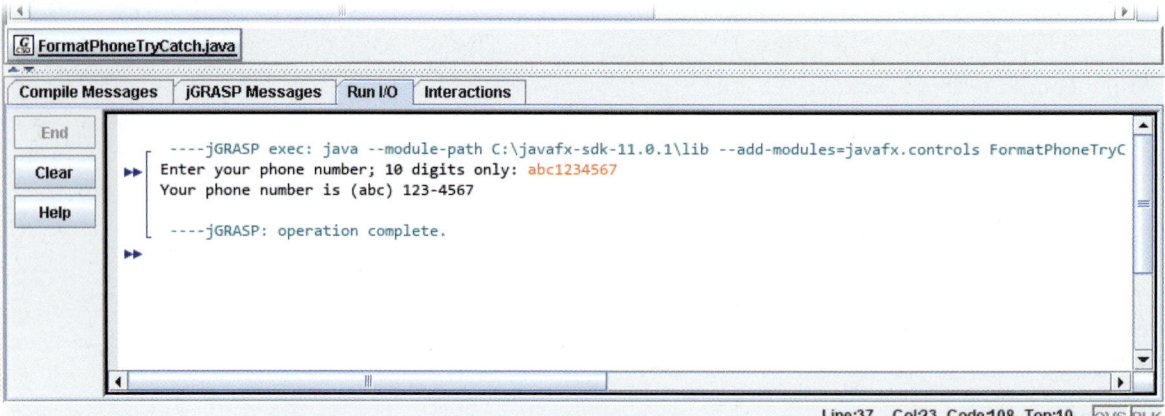

Figure 11-3. If nonnumeric characters are entered for the program shown in Figure 11-2, the input will be accepted if the total number of characters is ten.

After the while loop finishes, check the value of numeric. If false, then at least one digit was not numeric. Output a message to the user.

```
45    if ( !numeric ) {
46        System.out.println( "Please enter numeric characters." );
47    }
```

Modify the while statement of the do/while loop to include the !numeric test. Apply De Morgan's laws yet again!

```
48 } while ( inputPhone.length( ) != NUMBER_OF_DIGITS || !numeric );
```

HANDS-ON EXAMPLE 11.2.1

Validating Input

It is important to validate the input for numeric digits. One such example of where validating numeric input is important is the credit card program created in the Try It! for Hands-On Example 10.2.1. In this exercise, you will modify that program to verify the user enters only numbers. Before beginning this exercise, download the chapter files from the student companion website. The CreditCardValidateSnippet.java file will be used as a starting point.

1. Launch jGRASP, and open the CreditCardValidateSnippet.java file.

2. Examine the code, and review the loop that accepts input of the 16 digits.

```
/***** 2. Write the do/while loop; the input card number must be
    16 characters. Use the length method to count the number
    of characters. */
do {
    System.out.println( "Enter your credit card number; 16 digits only" );
    inputCard = input.nextLine( );
    if ( inputCard.length( ) != NUMBER_OF_DIGITS ) {
        System.out.println( "Please enter " + NUMBER_OF_DIGITS
                + " digits!" );
    }
} while ( inputCard.length( ) != NUMBER_OF_DIGITS );
```

3. Modify comment 2 to reflect the changes made to the program. Add this to the end of the comment: Ensure characters are numeric.

4. Define the boolean variable numeric before the validation loop. Add this on a new line directly after comment 2.

```
boolean numeric;
```

5. Add the while loop to validate the digits. Add this within the existing do/while loop after the if block that checks for ten digits. Note: it could be inserted before that check, if desired.

```
// Validate the characters are digits
numeric = true;
int i = 0;

while ( i < inputCard.length( ) && numeric ) {

    char digit = inputCard.charAt( i );
    if ( !Character.isDigit( digit ) ) {
        numeric = false;
        System.out.println( "Please enter numeric characters." );
    }
    i++; // get next character in String

} // end while check for numeric
if ( !numeric ) {
        System.out.println( "Please enter numeric characters." );
}
```

6. Add the nonnumeric test to the do/while loop.

```
} while ( inputCard.length( ) != NUMBER_OF_DIGITS || !numeric );
```

7. Compile and run the program. On the first run, test for too many digits. On the second run, test for too few digits. These checks ensure you did not break what was working in the previous code. Then, run the program a third time and test for nonnumeric characters. You have validated input for the credit card number!

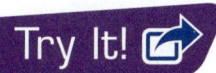

Before beginning this exercise, download the chapter files from the student companion website. The FormatZipCodeValidateTryItSnippet.java file will be used as a starting point. New ZIP codes are nine digits with a hyphen in the middle, such as 20050-4523. Write a Java program that accepts a nine-character ZIP code with a hyphen separating the two parts (ten total characters). Validate that ten characters are entered. Also, validate that the nine digits are numeric and a hyphen separates the two parts of the zip code. Remember, a String index begins at zero. Name the application FormatZipCodeValidate, and save the file in your working folder.

Data Type Validation

When accepting input from a user, care must be taken to ensure the correct data type is entered. Otherwise, the Scanner input method generates an InputMismatchException. For example, you can use a try/catch to avert disaster by checking that the entry is an integer. Or, you can take the input as a String and use String methods to investigate the value.

Try/Catch Validation

The code below is a one-time chance for the user to enter the integer. If the program must continue, you would use a loop to keep asking the user for the correct input.

```
import java.util.*;

public class ValidateDataTypeInput {

  public static void main( String [ ] args ) {

    Scanner input = new Scanner( System.in );
    String s = "";
```

(Continued)

```
System.out.print( "Enter an integer: " );
try {
  int integer = input.nextInt( );
  s = String.format( "The integer is %d." , integer );
}
catch ( InputMismatchException e ) {

  s = ( "There is a problem with your entry. It was not an "
  + "integer." );
}
finally {

  System.out.println( s );
} // end try/catch/finally

} // end main
} // end class
```

This is a general strategy for any data type to validate that the correct data type has been entered.

String Input Validation

Some programmers opt for accepting the input as a String. Then, they manipulate the String to get the numeric value of the required data type into the variable. Getting the input as a String provides the opportunity to perform validation.

For example, suppose the hours worked were accepted into the String stringHours. Using String methods, the program could validate that the string represented a double number. If the validation is complete with a positive result, then convert the String to the number and use the valueOf( ) method of the Double wrapper class.

```
double doubleHours = Double.valueOf( stringHours );
```

Otherwise report the problem to the user and provide options for the resolution of the input.

Output to a File

File output, or writing to a file, is a generic requirement for saving output. Every language has its own syntax and library support for this activity. In fact, Java has several ways to save to a file. The BufferedWriter, PrintWriter, FileOutputStream, DataOutputStream, RandomAccessFile, and FileChannel classes all support file output. Each has its special uses.

In order to focus on the task of outputting to a file, this discussion looks at just the FileWriter and PrintWriter classes. FileWriter can create new files and append to existing files. To *append* means to keep the data in an existing file and add new data to the end of the file. There are five steps to writing Strings to a file.

1. Identify the path where the file will be saved.
2. Determine whether a new file is needed or the data is being appended to an existing file; in either case, open the file. Opening the file consists of creating a FileWriter object by identifying the file name and mode. That FileWriter object is then fed to the PrintWriter constructor.
3. Create the output using the String data type and the String.format method.
4. Print the String to the file.
5. Close the file.

In Java, the objects are created from the FileWriter and PrintWriter classes. Import the java.io package to use these classes. The APIs for these classes are shown in **Figure 11-4.** If a new file is desired, pass false as a boolean to the FileWriter constructor. To append the output to an existing file, pass true as a boolean.

FYI

If you use FileWriter with a boolean argument of false, an existing file will be overwritten.

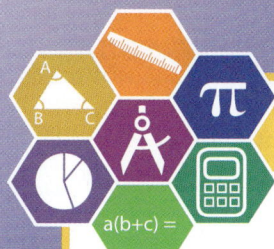

Math and Java

Quadratic Formula

Possibly the most exciting concept of all algebra is the quadratic formula. The sheer magnitude of power in one formula is that *any* quadratic equation in one variable can be solved using it. This is not paralleled anywhere else in mathematics. The quadratic formula is a ready-made algorithm to use, and it always yields an exact answer. The pattern of the formula lends itself to automation with computer programming.

With a little prep work, any second-degree equation in one variable can be written in standard form:

$$ax^2 + bx + c = 0$$

Then, the coder can solve the rest of the formula. The coefficients *a*, *b*, and *c* can be real numbers. That means the data type should be double.

Suppose a confused user enters the characters *a*, *b*, or *c* instead of real numbers. The program will crash because the data type does not match.

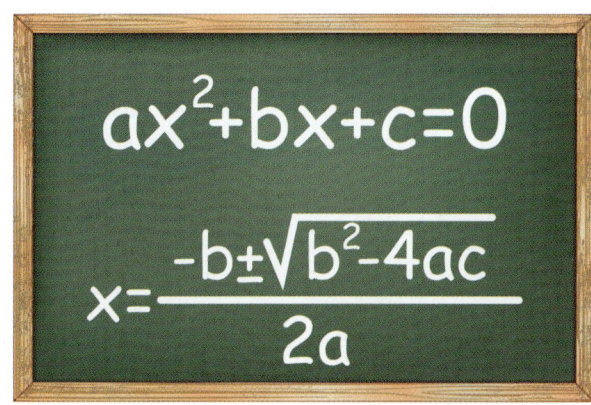

Imagestockdesign/Shutterstock.com

To avoid this, catch the InputMismatchException. Use a try/catch in a loop to ensure good input. Then, use the double values for *a*, *b*, and *c* in the formula. Below is an example for getting the first coefficient, *a*.

```java
Scanner input = new Scanner( System.in );
String s = "";              // output string
double a, b, c;             // coefficients of quadratic equation

boolean inputValidated = false;  // flag to repeat to get valid input

while ( !inputValidated ) {
  try {
    System.out.print( "Enter a: " ); // prompt user
    a = input.nextDouble( );       // if not a real number -> Exception!
    inputValidated = true;       // if try block completes, exit loop
  } catch ( InputMismatchException e ) {
    s = input.nextLine( );       // finish the aborted input
    s = "There is a problem with your entry for a. \n" +
      " a must be a real number." ;
    System.out.println( s );
  } // end try/catch block
} // end validate a while loop
```

Assignment

Before beginning this exercise, download the chapter files from the student companion website. The QuadraticFormulaSnippet.java file will be used as a starting point. Write the code to prompt the user for *a*, *b*, and *c*. Perform the calculation for the quadratic formula. Apply what you learned in Chapter 6 about the Math class method sqrt( ). That will come in handy for calculating the square root of the discriminant. Print out the original equation and the two roots. For an extra challenge, output any complex roots in the *a + bi* format.

Class	Explanation
FileWriter(String filename, boolean append) throws IOException	Writes characters to the specified file. If append is true, characters are appended to the file. Otherwise, the current contents of the file, if any, are deleted and data is written to the beginning of the file. If the file does not exist, it is created.
PrintWriter(Writer writer)	Prints formatted text to the specified file.

Goodheart-Willcox Publisher

Figure 11-4. The FileWriter and PrintWriter classes are used to create objects for writing to a file.

Be sure to check that the drive to which you are attempting to write actually exists and is writable.

The methods in the PrintWriter class used to write to the file and to close the file are shown in **Figure 11-5.** It is critical to close the file. Otherwise, Java never completes the printing. That results in nothing being added to the file.

Another important task is to ensure the path is correct. For example, the path provided in the FileWriter argument points to a specific folder. If a file cannot be written to this location, the FileWriter constructor will throw an IOException.

The code below shows how to create a new file and write the String "Hello" to it. Because exceptions are thrown, apply what you learned about using try/catch blocks. An IOException error will be thrown if the file or path cannot be found. This code is writing to the F: drive, which may be a virtual drive, mapped network location, or external drive such as a flash drive.

```
import java.util.*;
import java.io.*;
public class WriteToFile {

   public static void main( String [ ] args ) throws IOException {

      try {
        // pass the file destination path and false to create a new
           file.
        // f: is the disk name assigned to a flash drive.
        FileWriter writer = new FileWriter( "f:\\FirstFile.txt",
           false );
        PrintWriter = new PrintWriter( writer );

        String s = "Hello";
        printWriter.print( s );

        printWriter.close( );

      }
      catch ( IOException e ) {
        System.out.println( "We cannot write to the path indicated."
           + e );

      } // end try/catch blocks
   } // end main
} // end class
```

Method	Explanation
void print(String s)	Writes a string to the file.
void println(String s)	Writes a string to the file followed by a new line character.
void close( )	Closes the file, writes any data still in memory, and releases any system resources associated with the file.

Goodheart-Willcox Publisher

Figure 11-5. These methods of the PrintWriter class are used to write to a file and to close a file.

The FirstFile.txt file now exists and contains "Hello." The code below shows how to append to this existing file and write the String "World!" to it. To append to the file, be sure to change the boolean argument to true.

```
import java.util.*;
import java.io.*;
public class WriteToFile {

   public static void main( String [ ] args )
      throws IOException {
```

(Continued)

```
        try {
            FileWriter writer = new FileWriter( "f:\\FirstFile.txt",
                true );
            PrintWriter = new PrintWriter( writer );

            String s = " World!";
            printWriter.print( s );

            printWriter.close( );

        }
        catch ( IOException e ) {
            System.out.println( "We cannot write to the path indicated."
                + e );
        }

    } // end main
} // end class
```

The FirstFile.txt file now contains "Hello World!" Notice that in the first example, we used the PrintWriter print method and in the second example we used the println method. The PrintWriter print and println methods can be used to write all Java's primitive types to a file. Each argument to a print or println method is converted to a String before being written to the file. In fact, System.out, which is used to write output to the RUN/IO window, is also an object of the PrintWriter class. You have been using the PrintWriter print and println methods throughout this course!

Hands-On Example 11.2.2

Writing to a Text File 📷

In this exercise, you will create a new text file and write a String into it. Before beginning this exercise, download the chapter files from the student companion website. The WriteToFileSnippet.java file will be used as a starting point. You will also need a flash drive for this exercise.

1. Launch jGRASP, and open the WriteToFileSnippet.java file.

2. Examine the code and the comments.

3. Find the import statement for the java.io package on line 6, which supplies the classes for this program. Apply what you have learned to identify the try/catch blocks for writing to a file.

4. Locate comment 1, and follow the instructions to create an instance of the FileWriter class. Name it outputfile. In the argument list, add the path to your flash drive to name the file FirstFile.txt. Be sure the drive letter matches your actual drive. Choose false for the boolean so a new file is created.

   ```
   /***** 1. Create new file to write only. */
   FileWriter outputFile = new FileWriter( "f:\\FirstFile.txt", false );
   ```

5. Locate comment 2, and follow the instructions to create an instance of the PrintWriter constructor named printWriter. Pass the argument for the FileWriter object, which was named outputFile above.

   ```
   /***** 2. Instantiate the writer object. */
   PrintWriter printWriter = new PrintWriter( outputFile );
   ```

6. Locate comment 3, and follow the instructions to define a String and populate it with a quotation from Omar Khayyam's *Rubaiyat*. Use the PrintWriter object's print method to write the String to the file.

   ```
   /***** 3. Define a String and write it to the file. */
   String s = "The Moving Finger writes; and, having writ, Moves on: " +
      "Quote from 'Rubaiyat of Omar Khayyam'" ;
   printWriter.print( s );
   ```

7. Locate comment 4, and follow the instructions to call the close method of the printWriter class.

```
/***** 4. Close the file and write remaining data to the file. */
printWriter.close( );
```

8. Compile and run the program. The result proves not very satisfying. The message is simply: ----jGRASP: operation complete. The user probably wants confirmation that the operation worked. The code to do this is included in the file, but it is commented out. Locate comment 5. Move the end comment marker (*/) from the end of the line } // end try/catch on read file to the end of comment 5. Be sure the drive (F:) matches your flash drive.

```
/***** 5. Echo contents of file. */
try {
    // open the appropriate file
    File inputFile = new File( "f:\\FirstFile.txt" );
    Scanner file = new Scanner( inputFile );

// Line 1 here:
    if ( file.hasNext( ) ) {
        String line1 = file.nextLine( );
        System.out.println( line1 );
    } else {
        System.out.println( "The file has no data. We are sorry for the omission." );
    }
} catch ( IOException e ) {
    System.err.println( "Caught IOException: " + e.getMessage( ) );
} // end try/catch on read file
```

9. Compile and run the program. Verify the String was properly written into the file named FirstFile.txt. You have created a new file and written a String into it. A concern is that the path for the new file may not be correct. That will throw an exception or write the file somewhere unexpected. It would a good practice to prompt the user for at least the drive letter if not the full path.

Try It!

Using the Hands-On Example as a model, create a new program to append a second string to the file. Be sure to open the FirstFile.txt and append it, not overwrite it. Also adjust the echo code to accommodate printing out a second line.

SECTION REVIEW 11.2

Check Your Understanding

1. What is input validation?
2. What are the two requirements for inputted information to be correct?
3. When will an InputMismatchException error occur?
4. It is important to _____ a file after writing to it; otherwise, the remaining data in memory will not be written to the file.
5. If a file already exists and the FileWriter constructor is passed a boolean argument of false, what happens to the file?

Build Your Vocabulary

As you progress through this course, develop a personal computer science glossary. This will help you build your vocabulary and prepare you for a career. Write a definition for each of the following terms and add it to your computer science glossary.

append   input validation

Science and Java

Dwarf Planets

A dwarf planet orbits a sun just like the other planets. However, dwarf planets are much smaller than regular planets. Dwarf planets have two facts in common with regular planets:

- They have enough mass and gravity to be nearly round, unlike irregularly shaped asteroids.
- They travel in an orbit around a sun.

The big difference between dwarf and regular planets is the path for a dwarf planet is full of asteroids. A regular planet has a clear path around the sun. This fact is what caused Pluto, shown here, to be reclassified as a dwarf planet in 2006.

NASA

There may be dozens of dwarf planets in the solar system. So far, NASA has classified just a few. Most of them are very far away in the Kuiper Belt, which is located beyond Neptune. Here are the names of several classified dwarf planets and when they were discovered:

Dwarf Planet Name	Year Discovered
Ceres	1801
Pluto	1930
Haumea	2003
Makemake	2005
Eris	2005
The Goblin	2015
Farout	2018

Pluto is the most famous since it used to have a more-important status, but closer to home is Ceres. It is located in the asteroid belt between Mars and Jupiter. Ceres is the first dwarf planet to be visited by a spacecraft, NASA's Dawn. The image shown here was taken by that spacecraft. Dawn began its journey in September 2007. In March 2018, the spacecraft was captured by the dwarf planet's gravity. To see a video of the bumpy surface of Ceres, visit images.nasa.gov, and search the video collection for Ceres. You can see the results of collisions between Ceres and asteroids.

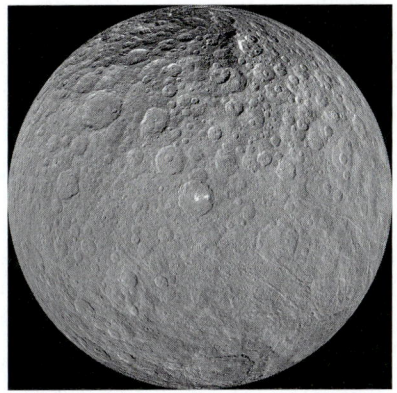

NASA

Assignment

Apply what you learned about creating text files to write a text file to your flash drive. Include in this file the names of the currently known dwarf planets and the year each was discovered. Include exception-handling code. Add each new line to the text file using the same String variable. Call the file dwarf.txt. Here is an example:

```
String s = "Ceres \t1801 ";
printWriter.println( s );
s = "Pluto \t1930 ";
printWriter.println( s );
```

Open the dwarf.txt file in Notepad or another text editor to see the file after it is written and verify the content. Print out the code as well as the text file.

Cooperative Coding

Homework Schedule Helper

Some days there is so much homework assigned you just do not know where to begin. It is a problem, a problem for *all* students! Planning the time to do the homework is part of the problem-solving process. Assess each homework task. Determine the amount of time and what resources you will need to complete the task. Think about your schedule for the remainder of the day. Determine what time is available for homework. You could use an online schedule to keep it all straight.

Dean Drobot/Shutterstock.com

Assignment

Your team can solve this problem by creating an interactive console program to accept a start and stop time to dedicate for each homework task, the subject, if this time slot will compete the homework assignment or if it is a subtask of a larger project, and any resources needed. Print this information as one string to a file using the student's name as the file name. At any time during the day, a student may display the day's schedule.

```
Homework Schedule Helper
Enter your name in CamelCase, no spaces: LoreneMcCall
    1 Add a new time slot to the schedule
    2 Display the schedule
    3 Exit
Enter 1, 2, or 3: 2

Assignment Schedule for LoreneMcCall
Time Slot      Course          Subtask     Resources Needed
1400–1430      Science         No          Baking soda, balloon, timer
1530–1700      Language Arts   Yes         Chapter 3, Mockingbird
```

Goodheart-Willcox Publisher

At the start of the program, prompt for a student name. If a file of that name does not exist, create a new file. Create a simple menu that is embedded in a loop. Ask if the student wants to add a homework block, display the day's schedule, or exit the program. Provide exception handling and input validation in the application. Think about how you will handle files for previous schedules.

Together with your group, determine the file format, and write the algorithm. Then, divide the tasks for coding the program. Assign students to handle the data collection code and others to do the display of the file. The HomeworkScheduleSnippet.java file on the student companion website can be used as a starting point.

1. What did you use for exception handling?
2. What were the criteria you established for valid input?

Chapter Summary

Section 11.1: Handling Exceptions

- When an exception is thrown, an exception handler must catch the exception and tell the runtime system what to do; otherwise, the program will crash.
- The try/catch/finally block can be used to handle both checked and unchecked exceptions by first trying an action, then catching the exception and finally telling the runtime system what to do whether the exception occurred or not.

Section 11.2: Data Validation and Output

- Input validation is the process of ensuring user input matches the expected data type and the information requested by the prompt; a loop is often used to accept input, validate it, and prompt the user to try again if needed.
- The FileWriter and PrintWriter classes can be used to write to a file. The process is to identify the path, determine if data should be written at the beginning of the file or appended to the existing data, create the output, print to the file, and close the file.

Chapter 11 Test

Multiple Choice

Select the best response.

1. If the program provides the meaningful output "File Not Found" instead of abruptly ending, this is an example of what?
   A. catching an exception
   B. crashing
   C. error writing to a file
   D. error in a println statement

2. In which section of a try/catch/finally block is the code IOException e to be found?
   A. try
   B. catch
   C. finally
   D. main

3. The code e.getMessage( ) would probably be an argument for which command statement?
   A. System.out.println
   B. System.in.println
   C. System.err.println
   D. File inputFile

4. Which exception is thrown when the nextInt method is waiting for a number and the user enters a letter?
   A. InputMismatchException
   B. OutOfBoundsException
   C. DivisionByZero
   D. DataTypeMismatch

5. A method for writing to a text file is:
   A. println( )
   B. printWrite( )
   C. writeFile( )
   D. readFile( )

Completion

Complete the following sentences with the correct word(s).

6. The use of a try/_____ block can prevent a program from crashing.

7. _____ an exception is when an exception object is sent to the runtime system.

8. The best way to avoid problems with user _____ is to provide meaningful prompts.

9. A(n) _____ exception will occur when the program cannot find the correct file when outputting to a file.

10. When appending a file, be sure the boolean argument to the FileWriter constructor is set to _____.

Matching

Match the correct term with its definition.

   A. exception
   B. overwriting
   C. appending
   D. try/catch/finally block
   E. input validation

11. Adding more data to an existing file.
12. Using an existing file name to create a new file.
13. Ensuring data is correct.
14. Object that contains information about an error.
15. Alerts the program about a set of errors and offers solutions to fix them.

Application and Extension of Knowledge

1. Write a Java program to calculate the simple interest on a $50,000 loan at 4.5 percent interest for a term of one year. The formula is: interest = principal × rate × term. Write the problem and the answer to a text file. Apply what you learned about the String.format method to create the Strings for output.

2. The Tompkins Company wants an application that calculates and displays an employee's gross pay. No one at the company works more than 40 hours per week, and everyone earns $10.65 per hour. Write a program that accepts and verifies that the number of hours entered is a number and is greater than or equal to zero, but less than or equal to 40. Getting the input as a String provides the opportunity to perform validation. To convert the String to the number, use the valueOf( ) method of the Double wrapper class. For example, double doubleHours = Double.valueOf(stringHours);. Then, calculate the gross pay. Prepare a series of tests that verifies your code solves the problem. Output the result to a text file.

3. Write a program that outputs chemical symbols for compounds to a text file named Compounds.txt. Include these compounds and their English names: NaCl sodium chloride, CO carbon monoxide, and KBr potassium bromide. Devise a way to indicate multiple atoms in a compound, such as H_2O water, and CO_2 carbon dioxide, and include these in the file as well.

4. Social Security numbers (SSN) are unique nine-digit numbers and were originally used for collecting taxes and distributing benefits. Since they were first issued in 1936, the SSNs have become used for other purposes. Now, having a SSN is necessary to open a bank account, obtain a credit card, or get a driver's license as well as being employed. It used to be the case you could tell in which state a person applied for a SSN by the three initial digits of the number. SSNs were assigned to each state starting with the northeastern states and moving to the southwest. Since 2011, SSNs have been assigned at random. Build an application that creates a Social Security number in this format based on numbers entered by the user: XXX-XX-XXXX. Verify the input for nine digits. The application should prompt the user for the nine digits and output the formatted Social Security number. Do not accept the dashes.

5. Look up the current population of the United States. What suggestions can you make for when the population exceeds the nine digits that are currently used for SSNs? How would you implement this in a Java program? Write one or two paragraphs to explain your suggestions.

Online Activities

Complete the following activities, which will help you learn, practice, and expand your knowledge and skills.

Vocabulary. Practice vocabulary for this chapter using the e-flash cards, matching activity, and vocabulary game until you are able to recognize their meanings.

Communication Skills

Reading. After you read this chapter, analyze how the author unfolds a series of ideas. Study the order in which the points are made as well as how they are introduced and developed. How are the individual ideas related or connected? Draw a conclusion about the author's purpose for writing this material.

Writing. A *concept map* is a type of graphic organizer used to show how facts, concepts, and ideas fit together. Carefully listen to your instructor as a lesson is presented. Note each main point by drawing a large circle on paper and writing the main point inside. Write supporting details around the circle and draw connecting lines to it.

Speaking. An effective strategy for committing information to memory is to recite the information you want to remember. After reading this chapter, choose important information to commit to memory. This may be definitions or information you anticipate to be included on a test. Then, recite the information aloud until you have committed it to memory.

Listening. A *barrier* is anything preventing clear, effective communication. *External barriers* are those that exist outside of a person, such as hearing other noises in the room or being distracted by looking out the window at something occurring outside. During class, attempt to recognize any external barriers to listening. How can you avoid external barriers?

Portfolio Development

College and Career Readiness

Talents. You have collected documents that show your skills and talents. Select a book report, essay, or poem you have written that demonstrates your writing talents. If you are an artist, include copies of your completed works. If you are a musician, create a video with segments from your performances.

1. Create a Microsoft Word document that lists your talents. Use the heading Talents along with your name. Next to each talent listed, write a description of an assignment or performance and explain how your talent is shown in it. If there is a video, state that it will be made available on request or identify where it can be viewed online. Indicate that sample screenshots are attached.

2. Scan hard-copy documents related to your talents to serve as samples. Save screenshots from a video in an appropriate file format. Place hard copies in the container for future reference.

3. Place the video file in an appropriate subfolder for your digital portfolio.

4. Update your master spreadsheet.

CTSOs

Case Study. A case study is an analysis of a person, group, or specific situation that is reviewed in order to learn information. A case-study event is an opportunity to demonstrate strategic analysis and decision-making skills. The activity may be a decision-making scenario for which your team will provide a solution. The presentation will be interactive with the judges. To prepare for a case-study event, complete the following activities.

1. Conduct an Internet search for case studies. Your team should select a case that seems appropriate to use as a practice activity. Look for a case that is no more than one printed page long. Read the case and discuss it with your team members. What are the important points of the case?

2. Make notes on index cards about important points to remember. Team members should exchange note cards so that each evaluates the other members' notes. Use these notes to study. You might be able to use these notes during the event.

3. Assign each team member a role for the presentation. Ask your instructor to play the role of competition judge as your team reviews the case.

4. Each team member should introduce himself or herself, review the case, make suggestions for the case, and conclude with a summary.

5. After the presentation is complete, ask for feedback from your instructor. You may also consider having a student audience listen and provide feedback.

Custom Classes and Methods

Sections

The advantages to writing your own classes are many. Bundling data and methods into a class helps to manage data for multiple items. For example, suppose a teacher is tracking grades for students. Creating a Student class and putting each student's name and grades into an object organizes each student's data as a single unit. This is much easier than managing multiple names and multiple grades as separate variables.

The ability to reuse classes is also an advantage. One programmer can write a class that can be used in many different programs. It can even be used by other programmers. For example, a Die class that simulates the roll of a die could be reused in many games of chance no matter who programs those games.

Classes also provide a level of protection for data. The data of an object can only be changed through the methods of the class. These methods can reject requested changes that would corrupt the data, thus ensuring the data is always valid.

College and Career Readiness

Reading Prep

Review questions serve as a self-assessment tool to help you evaluate your comprehension of the material. As you read this chapter, stop at the Check Your Understanding questions. Try to think of potential answers to these questions without referring to the chapter content.

While studying, look for the activity icon for:

- Vocabulary terms with e-flash cards and matching activities.
- Starter files for hands-on examples and other exercises.

These activities can be accessed at
www.g-wlearning.com/informationtechnology/1773

Chapter Glossary

access modifier: Specifies the scope of a class, method, or instance variable across an application.

accessor method: Allows a client to view the data in an object by returning the value of an instance variable; also called a *getter*.

actual parameter: Value sent to the method from the client when the method is called.

class scope: Means a variable can be used throughout an entire class.

client: Application that instantiates objects and calls methods of another class.

default constructor: Constructor that takes no arguments and assigns default values to instance variables.

formal parameter: Parameter name listed in the method header.

getter: Allows a client to view the data in an object by returning the value of an instance variable; also called an *accessor method*.

hash code: Sequence of characters that uniquely identifies an object; usually consists of the class name followed by @ and a hexadecimal address.

inherit: To take on characteristics of a parent class.

instance variable: Indicates each object (instance) of the class will have that data and the values of the data will vary from object to object.

is-a relationship: Object of a subclass is also an object of the superclass.

mutator method: Allows a client to change the values of the data in an object; also called a *setter*.

overloaded constructor: Has the same class name as another constructor, but takes a different number or data type of arguments.

package access: Means the item is available to any method or application in the same package or folder.

parameter: Any argument sent by the client to the method.

setter: Allows a client to change the values of the data in an object; also called a *mutator method*.

subclass: Class that inherits methods and data by extending a superclass.

superclass: Original class from which another class inherits.

this object reference: Special reference to the object on which the method should operate.

Creating a Class

All the applications written in this text so far have been defined as classes. These classes have been a special kind of class: applications. The classes in the Java Class Library are different. Those classes provide services to applications. The applications written so far have been clients of the service classes, such as String, Scanner, and Turtle.

It is possible for programmers to write their own classes that provide services to applications. Just like the classes in the Java Class Library, these service classes will not be executable. That is, these classes will not have a main method. Instead, a client application, which will have a main method, will create objects and call the methods of the service classes. As you continue your computer science education, you will find that applications will become more complex. Dividing the functions of the application into classes will simplify the code. Each class will manage its own data.

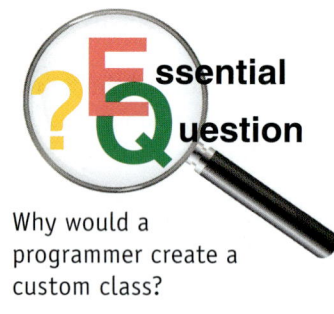

Essential Question

Why would a programmer create a custom class?

Learning Goals

- Explain the process for defining a class.
- Compare and contrast default and overloaded constructors.

Terms

access modifier	hash code
actual parameter	instance variable
class scope	overloaded constructor
client	package access
default constructor	parameter
formal parameter	

Class and Data Definition

Classes are usually created to perform some function in an application. Often, an application consists of many classes. Before writing an application, it is important to decide which classes are needed and what functions the objects of the classes will perform.

For example, in a game application, you might define a Player class to manage the players and their scores. You might also want to define a Die class to manage the random roll of a die. You might want a Manager class to alternate turns and determine when the game is over and which player won the game. Defining a class is performed in these steps:

1. Determine the name of the class.
2. Define the class.
3. Define the data of the class.
4. Create one or more constructors.
5. Write the methods of the class.

Class Name and Definition

The first step in creating a class is to decide on its name. By convention, classes are named as nouns, such as String, Scanner, and Turtle. Further, to follow Java naming style, the class name should begin with a capital letter and use CamelCase.

The syntax for defining a class should be familiar. The syntax is the same for service classes as for application classes:

```
public class ClassName {
}  // end class
```

Notice the keyword public. This keyword is called an access modifier because it determines who can use this class. An *access modifier* specifies the scope of a class, method, or instance variable across an application. Possible access modifiers are shown in **Figure 12-1.** The protected access modifier is included in the figure for completeness, but its use is beyond the scope of this text.

Classes and methods are generally defined to be public. This allows any other method or application to use the class and call the methods. For example, all the methods in the String class discussed in Chapter 10 have public access.

The data of the class are defined to have private access by convention. This means only methods of the class in which the data is defined can directly access the data. So, how can the object data be changed? A class usually provides methods to set and modify the data of the class. Because only the methods of the class can access the data directly, defining data to be private helps ensure the data is always valid. For example, for a Student class, the method to set a grade would require the value to be between 0 and 100. A call to the method with a request to set the grade to −1 would be rejected by the method.

If the modifier is not specifically given, the method or instance variable will receive package access automatically. *Package access* means the item is available to any method or application in the same package or folder. Package access widely exposes the data to many classes. This can cause unintended consequences. Therefore, it is important to remember to specify private access for data in a class.

As mentioned earlier, a Die class that simulates the roll of a die could be useful for many applications. To illustrate defining a class, we will create a Die class. Following the syntax for defining a class, here is the beginning of the Die class.

```
public class Die {

} // end class
```

FYI

Methods are usually defined to be public so they can be called from a range of classes. Data is defined to be private so only methods of the same class can change their values.

Access Modifier	Meaning
public	Available for use by any class.
private	Available only to methods of the class in which the item is defined.
(no modifier)	Available to methods in which the item has been defined as well as to methods in other classes in the same package or folder; also called *package access*.
protected	Available for direct use by subclasses, which are classes that inherit from another class.

Goodheart-Willcox Publisher

Figure 12-1. These are the access modifiers in Java.

Determine Data

The next step in writing a class is to determine the data for it. Each object created from a class has its own copy of the data items with values that are appropriate for that object. The term *instance variable* indicates each object (instance) of the class will have that data and the values of the data will vary from object to object. In contrast, all objects of a class share the methods of the class. Thus, one copy of the methods of a class are stored in memory, while each object has its own copy of data.

Keeping that in mind, instance variables should be chosen for data common to all objects of the class. For example, each object for a Student class could have a name and a test grade. Other possible data could be the grade level, such as freshman or sophomore, or the grade average for the year. For a Player class, each object would have a name, score, and set of abilities.

To define an instance variable, give the data type and a name. Optionally, an initial value can be assigned. Instance variables are defined using this syntax:

```
accessModifier datatype instanceVariableName;
```

or

```
accessModifier datatype instanceVariableName = initialValue;
```

As mentioned, the access modifier for instance variables is usually private so a client cannot directly change the value. A *client* is an application that instantiates objects and calls methods of another class.

Typically, the instance variables are defined at the top of the class. Regardless of where the instance variables are defined, however, they have class scope. *Class scope* means the variable can be used throughout the entire class.

For the example Die class, these instance variables come to mind:
- roll, which is the current value of the die
- numberOfSides, which is the number of sides for the die

Making the number of sides an instance variable allows the class to manage rolling a die with a nonconventional number of sides, as shown in **Figure 12-2.** Two other candidates for instance variables, which actually support the instance variables above, are:
- randGen, which is a random-number generator object to generate a random roll for the die
- DEFAULT_SIDES, which is a constant set to 6 to manage the most likely number of sides for the die

The Random class will need to be imported. The random number generator object can be initialized by instantiating the Random object in the definition of the instance variable. The Die class definition now becomes:

```java
/** Die class
    simulates the roll of a die
*/
import java.util.Random;

public class Die {

    // instance variables
    private int numberOfSides;
    private int roll;

    private Random randGen = new Random( );
    private final int DEFAULT_SIDES = 6;

} // end class
```

As in other blocks of code, indent any statements in the body of the class.

This Die class is not executable. Any attempt to run the class will generate an error saying the main method is not found. A separate client class that has a main method is needed to create Die objects. This is a possible client class that instantiates two Die objects:

```
/** Game class client of the Die class
*/

public class Game {

    public static void main( String [ ] args ) {

        Die die1 = new Die( );
        System.out.println( die1 );
        Die die2 = new Die( );
        System.out.println( die2 );

    } // end main
} // end class
```

Blackregis/Shutterstock.com

Figure 12-2. Some dice have more or fewer than six sides. By including an instance variable in the class for the number of sides, the class can be used to simulate the roll of a die with any number of sides.

In the Game class, first a Die object named die1 is instantiated. Die is the data type, and die1 is the object reference. The next statement outputs the die1 object. Then, another Die object named die2 is instantiated and also outputted. The Game class is executable. When run, the output looks like this:

```
Die@2669b199
Die@2344fc66
```

What you see is called the hash code of the object. A ***hash code*** is a sequence of characters that uniquely identifies an object. The hash code begins with the class name (Die), followed by @, followed by a hexadecimal number that is usually the location in memory of the object. The specific hexadecimal value depends on available memory and can vary from computer to computer, or from one execution to the next. You can see from the different hexadecimal values that die1 and die2 are different objects. Granted, the hash code is not terribly useful. After writing some methods, it will be possible to see the data in each object.

HANDS-ON EXAMPLE 12.1.1

Starting a Class

In this exercise, you will begin to write a class and a client for that class. As the chapter progresses, you will add to this class until the code is complete. Specifically, you will develop a class for tracking the balance of a fund as money is deposited and withdrawn. This fund could represent an allowance, a savings account, a scholarship fund, or any other account where money is collected. Obviously, this class has many opportunities for reuse. Two possible names for the class are Account or Fund. Either name would be appropriate, but the short yet meaningful name wins: Fund. It is time to start defining the Fund class.

1. Launch jGRASP, and begin a new Java file.

2. Write the header comment code.

```
/** A Fund class to track money in an account
    <your name here>
*/
```

3. Write the code to define the Fund class, leaving some empty lines to add the body of the class later.

```
public class Fund {


} // end class
```

4. What data should a Fund class have? It would make sense to have a name for the fund (a String) and a balance (a double). Write the code to define these two instance variables inside the class. Remember to define the instance variables as private. Each object created from this class will have a fundName and a balance. The class code should now look like this:

```
/** A Fund class to track money in an account
    <your name here>
*/

public class Fund {

    // define instance variables
    private String fundName;
    private double balance;

} // end class
```

5. Save the file as Fund.java. Remember, this class is not executable. The class does not have a main method. To illustrate this, compile and attempt to run the file. You will receive a warning, as shown. Click the **OK** button to close the warning.

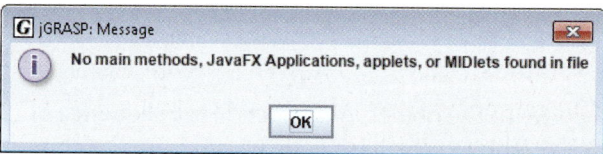

Goodheart-Willcox Publisher

6. Start a new Java file. To use the Fund class requires a client application class to have a main method and to create Fund objects. Name this class FundClient.

```
/** FundClient
    Creates Fund objects and calls Fund methods.
    <your name here>
*/

public class FundClient {

    public static void main( String [ ] args ) {



    } // end main
} // end class
```

7. Save the file in the *same* folder as the file for the Fund class.

8. In the FundClient file, create a Fund object named allowance and attempt to output the object. Put this code under the main method header and before the closing brace that ends the main method.

```
Fund allowance = new Fund( );
System.out.println( allowance );
```

9. Compile and run the FundClient file. You should see the hash code of the allowance object. Note: the hexadecimal value you see may be different from what is shown here.

```
Fund@4590c9c3
```

10. Save both the Fund.java and FundClient.java files. You may wish to keep these files open in jGRASP as you will continue coding them throughout the chapter.

Try It!

- Begin to write a Student class that can be used to track grades. This class will have two instance variables, studentName and grade. The studentName variable is a String that will hold the name of the student. The grade variable is an int to hold the student's grade. Also write a Teacher class that will be a client of the Student class. The Teacher class should create a Student object and output the object's hash code. Compile both files and run the Teacher class. You will continue to build these two files as you progress through the chapter.

Constructors

For the client to be able to create objects, a class provides one or more constructors. As defined in Chapter 6, a constructor is a special method called when an object is created (instantiated) using the new keyword. The job of any constructor is to ensure all instance variables of the class have valid values. A constructor has the same name as the class. A class can provide more than one constructor, as long as each constructor has a different argument list.

Default Constructor

Earlier, you may have wondered how the Game client could instantiate an object when no constructor had been written in the Die class. The answer is, the compiler automatically created a constructor. If a class does not have a constructor, the compiler creates a default constructor to allow objects to be created. A *default constructor* does not take any arguments and assigns default values to the instance variables. The values assigned depend on the data type of the instance variable, as shown in **Figure 12-3**. Numeric data are assigned the value 0, boolean data are assigned the value false, char data are assigned the lowest possible char value (0000), and object references are assigned the value null.

When writing a class, you are free to write your own default constructor to replace the one created by the compiler. Writing a default constructor is advisable if any of the instance variables are object references. Remember, many commonly used data types, such as Strings, are object references. Object references are automatically given the value null by the compiler's default constructor. Any attempt by the client to call a method using a null value will result in a NullPointerException error. Class authors often provide their own default constructor in order to assign a valid value other than null to object references. Another reason to write a default constructor is that the compiler no longer provides a default constructor if a class has any other constructors.

All default constructors have the same syntax:

```
public ClassName( ) {

}
```

The braces are required. Notice that the default constructor takes no arguments, and that no return value is given. Nevertheless, the constructor returns a reference to the newly created object. The body of the constructor should assign default values to the instance variables.

Data Type	Default Value
byte, short, int, long	0
float, double	0.0
boolean	false
char	0000 (hexadecimal)
object reference	null

Goodheart-Willcox Publisher

Figure 12-3. The values assigned to instance variables by the compiler's default constructor depends on the data type of the instance variable.

Language Arts and Java

Music Library Classes

Alejandro has an extensive music library. It spans many artists, genres, media formats, and locations. When he wants to listen to a specific tune, he must locate it among his many different media collections. He thinks if he catalogs his music, it will be easier to find each song.

Alejandro is taking Java programming in school. The current assignment is to code a Java class for a music library. He decides to take advantage of this assignment. Alejandro will take charge of his varied collection of music and create an electronic music library for himself.

wong yu liang/Shutterstock.com

He decides that the important information about each item is the title, artist, genre, album or single, language, and location. For example, he listens to the single "breathin" by Ariana Grande. The genre for this song is pop, the language is Spanish, and he has it stored in the cloud. An oldie favorite of his is "Layla" by Eric Clapton. It is on his phone in iTunes and it is a classic blues-rock song sung in English. Other types of media he owns are DVDs, flash drives, and files on his PC.

Assignment

Applying what you know about building service classes, in a word-processing document, write the directions in English for Alejandro defining the instance variables in the class. Identify the instance variables and the data type of each instance variable. Also, give the default value to which the default constructor should assign each instance variable.

The default constructor for the Die class could look like this:

```
public Die( ) {
   numberOfSides = DEFAULT_SIDES;                    // use default

   roll = randGen.nextInt( numberOfSides ) + 1;      // generate the
                                                     // random roll

}
```

This constructor assigns the default number of sides (6) to the numberOfSides instance variable, then rolls the die to assign a value to the roll instance variable. Now all the instance variables have valid values. Note that this default constructor is important. Without giving the class its own default constructor, the compiler's version would assign both numberOfSides and roll the value 0, which are both invalid values for a die.

It is important *not* to specify a return value for a constructor to avoid a compiler error. The tricky part about this error is that the compiler does not flag a constructor with a return value to be an error. The compiler just assumes the method is not a constructor. The problem shows up when a client attempts to call the constructor. At that time, the compiler reports there is no constructor taking the number and type of arguments the client is sending.

Multiple Constructors

To give clients more options for creating objects, classes often provide multiple, overloaded constructors. Like any overloaded method, ***overloaded constructors*** have the same class name, but take a different number or data type of arguments. Overloading constructors gives clients options in instantiating objects. Usually, at least one constructor will allow the client to set the initial values for the instance variables.

You have seen overloaded constructors in the Scanner class. One constructor, which was used for user input, takes an InputStream object as an argument. The other constructor, which was used for reading from a file, takes a File object as an argument:

```
Scanner( InputStream in )
Scanner( File inputFile )
```

When a class provides multiple constructors, the compiler figures out which constructor to call by looking at the data type of the arguments.

The syntax for a constructor that overloads the default constructor is:

```
public ClassName( dataType1 param1, dataType2, param2, …) {

}
```

Again, the braces are required and no return value is specified.

Clients send arguments to a method. To the method, any argument sent by the client to the method is called a ***parameter.*** A ***formal parameter*** is one of the names listed in the method header. An ***actual parameter*** is a value sent to the method from the client when the method is called. In the method body, the method refers to the parameters by using the formal parameter names. The scope of a method's parameter is limited to the method body.

Here is the second constructor for the Die class, which takes a parameter for the number of sides for the die. This overloaded constructor gives the client application the choice of setting the number of sides for the Die object. Remember, the default constructor sets the numberOfSides to 6.

```
// second constructor, accepts a parameter for number of sides
public Die( int sides ) {
    if ( sides >= 1 ) {                   // sides must be at least 1
        numberOfSides = sides;
    } else {
        numberOfSides = DEFAULT_SIDES;
    }

    roll = randGen.nextInt( numberOfSides ) + 1;   // generate the
                                                    // random roll
}
```

This constructor checks that the sides value sent by the client is at least 1. It would not make sense to have a die with no sides or a negative number of sides. If the parameter value is at least 1, then the parameter value is assigned to the instance variable. If the parameter value is less than 1, the default value for the number of sides is assigned to the instance variable instead. In this way, the constructor assures a valid value is assigned to the instance variable. After the number of sides has been established, the code rolls the die to put a valid value into the roll instance variable. This is an example of how the class methods ensure the data is always valid.

The code in main for the Game client to call each constructor is:

```
// call default constructor
Die die = new Die( );   // six-sided die
System.out.println( die );

// call second constructor to instantiate an object named bigDie
Die bigDie = new Die( 10 ); // 10-sided die
System.out.println( bigDie );
```

Notice that when the client calls the second constructor, the argument is the value 10. When the constructor in the Die class starts running, its parameter named sides will have been given the value 10. In this case, sides is the formal parameter and 10 is the actual parameter.

HANDS-ON EXAMPLE 12.1.2

Writing Constructors

In this exercise, you will continue writing the Fund and FundClient classes. Specifically, you will write two constructors for the Fund class and write code in the FundClient class to instantiate Fund objects by calling those constructors.

1. Launch jGRASP, and open the Fund.java and FundClient.java files created in the Hands-On Example 12.1.1.
2. The default constructor provided by the compiler does not do everything needed for the Fund class. The problem is that it will give the fundName instance variable, which is a String object reference, the value null. If a client attempted to call a method using fundName, a NullPointerException error would occur. To ensure that fundName is not null, you need to assign a default value. What the default value is for any class is a design decision made by the class author. With some classes, a meaningful default value is clear. For the Fund class, however, the value could be many things. Lacking a better option, assign an empty String.

Although an empty String is not meaningful, at least the value is not null. In the Fund class, write the default constructor after the definition of the instance variables. Also add the comment to indicate the default constructor.

```
// default constructor
public Fund( ) {

    fundName = ""; // makes an empty String to avoid null value
    balance = 0.0; // optional

}
```

In this case, assigning 0.0 to balance is optional. If any instance variable is not assigned a value, the compiler includes code to assign the default values shown in **Figure 12-3.** If you do not specifically assign a value to balance, the compiler will assign 0.0.

3. Write the code to define the second overloaded constructor. This constructor provides parameters so the client can send initial values for the instance variables. This code should be added after the closing brace of the first constructor, but before the end of the public class.

```
// second constructor with parameters
public Fund( String name, double startBalance ) {

    fundName = name;
    if ( startBalance > 0.0 ) {      // check for valid value
        balance = startBalance;
    } else {
        balance = 0.0;
    }

}
```

In this code, the formal parameters are name and startBalance. The actual parameters will be the values sent to the method by the client. Notice that the code checks whether the initial balance from the client is positive. If not, the balance is assigned 0.0. It would not make sense to start a fund with a negative balance.

4. Compile the Fund.java file.

5. Switch to the FundClient.java file. The code to create a Fund object by calling the default constructor is already in place. That code will now call the default constructor you wrote in the Fund class. Add the following code to call the second constructor to create an object named scholarship.

```
// call second constructor
Fund scholarship = new Fund( "College fund", 1000.00 );
System.out.println( scholarship );
```

When the second constructor in the Fund.java class begins executing, the formal parameter name will have the value of the actual parameter: College fund. The formal parameter startBalance will have the value of the actual parameter: 1000.00.

6. Compile and run the FundClient class. You should see output something like this, although the hexadecimal values may be different:

```
Fund@4590c9c3
Fund@32e6e9c3
```

7. You may wish to keep the Fund.java and FundClient.java files open in jGRASP as you will continue coding them throughout the chapter.

Try It!

Continue writing the Student class. Add a default constructor that sets the name of the student to "New student" and the grade to 0. Add a second constructor that accepts a name parameter for the student name and a beginGrade parameter for the grade. Ensure the beginGrade value is between 0 and 100. If not, set the value to 0. Add code to the Teacher class to call each constructor. Compile both files and run the Teacher class.

SECTION 12.1 REVIEW

Check Your Understanding

1. A method's parameters are given the value of the _____ in the method call.
2. Methods in a class are usually given the _____ access modifier, and instance variables are usually given the _____ access modifier.
3. What is a constructor?
4. What will the value of an object reference be if a constructor does not assign a value to the reference?
5. Why would an overloaded constructor be created in a class?

Build Your Vocabulary

As you progress through this course, develop a personal computer science glossary. This will help you build your vocabulary and prepare you for a career. Write a definition for each of the following terms and add it to your computer science glossary.

access modifier
actual parameter
class scope
client

default constructor
formal parameter
hash code
instance variable

overloaded constructor
package access
parameter

Writing Methods

You have created a class and defined the instance variables. However, a class is not complete without methods. Usual methods in a class allow the client to view and to change the values of the instance variables. Additionally, methods do the work of the class.

For example, for a Die class, the rollDie method would generate a new roll of the die determined by a random-number generator. For the Fund class, deposit and withdraw methods would allow the client class to add and remove amounts from the fund. For the Student class, a letterGrade method would convert a number grade to a letter. Finally, each class should provide a toString method that returns the data of the object in a formatted String.

Learning Goals

- Diagram the pattern of a value-returning method.
- Design accessor and mutator methods.
- Explain the use of the keyword this.
- Identify required methods for a given custom class.

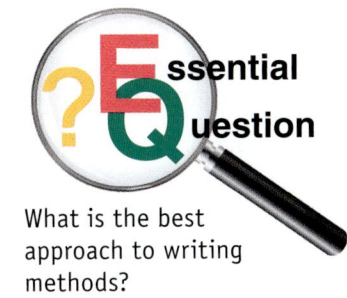

Essential Question

What is the best approach to writing methods?

Terms

accessor method

getter

mutator method

setter

this object reference

Accessor Methods

Some methods return a value. Other methods just perform their job without returning a value. Method definitions follow this syntax:

```
accessModifier returnType methodName( datatype1 param1, dataType2
param2, … ) {
   // method body
}
```

The access modifier is usually public. The braces are required.

Like instance variables, methods have class scope. This means any method can call any other method within the class. Also, because instance variables have class scope, any method can use any instance variable.

Classes usually provide accessor methods, which are value-returning methods. An *accessor method* allows a client to view the data in an object by simply returning the value of an instance variable. An accessor method is also called a *getter*. An accessor method uses the return statement to provide the value of the instance variable. The return statement has the syntax:

```
return expression;
```

FYI

Because instance variables and methods both have class scope, methods and instance variables can be defined in any order within the class definition.

When this statement is executed, the method stops, and the value of **expression** is passed back to the client. The value of **expression** effectively replaces the method call in the client program. Accessor methods follow a simple pattern:

- The method name usually starts with get followed by the instance variable name in CamelCase.
- The method does not take any parameters.
- The return type is the same data type as the instance variable.
- The method body consists of a return statement with the expression being the instance variable.

Here is the accessor method for the roll instance variable in the Die class:

```
// accessor method for roll
public int getRoll( ) {
    return roll;   // pass the value of roll to the client
}
```

This method is public so the client can call the method. The roll instance variable is of the int data type, so the return value from the method is int. The name of the method is getRoll. The empty parentheses in the method header indicate that the method does not take any parameters. The method body contains just one statement to return the current value of the instance variable to the client. Following this same pattern is the accessor for the numberOfSides instance variable:

```
public int getNumberOfSides( ) {
    return numberOfSides; // pass the number of sides to the client
}
```

If these statements are added to the client Game class:

```
System.out.println( "For the default die: " );
System.out.println( "\tThe roll is " + die.getRoll( ) );
System.out.println( "\tThe number of sides is "
          + die.getNumberOfSides( ) );
System.out.println( "For the bigDie: " );
System.out.println( "\tThe roll is " + bigDie.getRoll( ) );
System.out.println( "\tThe number of sides is "
          + bigDie.getNumberOfSides( ) );
```

the values of the data can now be seen after the class is compiled and run:

```
For the default die:
    The roll is 5
    The number of sides is 6

For the bigDie:
    The roll is 7
    The number of sides is 10
```

Because the value of roll is random, your output values for roll may be different. Notice that the \t escape character inside the output String inserts a tab into the output. This moves the text to the right by one tab stop.

HANDS-ON EXAMPLE 12.2.1

Writing Accessor Methods

In this exercise, you will continue writing the Fund and FundClient classes. Specifically, you will write two accessors for the Fund class and write code in the FundClient class to call those methods to view the data in the Fund objects.

1. Launch jGRASP, and open the Fund.java and FundClient.java files from Hands-On Example 12.1.2.

2. In the Fund.java file, write accessor methods for the fund name (getFundName) and the fund balance (getBalance). The fund name is a String, so the return value from getFundName should be a String. The balance is a double, so the return value from getBalance should be a double.

```java
// accessor methods
public String getFundName( ) {
    return fundName; // pass name of fund to the client
}

public double getBalance( ) {
    return balance;      // pass fund balance to the client
}
```

3. Compile the Fund.java file to check for errors, but do *not* attempt to execute the Fund class.

4. Switch to the FundClient.java file. Delete the code to output the object name. In its place, add code to call the accessors and output the return values. This code should be in the main method.

```java
System.out.println( "The allowance fund:" );
System.out.println( "\tThe fund name is "
                + allowance.getFundName( ) );
System.out.println( "\tThe balance is "
                + allowance.getBalance( ) );
System.out.println( "The scholarship fund:" );
System.out.println( "\tThe fund name is "
                + scholarship.getFundName( ) );
System.out.println( "\tThe balance is "
                + scholarship.getBalance( ) );
```

5. Compile and run the FundClient class. You should now be able to see and verify the data in each object.

```
The allowance fund:
    The fund name is .
    The balance is 0.0.

The scholarship fund:
    The fund name is College fund.
    The balance is 1000.0.
```

6. Notice the balance is not formatted in the output. That is good. The accessor's job is simply to return the value. The client chooses how to use the value. Perhaps the client wants to add the balances of the two funds to find the total in both funds. If the balances were formatted as Strings with a dollar sign, the client would not be able to use the balances in a calculation. To demonstrate, add the following code to the main in the client.

```java
// add balances to get a total
double totalFunds = allowance.getBalance( )
                + scholarship.getBalance( );
// output raw total
System.out.println( "The total in both funds is "
                + totalFunds );
// output formatted total
String s = String.format( "The total in both funds is $%.2f.",
                totalFunds );
System.out.println( s );
```

7. Compile and run the FundClient class. The additional output should look like this:

```
The total in both funds is 1000.0.
The total in both funds is $1000.00.
```

8. You may wish to keep the Fund.java and FundClient.java files open in jGRASP as you will continue coding them throughout the chapter.

Try It!

Continue writing the Student class. Add two accessors, getStudentName and getGrade. In the Teacher class, delete the code that outputs the student objects. Add code to the Teacher class to call each accessor for each object. Compile both files and run the Teacher class.

Mutator Methods

Classes usually provide mutator methods. A *mutator method* allows a client to change the values of the data in an object. These methods are also called *setters.* Mutators usually do not return a value. Rather, the return type is void. The scope of any method's parameters is the method body. Mutator methods typically follow this pattern:

- The method name usually starts with set followed by the instance variable name in CamelCase.
- The method takes one parameter, which has the same data type as the instance variable and is the new value for the instance variable.
- The return type is void, so the body of the method does not return a value.
- The body of the method validates that the new value for the instance variable is valid. If so, the method assigns the new value to the instance variable.

For the Die class, only one mutator is needed, which is to change the number of sides of the die. Another technique will be used later in the chapter to change the roll instance variable. Here is a mutator for the numberOfSides instance variable:

```
public void setNumberOfSides( int sides ) {
   if ( sides >= 1 ) {        // sides must be at least 1
      numberOfSides = sides;
   } else {           // parameter is invalid
      numberOfSides = DEFAULT_SIDES; // assign default value
   }
}
```

The method is public to allow the client to call the method. The method's return type is void, so the method does not return a value. The method's name is setNumberOfSides. The method takes one parameter, sides, which is an int value for the new number of sides for the die. Since a parameter's scope is the method body, the scope of the sides parameter is the body of the setNumberOf-Sides method.

Much of the code in the setNumberOfSides method should look familiar. The mutator is performing the same checks on the sides parameter as the second constructor discussed earlier. To avoid such duplication of code, it is advisable to change the second constructor to call this method instead of doing the checking itself. Remember that any method can call any other method in the class. Thus, the modified second constructor becomes:

FYI

Write code to validate the value of an instance variable in a mutator, and have the constructor call the mutator. In that way, code is not duplicated.

```
// second constructor, accepts a parameter
public Die( int sides ) {
   setNumberOfSides( sides );     // call mutator to verify value

   roll = randGen.nextInt( numberOfSides ) + 1;   // generate the
                                                  // random roll
}
```

Now the code to verify the sides parameter is only in the mutator. If later it is decided to change the rules for the number of sides, perhaps to impose an upper limit on the number of sides for a die, then the code needs to be changed in only one place.

Special Object Reference This

It is possible to name a parameter the same name as the instance variable. Doing so, however, could cause unintended consequences. A parameter with the same name as an instance variable "hides" the instance variable. With this *incorrect* mutator code:

```
public void setNumberOfSides( int numberOfSides ) {
  if ( numberOfSides >= 1 )
    numberOfSides = numberOfSides;    // code has no effect on
                                      // instance variable
  } else {
    numberOfSides = DEFAULT_SIDES;
  }
}
```

nothing happens to the numberOfSides instance variable because all mentions of numberOfSides refer to the parameter.

To "reveal" the instance variable, precede its name with a this object reference. The ***this object reference*** is a special reference to the object on which the method should operate. A name preceded with the this object reference refers to an instance variable. Using the keyword, the setNumberOfSides method becomes:

```
public void setNumberOfSides( int numberOfSides ) {
  if ( numberOfSides >= 1 ) {
    this.numberOfSides = numberOfSides;   // instance variable
                                          // gets parameter value
  } else {
    this.numberOfSides = DEFAULT_SIDES; // instance variable gets
                                        // default
  }
}
```

The line:

```
this.numberOfSides = numberOfSides;
```

means "assign the parameter value to the instance variable." This line:

```
this.numberOfSides = DEFAULT_SIDES;
```

means "assign the default number of sides to the numberOfSides instance variable."

The special object reference this solves this naming problem because when a method starts executing, the JVM sets the object reference this to the address of the object. Then, the method knows which object's data should be used for the method. Remember that all objects share one copy of the method code. By specifying this.numberOfSides, it is made clear to the method that the reference is to the instance variable of the object.

This code in the Game class will test the new code:

```
Die testerDie = new Die( 0 );  // invalid number of sides, default
                               // value is assigned

System.out.println( "For the testerDie:" );
```

(Continued)

```
System.out.println( "\tThe number of sides is "
                + testerDie.getNumberOfSides( ) );

System.out.println( "Calling the mutator with value 12" );
testerDie.setNumberOfSides( 12 );
System.out.println( "\tThe number of sides is now "
                + testerDie.getNumberOfSides( ) );
```

The output from these statements is:

```
For the testerDie:
    The number of sides is 6
Calling the mutator with value 12
    The number of sides is now 12
```

As expected, when the constructor was called with an invalid number of sides (0), the constructor called the mutator, and the mutator set the numberOfSides instance variable to the default value (6). Calling the mutator with a valid parameter of 12 *did* change the value.

Hands-On Example 12.2.2

Writing Mutator Methods

In this exercise, you will continue adding to the Fund and FundClient classes. Specifically, you will write one mutator for the Fund class, write code in the FundClient class to call that mutator method, and then call the accessor to view the modified data in the Fund objects.

1. Launch jGRASP, and open the Fund.java and FundClient.java files from Hands-On Example 12.2.1.

2. Write the mutator method for the fund name (setFundName). Use the fundName instance variable name as the parameter name. No validation of the parameter value is needed in this method because all fund names would be valid.

```
public void setFundName( String fundName ) {
    this.fundName = fundName;
}
```

3. In the FundClient.java file, add code to call the mutator for the scholarship object with the new fund name of "Medical School," and then call the accessor to output the new value.

```
scholarship.setFundName( "Medical School" );
System.out.println( "\tThe fund name is now "
                + scholarship.getFundName( ) + "." );
```

4. Compile and run the FundClient class. Verify the change to the fund name was made by viewing the additional output.

```
The fund name is now Medical School.
```

5. You may wish to keep both the Fund.java and FundClient.java files open in jGRASP as you will continue coding them throughout the chapter.

Try It!

Continue adding to the Student class. Write two mutators, setStudentName and setGrade. No validation is required for the student name. For the setGrade method, perform the validation of the parameter in this method. Then, remove the validation of the grade value from the second constructor and instead have the second constructor call the setGrade mutator. Add code to the Teacher class to call each mutator, then call the accessors to verify the change was made. Compile both files and run the Teacher class.

Math and Java

Line Segments

Line segments are a big topic in algebra and geometry. A Segment class can be written to provide information about any line segment given its endpoints. The class instance variables are the coordinates of the two endpoints: x_1, y_1, x_2, y_2. The constructor for this class uses these endpoints as parameters. Methods can find slope, whether or not a slope exists, length of the segment, y-intercept, and equation of the line containing the segment. Also, mutator and accessor methods can be written.

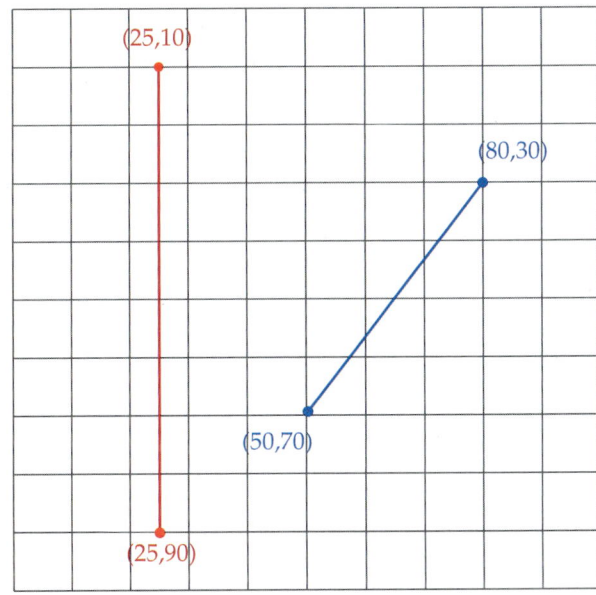

(25,10)

(80,30)

(50,70)

(25,90)

Goodheart-Willcox Publisher

Apply what you learned in math class to find slope, length, y-intercept, and the equation of the line.

To find slope:

$$m = (y_1 - y_2) \div (x_1 - x_2)$$

Recall that a problem arises if the x-coordinates are the same. The slope does not exist, and the line segment is vertical. The length of the line segment is calculated using the distance formula:

$$y = \sqrt{(x_1 - x_2)^2 + (y_1 - y_2)^2}$$

If the slope exists, you can find the y-intercept:

$$y\text{-intercept: } b = y_1 - mx_1$$

The equation of the line containing the segment can be written in slope-intercept form if the slope exists. Otherwise, the equation method will return the equation of the vertical line:

$$y = mx + b$$
or
$$x = x_1$$

Assignment

Before beginning this exercise, download the chapter files from the student companion website. Two starter files are provided: Segment.java and SegmentClient.java. Follow the prompts to complete these two files and verify the output. When the output is correct:

1. Copy the output into a word-processing document, and submit it along with the two completed class files.

2. Change the values of the coordinates in the client program, compile, run, and verify the output.

3. See what happens if you enter two endpoints with the same *x* value. Explain what you observe.

For an additional challenge, use Scanner to get the coordinates by user input.

Finishing the Class

The reason-for-being of a class is to perform some action with the data beyond allowing the client to see or change the values. Whether or not these methods are void or return a value and whether or not the methods take parameters depend on the purpose of the method. In addition, all classes should have a toString method that returns the data of the class as a String.

Purpose of the Class

Consider the Die class example presented in this chapter. What is the purpose for this class? The main purpose of the Die class is to simulate a roll of a die. To that end, an essential method for the class is rollDie. This method should generate a random number to represent the roll of the die. The rollDie method should store that number in the roll instance variable and also return the number to the client. No parameter is needed because the method works with only instance variables: roll, randGen, and numberOfSides. Remember that instance variables have class scope. This means all methods of the class can access the instance variables.

Thus, the rollDie method is public, takes no parameters, and returns an int:

```java
public int rollDie( ) {
    roll = randGen.nextInt( numberOfSides ) + 1; // generate the
                                                 // random roll
    return roll;
}
```

Technically, the rollDie method is a mutator because it changes the value of an instance variable.

toString Method

It is a best practice to include a toString method in every custom class so the object data can be returned as a String. This method is especially important because the JVM automatically calls the toString method whenever an object reference needs to be converted to a String. This conversion happens when an object reference is concatenated with a String or when the object reference is the only parameter to System.out.println. The toString method takes no parameters and returns a String.

The format of the return value is a design decision. Usually, however, the String contains some description along with the instance variable values. For example, a possible toString method for the Die class is:

```java
public String toString( ) {
    return "The roll for a " + numberOfSides + "-sided die is " +
        roll;
}
```

FYI

Notice that the toString method returns a String to the client. The method does not *print* the value. It is the client's decision what to do with the return value. For example, it may print the value or write the value to a file.

Without a toString method, the hash code of an object is used for the String representation. You saw this in earlier examples in this chapter. This code in the client would output the hash code if there was not a toString method:

```java
// without toString
Die myDie = new Die( );
System.out.println( myDie ); // outputs hash code
```

After writing a toString method, the same client code outputs the message from the toString method instead:

```java
// with toString
Die myDie = new Die( );
System.out.println( myDie );   // outputs: The roll for a 6-sided
                               // die is 6
```

It is also possible to directly call the toString method. The output is the same as the automatic call:

```
Die myDie = new Die( );
System.out.println( myDie.toString( ) );   // outputs: The roll for
                                            // a 6-sided die is 6
```

Publishing the API

Now that the Die class is finished, the API can be published. **Figure 12-4** shows the complete API for the Die class. Using this API, other programmers can include the Die class in their applications.

Constructors	Explanation
Die( )	Default constructor; assigns the default of 6 to the number of sides of the die and rolls the die.
Die(int sides)	Assigns sides to the number of sides of the die and rolls the die.
Methods	**Explanation**
int rollDie( )	Rolls the die and returns the value of the roll.
int getRoll( )	Returns the current roll of the die.
int getNumberOfSides( )	Returns the number of sides of the die.
void setNumberOfSides(int sides)	Sets the number of sides to sides if sides is at least 1; otherwise, sets the number of sides to the default value of 6.
String toString( )	Returns a String containing the number of sides of the die and the current roll.

Goodheart-Willcox Publisher

Figure 12-4. This is the API for the Die custom class.

HANDS-ON EXAMPLE 12.2.3

Completing the Class and the API

In this exercise, you will complete the Fund and FundClient classes. Specifically, you will write two methods in the Fund class to deposit and withdraw money and also the toString method. In the FundClient class, you will write code to call those methods.

1. Launch jGRASP, and open the Fund.java and FundClient.java files from Hands-On Example 12.2.2.
2. Write the deposit method to allow the client to add money to the fund. Technically, the deposit method is a mutator method because it alters the value of the balance instance variable.

```
public void deposit( double amount ) {
   balance += amount; // add deposit amount to balance
}
```

3. Write the withdraw method to allow the client to take money from the fund. Technically, withdraw is also a mutator method.

```
public void withdraw( double amount ) {
   balance -= amount; // subtract withdrawal amount from balance
}
```

4. Write the toString method to return a String describing the data in the object. For this method, you *do* want to format the balance because the return value is a String meant for displaying or printing, not for further calculations.

```
// toString method; returns fund name and formatted balance
public String toString( ) {
    String formattedBalance = String.format( "$%.2f", balance );
    return "The balance for the " + fundName + " fund is "
        + formattedBalance + ".";
}
```

5. Switch to the FundClient.java file. Add code before the end of the main method to create a new Fund object, call the deposit and withdraw methods, and then output the state of the object by calling the toString method.

```
Fund piggyBank = new Fund( "Savings", 56.75 );

piggyBank.deposit( 100.00 );
piggyBank.withdraw( 50.00 );
System.out.println( piggyBank ); // output the fund name and balance
```

6. Compile and run the FundClient class. The four statements above should generate this additional output:

```
The balance for the Savings fund is $106.75
```

7. Now that the Fund class is complete, write the API so other programmers can use your class.

Try It!

Complete the Student class. Write a letterGrade method that returns a String representing the letter grade for the numeric grade. The letterGrade method takes no parameters. Also, write the toString method that outputs the name, grade, and letter grade. Compile both files and run the Teacher class. Notice that without changing any code, the previous statements to output the Student object references now output the return value from your toString method instead of the hash code. Write the API for the Student class.

SECTION REVIEW 12.2

Check Your Understanding

1. Which type of method allows a client program to *view* the value of an instance variable?
2. Which type of method allows a client program to *change* the value of an instance variable?
3. What is the scope of a method's parameters?
4. Which keyword should be used with an instance variable to avoid it being hidden by a method's parameter?
5. Which method is automatically called by the JVM when an object reference is used as a String and, therefore, should be included in every custom class?

Build Your Vocabulary

As you progress through this course, develop a personal computer science glossary. This will help you build your vocabulary and prepare you for a career. Write a definition for each of the following terms and add it to your computer science glossary.

accessor method	mutator method	this object reference
getter	setter	

Science and Java

Sleep Deprivation among Teenagers

Biological clocks produce and regulate the timing of circadian rhythms. Humans, fruit flies, mice, fungi, and *all* organisms have innate timing devices. Biological clocks consist of specific protein molecules that interact in cells throughout the organism. Biological clocks are found in nearly every tissue and organ.

Natural factors within the body produce circadian rhythms, but the rhythms can be changed by signals from the environment. The main cue influencing circadian rhythms is daylight. Changing the light-dark cycles can speed up, slow down, or reset biological clocks as well as circadian rhythms. For example, traveling to a significantly different time zone can change your circadian rhythms, but that is temporary. Another more permanent factor that influences circadian rhythms is puberty. This is due to the increase in hormones.

According to researchers at Johns Hopkins University, teens need at least nine hours of sleep every night. However, children at age 10 need only eight hours. That is because teens are experiencing another stage of cognitive development. It will be impeded if they are awake too long. More sleep is vital to their maturing brains. In addition, they are going through physical growth spurts.

Early school start times, an increase in homework, extracurricular activities, and sometimes a part-time job contribute to lack of sleep. When Seattle, Washington, moved the start times for high school later, a study by scientists at the University of Washington and the Salk Institute found that most teens got at least a half hour more sleep. Median grades improved 4.5 percent and first-period absences declined 12 percent.

Assignment

Consider the scenario in which your state has decided to begin the high school day one hour later. You are part of a team of researchers monitoring the change in grades of five different counties in your state over one semester. Build a SleepIn Java class that has instance variables for the reporting period, percentage of students late, percentage of students absent, and the percentage change in grades. Write the SleepInClient client file to create five objects, echo the variables with the toString method, and calculate the average percentage change in grades.

Iurii Stepanov/Shutterstock.com

Graphical Classes

A graphical class is like any other class in that it can have constructors, accessors and mutators, and other methods. One difference is that a graphical class includes a method to draw a graphical object. Another difference is the client for the object is a JavaFX application.

The Turtle object that is used throughout this text is a graphical object. This section demonstrates how to create an object that will act like a Turtle object, but the object will have a different appearance. The graphical object will respond to all the Turtle class commands. In other words, it is possible to make a graphical object composed of shapes that can be told to move forward or backward, turn, or perform other tasks the Turtle object performs.

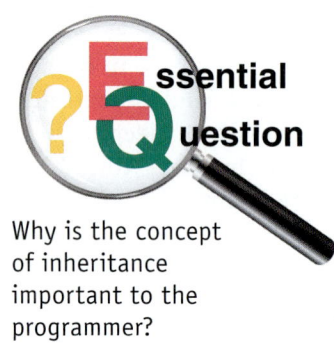

Why is the concept of inheritance important to the programmer?

Learning Goals

- Explain the concept of inheritance from superclass to subclass.
- Construct a graphical class.

Terms

inherit   subclass

is-a relationship   superclass

Brief Introduction to Inheritance

The Turtle class is quite simple. It consists of a few instance variables, a constructor, and one method (buildTurtle) that draws a turtle using six Circle shapes. So how can the Turtle object respond to all its methods: forward, turnRight, backward, penUp, getX, and others? The answer is that the Turtle class *extends* the Sprite class. All the methods mentioned above are defined in the Sprite class. When the Turtle class extends the Sprite class, it inherits all those methods. To *inherit* means to take on the characteristics of the superclass. Therefore, all the Sprite class methods can be called on a Turtle object. In this case, the Sprite class is the superclass and the Turtle class is a subclass of Sprite.

A *superclass* is the original class from which another class inherits. A *subclass* is a class that inherits methods and data by extending a superclass. A superclass is often called a parent class and the subclass is often called a child class. In effect, a Turtle object becomes a Sprite object. In fact, inheritance is defined as an "is-a" relationship. An *is-a relationship* means an object of a subclass "is an" object of the superclass. A Turtle object "is a" Sprite object.

An annotated version of the Turtle class is shown in **Figure 12-5.** Some code dealing with the Turtle color uses arrays, which have not been covered yet. Arrays are covered in Chapter 13. Those statements dealing with arrays are not discussed in depth here.

On line 9, where the Turtle class is defined, the phrase extends Sprite indicates that the Turtle class will inherit methods from the Sprite class. The keyword is extends, and Sprite is the superclass. The instance variables from lines 10 to 17 are defined to support drawing and coloring the Turtle object.

```
1  /** Turtle
2     inherits from Sprite
3  */
4
5  import javafx.scene.*;
6  import javafx.scene.paint.*;
7  import javafx.scene.shape.*;
8
9  public class Turtle extends Sprite { // inherits from Sprite
10    // instance variables
11    public final double WIDTH = 15;
12
13    // variables to choose Turtle color
14    private Color[ ] turtleColors = { Color.GREEN, Color.CORNFLOWERBLUE,
15                Color.TOMATO }; // array of colors
16    private static int colorIndex = -1;
17    private Color bodyColor;
18
19
20    // constructor accepts parameters for root and starting x and y location
21    public Turtle( Group root, double x, double y ) {
22      super( root, x, y ); // pass parameters to constructor in Sprite class
23
24      // choose Turtle color
25      bodyColor = turtleColors[ ( ++colorIndex ) % turtleColors.length ];
26      setDrawingColor( bodyColor.darker( ) );
27
28      buildTurtle( ); // call method that draws Turtle
29    }
30
31    // method to draw Turtle
32    public void buildTurtle( ) {
33      // draw legs
34      Circle leg1 = new Circle( 0 - WIDTH/2, 0 - WIDTH/3, WIDTH/6,
35                                bodyColor.brighter( ) );
36      Circle leg2 = new Circle( 0 - WIDTH/2, 0 + WIDTH/3, WIDTH/6,
37                                bodyColor.brighter( ) );
38      Circle leg3 = new Circle( 0 + WIDTH/2, 0 - WIDTH/3, WIDTH/6,
39                                bodyColor.brighter( ) );
40      Circle leg4 = new Circle( 0 + WIDTH/2, 0 + WIDTH/3, WIDTH/6,
41                                bodyColor.brighter( ) );
42
43      // draw head and body
44      Circle head = new Circle( 0, 0 - WIDTH/2, WIDTH/5,
45                                bodyColor.brighter( ) );
46      Circle body = new Circle( 0, 0, WIDTH/2, bodyColor );
47
48      // add all shapes to Group
49      this.getChildren( ).addAll( leg1, leg2, leg3, leg4, head, body );
50    }
51 }
```

Figure 12-5. This is the full code for the Turtle custom class.

The class has one constructor, which is on lines 20 to 29. The constructor accepts three parameters: root and the starting x and y values (locations) for the Turtle object. In fact, those three parameters are meant for the constructor of the Sprite class. The first statement in the constructor on line 22:

```
super( root, x, y );
```

uses the keyword **super** to call the constructor of the Sprite class, passing those parameters through to the Sprite class constructor.

The constructor then sets the color for the Turtle object, rotating through three colors. It then calls the buildTurtle method to draw the Turtle object. Drawing the graphical object is performed in the buildTurtle method discussed next.

Graphical Object Construction

Drawing the graphical object can best be accomplished by putting the code in a separate method that the constructor calls. In the case of the Turtle object, the drawing code is isolated to the buildTurtle method. Having a separate drawing method simplifies the code in the constructor. The method to draw the graphical object becomes a method of the graphical class.

Drawing shapes is covered in Chapter 7. Those concepts of combining circles, lines, rectangles, and other shapes to draw a figure also apply here. One difference is that the figure should be drawn assuming that its x and y values define its origin (0,0). All shapes composing the figure should be drawn with their x and y values relative to that (0,0) reference point. Exactly where on the drawing the origin (0,0) will be placed is a design decision. It may depend on whether or not the graphical object will be rotated in the window. Perhaps (0,0) is the top left of the drawing. Perhaps (0,0) is the center of the drawing. In the buildTurtle method above (lines 31 to 50), the origin is the center of the body, as shown in **Figure 12-6.**

The x and y parameter values sent to the constructor determine where the drawing will first appear in the window. The (0,0) point on the drawing becomes the (x, y) point in the window. Then, the Sprite methods rotate and move the drawing around the window as requested by the client method calls. Because the drawing is specified as a Group, all the elements move together.

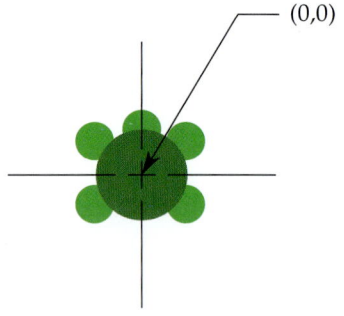

(0,0)

Goodheart-Willcox Publisher

Figure 12-6. The origin (0,0) of the Turtle object is the center of the body.

Coding Conundrum

You are writing a custom Player class. The class begins:

```
public class Player {

    private String name;
    private int score;

    public Player( ) {
        name = "Player";
        score = 0;
    }
...
```

1. For the second constructor, you have written this code:

```
public void Player( String pName ) {
    name = pName;
    score = 0;
}
```

(Continued)

Coding Conundrum

Continued

The compiler does not report any errors, but when a client program attempts to create a **Player** object using this code:

```
Player two = new Player( "Jen" );
```

The compiler reports an error for the client:

```
constructor Player in class Player cannot be applied to given types;
Player two = new Player( "Jen" );
                 ^
required: no arguments
found: String
reason: actual and formal argument lists differ in length
```

It is a conundrum! How can you fix this code?

2. For the accessor method for **playerName**, you have written:

```
public int getName( ) {
   return name;
}
```

The compiler reports this error:

```
Player.java:21: error: incompatible types: String cannot be converted to int
   return name;
          ^
```

It is a conundrum! How can you fix this code?

3. You have written the **toString** method as:

```
public String toString( ) {
   System.out.println( name + ": " + score );
}
```

The compiler reports this error:

```
error: missing return statement
   }
   ^
```

It is a conundrum! How can you fix this code?

4. For the mutator method, you have written:

```
public void setName( String name ) {
   name = name;
}
```

The compiler does not report an error, but when a client program calls this method, the name is not changed. It is a conundrum! How can you fix this code?

5. You have finished the class definition. You compile and run the **Player** class, but receive this message at runtime:

```
No main methods
```

It is a conundrum! What is wrong?

HANDS-ON EXAMPLE 12.3.1

Running a Race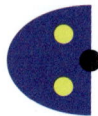

In this exercise, you will create a graphical class. Then, you will "run a race" with two graphical objects instantiated from that class. Before beginning this exercise, download the chapter files from the student companion website. You will use the Turtle files as well as the Die class developed throughout this chapter. You will also write code for the graphical class in the file (Runner.java) and in the executable JavaFX file (RunRace.java).

The graphical object you will create is shown here. The left-hand image shows the object as originally drawn. The small black circle indicates where the origin (0,0) is on the drawing. The black circle is drawn for illustration only and is not part of the final figure. Remember that when a Turtle or any Sprite starts, the figure is facing up. For the figure to run a race from left to right, the figure needs to be turned to face right. The image on the right shows the figure rotated 90 degrees to the right. The origin is now on the leading edge of the figure. This will help in the race because you can find there is a winner when the (x, y) location of a figure crosses the finish line.

Goodheart-Willcox Publisher

1. Launch jGRASP, and open the Runner.java file. This is the skeleton of the graphical class.

2. Locate comment 1, and follow the instructions to specify that this class inherits from the Sprite class. Add the phrase extends Sprite to the existing code.

   ```
   /***** 1. Insert extends Sprite in the line below. */
   public class Runner extends Sprite { // inherits from Sprite
   ```

3. Locate comment 2, and follow the instructions to complete the constructor.

   ```
   /***** 3. Write the constructor. The constructor accepts parameters for Group root,
         and x and y locations, both of which are doubles.
         a. Call super and pass the parameters on to the Sprite constructor.
         b. Call the buildRunner method, which will draw the runner.
   */
   public Runner( Group root, double x, double y ) {

      super( root, x, y );  // pass parameters to Sprite constructor

      buildRunner( );    // call method to draw the runner
   }
   ```

4. Locate the buildRunner method. The code to draw the runner's body, an Arc, is already written. The two eyes of the runner still need to be drawn. Locate comment 3, and follow the instructions to draw the two eyes.

   ```
   public void buildRunner( ) {
      // code to draw the body is here

      /***** 3. Complete the figure by drawing the eyes.
          The eyes are yellow circles with a radius of 4.
          One eye's center is at (-10, 10) and the second eye is at (-10, -10). */
      Circle eye1 = new Circle( -10, 10, 4 );
      eye1.setFill( Color.YELLOW );

      Circle eye2 = new Circle( -10, -10, 4 );
      eye2.setFill( Color.YELLOW );
   ```

5. Locate comment 4, and follow the instructions to add the three shapes to the Group.

   ```
   /***** 4. Add the three shapes to the Group. */
   this.getChildren( ).addAll( body, eye1, eye2 );
   ```

6. Compile the Runner class to check for errors. Remember that this class is not executable.

7. Open the RunRace.java file, and examine the code. This is the JavaFX client for the Runner graphical class. All code to set up and run the race will be in the buildScript method.

8. Locate comment 5, and follow the instructions to define a constant for the finish line to be 100 pixels to the left of the right edge of the window.

```
/***** 5. Define the finish line as a constant. */
final int FINISH_LINE = windowWidth - 100;
```

9. Locate comment 6, and follow the instructions to use a Turtle object to draw the finish line.

```
/***** 6. Use a Turtle to draw the finish line.
   a. Instantiate a Turtle at (FINISH_LINE, 10 ).
   b. Hide the Turtle.
   c. Set the speed to no animation.
   d. Turn the Turtle to face down.
   e. Draw a line windowHeight long.
*/
Turtle finish = new Turtle( root, FINISH_LINE, 10 );
finish.hide( );
finish.setSpeed( Turtle.NO_ANIMATION );
finish.turnRight( 180 );
finish.forward( windowHeight );
```

10. Locate comment 7, and follow the instructions to instantiate a Die object using the default constructor. The Die will be used in the race to determine each runner's forward movement.

```
/***** 7. Instantiate a Die object using the default constructor. */
Die die = new Die( );
```

11. Locate comment 8, and follow the instructions to instantiate two Runner objects. Remember that a newly instantiated Sprite is facing up. Turn each runner right to face toward the finish line, and lift up the pen so the runners will not leave a trail.

```
/***** 8. Instantiate two Runner objects at (100, 100) and at (100, 200).
      Turn each Runner right and lift the pen. */
Runner racer1 = new Runner( root, 100, 100 );
racer1.turnRight( );
racer1.penUp( );

Runner racer2 = new Runner( root, 100, 200 );
racer2.turnRight( );
racer2.penUp( );
```

12. The runners are ready to run the race. Locate comment 9, and follow the instructions to write a while loop that rolls a Die two times for each racer and moves each racer forward the amount of the roll until one of the racers has passed the finish line.

```
/***** 9. Write a while loop that rolls the Die twice for each racer
      and moves each racer forward the amount of the roll
      until one of the racers has passed the finish line. */
while ( racer1.getX( ) < FINISH_LINE && racer2.getX( ) < FINISH_LINE ) {

    int moveRacer1 = die.rollDie( ) + die.rollDie( );
    racer1.forward( moveRacer1 );

    int moveRacer2 = die.rollDie( ) + die.rollDie( );
    racer2.forward( moveRacer2 );
}
```

13. Compile all files, and run the RunRace.java file. The graphics window should appear, and the two runners should move across the screen until one reaches the finish line.

Goodheart-Willcox Publisher

Try It!

Modify the Runner class to create another graphical class named Dancer. Create a simple drawing for the dancer. Modify the RunRace client program to instantiate two Dancer objects and have them "dance." For example, in a for loop that executes ten times, you could have one Dancer object move up and down while the second Dancer object moves left and right. The distance each Dancer moves can be determined by the roll of a 50-sided Die object.

SECTION REVIEW 12.3

Check Your Understanding

1. The "is-a" relationship means an object of the _____ is also an object of the _____.
2. Which keyword is used to indicate methods will be inherited from another class?
3. Which keyword calls the constructor of a superclass?
4. Why should the code to draw a graphical object be placed in a separate method called by the constructor instead of in the constructor itself?
5. What two factors determine where the constructor will place a graphical object in the JavaFX window?

Build Your Vocabulary

As you progress through this course, develop a personal computer science glossary. This will help you build your vocabulary and prepare you for a career. Write a definition for each of the following terms and add it to your computer science glossary.

inherit   subclass
is-a relationship   superclass

Cooperative Coding

Program Modification

Often, existing programs are modified to create solutions to different problems. In some cases, an individual coder may modify the program. However, in many cases modifying a program is a group activity. The purpose for the modification may be identified by a group. This may be the coders, but the group is likely to be management.

Jacob Lund/Shutterstock.com

Assignment

Once the purpose for the modification has been specified, identify what needs to change in the program. This is often a group activity during which the coders who will work on the project each share input. This may also be the time when tasks are assigned to each coder in the group.

If you and your group have not completed Hands-On Example 12.3.1, Running a Race, do so now. You will need the complete game running without errors before beginning this exercise. This exercise edits the race in the following ways.

- Add two more racers.
- Modify the Runner constructor method to assign different colors to the racers.
- Use the Turtle object to draw lane lines for each racer.

Work together as a group to solve these problems. Run the race several times to test your edits.

Mark Herreid /Shutterstock.com

1. As a group, discuss if the modifications improve the game. Develop a position, and be prepared to defend the group's position in class discussion.
2. What intended and unintended consequences of your edits were encountered?

Chapter Summary

Section 12.1: Creating the Class
- A class consists of data and methods, and each object created from a class has its own copy of the data stored in the class's instance variables, but all objects of a class share the class methods.
- A class should provide one or more constructors for the client to be able to create objects; otherwise, the compiler's default constructor is used, which assigns default values to the instance variables based on the data type of the variable.

Section 12.2: Writing Methods
- An accessor, which is a value-returning method, allows the client to view data in an object and follows the basic pattern of having a name starting with get followed by the name of the instance variable, does not take parameters, and returns the same data type as the instance variable.
- A mutator method allows a client to change the values of the data in an object and follows the basic pattern of having a name starting with set followed by the name of the instance variable, taking one parameter that is the new value for the instance variable, and is usually a void method.
- Preceding a name with this refers to an instance variable, and using this object reference is done to prevent hiding the instance variable if a parameter has the same name as the variable.
- A custom class should include at least one method that completes the intended function of the class, and it should also include a toString method to return the data of the class as a String.

Section 12.3: Graphical Classes
- A subclass inherits public methods from a superclass; inheriting from a superclass is defined as an "is-a" relationship because each object of the subclass also "is an" object of the superclass.
- Place the code for constructing a graphical class in a separate method the constructor calls. This will clean up the constructor code; the shapes composing the figure should be drawn relative to an origin (0,0) for the figure, which is used by the JavaFX runtime system to place the figure in the graphics window.

Chapter 12 Test

Multiple Choice
Select the best response.

1. What describes a constructor when the class provides more than one constructor?
   A. overloaded
   B. inherited
   C. subclass
   D. superclass

2. What are arguments from a client class called when inside the method?
   A. constructors
   B. parameters
   C. modifiers
   D. variables

3. Instance variables can be unhidden from parameters with the same name by using the _____ object reference.
   A. this
   B. default
   C. set
   D. get

4. When creating an accessor method, what precedes the instance variable name in the method header?
   A. catch
   B. send
   C. set
   D. get

5. What is meant by an "is-a" relationship?
   A. The method is a means of constructing an object.
   B. The class is a member of the client application.
   C. The subclass object is also an object of the superclass.
   D. The superclass object is a valid object.

Completion

Complete the following sentences with the correct word(s).

6. By convention, _____ are nouns beginning with a capital letter and following CamelCase.

7. In a class definition, the data for an object are stored in _____.

8. Methods and instance variables have _____ scope.

9. It is a best practice to include a(n) _____ method in every custom class so the object data can be returned as a String.

10. The keyword _____ specifies inheritance from a superclass.

Matching

Match the correct term with its definition.

A. Scanner, String, Turtle

B. public, private, protected

C. 0, 0.0, false, null

D. a and b in methodName(double a, int b)

E. toString( )

11. Takes no parameters.

12. Formal parameters.

13. Service classes.

14. Access modifiers.

15. Default values of instance variables.

Application and Extension of Knowledge

1. The manager of the Baltimore Book Store wants an application that calculates and displays the total amount a customer owes. A customer may purchase one or more books. The application should read the book title (one word), ISBN, and price from a file. For each line in the file, create a Book object, and ask the user if he or she wants to buy the book. If so, add the title to the shopping cart and the price to the total price. At the end, the application should display the books purchased and the total amount of the purchase. Create two classes to provide this solution.

2. Write an application that allows a user to compare the mileage for two vehicles. Create a class called Vehicle that has these instance variables: milesTraveled(double), gasInGallons(double), and costPerGallon(double). Write the necessary constructors, accessors, mutators, and toString methods. Also, write a method to calculate miles per gallon and another method to calculate cost per mile. Create a client application that prompts the user for miles traveled, gas used, and the cost of gas for two vehicles. Create two Vehicle objects, and output a comparison of the mileage and cost.

3. The Accounts Payable department at a company has a new supervisor who wants to keep track of the checks. The specifics she desires are the check number, payee name, and amount of each check. Create a Check class that has instance variables for the check number, payee name, and check amount. Write a client application that creates three Check objects using this data: 123, Lorraine Jackson, $1045.67; 124, Kathleen Marsiglia, $2040.12, and 125, Julie Austin, $3214.89. Write the data (call toString) in the three objects to a text file.

4. Write a class definition for a class named Employee. The class should have three instance variables: name (String), salary (double), and a MINIMUM_SALARY constant for the minimum salary, which is $10.00 per hour. Create a constructor that accepts the name and salary as parameters and a default constructor that assigns the minimum salary.

5. Video game developers often create a game object from a class that keeps track of the instance's image, dimensions, location on the screen, and other game-related properties. Write a class called GameObject with instance variables length, height, xLoc, and yLoc. The constructor will accept arguments for initial values for all of these properties. Define getter and setter methods for each instance variable. A game mechanic allows the size of the GameObject to be changed based on gameplay. Also, create a resize method that accepts a scale parameter and resizes the GameObject by multiplying the length and height by the scale factor. In the client, create a GameObject, print the original length and height, call the resize method, and print the new length and height.

Online Activities

Complete the following activities, which will help you learn, practice, and expand your knowledge and skills.

Vocabulary. Practice vocabulary for this chapter using the e-flash cards, matching activity, and vocabulary game until you are able to recognize their meanings.

Communication Skills

Reading. Being able to respond to questions about what you read demonstrates that you comprehend the information. After reading each section of the chapter, stop to answer the Check Your Understanding questions and complete the Build Your Vocabulary activity without looking back at the information covered in the chapter. Then, look back at the information to check your work.

Writing. Being able to *retell* or *summarize* what you read can help you confirm your understanding of the material and demonstrate that you comprehend the subject. Summarizing is a technique that entails writing about the main points that you read. Summarizing involves identifying the most important ideas in the material and retelling them in your own words. Reread the first section in this chapter. Summarize the information in several paragraphs.

Speaking. A *presentation* is usually a speech given to a group of people. Plan and deliver a speech about a topic covered in this chapter. Be clear in your perspective for the idea and demonstrate solid reasoning.

Listening. When taking notes in class, discretion must be used to decide which key words are important to main ideas so they can be recorded. It is not necessary to write down every word you hear. A first rule of notetaking is to practice active listing. You must not only hear what is said, but comprehend the information. One way to determine importance is to listen to repetition by the instructor. If information is repeated, it must be important. Listen closely to the lectures by your instructor. What have you noticed that he or she repeats?

Portfolio Development

College and Career Readiness

Diversity Skills. As part of an interview with an organization, you may be asked about your travels or experiences with people from other cultures. Many different organizations serve people from a variety of geographic locations and cultures. Some have offices or other types of facilities in more than one region or country. You may need to interact with people from diverse cultures or travel to facilities in different countries. Speaking more than one language and having traveled, studied, or worked in other countries can be valuable assets. You may be able to help an organization understand the needs and wants of diverse people. You may also be better able to communicate and get along with others.

1. Identify travel or other educational experiences that helped you learn about another culture, such as foreign languages studied or trips taken.

2. Create a Microsoft Word document that describes the experience. Use the heading Diversity Experience and your name. Explain how the information you learned might help you better understand classmates, customers, or coworkers from this culture. Save the document in an appropriate folder.

3. Place a printed copy in the container for future reference.

4. Update your checklist to reflect the file format and location of the document.

CTSOs

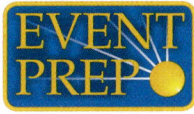

Communications Skills. Competitive events may judge communications skills of the participants. Presenters must be able to exchange information with the judges in a clear, concise manner. The evaluation will include all aspects of effective writing, speaking, and listening skills. To prepare for the business communications portion of an event, complete the following activities.

1. Visit the organization's website and look for specific communication skills that will be judged as a part of a competitive event.

2. Spend time to review the essential principles of business communication, such as grammar, spelling, proofreading, capitalization, and punctuation.

3. If you are making a written presentation, ask an instructor to evaluate your writing. Review and apply the feedback so your writing sample appears professional and correct.

4. If you are making an oral presentation, ask an instructor to listen for errors in grammar or sentence structure. After you have received comments, review the feedback and adjust the presentation until you are satisfied and comfortable with your presentation.

5. Review the Communication Skills activities that appear at the end of each chapter of this text as a way to practice your reading, writing, speaking, and listening skills.

6. To practice listening skills, ask your instructor to provide a set of directions. Then, without assistance, repeat those directions to your instructor. Did you listen closely enough to be able to do what was instructed?

Working with Arrays

Sections

So far, you have worked with single data items. The user entered the items at the keyboard, the program read the data from a file, or you generated random values within the program. These data items were handled one at a time in a loop, either a while loop, for loop, or do/while loop. Processing data—one item at a time—has limitations. When an item is entered, read, or generated, all processing of the item must be done at that time. The next item that is entered, read, or generated replaces the previous item, so only one item is available at any time.

What if the eventual processing of the data is not known at the time it is available? Would the program be better if the user could choose what happens to the data? Arrays make that possible.

Reading Prep

The summary at the end of the chapter highlights the most important concepts presented in the chapter. Before reading this chapter, review the summary. Based on this information, write down two or three items you think are important to note while you are reading.

College and Career Readiness

While studying, look for the activity icon ⟶ for:

- Vocabulary terms with e-flash cards and matching activities.
- Starter files for hands-on examples and other exercises.

These activities can be accessed at
www.g-wlearning.com/informationtechnology/1773

Chapter Glossary

array: Ordered sequence of data of the same data type.

bar graph: Consists of side-by-side lines and is commonly used to compare individual values.

element: Each data item in an array.

scale: To multiply each value by a set amount.

search key: Value to find in a search.

sequential search: Algorithm for searching an array by comparing the search key to each element in order.

sort: To arrange elements in an array in some order based on their values.

Introduction to Arrays

Y ou have used the Turtle class several times to this point. In Chapter 12, you saw an example that illustrated inheritance. Some of the code in the example for controlling the turtle's color was not discussed at that time because it involved arrays. In this section, you will learn what an array is. You will also learn how to create arrays and how to access elements within arrays.

Essential Question

What is the most important concept related to arrays for the programmer?

Learning Goals

- Explain the process for creating an array.
- Modify array elements.

Terms

array
element

Creating Arrays

An *array* is an ordered sequence of data of the same data type. That means a group of data can be stored next to each other in memory. Each piece of data has a position in the array. The data in an array must all be the same data type: all ints, all doubles, all Strings, and so on. The array can be named, and all data within the array can be accessed using the array name.

Consider the Very Large Array radio telescope, as shown in **Figure 13-1.** This is a group of identical antennae. Each one is individually accessed and operated, but it has properties and methods identical to all the other antennae. This is a physical example of an array and conceptually is similar to the Java programming array. In Java, an array is a group of data that is stored consecutively in memory and is accessed using the array name and an index.

Arrays are objects, so arrays must be instantiated. This is the syntax for creating an array:

```
datatype[ ] arrayName = new datatype[size];
```

An alternative is to declare the array name on one line and instantiate the array on another line, using this syntax:

```
datatype[ ] arrayName;
arrayName = new datatype[size];
```

The arrayName is actually a reference to the array. Thus, its initial value is null before the array is instantiated. The keyword new should be familiar for creating objects. Notice that empty square brackets follow the data type when defining the name of the array. On the right side of the assignment

NRAO/AUI/NSF

Figure 13-1. This is the Very Large Array (VLA), which is located on the Plains of San Agustin in New Mexico. It consists of 25 antennae, each 82 feet across and weighing 230 tons.

operator, the number of data items in the array is given as the value size inside the square brackets.

For example, this statement creates an array to store the seven high temperatures for a week:

```
int[ ] temperatures = new int[7];
```

The following statement creates an array to store the magnitudes of ten earthquakes.

```
double[ ] magnitudes = new double[10];
```

Each data item in an array is called an *element.* Thus, the temperatures array has seven elements, and the magnitudes array has ten elements. Like other objects, the elements in an array are assigned default values when the array is created. Refer to **Figure 13-2.** This means that all the elements in the temperatures array initially have the value 0, while all the elements in the magnitudes array initially have the value 0.0. Strings are objects, so the elements in an array of Strings are object references. The elements of a String array would be given the default value null.

Alternately, an array can be instantiated by providing the values within braces. The values are separated by a comma using this syntax:

```
datatype[ ] arrayName = { element0, element1, element2, … };
```

For example, this statement creates an array of chars containing eight special characters:

```
char[ ] specialCharacters = { '!', '@', '#', '$', '%', '^', '&',
                              '*' };
```

When providing values in this manner, the size of the array is determined by the number of values in the list. In this case, the size of the specialCharacters array is eight.

Figure 13-3 illustrates the three example arrays presented above. The reference temperatures points to the location of seven consecutive int elements initialized to 0. The reference magnitudes points to the location of ten consecutive double elements initialized to 0.0. The reference specialCharacters points to the location of eight consecutive char elements containing the special characters that were provided in the initialization list.

FYI

The *magnitude* of an earthquake measures its severity.

Data Type	Default Value
byte, short, int, long	0
float, double	0.0
boolean	false
char	0000 (hexadecimal)
Object reference	null

Goodheart-Willcox Publisher

Figure 13-2. These are the default values for array elements.

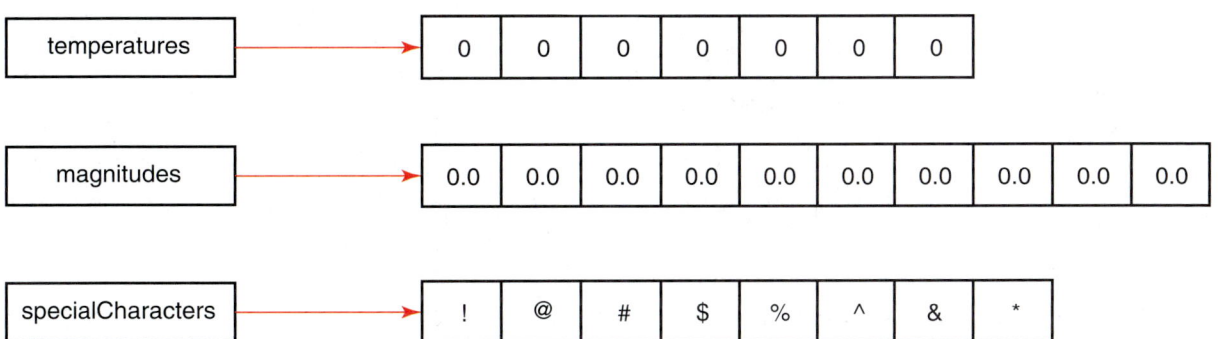

Goodheart-Willcox Publisher

Figure 13-3. After instantiation, each array name refers to the memory address of the elements within the array.

FYI

Note that the length *variable* for arrays is different from the length( ) *method* for finding the number of characters in a String.

The size of an array can be accessed through code. To find the number of elements in an array, use the instance variable length. For example:

```
temperatures.length        // value is 7
magnitudes.length          // value is 10
specialCharacters.length   // value is 8
```

Once an array is created, its size is fixed. To add an element to an array requires creating a new, larger array and copying the values from the original array into the new array. Similarly, deleting an element in an array requires creating a new, smaller array and copying all elements to keep into the new array.

Accessing Array Elements

Similar to Strings, each element in an array can be accessed through an index. As you have previously learned, an index is a number representing the relative position of an element in a String. Like Strings, the first index in an array is 0. To access a single element in an array, use square brackets with this syntax:

```
arrayName[index]
```

which is read as "*arrayName* sub index." For example, the second element of the temperatures array would be accessed as:

```
temperatures[1]      // temperatures sub 1
```

The first element of an array is always available at index 0:

```
arrayName[0]
```

and the last element of an array is always available at the index length − 1:

```
arrayName[arrayName.length - 1]
```

Thus, this code outputs the first element in the specialCharacters array:

```
System.out.println( specialCharacters[0] ); // outputs !
```

and this code outputs the last element in the specialCharacters array:

```
System.out.println( specialCharacters[specialCharacters.length -
                    1] ); // outputs *
```

Values can be assigned to individual elements in an array. For example, the following code gives a value to each element in the magnitudes array. **Figure 13-4** shows the magnitudes array after this code is executed.

```
magnitudes[0] = 7.5;
magnitudes[1] = 4.8;
magnitudes[2] = 7.1;
magnitudes[3] = 4.4;
magnitudes[4] = 6.3;
magnitudes[5] = 7.3;
magnitudes[6] = 5.0;
magnitudes[7] = 5.5;
magnitudes[8] = 7.0;
magnitudes[9] = 5.0;
```

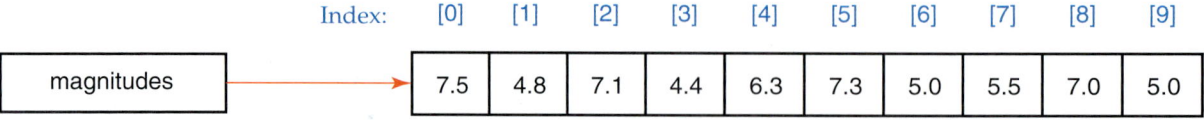

Goodheart-Willcox Publisher

Figure 13-4. The array has been populated with values. Note the index of each element.

HANDS-ON EXAMPLE 13.1.1

Creating an Array and Adding Values

There are many applications for arrays in computer programs. In this exercise, you will create an array to track the number of steps you take each day for a week.

1. Launch jGRASP, and start a new Java file. Write the header comment code and the main structure of the program.

```
/* An array to track steps
   your name here
*/
import java.util.*;

public class CreateArray {
    public static void main( String [ ] args ) {

        Scanner input = new Scanner( System.in );

        // add the remainder of the code from this exercise here

    } // end main
} // end class
```

2. Save the file in your working folder under the correct name.

3. Define a constant for the size of the array, and then create an array named dailySteps with seven int elements.

```
final int NUMBER_OF_DAYS = 7;
int[ ] dailySteps = new int[NUMBER_OF_DAYS];
```

4. Output the elements in the array on one line, separating each element by a space. Note that when outputting each element, use the print method rather than println so all elements are displayed on the same line.

```
System.out.println( "The array values are:" );
System.out.print( dailySteps[0] + " " );
System.out.print( dailySteps[1] + " " );
System.out.print( dailySteps[2] + " " );
System.out.print( dailySteps[3] + " " );
System.out.print( dailySteps[4] + " " );
System.out.print( dailySteps[5] + " " );
System.out.print( dailySteps[6] + " " );
System.out.println( );
```

5. Compile and run the program. You will see that each element has been initialized to 0.

6. Add code to allow the user to change one value in the array. Ask the user for the index of the element to change and for the new value for the steps. Ensure the index is valid!

```
// a valid index is between 0 and length - 1
int index;

do {
    System.out.print( "Enter the index of the element to change: " );
    index = input.nextInt( );

} while ( index < 0 || index >= dailySteps.length );

// index is valid; read the new value
System.out.print( "Enter the new number of steps for element " + index + ": " );
int value = input.nextInt( );

dailySteps[index] = value; // assign new value
```

7. Output the elements in the array so you can see the value changed. This is the same code as used earlier. You can copy and paste the code from there.

```
System.out.println( "The array values are:" );
System.out.print( dailySteps[0] + " " );
System.out.print( dailySteps[1] + " " );
System.out.print( dailySteps[2] + " " );
System.out.print( dailySteps[3] + " " );
System.out.print( dailySteps[4] + " " );
System.out.print( dailySteps[5] + " " );
System.out.print( dailySteps[6] + " " );
System.out.println( );
```

8. Compile and run the program. Check that the appropriate element has been changed to the new value.

Try It!

Create a Java program that instantiates an array of chars containing the first ten letters of the alphabet (A–J). Use the comma-separated list syntax to create the array. Output the elements in the array on one line, separating each element by a space.

SECTION 13.1 REVIEW

Check Your Understanding

1. What is true of the data type(s) of the elements in an array?
2. What is the syntax for creating an array?
3. Which variable returns the number of elements in an array?
4. What does this code return: movies[8] ?
5. What is the syntax for changing the value of index 4 of the price array to 1.5?

Build Your Vocabulary

As you progress through this course, develop a personal computer science glossary. This will help you build your vocabulary and prepare you for a career. Write a definition for each of the following terms and add it to your computer science glossary.

array   element

Processing Arrays

In the last section, you learned how to output each element in an array using separate statements. Did you get tired of entering the same statement with only the index changed? Imagine how tedious it would be to write the code for an array with a large number of elements, say 100 or even 1,000. Fortunately, the for loop is a helpful tool for traversing and processing arrays element by element.

As in traversing Strings, the for loop counter variable can be used to reference each index in the array. The for loop body will contain the statements to output or manipulate a single array element. The for loop is essential in processing all the elements of an array because an attempt to output the array name itself will output the hash code, not the array elements. Remember, the array name is a reference that points to where the array is located in memory.

Learning Goals

- Process an array element-by-element.
- Explain how to construct a bar graph from an array.
- Identify the steps for processing an array of objects.

Terms

bar graph   scale

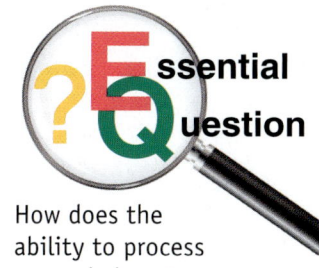

Essential Question

How does the ability to process arrays help a programmer manipulate the data in an array?

Processing Arrays Element-by-Element

To use a for loop to traverse an array, this pattern does the job:

```
for ( int i = 0; i < arrayName.length; i++ ) {
   // use arrayName[i]
}
```

The loop counter (i) is used to represent each index in the array. The loop counter's initial value is 0, which is the index of the first element. The for loop's condition is that the index is less than the length of the array because the index of the last element is always 1 less than the length of the array. The update statement adds 1 to the loop counter to move to the next element in the array. Inside the loop, the current element is addressed as arrayName[i].

Be careful not to use the less than or equal to operator (<=) in the for loop condition. The condition:

```
i <= arrayName.length
```

will cause the for loop to try to access an element at arrayName.length, which does not exist. This will cause an ArrayIndexOutOfBoundsException error at runtime.

Filling an Array

The values in an array can come from a variety of sources. If the array is small, the user can be asked to enter the values. If the array is larger, the values can be read from a file or generated randomly, if appropriate. Regardless of where the array values come from, the general form is to assign a value to one array element in each iteration of the for loop.

Consider the text file shown in **Figure 13-5.** This text file is named earthquakes.txt. This for loop reads values from the file into the magnitudes array:

```
double[ ] magnitudes = new double[10]; // instantiate the array

File earthquakes = new File( "earthquakes.txt" );
Scanner file = new Scanner( earthquakes );

for ( int i = 0; i < magnitudes.length; i++ ) {
   magnitudes[i] = file.nextDouble( );
}
```

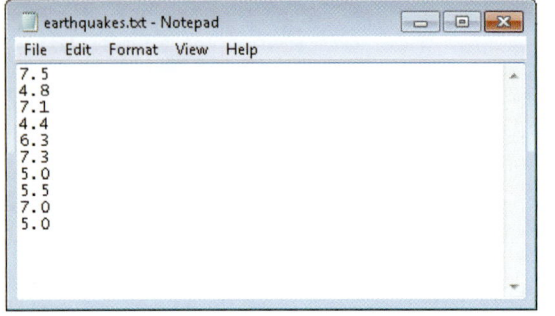

Goodheart-Willcox Publisher

Figure 13-5. The contents of this earthquakes.txt file will be used to populate an array.

When reading values from a file into an array, a for loop is better than a while loop with the condition file.hasNext( ). There are two reasons for this.

The first reason is while loops are used when the number of items in a file is not known, but in this case you know the file has ten items. Whenever a program is reading from a file, a contract of sorts exists between the creator of the file and the program. The creator guarantees the file will have a specific format. The programmer then writes the code to read the file in that format. In this case, the file creator guarantees that ten values are in the file, one per line.

The second reason is that with a for loop, the index for storing the value in the array is readily available as the loop counter. With a while loop, the array index would need to be handled as a separate variable.

Printing an Array

When the code above is executed, the magnitudes array will contain the values shown in the file in **Figure 13-5.** That can be verified by printing the contents of the array. Again, a for loop is an efficient tool for outputting values in an array.

```
System.out.println( "The array values are:" );
for ( int i = 0; i < magnitudes.length; i++ ) {
   System.out.print( magnitudes[i] + "  " ); // two spaces for
                                             // readability
}
System.out.println( ); // output a new line
```

This code should look similar to the code in Section 13.1 that outputted each element in a separate statement. As you can see, this technique using a for loop is easier and more efficient to code. The output from this code verifies the contents of the array are as expected:

```
The array values are:
7.5  4.8  7.1  4.4  6.3  7.3  5.0  5.5  7.0  5.0
```

Finding the Average in an Array

The average of the elements in an array is found by calculating the sum of all the values, and then dividing by the number of elements in the array. There is no need to count the number of values because the length variable holds the size of the array. Again, the standard for loop is a good tool.

```
double sum = 0.0; // initialize sum

// use for loop to calculate the sum
for ( int i = 0; i < magnitudes.length; i++ ) {
   sum += magnitudes[i];
}

double average = sum / magnitudes.length;  // sum divided by number
                                           // of values
// output the average to two decimal places
String s = String.format( "The average earthquake magnitude is %.2f",
                          average );
System.out.println( s );
```

When this code is run, the output is:

```
The average earthquake magnitude is 5.99
```

Finding the Minimum and Maximum Values in an Array

To find the minimum value in an array, it is a good idea to initialize a minimum variable to the value of the first element. Then, traverse the rest of the elements and compare each value to the current value of minimum. If the value of any element is less than the current value of minimum, store that value as the new minimum. In this way, once the whole array has been traversed, the minimum variable will hold the lowest value in the array.

The for loop below has one change from the standard pattern. Because minimum is initialized to the first element, there is no need to compare the minimum to the first element. So, the for loop initializes the loop counter i to 1, rather than to 0. Here is the code:

```
double minimum = magnitudes[0]; // initialize to first element

for ( int i = 1; i < magnitudes.length; i++ ) {// start at 1
   if ( magnitudes[i] < minimum ) {       // compare current value
                                          // to minimum
      minimum = magnitudes[i];
   }
}
System.out.println( "The minimum magnitude is " + minimum );
```

When this code is run, the output is:

```
The minimum magnitude is 4.4
```

Finding the maximum value requires similar logic. The only difference, other than naming the variable maximum, is that the if statement inside the for loop compares if the value of the element is greater than the value of maximum:

```
if ( magnitudes[i] > maximum ) { // compare current value to
                                 // maximum
   maximum = magnitudes[i];
}
```

HANDS-ON EXAMPLE 13.2.1

Processing Elements in an Array

The program you will create in this exercise is similar to the program created in Hands-On Example 13.1.1. You will insert values into the dailySteps array. Then, you will output the array values, the average number of steps, and the highest and lowest number of steps for the week.

1. Launch jGRASP, and begin a new Java file. Write the header comment code and the main structure of the program.

```
/* An array to analyze daily steps counts
   your name here
*/
```

2. Add code to create the dailySteps array with seven elements.

```
import java.util.*;

public class AnalyzeSteps {

    public static void main( String [ ] args ) {

        Scanner input = new Scanner( System.in );
        Random randInt = new Random( ); // this code uses random numbers

        final int NUMBER_OF_DAYS = 7;
        int[ ] dailySteps = new int[NUMBER_OF_DAYS];

        // add the remainder of the code from this exercise here

    } // end main
} // end class
```

3. Save the file in your working folder under the correct name.

4. Fill the dailySteps array with random numbers between 2,000 and 12,000. These numbers will represent the number of steps taken each day since the user is not being asked to enter that data. An easy way to calculate the number of values needed for the nextInt method is subtracting the start value (2,000) from the end value (12,000) and adding 1. Then add 2000 to that value returned from nextInt to start the random values at 2000.

```
// generate random values between 2000 and 12000
for ( int i = 0; i < dailySteps.length; i++ ) {
    dailySteps[i] = randInt.nextInt( 12000 - 2000 + 1 ) + 2000;
}
```

5. Output the array values on one line. Label each value with the corresponding day. Note that 1 is added to the index to get the day number because humans start counting at 1.

```
// output the array values
for ( int i = 0; i < dailySteps.length; i++ ) {
    System.out.println( "Day " + ( i + 1 ) + ":\t"  // add 1 to i
              + dailySteps[i] + " steps." );
}
```

6. Calculate the average number of steps to one decimal place.

```
// find the average number of steps
int sum = 0;

for ( int i = 0; i < dailySteps.length; i++ ) {
    sum += dailySteps[i];
}

double average = (double) sum / dailySteps.length;
String s = String.format( "The average number of steps per day is %,.1f", average );
System.out.println( s );
```

7. Find the maximum number of steps by initializing the maximum to the first element value.

```
// find the maximum number of steps
int maximum = dailySteps[0];

for ( int i = 1; i < dailySteps.length; i++ ) {
    if ( dailySteps[i] > maximum ) {
        maximum = dailySteps[i];
    }

}

System.out.println( "The maximum number of steps is " + maximum );
```

8. Compile and run the program. Verify all calculations are correct.

Try It!

Continue working in the **AnalyzeSteps** program. Add code to find the minimum number of steps. Also, add code to ask the user for his or her target number of daily steps. Then, count the number of days the user met or surpassed the target and report that number.

Graphing an Array

Graphing the values in an array helps to visualize how the values are related. For example, **Figure 13-6** shows a bar graph created to display the values in the magnitudes array. A *bar graph* consists of side-by-side lines and is commonly used to compare individual values. A bar graph of array values can be made using a for loop in a JavaFX application. The for loop creates a bar for each value in the array.

A close look at **Figure 13-6** reveals that each bar is simply a Rectangle object and the values are displayed as Text objects. **Figure 13-7** shows how each rectangle is created from the values in the array. Remember from Chapter 7 that to draw a rectangle, you need to know the (x, y) value of its upper-left corner, its width, and its height. The width is a constant. The base y value of each rectangle is also a constant so the bars align horizontally. The height of each rectangle is the value of the array element, giving each bar a varied height.

For the top-left corner, the x value is a variable. In this example, it is named barX. The y value of the top-left corner is the base of the rectangle minus the value of the array element. As the for loop traverses the array, the value of barX moves to the right, positioning to the next bar. After drawing each bar, the width of the bar and a constant spacing is added to the barX value.

A Text object is used to output the value of the array element. A constant is used to position the text. In this example, the constant is named BAR_VALUE.

The first step is to define the array. For efficiency, use the list of values technique to create the array. Here is the code inside the start method of a JavaFX application to create the bar graph:

```
// 1. define the magnitudes array and fill with values
double[ ] magnitudes = { 7.5, 4.8, 7.1, 4.4, 6.3, 7.3, 5.0, 5.5,
7.0, 5.0 };
```

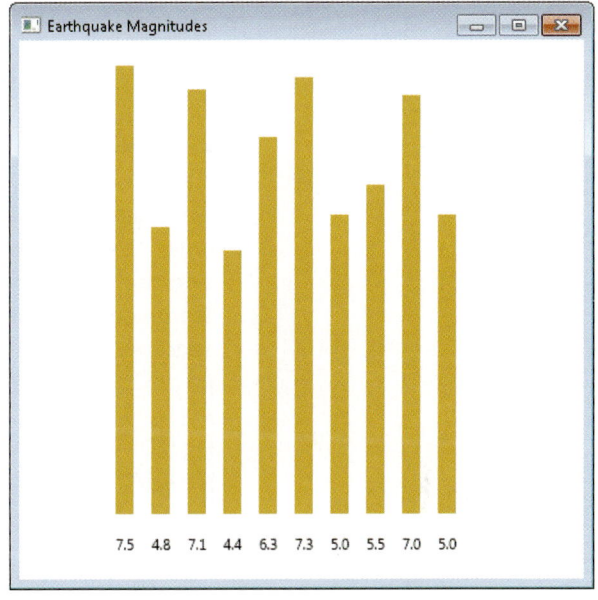

Goodheart-Willcox Publisher

Figure 13-6. This bar graph of earthquake magnitudes was created from an array.

FYI

Remember, y values increase from the top of the window to the bottom.

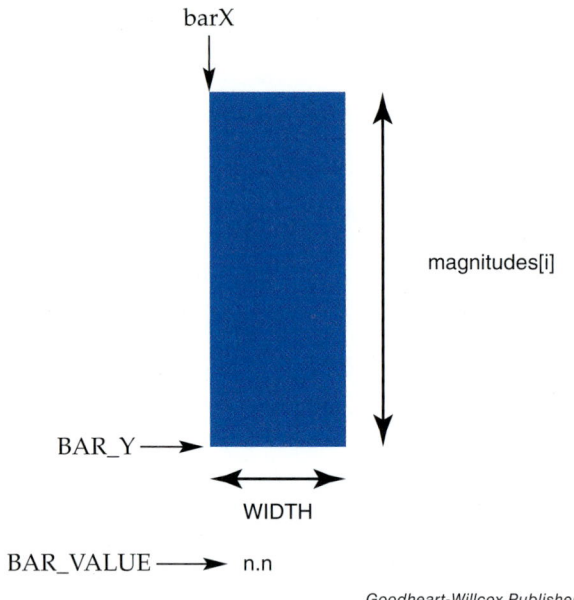

barX

magnitudes[i]

BAR_Y

WIDTH

BAR_VALUE ⟶ n.n

Figure 13-7. The bars in a bar graph are formed by Rectangle objects.

The next step is to define the drawing variables. The barX variable starts at 100 and will increase in value (move to the right) as each bar is drawn. The rest of the constants were chosen by considering the number of values and the values themselves.

One new item is the SCALE constant. The values in the magnitudes array range from 4.4 to 7.7. If Rectangle objects are drawn with heights ranging from 4.4 to 7.7 pixels, the bar graph would be very small and the relative differences between the bars would be difficult to see. To fix this problem, each bar is scaled. To *scale* means to multiply each value by a set amount. In this case, that amount is 50, which is the value of the SCALE constant. In this way, the relative sizes of the bars are maintained, but the graph is larger and thus easier to see.

```
// 2. Define variables needed for bars.
int barX = 100;
final int BAR_Y = 400;
final int VALUE_Y = BAR_Y + 30;
final int WIDTH = 15;
final int SPACING = 15;
final int SCALE = 50;
```

Next, the for loop begins with the standard loop header. The first task in the for loop is to multiply the array value by the scale, creating the actual height of the bar in pixels. This is indicated by comment 4. The Rectangle object is then created using (barX, BAR_Y − height) as the top-left corner, the constant WIDTH, and the height just calculated. See comment 5. To display the magnitude value, a formatted String is passed to the Text constructor along with the (*x*, *y*) value of (barX, VALUE_Y). See comment 6. Both the Rectangle and the Text are then added to the scene. See comment 7. Finally, the barX value is moved to the next bar location by adding the width of the bar and a constant spacing. See comment 8.

```
// 3. The for loop to draw the bars.
for ( int i = 0; i < magnitudes.length; i++ ) {

    // 4. Scale the bar.
    double height = magnitudes[i] * SCALE;

    // 5. Create the Rectangle.
    Rectangle bar = new Rectangle( barX, BAR_Y - height, WIDTH,
        height );
    bar.setFill( Color.GOLDENROD );

    // 6. Create the Text for the value.
    String s = String.format( "%.1f", magnitudes[i] );
    Text value = new Text( barX, VALUE_Y, s );

    // 7. Add Rectangle and Text to Group.
    root.getChildren( ).addAll( bar, value );

    // 8. Move to next bar location.
    barX += WIDTH + SPACING;

} // end for
```

Science and Java

Sinkholes

Sinkhole development is a geological hazard found all over the world. One type of sinkhole usually occurs in areas where the underground bedrock is made of soft materials such as salt, gypsum, or limestone. Rainwater that has become slightly acidic seeps down and reacts with the material. Water flowing through these spaces carries sediment away, causing underground spaces to form. Then, soil that was above the space drops to fill the gap. Human activity such as drilling, mining, fracking, vehicle traffic, and leaking pipes can also create sinkholes. Sinkholes in urban areas are increasing around the world caused by aging water and sewer infrastructure.

a katz/Shutterstock.com

According to the US Geological Survey (www.usgs.gov), one area prone to sinkholes is northwest Florida. Additionally, in regions where there are underground caves, sinkholes can open without any warning. This is due to the ceiling of the cave collapsing because it can no longer support the weight of the surface soil.

Sinkholes are dangerous. They can also cause millions of dollars in damage. YouTube has many videos where you can see sinkholes happening, particularly on roadways.

Governmental agencies routinely use geophysics as part of a geotechnical evaluation of sinkholes after a collapse has occurred. Ground penetrating radar (GPR) and electrical resistivity imaging (ERI) are important tools for detecting and evaluating the subsurface dimensions of buried depressions, caves, or leaking pipes associated with ground surface collapse. These methods create an underground picture of where the spaces are.

Assignment

Before beginning this exercise, download the chapter files from the student companion website. Open the USSinkholes.txt file, and review the data it contains. On each line of the file is the diameter of a sinkhole found in Florida, Louisiana, Utah, Texas, New Mexico, or Alabama. The data was collected by the US Geological Survey (www.usgs.gov). Start a new Java application. Code an array that will hold sinkhole diameters. Read the diameter (in feet) of each sinkhole from the file. Output the contents of the array. Also output the maximum, minimum, and average diameter of the sinkholes.

Math and Java

Deck of Cards

Card games are popular video games. They can be found on all types of computers, from desktop or laptop home computers to mobile devices such as tablets and smartphones. Arrays are useful in managing a deck of cards. There are 52 cards in a standard deck; there are 54 cards if you count the Jokers. The cards occur in patterns, as shown.

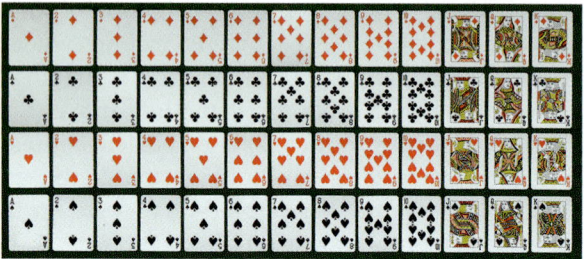

gmlykin/Shutterstock.com

There are four suits: diamonds, clubs, hearts, and spades. Each suit has 13 cards: Ace, 2, 3, 4, 5, 6, 7, 8, 9, 10, Jack, Queen, and King. An array of 52 integers can be filled with 52 unique random numbers between 0 and 51.

First, instantiate the array, and fill it with the numbers 0–51.

```
final int NUMBER_OF_CARDS = 52;
int arrayDeck[ ] =
new int[NUMBER_OF_CARDS];
for ( int i = 0; i < arrayDeck.length;
    i++ ) {
  arrayDeck[i] = i;
}
```

The array can be separated, or parsed, into the four suits and 13 cards using the mod operator (%).

```
int card = arrayDeck[i];
int intSuit = card / 13;
int intNumber = card % 13;
```

Any card in the array can be identified by its value. The card faces are numbered 0 to 12, and the suits are numbered 0 to 4. For the card faces, 0 is the Ace, 1 is a 2, 9 is a 10, 10 is a Jack, 11 is a Queen, 12 is a King, and so on. For the suits, 0 is diamonds, 1 is clubs, 2 is hearts, and 3 is spades. A JavaFX application can be used to draw the card. An array of Unicode characters provides graphic values for the suit symbols.

```
String suits[ ] = new String[4];
suits[0] = "\u2666"; // diamonds
suits[1] = "\u2663"; // clubs
suits[2] = "\u2665"; // hearts
suits[3] = "\u2660"; // spades
```

For example, if the value of arrayDeck[i] = 40, then the card face number would be 40 mod 13 = 1. The suit would be 40 / 13 = 3. Therefore, the card is the 2 of spades, as shown.

arrayDeck[i] = 40
Card number: 40 mod 13 = 1
Card suit: 40 / 13 = 3

2 ♠

Goodheart-Willcox Publisher

To deal the deck, assign random numbers between 0 and 51. Search the array to verify a card is not dealt more than once.

Assignment

Before beginning this exercise, download the chapter files from the student companion website. The DealCardsSnippet.java file will be used as a starting point. Open the file, and follow the algorithm to fill the array with the numbers 0 to 51. To shuffle the cards, use a for loop from 0 to 51. For each index number i, generate a random number and swap the element at i with the element at the random number.

HANDS-ON EXAMPLE 13.2.2

Drawing a Bar Graph 📲

In this exercise, you will draw a bar graph based on the dailySteps array created in Hands-On Example 13.2.1. Before beginning this exercise, download the chapter files from the student companion website. The GraphDailyStepsSnippet.java file will be used as a starting point. Do *not* use the program you created in Hands-On Example 13.2.1 as it will have different values.

1. Launch jGRASP, and open the GraphDailyStepsSnippet.java file.

2. Locate comment 1, and follow the instructions to create the dailySteps array with some values.

```
/***** 1. Assign arbitrary values to array. */
int[ ] dailySteps = { 6000, 10000, 7526, 2345, 12054, 8756, 7890 };
```

3. Locate comment 2, and follow the instructions to define the variables and constants used in the drawing. These numbers are slightly different from the magnitudes array presented in the text because of the range of the values and the number of elements in the dailySteps array.

```
/***** 2. Define variables used in the bars. */
int barX = 100; // start the first bar at 100
final int BAR_Y = 400;
final int BAR_VALUE = 430;
final int WIDTH = 30;
final int SPACING = 20;
final double SCALE = .02;
```

4. Locate comment 3, and follow the instructions to scale the value inside the for loop. The scaling for the dailySteps array is opposite to the scaling of the magnitudes array. In this case, the bars need to be *smaller* than the actual values to fit in the window. Thus, considering the values, the scale factor .02 works well. This value is set in the SCALE constant.

```
/***** 3. Scale the value. */
double height = dailySteps[i] * SCALE;
```

5. Locate comment 4, and follow the instructions to create the Rectangle and Text and add both to the scene. This code is almost identical to that used in the magnitudes array example. The difference is the bar color and the formatting of the value. The format specifies adding a comma every three digits.

```
/***** 4. Create the Rectangle and Text and add to scene. */
Rectangle bar = new Rectangle( barX, BAR_Y - height, WIDTH, height );
bar.setFill( Color.CORNFLOWERBLUE );

String s = String.format( "%,d", dailySteps[i] ); // format with comma
Text value = new Text( barX, BAR_VALUE, s );

root.getChildren( ).addAll( bar, value );
```

6. Locate comment 5, and follow the instructions to move xBar to the next bar location. This is the last code inside the for loop.

```
/***** 5. Move to next bar location. */
barX += WIDTH + SPACING;
```

7. Compile and run the program. Verify all bars are displayed and the relative values are correct.

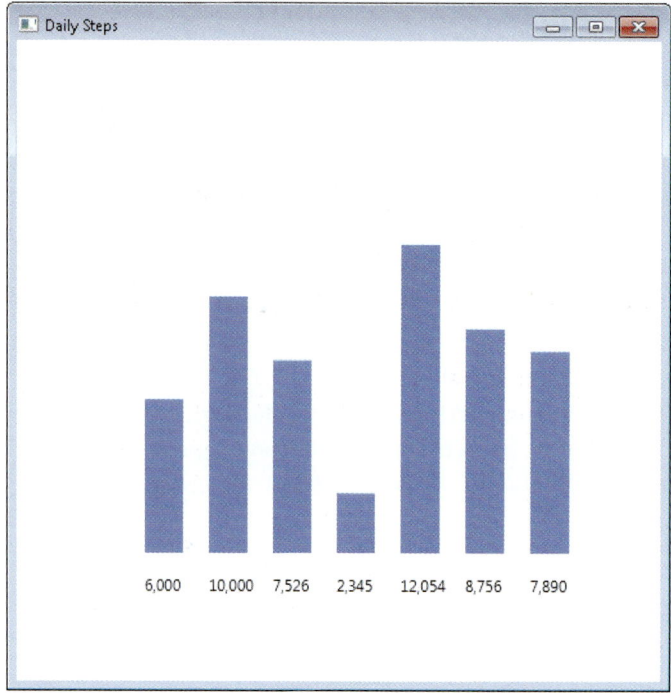

6,000   10,000   7,526   2,345   12,054   8,756   7,890

Goodheart-Willcox Publisher

Try It!

Continue working with the GraphDailyStepsSnippet program. Add code to prompt the user for his or her target number of daily steps. Then, add code to determine which days the user did not meet the goal. For each day the daily steps are less than the target, color the bar and text red.

Processing Arrays of Objects

Arrays can contain objects as well as primitive types. Remember that classes are a data type. Arrays of objects differ from arrays of primitive types (int, double, etc.) because each element of the array is a reference to an object, rather than storing the object itself. Creating an array of objects takes two steps:

1. Instantiate the array.
2. Instantiate each object.

The code below instantiates an array of Strings called students. Because Strings are objects, each element will initially have the value null.

```
String[ ] students = new String[4]; // instantiate the array
```

These statements instantiate three Strings in the array:

```
// instantiate each object
students[0] = "Lily";
students[2] = "Jane";
students[1] = "Roberto";
```

Figure 13-8 shows the resulting students array.

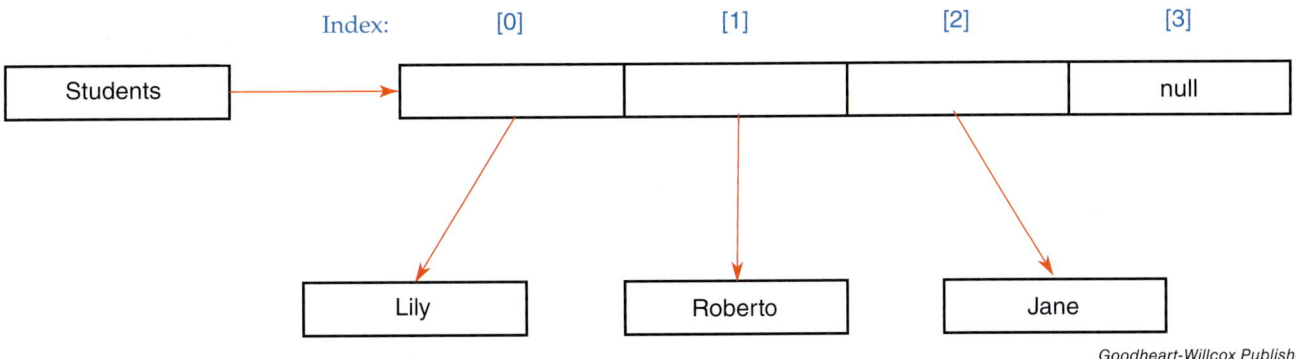

Index: [0] [1] [2] [3]

Figure 13-8. This shows the students array after instantiating some elements.

As you can see, each array element is a reference to the location of the object (the String) in memory. Notice that the last element (students[3]) is still null because no String has been instantiated and assigned to that element. Outputting the array verifies **Figure 13-8:**

```
for ( int i = 0; i < students.length; i++ ) {
   System.out.print( students[i] + "\t" );
}
System.out.println( );
```

The output is:

```
Lily Roberto Jane null
```

To call a method on a student name in the array, use the dot operator and the name of the array with an index. For example:

```
System.out.println( students[2].toUpperCase( ) ); // outputs JANE
```

Think of the array name with an index (students[2]) as the name of the String that has the value "Jane".

To illustrate this, consider again the magnitudes array. The previously presented code calculated the maximum magnitude of the earthquakes listed in the file. However, the magnitudes array held no information on the location of the earthquake. The location of each earthquake can be captured and used by creating an Earthquake class that has both the magnitude and location as instance variables. An abbreviated version of the class is shown below. This version illustrates only the instance variables, a constructor, accessors, and a toString method.

```
/** Earthquake class */

public class Earthquake {
   // instance variables
   private String location;
   private double magnitude;

   // constructor
   public Earthquake( double magnitude, String location ) {
      this.magnitude = magnitude;
      this.location = location;
   }

   // accessors
   public double getMagnitude( ) {
      return magnitude;
   }
```

(Continued)

```java
public String getLocation( ) {
   return location;
}

// ... other methods

// toString
public String toString( ) {
   return "magnitude " + magnitude + " at " + location;
}

} // end class
```

The next step is to alter the input file to contain the locations of the earthquakes as well as the magnitudes. The earthquakeLocations.txt file is shown in **Figure 13-9.** Each line in the file contains the magnitude of an earthquake followed by location of that earthquake. The information in this file represents earthquakes that occurred on January 1 and 2, 2019. This data is taken from the US Geological Survey website (www.usgs.gov).

Finally, the data is analyzed. The code is shown below. The code first creates an array that will hold Earthquake objects. See comment 1. For each line in the file, the code reads the magnitude and the location and then uses these values to create an Earthquake object. The object is then stored in the earthquakes array. See comment 2. The last job is to find the earthquake with the maximum magnitude. This code uses a slight variation of the pattern for finding the maximum. Instead of finding the maximum magnitude, the code finds the location in the array of the object that has the maximum magnitude. This is accomplished by starting with an index of 0 (the first element) and saving the *index* of the object that has the maximum magnitude. See the code in comment 4. Instead of comparing the values directly, the code compares the values of the magnitudes of the objects at i and at maxIndex. Having the index of the earthquake with the maximum magnitude makes reporting the earthquake location and magnitude easy by calling the toString method of that element. See comment 5.

```java
/* Earthquake analysis */
import java.util.*;
import java.io.*;

public class EarthquakeAnalysis {

   public static void main( String [ ] args ) {

      // 1. Instantiate array of Earthquake objects.
      final int NUMBER_OF_EARTHQUAKES = 10;
```

(Continued)

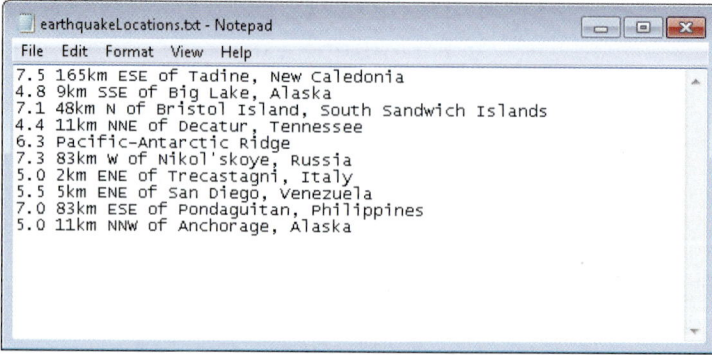

Figure 13-9. The earthquakeLocations.txt file can be read to extract all data.

```java
Earthquake[ ] earthquakes
              = new Earthquake[NUMBER_OF_EARTHQUAKES];

try {
   File inFile = new File( "earthquakeLocations.txt" );
   Scanner file = new Scanner( inFile );

   // 2. Read file, instantiate Earthquake objects, add to
   // array.
   for ( int i = 0; i < earthquakes.length; i++ ) {

      double magnitude = file.nextDouble( );
      String location = file.nextLine( );
      location = location.trim( );
          // instantiate the object
      Earthquake e = new Earthquake( magnitude, location );
      earthquakes[i] = e; // assign the object to the array
   }

}
catch ( FileNotFoundException fnf ) {
   System.out.println( "Unable to find the file." );
   System.exit( 1 ); // exit the program
}

// 3. Output the earthquakes by calling toString.
System.out.println( "The earthquakes are: " );
for ( int i = 0; i < earthquakes.length; i++ ) {
   System.out.println( earthquakes[i].toString( ) );
}

// 4. Find earthquake with the maximum magnitude.
int maxIndex = 0; // start with index 0

for ( int i = 1; i < earthquakes.length; i++ ) {
   // compare magnitude of earthquake at [i] and [maxIndex]
   if ( earthquakes[i].getMagnitude( ) >
      earthquakes[maxIndex].getMagnitude( ) ) {
      maxIndex = i; // if greater, save the index
   }
}

// 5. Call toString of object at maxIndex.
System.out.println( "\nThe strongest earthquake is "
             + earthquakes[maxIndex].toString( ) );
} // end main
} //end class
```

After reading the file, the code outputs the earthquakes in the array by calling toString on each object. Remember the toString method is automatically called when an object is used as a String. So this statement would have worked just as well.

```java
System.out.println( earthquakes[i] );   // automatically call
                                         // toString
```

Then, the code outputs the earthquake with the greatest magnitude. The output of the program is:

```
The earthquakes are:
magnitude 7.5 at 165km ESE of Tadine, New Caledonia
magnitude 4.8 at 9km SSE of Big Lake, Alaska
magnitude 7.1 at 48km N of Bristol Island, South Sandwich Islands
magnitude 4.4 at 11km NNE of Decatur, Tennessee
magnitude 6.3 at Pacific-Antarctic Ridge
magnitude 7.3 at 83km W of Nikol'skoye, Russia
```

(Continued)

```
magnitude 5.0 at 2km ENE of Trecastagni, Italy
magnitude 5.5 at 5km ENE of San Diego, Venezuela
magnitude 7.0 at 83km ESE of Pondaguitan, Philippines
magnitude 5.0 at 11km NNW of Anchorage, Alaska

The strongest earthquake is magnitude 7.5 at 165km ESE of Tadine,
New Caledonia
```

HANDS-ON EXAMPLE 13.2.3

Coding a World Capitals Quiz 📲

In this exercise, you will create a quiz on world capitals. You will write a Country class that has instance variables for a country name and its capital. The client program reads a list of selected country names and capitals from a file. Using that data, the program instantiates Country objects and stores each Country object in an array. The client will then quiz the user by choosing a country at random from the array and asking the user to name the capital of that country. Before beginning this exercise, download the chapter files from the student companion website. The Country.java, CountryCapitalQuiz.java, and worldCapitals.txt files will be used as a starting point.

1. Launch jGRASP, and open the Country.java file.

2. Locate comment 1, and follow the instructions to define the two instance variables.

   ```java
   /***** 1. Define the instance variables: countryName and capital, both Strings. */
   private String countryName;
   private String capital;
   ```

3. Locate comment 2, and follow the instructions to complete the constructor.

   ```java
   /***** 2. Complete the constructor. It accepts the country name and capital as parameters. */
   public Country( String countryName, String capital ) {
      this.countryName = countryName;
      this.capital = capital;
   }
   ```

4. Locate comment 3, and follow the instructions to write the accessor methods for countryName and capital.

   ```java
   /***** 3. Write the accessors for the country name and capital. */
   public String getCountryName( ) {
      return countryName;
   }

   public String getCapital( ) {
      return capital;
   }
   ```

5. Locate comment 4, and follow the instructions to write the toString method.

   ```java
   /***** 4. Write toString. It returns "The capital of ... is ...". */
   public String toString( ) {
      return "The capital of " + countryName + " is " + capital;
   }
   ```

6. Compile the file. Correct any errors reported.

7. Open the CountryCapitalQuiz.java file. This is the client program that will run the quiz.

8. Locate comment 1, and follow the instructions to create an array to hold 82 Country objects. The size of the array is determined by the worldCapitals.txt file, which contains a selection of 82 countries and capitals. Place this code before the try/catch block.

   ```java
   /***** 1. Instantiate an array to hold 82 Country objects. */
   final int NUMBER_OF_COUNTRIES = 82;
   Country [ ] countries = new Country[NUMBER_OF_COUNTRIES];
   ```

9. The try/catch block for opening the worldCapitals.txt file is coded for you. The file contains 164 lines, alternating between country names and capitals. In other words, the first line contains a country name, the second line contains its capital, the third line contains another country, the fourth line contains that country's capital, and so on. Locate comment 2, and follow the instructions to read the file, create Country objects, and store them in the array. Notice that the code sends the input values from reading the country name and capital to the Country constructor.

```
/***** 2. Using a for loop, read the countries and their capitals,
       then create a Country object and store it in the array */
for ( int i = 0; i < countries.length; i++ ) {
   Country c = new Country( file.nextLine( ), file.nextLine( ) );
   countries[i] = c;
}
```

10. Create the quiz. Use what you have learned about do/while loops to prompt the user for the capital of a country chosen at random until the user makes three incorrect guesses. The Scanner object for the keyboard and a Random object have been created for you.

```
/***** 3. Play the game until the user makes three incorrect guesses:
          Set up counters for the number of correct and incorrect guesses.
          Using a do/while loop, generate a random number between 0 and the size of
            the countries array.
          Using that number, get the countryName for that Country object in the array
            and ask the user to guess the capital.
          If the user is incorrect, call the toString method to display the correct capital.

          When the user makes three incorrect guesses, the quiz ends.
          Display the number of correct and incorrect guesses. */

Scanner input = new Scanner( System.in );
Random rand = new Random( );

int correct = 0; // counter for correct answers
int incorrect = 0; // counter for incorrect answers

do {
   // choose a random Country index
   int randomIndex = rand.nextInt( countries.length );

   // ask the user to guess
   System.out.println( "\nWhat is the capital of "
   + countries[randomIndex].getCountryName(
      ) + "?" );
   String userGuess = input.nextLine( );
   userGuess = userGuess.trim( ); // remove leading and trailing spaces

   // compare user guess to the capital in the Country object
   if ( countries[randomIndex].getCapital( ).equalsIgnoreCase( userGuess ) ) {
      System.out.println( "Correct!" );
      correct++;
   } else { // output correct answer using toString
      System.out.println( "Sorry. " + countries[randomIndex] );
      incorrect++;
   }

} while ( incorrect < 3 ); // 3 incorrect guesses ends the game

System.out.println( "\nGame over. You got " + correct + " answers correct." );
```

11. Compile and run the program to test your knowledge of world capitals!

Try It!

Create a quiz for the state capitals in the United States. Create a text file containing each state and its capital. Then, create a State class. Finally, create a StateCapitalQuiz quiz client program.

Language Arts and Java

William Shakespeare

William Shakespeare was born in Stratford-upon-Avon, England in 1564. In 1582, he married Anne Hathaway and had three children. Shakespeare lived in London for 25 years, where he wrote most of his plays. He died at his home in Stratford-upon-Avon in 1616 at age 52.

Georgios Kollidas/Shutterstock.com

Shakespeare wrote 37 plays. He wrote three different types of plays: histories, about the lives of kings and famous figures of the past; comedies, which all ended with a marriage; and tragedies, which all closed with the death of the main character. Shakespeare also wrote poetry and, in 1609, published a book of 154 sonnets. His work was very popular, and he earned enough income to live in an affluent area of London.

Shakespeare had an incredible influence on the English language. According to the Folger Shakespeare Library, he invented more than 1,700 words that are still in use today. Here are a few: amazement, bedroom, champion, dawn, eyeball, fashionable, gossip, moonbeam, and Olympian.

William Shakespeare remains the world's most romantic dramatist and poet. Many trivia games and Jeopardy-like activities often have a Shakespeare category. The contestant will pick 100, 200, 300, 400, or 500 representing dollars or points. The program will print a Shakespeare quote. The contestant will read the quote and say from which Shakespeare work the quote originated. The moderator, who has a copy of the answers, will say "correct" or "not correct" and give the correct answer to the contestant.

Assignment

Start a new Java application. Create an array that contains the quotes and another array that contains the Shakespeare work. Refer to the table shown. Prompt the user for the value desired and display the associated quote as shown below. Use a switch statement to choose the appropriate quote. Prompt the user for the name of the work. Tell the user if the answer is correct.

Value	Quote	Shakespeare Work
100	If music be the food of love, play on.	*Twelfth Night*
200	Shall I compare thee to a summer's day? Thou art more lovely and more temperate.	"Sonnet 18"
300	Love sought is good, but given unsought is better.	*Twelfth Night*
400	I love you with so much of my heart that none is left to protest.	*Much Ado About Nothing*
500	Love is a smoke raised with the fume of sighs.	*Romeo and Juliet*

SECTION 13.2 REVIEW

Check Your Understanding

1. Which type of loop is best to traverse an array?
2. How many times is the loop that traverses an array repeated?
3. When creating a bar graph from an array, what determines the height of each rectangle?
4. An array of a primitive data type stores the object itself. How does an array of objects differ from this?
5. What are the two steps needed to create an array consisting of objects?

Build Your Vocabulary

As you progress through this course, develop a personal computer science glossary. This will help you build your vocabulary and prepare you for a career. Write a definition for each of the following terms and add it to your computer science glossary.

bar graph   scale

Searching Arrays

O ften, it is useful to determine if a particular value is in an array. For example, in the earthquakes array, you might want to know if any earthquakes occurred in Alaska. For a students array, you might want to know if a student named "Roberto" is in the array. It can also be useful to know if a certain value is *not* in the array. Perhaps you want to search the earthquakes array to confirm that no earthquakes have occurred in your state.

All the values in an array are in memory at once. Therefore, these searches can be done dynamically. The program can prompt the user at runtime for a value to find. That same code can be used for multiple searches. Over time, computer scientists have developed many search algorithms. In this section, you will use the sequential search algorithm on an unsorted array and then on a sorted array.

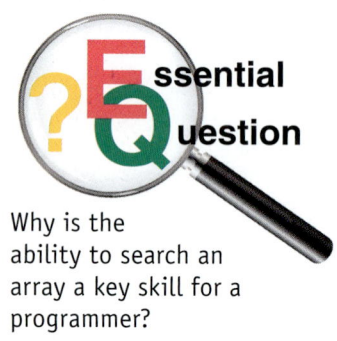

Essential Question

Why is the ability to search an array a key skill for a programmer?

Learning Goals

- Construct a search of an unsorted array.
- Explain how to sort an array and then search it.

Terms

search key
sequential search

sort

Searching an Unsorted Array

One algorithm to search an array for a particular value is to traverse the array in order. At each element, check whether the element's value matches the search key. The *search key* is the value to find in a search.

For example, consider an array of six ints named intArray, as shown in **Figure 13-10.** Suppose the user wants to find if the array contains the number 12. In this case, 12 is the search key. The algorithm is to start at the first element and compare each element to the search key. If there is a match, the value is found. To indicate that the value is found, the search code saves the index of the matching element. If the value does not match, move on to the next element.

How do you know if the search key value is not in the array? The answer is that the end of the array is reached without finding a match. To indicate the value is not found, the search code reports an index of –1.

This behavior should be familiar. The indexOf method in the String class returns the same values. The index of the first character of the search String is returned if the value is found. If not found, –1 is returned.

This algorithm is called a *sequential search* because the code searches an array by comparing the search key to each element in order. Here are the steps to the algorithm. Starting with the first element in the array:

1. Compare the element to the search key.
2. If a match, save the index.
3. Go to the next element.
4. Repeat step 1.

Code to sequentially search an array is shown below. The variable index will hold the results of the search. Before the for loop begins, index is initialized to –1. As the for loop is executed, if a match is found, the index of the matching element is assigned to index. After the for loop ends, the results are checked. If index is still –1, the search key was not found. If the index is not –1, then a match was found, and the value of index is the location in the array where the search key was found.

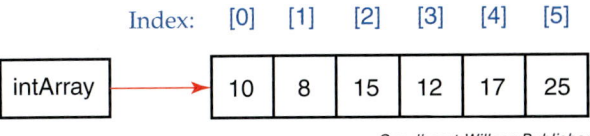

Figure 13-10. This array of ints will be searched.

Index: [0]   [1]   [2]   [3]   [4]   [5]

intArray   →   10   8   15   12   17   25

Goodheart-Willcox Publisher

```
System.out.println( "Enter the value to find" );
int searchKey = scan.nextInt( );

int index = -1; // start with not-found indication

for ( int i = 0; i < intArray.length; i++ ) {
   if ( intArray[i] == searchKey ) { // compare element and search key
      index = i; // if found, save the index
   }
}

// check results
if ( index == -1 ) {
   System.out.println( searchKey + " was not found." );
} else {
   System.out.println( searchKey + " was found in index " + index );
}
```

How many comparisons must the code perform to discover the search key is not in the array? The answer is that each element must be checked. So, the number of comparisons is equal to the number of elements in the array. How many comparisons must the code make if the search key is found? Unfortunately, the answer is the same. All elements are still checked even after the search key is found. Because the entire array is searched, this code finds the last occurrence of the search key. If the search key appears multiple times in the array, when the for loop ends, the index variable will contain the index of the last place the search key was found.

This code can be improved by adding an early exit from the for loop when the search key is found. This can be accomplished by modifying the loop condition to check if index is still –1. Once the search key is found, the value of index will be changed. When index is no longer –1, the loop can be exited.

The more-efficient code is shown below, including a test to see that the new for loop does indeed end early. A statement is inserted at the beginning of the for loop to output the value of i. Notice the change in the loop header:

```
for ( int i = 0; i < array.length && index == -1; i++ ) {
   System.out.println( "Checking element " + i );
   if ( array[i] == searchKey ) { // compare element and search key
      index = i; // if found, save the index
   }
}
```

With a search key of 13, which will be found in index 2, this code outputs:

```
Checking element 0
Checking element 1
Checking element 2
```

As you can see, the loop ends after finding the search key at index 2.

Note that the new for loop header does nothing to improve the number of comparisons when the search key is *not* found. The entire array still needs to be checked if there are no matches. However, there is a change in the results for multiple occurrences of the search key. Because this code exits the loop as soon as the search key is found, the code will find the first occurrence of the search key in the array. This behavior more closely matches the behavior of the indexOf method in the String class.

HANDS-ON EXAMPLE 13.3.1

Processing Movie Theater Show Times

In this exercise, you will write a sequential search for a multiplex movie theater. The user will ask to see a movie. The program will tell the user in what theater the movie is playing or will inform the user the movie is not playing. The names and locations of eight movies being shown are contained in a text file. Before beginning this exercise, download the chapter files from the student companion website. The MovieTheaterSnippet.java and movies.txt files will be used as a starting point. The text file contains the names of eight classic Disney movies, one per line.

1. Launch jGRASP, and open the MovieTheaterSnippet.java file.

2. Locate comment 1, and follow the instructions to define the array of Strings named movies.

```
/***** 1. Declare and instantiate an array of Strings named movies. */
final int NUMBER_OF_MOVIES = 8;
String[ ] movies = new String[NUMBER_OF_MOVIES];
```

3. Locate comment 2, and follow the instructions to open the text file and read the movies into the array.

```
/***** 2. Open the file and read the movies into the array; movies are one per line in the
    file. */
try {
    File inFile = new File( "movies.txt" );
    Scanner file = new Scanner( inFile );

    for ( int i = 0; i < movies.length; i++ ) {
        movies[i] = file.nextLine( );
    }
}
catch ( FileNotFoundException fnf ) {
    System.out.println( "Unable to find the file." );
    System.exit( 1 );
}
```

4. Locate comment 3, and follow the instructions to prompt the user for the movie name.

```
/***** 3. Prompt the user for the movie they want to see. That is the search key. */
Scanner input = new Scanner( System.in );

System.out.println( "Welcome to Multiplex Movies" );
System.out.println( "What movie would you like to see today?" );
String userMovie = input.nextLine( );
```

5. Locate comment 4, and follow the instructions to perform the sequential search.

```
/***** 4. Search for user's title. */
int index = -1;
for ( int i = 0; i < movies.length && index == -1; i++ ) {
    if ( movies[i].equalsIgnoreCase( userMovie ) ) {
        index = i;
    }
}
```

6. Locate comment 5, and follow the instructions to report the results of the search by checking the value of index. Assume the position of the movie in the array corresponds to the theater number where the movie is showing. You need to make one adjustment, however. Because array indexes start at 0 and movie theater numbers start at 1, add 1 to the index to report the theater number.

```
/***** 5. Report the "theater number" where the movie is showing
    or inform the user if the movie is not playing.
```

(Continued)

```
         Because theaters start at 1 and indexes start at 0,
         the theater number, if found, should be reported as ( index + 1 ). */
   if ( index == -1 ) {
      System.out.println( "Sorry, " + userMovie + " is not playing today." );
   } else {
      System.out.println( userMovie + " is playing in theater " + ( index + 1 )
                    + ". \nBe sure to visit our concession stand!" );
   }
```

7. Compile and run the program. Test the program by entering a movie name that is in the text file. Be sure the number of the theater is correct (array index plus 1). Then, run the program again and enter a movie name that will not be found in the file.

Try It!

Write a program that allows the user to search a students array for the name of a student. Create an array of Strings with about ten names of your choice. Populate the array within code, not in a separate text file. Then, prompt the user for the search key, search the array, and report whether or not the student name is found.

Searching a Sorted Array

The intArray in the example discussed above contained numbers in no particular order. For some applications, it is helpful if the numbers are in either ascending or descending order. To *sort* an array means to arrange the elements in some order based on their values. For example, an array with numeric values would be sorted if the numbers increased from the beginning of the array to the end. If the numbers decreased from the beginning of the array to the end, the array would be in reverse sorted order.

Sorting

Over time, computer scientists have developed many sophisticated sort algorithms. For applications in this text, the static sort method in the Arrays class can be used. The Arrays class is in the java.util package with the other utility classes, such as Scanner and Random. The sort method's API is shown in **Figure 13-11.**

Starting with the intArray presented earlier:

```
int[ ] intArray = { 10, 8, 15, 12, 17, 25 };
```

the array is sorted with this statement:

```
Arrays.sort( array );   // sort the array
```

Then, to verify the sort was successful, this code prints the array:

```
for ( int i = 0; i < intArray.length; i++ ) {
   System.out.print( intArray[i] + " " );
}
System.out.println( );
```

The output from this code is indeed a sorted array:

```
8 10 12 15 17 25
```

An array of Strings or chars is sorted by arranging the elements in order by their Unicode values. Given this array of Strings:

```
String[ ] flowers = { "Violet", "Rose", "carnation", "Gardenia" };
```

Method	Action
static void sort(datatype[ ] array)	Arranges the elements in the array in ascending order. Overloaded versions of this method are available for all primitive types except boolean as well as for an array containing objects.

Goodheart-Willcox Publisher

Figure 13-11. This is the API for the static sort method in the Arrays class.

Coding Conundrum

1.   The programmer has written this code to define an array of integers:

```
int scores = new int[10];
```

but the compiler reports this error:

```
error: incompatible types: int[ ] cannot be converted to int
int scores = new int[10];
            ^
```

It is a conundrum! How can you fix this code?

2.   The programmer has written this code to define an array and output its elements:

```
boolean[ ] bArray = new boolean[5];
System.out.println( bArray );
```

The code compiles and executes, but produces this unexpected output:

```
[Z@b97c004
```

It is a conundrum! How can you fix this code?

3.   The programmer has written this code to define an array and print its elements:

```
double[ ] array = new double[3];
for ( int i = 0; i <= array.length; i++ ) {
   System.out.println( array[i] );
}
```

The code compiles without errors, but when the code runs, an exception occurs:

```
Exception in thread "main" java.lang.ArrayIndexOutOfBoundsException:
Index 3 out of bounds for length 3
```

It is a conundrum! How can you fix this code?

4.   The programmer has written this code to define an array and find the minimum value. When the code runs, it outputs 100, rather than the correct minimum of 5.

```
int[ ] iArray = { 42, 23, 100, 5, 10 };
int minimum = iArray[0];
for ( int i = 1; i < iArray.length; i++ ) {
   if ( iArray[i] > minimum ) {
      minimum = iArray[i];
   }
}
System.out.println( minimum );
```

It is a conundrum! How can you fix this code?

5.   The programmer has written this code to define an array of Strings and convert the first element to uppercase:

```
String[ ] names = new String[5];
String uppercaseName = names[0].toUpperCase( );
```

The code compiles, but when it runs, a NullPointerException error occurs. It is a conundrum! What is wrong with the code?

that is then sorted with this code:

```
Arrays.sort( flowers );
```

the sorted array outputted is:

```
Gardenia Rose Violet carnation
```

This result may be surprising, considering that "carnation" is listed last. Remember, however, that lowercase letters have a higher Unicode value than uppercase letters. Thus, words starting with a lowercase letter will follow any words starting with an uppercase letter.

Searching

If an array is sorted, the search algorithm can be made even more efficient. With the elements in ascending order, as soon as the for loop finds an element larger than the search key, then all following elements will also be larger than the search key. At that point, you know the search key will not be found, and the code can exit the loop. This is more efficient than searching an unsorted array until the end to determine the search key is not found.

Here is the code for searching a sorted array. The only difference between searching an unsorted array is the for loop condition, which now becomes:

```
i < intArray.length && index == -1 && intArray[i] <= searchKey;
```

An added bonus to a sorted array is that finding the minimum or maximum values is trivial. When sorted in ascending order, the minimum is always in the first element, and the maximum is always in the last element. For the sorted array of int values presented above:

```
System.out.println( "The minimum is " + intArray[0] );
// outputs: The minimum is 8

System.out.println( "The maximum is "
                    + intArray[intArray.length - 1] );
// outputs: The maximum is 25
```

SECTION 13.3 REVIEW

Check Your Understanding

1. What is the term for the value to find in an array?
2. What does the indexOf method return if the value being searched for is not found?
3. Describe a sequential search.
4. What does it mean to sort an array?
5. If an array is sorted in ascending order, what is true of the first element?

Build Your Vocabulary

As you progress through this course, develop a personal computer science glossary. This will help you build your vocabulary and prepare you for a career. Write a definition for each of the following terms and add it to your computer science glossary.

search key   sequential search   sort

Cooperative Coding

Character Movement

An array can be used to synchronize the movement of animated characters. By using an array, multiple characters can be made to have the same movements. Additionally, if the movements are saved in a text file, it is easy to modify the movement by simply editing the file. In this activity, your team will create a dance routine for multiple Turtle objects.

Step	Turtle Movement
f	forward
b	backward
l	turn left
r	turn right
sl	make a square turning left
sr	make a square turning right
fb	forward/back

As a team, choreograph the dance steps for the turtles. Consider the possible dance steps shown here. You can add your own steps to these to do more creative movements.

Assignment

Create a text file containing the dance steps. The file should contain one step per line. Each step should be one- or two-letter keys followed by an integer amount. For example, with this file:

```
f 20
l 45
b 15
sr 20
fb 15
```

a Turtle "dancer" would move forward 20 pixels, then turn left 45 degrees, move backward 15 pixels, make a square to the right 20 pixels in size, then move forward and back 15 pixels.

Once the choreography is done, create a JavaFX application to execute the dance. In the JavaFX application, create an array of Turtle objects and place each Turtle object in the window. Then, read the file containing the dance steps. For each step, traverse the array of Turtle objects and move each Turtle object accordingly.

1. How did you use the letter keys to assign movement to the objects?
2. How would you change the code to have Turtles in odd-numbered indexes perform one dance while Turtles in even-numbered indexes perform a different dance?

Chapter Summary

Section 13.1: Introduction to Arrays

- An array is an ordered sequence of data of the same data type, called elements; arrays are objects, so arrays must be instantiated.
- Array elements are referenced by an index, which can be used to retrieve the element value or change it.

Section 13.2: Processing Arrays

- A for loop is used to traverse an array with a loop condition that the index is less than the length of the array, and this can be used to process each element of the array, for example, to fill an array, print an array, find the average of the array, or find the minimum and maximum values in the array.
- A bar graph can be created from an array using a for loop and Rectangle objects; the value of each element is used to draw the height of each bar in the graph.
- Each element in an array of objects is a reference to the object, not the object itself; creating an array of objects requires instantiating the array and then instantiating each object in the array.

Section 13.3: Searching Arrays

- The search key is the value to find, and a sequential search processes the array in order; an unsorted array must be searched in its entirety if the search key is not found, but can be exited early if the key is found.
- The sort method of the Arrays class can be used to return the values in an array in a sequential order; searching a sorted array is more efficient than sorting an unordered array.

Chapter 13 Test

Multiple Choice

Select the best response.

1. Which of the following statements is *not* true about arrays?
   - A. An array can contain any number of elements of the same data type.
   - B. The number of elements in an array can be changed after the array is created.
   - C. An array can be populated in the same statement in which the array is created.
   - D. Each element in an array can be accessed by its index.

2. The following code creates an array. What is the value of specialChars[3]?

```
char[ ] specialChars = { '!', '@', '#',
'$', '%', '^', '&', '*' };
```

   - A. !
   - B. @
   - C. #
   - D. $

3. What does this code do?
specialChars[specialChars.length − 1]
   - A. Accesses the last element in the array named specialChars.
   - B. Creates an array named specialChars and populates it with the value −1.
   - C. Reduces the size of the array named specialChars by 1.
   - D. Removes (deletes) the array named specialChars and frees the memory it occupied.

4. Which for loop header is correct for traversing an array named myArray from the first to the last element?
   - A. for (int i = 0; i <= myArray.length; i++)
   - B. for (int i = 0; i == myArray.length; i++)
   - C. for (int i = 0; i < myArray.length; i++)
   - D. for (int i = 0; i > myArray.length; i++)

5. Which of the following statements sorts the sales array in ascending order?
   - A. Arrays.sort(sales)
   - B. Sales.Sort( )
   - C. Sort(sales)
   - D. SortArray(sales)

Completion

Complete the following sentences with the correct word(s).

6. An array of the names of the states in the United States would contain 50 _____.

7. The number of elements in an array is available through the _____ variable.

8. Outputting all values in an array requires a(n) _____ loop to traverse the data.

9. In a sequential search, the value _____ for the index means no match was found.

10. For an array sorted in ascending order, the minimum is always in the _____ element.

Matching

Match the correct term with its definition.

A. length

B. sequential search

C. sort

D. element

E. index

11. Algorithm for searching an array by comparing each element in order to the search key.

12. Data item in an array.

13. Position of an element in an array.

14. To arrange elements in an array in some prescribed order.

15. Variable holding the size of an array.

Application and Extension of Knowledge

1. Review the example of an array created with Earthquake objects presented in this chapter and the text file shown in **Figure 13-9.** Write an application that searches this array. Allow the user to search for a location. Instead of requiring an exact match between the location in the Earthquake object and the search key, consider the search key found if the search key is part of the location. Hint: use the indexOf method of the String class.

2. Before beginning this activity, download the chapter files from the student companion website. The halogens.txt file is provided for use in this activity. Review the World Capitals Quiz created in Hands-On Example 13.2.3. Apply what you learned in that exercise to make a quiz on the scientific abbreviations of the halogen group of the periodic table. Create a Halogen class with instance variables for halogen names and abbreviations. Read the data from the halogens. txt file. Create Halogen objects and store them in an array. Display the name of the element to the user, and then ask for the abbreviation. After two wrong answers, the program should stop.

3. The local juice bar sells varying amounts of bananas over the year. The monthly amounts were 325.6, 200.4, 456, 324.7, 256.8, 431, 509.5, 445, 333.5, 451, 421.5, and 476 pounds. Apply what you learned about creating a bar graph in Hands-On Example 13.2.2 to create a horizontal bar graph of the amounts of bananas used over the year. Define an array of Rectangle objects and use the monthly amounts for the width of each rectangle. Advance the *y* coordinate for each month. Display the number of the month, the amount, and the rectangle for each element in the array.

4. The store manager of the juice bar wants to know the total amount of bananas used during the year as well as the average per month. Using the array from the previous activity, output the total for the year and the monthly average.

5. Identify at least five flowers native to your area. Write the statements to create an array of flower names called flowerTypes, and store the flower names you identified. Allow the user to search for a flower. The program should output whether or not the flower was found.

Online Activities

Complete the following activities, which will help you learn, practice, and expand your knowledge and skills.

Vocabulary. Practice vocabulary for this chapter using the e-flash cards, matching activity, and vocabulary game until you are able to recognize their meanings.

Communication Skills

Reading. Good readers are able to reflect on their own reading abilities and make changes to improve these skills. They often monitor their comprehension, adjust their speed to match the difficulty of the text, and self-evaluate any comprehension problems they encounter. Students who are able to self-evaluate their reading skills ultimately comprehend more in all areas of communication. Review your own reading habits to identify where you can improve. This can include timing yourself as you silently read this chapter, paying attention to how often you interrupt yourself, or staying on topic when your thoughts begin to wander. What did your self-assessment reveal about your reading skills?

Writing. Critical thinking is necessary for success as a student and in your future career. Learning how to apply critical-thinking skills now will help you develop the ability to use them to handle challenges throughout your life. Recall a problem you needed to solve that was important to your success at school or work. Write several paragraphs to describe the problem and explain how you applied critical-thinking skills to arrive at a solution. Summarize the pros and cons of your choice.

Speaking. Demonstrating leadership qualities is a way to make a helpful contribution to a team. Identify leadership characteristics you believe all team members should have. Create a graphic organizer to present your ideas visually. Develop a short oral presentation that focuses on the use of your leadership graphic to explain the topic. As you are presenting, adjust your presentation length to fit the attention of the audience.

Listening. Practice active-listening skills while listening to your teacher present a lesson. Focus on the message and monitor it for understanding. Were there any barriers to effective listening? How did you use prior experiences to help you understand what was being said? Evaluate your teacher's point of view and use of material in the presentation.

Portfolio Development

College and Career Readiness

Hard and Soft Skills. Employers reviewing candidates for various positions and colleges are always looking for qualified applicants. When listing your qualifications, illustrate both hard and soft skills. For example, you might discuss programming languages you know or the computer platforms in which you are versed. These abilities are often called *hard skills.* The ability to effectively communicate, get along with customers or coworkers, and solve problems are examples of *soft skills.* These are also important skills for many jobs. Make an effort to learn about and develop the hard and soft skills needed for your chosen career field.

1. Conduct research about hard and soft skills and their value in helping people succeed.
2. Create a Microsoft Word document and list the hard skills you possess that are important for a job or career that interests you. Use the heading Hard Skills and your name. Next to each skill, write a paragraph that describes the skill and give examples to illustrate it. Save the document.
3. Create a Microsoft Word document and list the soft skills you possess that are important for a job or career that interests you. Use the heading Soft Skills and your name. Next to each skill, write a paragraph that describes the skill and give examples to illustrate it. Save the document.
4. Update your master spreadsheet.

CTSOs

Job Interview. Job interviewing is an event you might enter with your CTSO. By participating in the job interview, you will be able to showcase your presentation skills, communication talents, and ability to actively listen to the questions asked by the interviewers. For this event, you will be expected to write a letter of application, create a résumé, and complete a job application. You will also be interviewed by an individual or panel. To prepare for a job interview event, complete the following activities.

1. Use the Internet or textbooks to research the job application process and interviewing techniques.
2. Write your letter of application and résumé, and complete the application form (if provided for this event). You may be required to submit this before the event or present the information at the event.
3. Make certain that each piece of communication is complete and free of errors.
4. Solicit feedback from your peers, instructor, and parents.

Graphical User Interface

Sections

14.1 JavaFX Graphical User Interfaces

14.2 GUI Input

14.3 JavaFX Games

As you learned in Chapter 2, a user interface is the collection of outputs and inputs in a computer program that provide communication between the application and the user. A command-line interface is text only and delivers procedural processing. The program requests input. The user provides it. The program processes the input and provides output. The interactions are sequenced by the program. However, humans prefer visual interactions. The graphical user interface provides this, and has become the preferred user interface.

Effective user experience and interface design are so important that entire college majors are devoted to these elements. Some of the important aspects include: know the user; understand the principles of good interface and screen design; select proper interaction devices and controls; write clear messages; provide effective feedback; create meaningful graphics, icons, and images; choose proper colors; organize the screen; and test heartily. While these seem like obvious considerations, not every graphical user interface meets these criteria. This chapter explores the design and effectiveness of good graphical user interfaces created with JavaFX that provide interactions and responses.

College and Career Readiness

Reading Prep

Special features focus on topics of interest related to the material presented in the chapter. Before reading this chapter, preview the Math and Java, Science and Java, and Language Arts and Java special features so you can relate the information to the main text as you read the content.

While studying, look for the activity icon for:

- Vocabulary terms with e-flash cards and matching activities.
- Starter files for hands-on examples and other exercises.

These activities can be accessed at
www.g-wlearning.com/informationtechnology/1773

Chapter Glossary

asset: External file used for multimedia.

avatar: Visual representation of a player character.

button: Control used to trigger an action when the user clicks the control.

check box: Control presented in a group in which all, some, or none of the controls can be selected.

event: Object created as a result of user action.

event handler: Code executed when an event is generated.

focus: Indicates which control is active to accept input.

graphical user interface (GUI): Screen displays visual options with which the user can interact.

horizontal box layout: Provides a side-by-side arrangement of the controls.

image control: Holds the reference to an image file as an Image object.

image view control: Displays an Image object.

Lambda expression: Shortcut from the full event handler statement.

label: Noneditable text used to provide information to the user.

password field: Text field that has a special property to hide the characters entered.

radio button: Control presented in a group in which only one control can be selected.

uniform resource locator (URL): Address that points to a specific document or other resource on a computer network.

user experience (UX): Feeling users get from interacting with a website or computer application.

vertical box layout: Aligns controls from top to bottom.

JavaFX Graphical User Interfaces

You can apply what you learned about JavaFX graphics to create scenes for a graphical user interface. A stage is a screen containing all the objects to be displayed. A group is a collection of these objects that make up the scene to be displayed. JavaFX components can be added to provide interactions. Each component serves a specific purpose in input or output. Proper use of these components is paramount in effective graphical user interface designs.

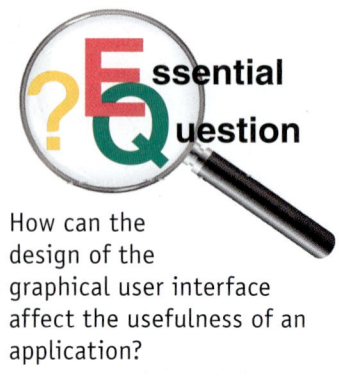

How can the design of the graphical user interface affect the usefulness of an application?

Learning Goals

- Evaluate a graphical user interface for effectiveness of the user experience.
- Identify the steps for including control objects in a graphical user interface.

Terms

button

event

event handler

graphical user interface (GUI)

label

Lambda expression

user experience (UX)

Effective GUI Applications

In Chapter 7, graphics and text were developed and colors changed in a scene, but no interactions were used. In essence, you were only at a graphical program stage. In a *graphical user interface (GUI),* the screen displays visual options with which the user can interact. These options include graphics, buttons, text input, display text, and many more.

Industry standards for GUI applications are informal, but many practitioners have published their own standards. Notable in the field of user experience and usability is the Nielsen Norman Group. Jakob Nielson and Don Norman have led the user experience field since 1998. An overarching concern for Nielson and Norman is that the user should not have to stop and figure out what to do next in a user interface. "Minimize learning" is a major guideline and helps with the user experience.

User experience (UX) is the feeling users get from interacting with a website or computer application. UX designers feel that if users do not like interacting with an interface, they will take their business elsewhere. To promote a good feeling from the interface, UX designers make the interface clear, easy to learn, attractive, and able to rapidly produce meaningful results.

Design for the User

Early in user interface design, many developers produced what they thought was cool. Often, the background was black and text was blue. It was common to find graphics that had little to do with the application or website. The color combination resulted in text that was difficult to read. The graphics, while perhaps fun, did not communicate a message to the user. Today, designers study which text colors users find easy to read and which graphics produce the most meaning.

Compare the two GUIs shown in **Figure 14-1.** The image on the left is considered to be a poor example of UX. The image on the right is better. Think about why this is so.

Notice first the color combinations in each image. The GUI on the left has a harsh contrast between the blue background and the red buttons. Additionally, the deep blue of the title provides too little contrast with the black background, making the text difficult to read. The menu bar is confusing because the label Menu looks like a button rather than the title of the three items below it. Should the user click Menu? Compare the colors orange and red on the menu items. This combination is a problem for people with vision-color deficiencies. On the other hand, the GUI on the right is bright and each section of the screen is clear. Contrast between elements and background colors is high. The colors used in the GUI complement one another.

Next, compare the graphics. The GUI on the left has a photograph of a bird of paradise flower. It is hard to figure out why that is there. This is a lovely plant, but it provides no information about the website or its products. Graphics, information, or other elements that are not related to the message or topic are called *gratuitous.* It is best to avoid them. The graphics in the GUI on the right are well placed and inviting. They also are informative and provide an indication of the website's purpose.

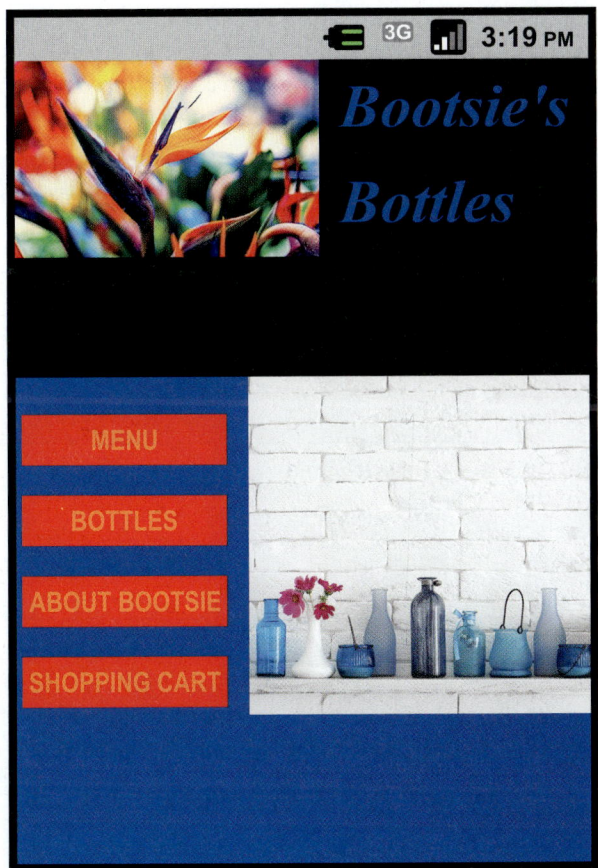
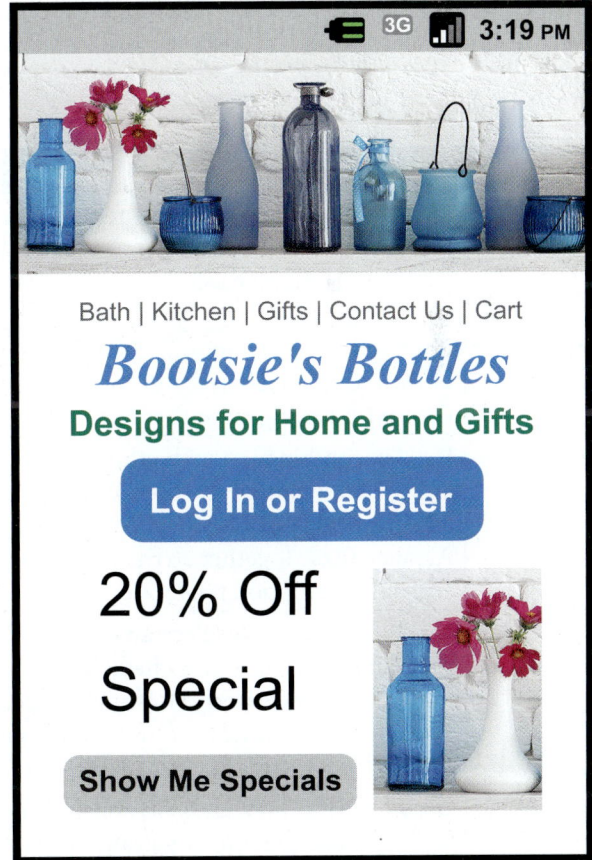

Goodheart-Willcox Publisher; photos: warasit phothisuk, Alena Ozerova/Shutterstock.com

Figure 14-1. The GUI on the left is hard to read and does not have a clear message. The GUI on the right is attractive, easy to view, and clear to the user how to use it.

Regarding the interactions, it is clear what to click in the GUI on the right, but not so much in the GUI on the left. Designing for the user means displaying what the user wants to see. Surely, a 20 percent-off sale is in the best interest of the user.

As you use software and websites, pay attention to your user experience. Take note of what works well for you, and consider if all users would feel the same way. Also note what does *not* work well. Take your observations and apply them to your own GUI designs.

Design the controls on the stage with the user in mind. Apply what you learned in the drawing chapter. Use graph paper to lay out the scene and document your design. Locate the controls where a user would likely look for them. Clearly label all controls. Once you design the layout of a scene, ask a potential user to review the design. In UI design, the guideline is, "All of us know more than any of us." The more users you can involve in the design and testing of your program, the better the user experiences will be.

Elements of GUI Applications

JavaFX provides a large number of objects to enhance the user experience. A good designer will select the best objects for each interface. The API for JavaFX GUI is located at the Oracle Docs website. Search the Internet using the string Oracle Docs JavaFX UI controls to locate the full API. Some of the more common UI controls are listed in **Figure 14-2.** This figure also describes the best practices for the implementation of these controls.

Using JavaFX UI Controls

The programs designed and coded in this text so far have been of the procedural type. The sequence of action was determined by the program. In a GUI, the flow of a program is determined by the user interacting with a selection of controls

UI Control	Best Practice	Example
Label	Used to display noneditable text or an image to provide information to the user. Do not make it look like it is clickable or changeable.	You rolled a 6.
Button	Used to trigger an action when the user clicks the button. Can be labeled with text and an image. Make the text meaningful to the user.	Start Game
Radio Button	Used in combination with other radio buttons to provide a single selection for the user. When one button in the group is selected, the others automatically are deselected. Arrange these together so the user sees they are a group and only one option is allowed.	New Game / Load Saved Game / Quit
Check Box	Used to select or deselect an option. May be used in combination with other check boxes. One or more or none may be selected. These are best used to toggle a setting. Useful to select or deselect program or game options.	✓ Advanced Play / Assign Avatar / ✓ Display Inventory
Text Field	Used for data entry. User may change the contents. Useful for forms completion, or getting a player's name. Do *not* use these for displaying text; use a label instead.	Player Name:

Figure 14-2. These are some of the more common UI controls in JavaFX.

presented on the stage. When the user interacts with a control, the control generates an event. An *event* is an object created as a result of user action. When a user interacts with a button, radio button, or check box control, an ActionEvent is generated. The programmer writes an event handler. An *event handler* is code executed when the event is generated. The event handler code responds to the user's action. For example, an event could be the user clicking a button to start a game. The event handler for the button would contain the code to start the game. The sequence for using a JavaFX control is similar for each control:

1. Import the JavaFX control class.
2. Define an object of the JavaFX control class.
3. Set the properties for the JavaFX control object.
4. Add the object to the list of nodes in the scene.
5. Write an event handler.

Button UI Control

A *button* is a control used to trigger an action when the user clicks the control. Apply the above sequence to program a button. The first step to use JavaFX controls is to import the control package. Use the asterisk (*) as a wild card to add all control classes to your application. The compiler will select the ones it needs based on your usage of the controls.

```
import javafx.scene.control.*;
```

The next step is to define a JavaFX control object. A button has three possible constructors, shown in **Figure 14-3.** You can create one with or without its label and optionally add an image. The following line of code defines a start button for a game.

```
Button btnStart = new Button( "Start Game" );
```

A good coding practice is to name controls with a prefix based on the type of control. This provides internal documentation and avoids careless errors such as changing the properties of the wrong control. Use btn for a button, lbl for a label, rb for a radio button, and cb for a check box.

Next, set the properties of the JavaFX control. All controls inherit from the Node class in the javafx.scene package. Locating the button on the stage is done using the setLayoutX( ) and setLayoutY( ) methods in the Node class:

```
btnStart.setLayoutX( 120 );
btnStart.setLayoutY( 140 );
```

Keep the UX in mind when locating the button. Place the button where an average user would expect to see it. For example, in a game there are some pregame activities, such as entering the player's name. Locate the button to start the game where it makes sense to find it after these activities are concluded.

After the control has been located, add it to the list of nodes in the scene:

```
root.getChildren( ).add( btnStart );
```

The control will not be visible without adding it to the list of nodes.

> **FYI**
>
> It is a good coding practice to name controls with a prefix indicating the type of control.

Constructor	Explanation
Button( )	Creates a button with an empty string for its label.
Button(String text)	Creates a button using the text for its label.
Button(String text, Node image)	Creates a button with the text as its label and the image as an icon.

Figure 14-3. These are the three possible constructors for a button control.

Language Arts and Java

Effective GUI Design

Mobile advertising is a very large industry. Advertising agencies make a lot of money developing pop-up ads for apps. People who make apps depend on revenue earned by selling links to their apps to mobile ad agencies. Companies depend on revenue earned from selling their products online. If you use apps, you have seen these ads.

Alex Ruhl/Shutterstock.com

Often, while using an app, a user is presented with a pop-up ad. To keep the user from closing the ad without reading it, the ad must be compelling in some way. Once the user stops to look at the ad, the ad must present a positive user experience, or the user closes the pop-up without engaging and ad revenue is lost. Think about the ads that pop up while you are using apps. What about them keeps you looking and not closing? Use these thoughts to perform this activity.

Assignment

You are the communications director for an advertising agency. Your team is tasked with creating a proposal for an ad campaign for a new product. As director, your leadership is critical in providing a design for the interface of a pop-up ad for a social media page. Using some of the controls listed in **Figure 14-2** or on the Oracle Java UI Controls web pages, create a GUI design for a pop-up ad for Bea's Bees Organic Local Honey. Key items to include in the campaign are listed in the request for proposal issued by Bea's Bees:

- Realistic instead of cartoonish
- Image of bees
- Image of honey
- Incorporate Bea's Bees logo
- Organic-looking typeface for a description of the products: Raspberry Honey, Wild Field Honey, and Bea's Best Honey
- Clean, crisp page with good color combinations
- Buttons to click to find out about honey, Bea's Bees company, and buying Bea's Bees honey
- Label and text field to capture e-mail addresses
- A close image (**X**) to return to the previous screen

Write a paragraph to describe your vision for the design for the pop-up ad. Then, draft the design in a word-processing document. Alternatively, you can draft the design on graph paper. Insert images, shapes, icons, 3D models, and other elements as needed to illustrate the design. Show the controls your team will program as they create this social media advertisement page.

Keep in mind the user's attributes and how they will interact with the page. The customer is expecting a great deal of information to be communicated on this page. Plan to make it easy for the user to figure out what to do.

Once you have made the design, share it with someone else to see if it is easy for them to figure out what to do. Make final updates and submit it to your instructor.

The final step is to write the event handler. Use the setOnAction( ) method for the button with the parameter event and an arrow constructed out of a dash and the right-angle bracket. This instruction is called a *Lambda expression* and is a shortcut from the full event-handler statement. You will learn more about this as you take further courses in computer programming.

```
btnStart.setOnAction(
   event -> {
      // statements to execute when button is clicked
   }
); // end btnStart event handler
```

Label UI Control

A *label* is noneditable text used to provide information to the user. As you progress through this explanation on creating a label control, pay attention to the similarity in the process of creating a button control. Use these commonalities to generalize and, therefore, simplify the process in your mind.

The first step is to define a JavaFX control object. Like buttons, the label control has three constructors, as shown in **Figure 14-4.** You can create a label with or without its text and optionally add an image. The following line of code defines a label to inform the player how to begin the game.

FYI

There are image view controls that can be used to add images.

```
Label lblStart = new Label( "When all options are selected, press "
+ "Start to begin the game." );
```

The next step is to set the properties of the JavaFX control. Locate the label on the stage using the setLayoutX( ) and setLayoutY( ) methods.

```
lblStart.setLayoutX( 100 );
lblStart.setLayoutY( 60 );
```

For the best UX, place the label where an average user would expect to see it. The label control inherits from the javax.scene.control.Labeled class. Therefore, it has methods to set the typeface (font) and color.

```
lblStart.setFont( Font.font( "Tahoma", 20 ) );
lblStart.setTextFill( Color.BLUE );
```

After the properties have been set, add the control to the list of nodes in the scene.

```
root.getChildren( ).add( lblStart );
```

Constructor	Explanation
Label( )	Creates an empty label.
Label(String text)	Creates a label with the text.
Label(String text, Node image)	Creates a label with the text and an image.

Goodheart-Willcox Publisher

Figure 14-4. These are the three possible constructors for a label control.

The control must be added to the list in order for it to appear and be functional.

Usually, the final step is to write the event handler. However, the user cannot interact with a label, and there is no event.

HANDS-ON EXAMPLE 14.1.1

Displaying a GUI Application ➦

In this exercise, you will program a button to make a cat drawing appear on the stage. This exercise is based on the JavaFXCatSnippet.java you built in the drawing chapter. Before beginning this exercise, download the chapter files from the student companion website. The JavaFXGUISnippet.java file will be used as a starting point.

1. Launch jGRASP, and open the JavaFXGUISnippet.java file.

2. Compile and run the application. The output is the cat drawing. This snippet is based on the one created in the drawing chapter. Close the graphics window.

3. Examine the code on lines 13–16. These are the import statements required for event handling.

```
// add GUI elements
import javafx.event.ActionEvent;
import javafx.event.EventHandler;
import javafx.scene.control.*;
```

4. Locate comment 5, and examine the code on lines 61–72. This is where the individual shapes of the cat are collected into a Group named cat, which allows properties and methods to be applied to the cat as a whole. Note that the individual shapes are no longer added to the scene. The cat is added as a group node.

```
Group cat = new Group( );
cat.getChildren( ).add( head );
cat.getChildren( ).add( body );
cat.getChildren( ).add( tail );
cat.getChildren( ).add( ear1 );
cat.getChildren( ).add( ear2 );

/***** 5. Add the shapes to the group. */
// get the list of current nodes and add the cat group
ObservableList<Node> effects = root.getChildren( );
effects.add( cat );
```

5. Locate comment 6 at the end of the cat group. Set the visible property to false to hide the cat. With this setting, the cat will not be visible when the application begins. The user will click a button to make the cat visible. Without the Group statement, you would have to set the visible property for each shape. Setting the visibility for a group of objects is much more efficient code.

```
/***** 6. Hide the cat. */
cat.setVisible( false );
```

6. Locate comment 7, and define the button. Assign the label See Cat to the button, and locate the button at coordinates (120,100);

```
/***** 7. NEW GUI: Button. */
Button btnSeeCat = new Button( "See Cat" );
btnSeeCat.setLayoutX( 120 );
btnSeeCat.setLayoutY( 100 );
```

7. Locate comment 8, and add the button to the scene.

```
/***** 8. Add the button to the scene. */
effects.add( btnSeeCat );
```

8. Compile and run the application. You should see the button, but no cat, as shown. Click the button. Nothing happens because the event handler has not been programmed. Close the graphics window. Note: clicking the button results in the button briefly changing colors to indicate it was clicked and released. That is a native part of the button control. The programmer does not need to do anything to make that happen.

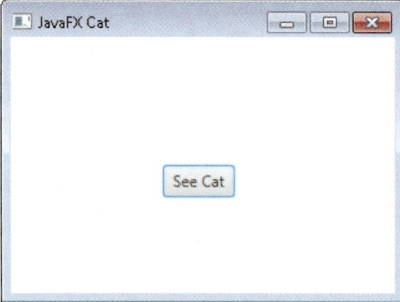

Goodheart-Willcox Publisher

9. When the user clicks the button, the cat should appear. For this example, the button should also disappear. Locate comment 9, and create the event handler for the button click.

```
/***** 9. Create the Button event handler. */
btnSeeCat.setOnAction(
    event -> {

        cat.setVisible( true );
        btnSeeCat.setVisible( false );

    }
); // end btnSeeCat event handler
```

10. Compile and run the application. Click the button to see the cat appear and the button disappear. You have created your first JavaFX control!

Try It!

Continue working in the JavaFXGUISnippet.java file. Create a label with text A JavaFX Shapes Cat. Apply what you learned about text in the drawing chapter. Use an appropriate typeface (font) and color. Import the Text class. Place the label in an appropriate location, say, right after the button is added to the scene. Make the label visible at the start of the application and invisible when the button is clicked.

SECTION REVIEW 14.1

Check Your Understanding

1. The acronym GUI stands for _____.
2. What is the user experience in relation to computer applications?
3. In Java, what is an event?
4. Which code is executed when the event is generated?
5. Which Java control is used to provide information, but is not interactive?

Build Your Vocabulary

As you progress through this course, develop a personal computer science glossary. This will help you build your vocabulary and prepare you for a career. Write a definition for each of the following terms and add it to your computer science glossary.

button	graphical user interface (GUI)	Lambda expression
event	label	user experience (UX)
event handler		

GUI Input

There are many uses for input in a GUI application. Providing personal information in an online form, entering a password, making selections for a search, choosing options for a game, and other situations require input from the user. When choices from a set of options are required, radio buttons and check boxes can be used. When unknown input such as usernames are required, text fields can be used. A great deal of thought about the user's needs must be done before selecting the appropriate options for input. Critical to the decision is both the needs of the user and the clarity with which you provide the input selection.

Essential Question

Which is a better method of getting user input, text fields or radio buttons?

Learning Goals

- Describe how to accept input from a text control.
- Compare and contrast radio button and check box controls.
- Explain how to incorporate multimedia into a JavaFX application.

Terms

asset	image view control
check box	password field
focus	radio button
horizontal box layout	uniform resource locator (URL)
image control	vertical box layout

Text Fields and Layout Controls

A common use of a text field is the log-in screen for an application. This is often a first line of defense to protect computer systems and databases from unauthorized use and tampering. Generally, a username and password are requested. Four JavaFX controls are required for this data entry. A label control tells the user what to do. A text control accepts the input. A password control hides the input from onlookers. A button provides the means to execute the log-in. Additionally, once text has been entered, it is accessible through code. Action can be taken on the text.

Log-in Screen

Suppose you are providing a login to a game. This is an example of using JavaFX controls to gather game information. One label is used to tell the user what input is required. Two input fields are used. One is for the player's name. The second input field is a *password field*, which is a text field that has a special property to hide the characters entered. To organize these controls on the screen, a layout is created. A *horizontal box layout* provides a side-by-side arrangement of the controls. A *vertical box layout* aligns controls from top to bottom. These layouts are groups. Apply what you know about groups to layouts. The controls added to the layouts are their children. The code for creating these controls is given in **Figure 14-5.**

Control	Explanation
TextField txtPlayerName = new TextField( );	Creates a text field control for input.
PasswordField pwdPassword = new PasswordField( ); pwdPassword.setPromptText("Enter password");	Creates a password field with a prompt displayed inside the input area.
HBox hbLogin = new HBox( ); hbLogin.setSpacing(10);	Creates a horizontal layout group with ten pixels between each control.
VBox vbLogin = new VBox( );	Creates a vertical layout group.

Figure 14-5. This code can be used to create controls for a log-in screen.

Use of these controls follows the pattern discussed in the last section: define the control, set its layout properties, and add it to the scene. In the case of the horizontal and vertical layout boxes, use the same code as is used when making a group to add the controls to the layout. The layout boxes group the controls so that setting the coordinates is required only for the box itself. To find out which other properties and methods these controls inherit, search the Internet using the string Oracle Docs JavaFX with the name of the control, such as Oracle Docs JavaFX HBox.

HANDS-ON EXAMPLE 14.2.1

Creating a Log-in Feature

In this exercise, you will create a log-in screen for a video game. A log-in screen requires multiple controls. You will group the controls into one layout. Before beginning this exercise, download the chapter files from the student companion website. The LoginScreenSnippet.java file will be used as a starting point.

1. Launch a web browser, navigate to a search engine, and search for Oracle Docs JavaFX VBox. Gather information from the Oracle Docs website about the VBox control. Keep this page open while you complete the following lines in the code.

2. Launch jGRASP, and open the LoginScreenSnippet.java file.

3. Locate comment 1, and follow the instructions to add the name-entry and password-entry controls: label to prompt for the name, text field for input of the name, and password field including a prompt.

```
/***** 1. Create password entry controls. */
Label lblPlayer = new Label( "Player Name:" );
TextField txtPlayer = new TextField( );
PasswordField pwdPassword = new PasswordField( );
pwdPassword.setPromptText( "Enter password" );
```

4. Locate comment 2, and follow the instructions to add a vertical layout control. Add the password-entry controls to the vertical layout control, set the spacing between lines to 10 pixels, locate the box at coordinates (50, 50), and add the vertical layout control to the scene. Notice that effects has been defined as the list of nodes in the application on line 27.

```
/***** 2. Add layout control for login controls. */
VBox vbLogin = new VBox( );
vbLogin.getChildren( ).addAll( lblPlayer, txtPlayer, pwdPassword );
vbLogin.setSpacing( 10 );
vbLogin.setLayoutX( 50 );
vbLogin.setLayoutY( 50 );
// add box to the scene
effects.add( vbLogin );
```

5. Locate comment 3, and follow the instructions to create the welcome message. After the user completes the login, the welcome message will be displayed.

```
/***** 3. Create Welcome label. */
Label lblWelcome = new Label( "Race to the Top\n  GET READY!" );
lblWelcome.setVisible( false ); // hide until after login
lblWelcome.setLayoutX( 40 );
lblWelcome.setLayoutY( 50 );
lblWelcome.setFont( Font.font( "arial", FontWeight.BOLD, 30 ) );
lblWelcome.setTextFill( Color.ORANGE );

// add label to scene
effects.add( lblWelcome );
```

6. Locate comment 4, and follow the instructions to create the log-in button for the user to click after filling in the fields. Notice how similar the statements are to those used for the label created under comment 3.

```
/***** 4. Create Log-in Button. */
Button btnLogin = new Button( "Log In" );
btnLogin.setLayoutX( 120 );
btnLogin.setLayoutY( 150 );
btnLogin.setVisible( true );

effects.add( btnLogin );
```

7. Locate comment 5, and follow the instructions to create the event handler for the button, hide the login controls, and show the welcome message. Note: for this exercise, there is no validation to be sure the user correctly entered the log-in information.

```
/***** 5. Create the Button event handler. */
btnLogin.setOnAction(
    event -> {

        // Hide login controls
        vbLogin.setVisible( false );
        btnLogin.setVisible( false );
        // Show Welcome
        lblWelcome.setVisible( true );

    }
); // end Button event handler
```

8. Compile and run the application. The user should first be presented with a screen to enter log-in information, and then when the button is clicked, the welcome message should be displayed, as shown. You have created a log-in feature for a video game! Note: no input validation is coded at this time. Any input will be accepted.

 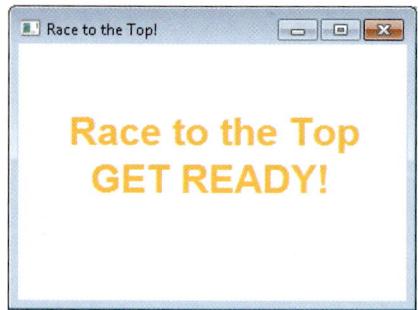

Goodheart-Willcox Publisher

Try It!

Continue working in the LoginScreenSnippet.java file. Rename the class to LoginScreenHorizontal, and save the file under the new name. Edit the scene size to 600 × 200 pixels. Change the layout to a horizontal box. Research the horizontal-box layout control as needed to update the code. Compile and run the application. Examine the layout to see if it looks balanced. Make any adjustments to the layout coordinates to present a balanced scene. Think about horizontal and vertical layouts. Which layout do you feel better suits the user's needs?

Text Input from a Text Field

Each of the controls in the LoginScreenSnippet file created in Hands-On Example 14.2.1 contains text. A coder may now use that information. For example, suppose there is a user file that contains the player's name and password. You could apply what you learned about text files to open that file. Remember to use the try/catch statement to ensure the file is available to the application. Then, you could apply what you learned about Strings to search the file for a match to the username and password.

In the LoginScreenSnippet file, the txtPlayer and pwdPassword fields hold what was entered in the snippet. The text property is what needs to be accessed. The getText accessor method performs that function. The code below accesses the text entered in the text field and password field defined in the LoginScreenSnippet file.

```
String playerName = txtPlayer.getText( );
String playerPassword = pwdPassword.getText( );
```

Using Methods to Simplify JavaFX UI Code

It is now obvious the JavaFX UI controls require many lines of code to set up. Also, the code is very similar for each control. It is possible to remove this clutter from the start method. To do this, define the control before the start method as an instance variable. The scope of the variable must be the entire program because it is used in the new method and in the start method. Then, apply what you learned about methods in a class to create a new method. Put the setup code in the new method, and call the method from the start method. Name the method something meaningful such as setupGetPlayerName( ). This code demonstrates the action:

```
/***** Define player name TextField before the start method. */
private TextField txtPlayerName;

/***** Call the method in the start method where this code would
   be executed. */
   setupGetPlayerName( );

/***** Write custom method to set up properties to get the
   player's name. */
public void setupGetPlayerName( ) {

   txtPlayerName = new TextField( );
   txtPlayerName.setLayoutX( 40 );
   txtPlayerName.setLayoutY( 50 );
   txtPlayerName.setFont
      ( Font.font( "arial", FontWeight.BOLD, 30 ) );

   // add TextField to scene
   effects.add( txtPlayerName);   // effects needs to be an instance
                                  // variable too
}
```

This technique reduces the number of lines of code in the start method to one instead of five. The readability of the start method is increased. You will learn how to do this in the next Hands-On Example.

Radio Buttons and Check Boxes

Sometimes there is a set number of known options for input. When this is the case, the options should be presented to the user as a group for selection.

Two JavaFX UI controls that provide selection from options are radio buttons and check boxes.

Radio buttons and check boxes can be focused. The *focus* indicates which control is active to accept input. It is usually indicated by a blue outline around the control. The user can press the [Tab] key to change focus between input controls. Focus can also be changed with a touch or mouse click. **Figure 14-6** shows the API for the setFocused method.

As you work through the following material, make note of the similarities among methods for UI controls. Setting layout, adding to the scene or layout control, and setting text are all remarkably similar for all controls.

Radio Button UI Control

A *radio button* is a control presented in a group in which only one control can be selected. Use radio buttons when only one selection in a set of options is allowed. The options must be mutually exclusive, meaning the user can choose only one option at a time. For example, in a welcome screen for a game, several options are mutually exclusive: play, set options, and quit.

Figure 14-7 shows the API for several actions with radio buttons. Radio buttons must be added into a ToggleGroup object for the exclusive selection to work. Group the radio buttons together in a vertical layout for easy comparison of the options.

Figure 14-8 shows a grouping of a label and radio buttons for the exclusive selection. The code to accomplish this set of options is below.

```
// Game Options Radio Buttons
Label lblActions = new Label( "Select an Action" );
RadioButton rbPlayGame = new RadioButton( "Play Game" );
RadioButton rbOptions = new RadioButton( "Options" );
RadioButton rbQuit = new RadioButton( "Quit" );
```

(Continued)

Method	Explanation
void setFocused(boolean value)	Sets the value of the property focused to true or false.

Goodheart-Willcox Publisher

Figure 14-6. This is the API for the setFocused method, which is used to indicate which control is active to accept input.

Code	Explanation
RadioButton rbOption1 = new RadioButton( );	Defines a radio button with no label.
RadioButton rb2Option2 = new RadioButton("Single Player");	Defines a radio button with a label.
rb2Option2.setSelected(true);	Shows the radio button as selected; used to indicate a default setting.
boolean isSelected( )	Method that gets the value of the property selected. A value of true means the radio button is selected; false means not selected.
ToggleGroup( ) toggleGroup = new ToggleGroup( )	Instantiates a ToggleGroup named toggleGroup.
rbOption1.setToggleGroup(toggleGroup) rbOption2.setToggleGroup(toggleGroup)	Adds two radio buttons to the ToggleGroup named toggleGroup.

Goodheart-Willcox Publisher

Figure 14-7. This is the API for radio buttons.

```
// create toggle group
ToggleGroup toggleGroup = new ToggleGroup( );
rbPlayGame.setToggleGroup( toggleGroup );
rbPlayGame.setSelected( true );
rbOptions.setToggleGroup( toggleGroup );
rbQuit.setToggleGroup( toggleGroup );

// align controls in vertical arrangement
VBox vbGameOptions = new VBox( );
vbGameOptions.getChildren( ).addAll( lblActions,
   rbPlayGame, rbOptions, rbQuit );
vbGameOptions.setSpacing( 10 );
vbGameOptions.setLayoutX( 30 );
vbGameOptions.setLayoutY( 20 );

// add box to the scene
root.getChildren( ).add( vbGameOptions );
```

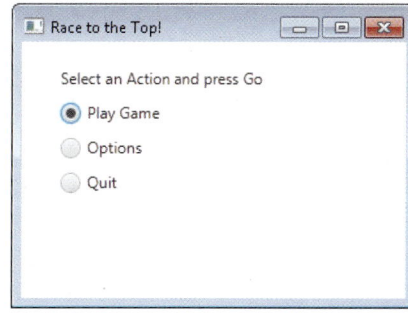

Goodheart-Willcox Publisher

Figure 14-8. This is an example of using radio buttons. Notice the blue outline is visible on **Play Game**. This indicates the focus.

Check Box UI Control

A *check box* is a control presented in a layout group in which all, some, or none of the controls can be selected. When one or a set of options are independent and may be selected or not, use the check box UI control. A check box can be used to turn a single option on or off. **Figure 14-9** shows the API for several actions with check boxes.

If more than one check box is used, group them in a layout control. **Figure 14-10** shows the vertical grouping of a label and check boxes for selecting more than one option. The code to accomplish this set of options is below.

```
// check boxes
Label lblGameOptions = new Label( "Select or Deselect Options" );
CheckBox cbAdvPlay = new CheckBox( "Advanced Play" );
CheckBox cbAvatar = new CheckBox( "Assign Avatar" );
CheckBox cbInventory = new CheckBox( "Display Inventory" );
cbInventory.setSelected( true );

VBox vbGameOptions = new VBox( );
vbGameOptions.getChildren( ).addAll( lblGameOptions, cbAdvPlay,
   cbAvatar, cbInventory );
vbGameOptions.setSpacing( 10 );
vbGameOptions.setLayoutX( 30 );
vbGameOptions.setLayoutY( 20 );
// add box to the scene
root.getChildren( ).add( vbGameOptions );
```

Code	Explanation
CheckBox cb1 = new CheckBox( );	Defines a check box without text.
CheckBox cb2 = new CheckBox("Fast Play");	Defines a check box with text.
cb1.setText("Option 1");	Sets the text for a check box.
cb1.setSelected(true);	Shows check box as selected; used to indicate a default setting.
boolean isSelected( )	Method that gets the value of the property selected. A return value of true means the check box is selected; false means not selected.

Goodheart-Willcox Publisher

Figure 14-9. This is the API for check boxes.

Math and Java

Degrees and Radians

The size of angles is measured using one of two units. Most likely, the first measure you learned was that a right angle has 90 degrees. That followed with 360 degrees in a circle. There is a second measure for angles that is defined using the radius of a circle. One radian is the number of degrees in the arc of a circle where the arc is the same length as the radius. The circumference of a circle is $2\pi r$, so there are 2π radians in a circle. Degrees and radians represent the same quantity of angle size.

Radians are used as arguments for most trigonometric and other mathematical methods in programming. At times, it is advantageous to be able to convert between the two measurement units. Writing an equation for these two measures allows conversion formulas to be developed.

$$360 \text{ degrees} = 2\pi \text{ radians}$$

Simplify by dividing both sides by 2:

$$180 \text{ degrees} = \pi \text{ radians}$$

$90° = \dfrac{3\pi}{6}$ Radians $= \dfrac{\pi}{2}$ Radians

$60° = \dfrac{2\pi}{6}$ Radians $= \dfrac{\pi}{3}$ Radians

$30° = \dfrac{\pi}{6}$ Radians

$0° = 0$ Radians

$\dfrac{\pi}{6}$

$\overline{OA} = 1$

Attaphong/Shutterstock.com

Therefore:

	Degrees to Radians	Radians to Degrees
Equation	180 degrees = π radians	π radians = 180 degrees
Explanation	Divide both sides by 180. $1 \text{ degree} = \dfrac{\pi \text{ radians}}{180}$	Divide both sides by π radians. $1 \text{ radian} = \dfrac{180°}{\pi \text{ radians}}$
	Each degree is $\pi \div 180$ radians. Multiply the number of degrees by $\pi \div 180$ to find the number of radians.	Each radian is $180 \div \pi$ degrees. Multiply the number of radians by $180 \div \pi$ to find the number of degrees.
Conversion	radians = degrees × π ÷ 180	degrees = radians × 180 ÷ π

Assignment

Before beginning this activity, download the chapter files from the student companion website. The DegreesAndRadiansSnippet.java file will be used as a starting point of this activity. Write a JavaFX application that uses radio button controls to determine which conversion is taking place, has a text field control to accept a number of degrees or radians, provides a label to prompt the user to enter an angle measure, and has a button control to perform the appropriate conversion. Output the result in another label. To begin, draw a layout of where each control will be located and what its identifier will be. Then, write the code.

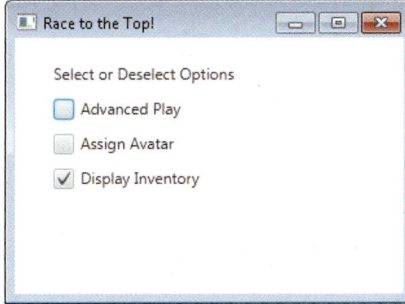

Figure 14-10. This is an example of using check boxes. Notice the blue outline is visible on **Advanced Play**. This indicates the focus.

HANDS-ON EXAMPLE 14.2.2

Creating Radio Buttons and Check Boxes

In this exercise, you will create groups of radio buttons and check boxes. This will create an interface similar to what you may find for starting a video game. Before beginning this exercise, download the chapter files from the student companion website. The InputExamplesSnippet.java file will be used as a starting point. This file is similar to the one created in Hands-On Example 14.2.1, but is slightly modified. If you attempt to run the file, it will compile, but will not execute. It will also throw an exception if you click a button because there are missing items. The Hands-On Example steps will fix this.

Launch jGRASP, and open the InputExamplesSnippet.java file. Review the code to see how it has changed from the last Hands-On Example. Code that performs a single function has been collected to create a class method. To do that, the JavaFX controls are defined, but not instantiated, in the snippet. This definition occurs as instance variables right before the start method. Notice three radio buttons (rb1, rb2, and rb3) and three check boxes (cb1, cb2, and cb3) are defined. This demonstrates the advantage of planning the application first: you know the controls required. Refer to Chapter 12 if you need to review creating custom classes and methods.

```
public class InputExamplesSnippet extends Application {

// define all controls here - private to the class
// but available to all methods within the class
private Scene scene;
private Circle head;
private Ellipse body;
private Arc tail;
private Path ear1, ear2;
private Group cat;
private ObservableList<Node> effects;
private Label lblGameOptions, lblPlayerName;
private TextField txtPlayerName;
private VBox vboxRB, vboxCB;
private HBox hboxPlayer;
private RadioButton rb1, rb2, rb3;
private CheckBox cb1, cb2, cb3;
private Button btnSeeCat;
```

The code for instantiation and setting properties is collected in a method after the end of the start method. This helps clean up the code since drawing the cat used many lines of code and took up much of the visual space of the beginning of this application.

```
public void drawCat( ) {
/***** Draw head: black circle, center (150, 50), radius 20. */
    head = new Circle( 150, 50, 20 );
    head.setFill( Color.BLACK );
```

(Continued)

```
        cat.setVisible( false );
        effects.add( cat );
} // end draw cat
```

The call to this code is placed in the start method.

```
// instantiate FX controls and set properties in methods
drawCat( ); // draw cat and hide it
```

Examine the rest of the file to see how the method creation organizes the file. Notice that the method calls make it easier to see the major steps in the code. When the user clicks the button, all the action that is within the button handler takes place.

```
// instantiate FX controls and set properties in methods
drawCat( );                // draw cat and hide it
setupOptionsLabel( );      // Show Game Options Label
setupGetPlayerName( );     // Show label and text field
setupRadioButtons( );      // Show new/load/quit game options
setupCheckBoxes( );        // Show advance/avatar/inventory options

setupButton( );            // Show button to accept input, show cat

// create the Button event handler
btnSeeCat.setOnAction(
    event -> {

        hideOptionsControls( );
        cat.setVisible( true );

        /***** 3. Show name input and selections. */


        /***** 4. Check for selections. */


    }
); // end btnSeeCat set On Action event handler
```

When you are comfortable with the changes in the file, follow the steps below to complete the application.

1. Locate comment 1, and follow the instructions to instantiate each radio button. Use the radio button constructor that accepts the display text as an argument. The three options that will be presented to the user are New Game, Saved Game, or Quit.

```
public void  setupRadioButtons( ) {

    /***** 1.  Radio buttons */
    rb1 = new RadioButton( "New Game" );
    rb2 = new RadioButton( "Saved Game" );
    rb3 = new RadioButton( "Quit" );
    // create toggle group
    ToggleGroup toggleGroup = new ToggleGroup( );
    rb1.setToggleGroup( toggleGroup );
    rb2.setToggleGroup( toggleGroup );
    rb3.setToggleGroup( toggleGroup );
    rb2.setSelected( true );
```

2. Rather than dealing with the layout for all three buttons individually, put them in a vertical box, and set the layout for just that box. Then, add the box to the scene. Place this code within the method for setting up the radio buttons.

```
    vboxRB = new VBox( );
    vboxRB.getChildren( ).addAll( rb1, rb2 ,rb3 );
    vboxRB.setLayoutX( 20 );
    vboxRB.setLayoutY( 50 );
    vboxRB.setSpacing( 10 );

    effects.add( vboxRB );
} // end radio buttons
```

3. Locate comment 2, and follow the instructions to add the method for the check boxes. Place these in a vertical layout as well.

```java
public void setupCheckBoxes( ) {

    /***** 2. Check boxes */
    cb1 = new CheckBox( "Advanced Play" );
    cb2 = new CheckBox( "Assign Avatar" );
    cb3 = new CheckBox( "Display Inventory" );
    cb3.setSelected( true );

    vboxCB = new VBox( );
    vboxCB.setLayoutX( 150 );
    vboxCB.setSpacing( 10 );
    vboxCB.setLayoutY( 50 );
    vboxCB.getChildren( ).addAll( cb1, cb2, cb3 );

    effects.add( vboxCB );

} // end set up check boxes
```

4. Now you need to add code to the button handler for when the player clicks the button to see the cat. Locate comment 3, and follow the instructions to get the text from the PlayerName field and to set the property of the GameOptions label to "*player name* wins a Cat!"

```java
/***** 3. Show name input and inventory. */
lblGameOptions.setText( txtPlayerName.getText( ) + " has a Cat!" );
```

5. Locate comment 4, and follow the instructions to create a new label control to show which of the radio buttons was selected. Use if-else statements to find the one selected.

```java
/***** 4. Check for and display selections. */
Label rbOption = new Label( );
if ( rb1.isSelected( ) ) {
    rbOption.setText( "New Game" );
} else if ( rb2.isSelected( ) ) {
    rbOption.setText( "Load Saved Game" );
} else {
    rbOption.setText( "Quit" );
}
rbOption.setFont( Font.font( "arial", FontWeight.BOLD, 16 ) );
rbOption.setTextFill( Color.RED );
rbOption.setLayoutX( 10 );
rbOption.setLayoutY( 175 );
effects.add( rbOption );
```

6. Compile and run the application. The output should look as shown. You have added functional radio buttons and check boxes to a JavaFX application!

Goodheart-Willcox Publisher

Try It!

Continue working in the InputExamplesSnippet.java file. Gather information from the check boxes, and create labels to display the text for each check box that is checked. Display this text above in an appropriate location on the stage. You may wish to change the dimensions of the stage to create more space for the check box text.

Multimedia Files

The word *multimedia* indicates that more than one medium is used to convey information. Text, images, sound, and video are all included in the set of media in the multimedia definition. The support files for multimedia are all external to the Java program. Each Java program file is simply text. The picture files for the images, the audio files for sound effects and music, and the video files must be located by the program. If these items cannot be located, a runtime exception is thrown. Managing the media is the responsibility of the programmer. Normally, there are experts who develop these multimedia files, and then the programmer incorporates the files into the application.

An external file used for multimedia is called an *asset.* All multimedia assets, regardless of file type, are simply patterns of 1s and 0s. Each type of file is created with its own internal format. The format of the asset determines how the information is stored and how to redisplay or play the information to generate the media. Examples of commonly used multimedia file formats are listed in **Figure 14-11.** The programmer must work with the asset developers to ensure the correct format is delivered for use in the application. The programming language being used to create the application may or may not be able to handle a particular multimedia file type.

It is the job of the programmer to locate each of these files for the application. A URL class is used. A *uniform resource locator (URL)* is an address that points to a specific document or other resource on a computer network. For example, for a short audio file saved in the same folder as the Java program, a URL object will create the path to that file so the application can access it.

```
// add audio: create resource URL to load file
URL resource = getClass( ).getResource( "click.wav" );
sndMove = new AudioClip( resource.toString( ) ); // sndMove is an
                                                  // AudioClip object
```

To use URL and AudioClip objects, import their classes:

```
import javafx.scene.media.AudioClip;
import java.net.URL;
```

File Format	File Extension	Media Type	Characteristics
Bitmap	.bmp	Image	Original Microsoft image format; intended for small images or clip art.
Graphics Interchange Format (GIF)	.gif	Image	Proprietary image format of CompuServe, which was an early Internet Service Provider; works best with clip art, and can be animated.
Joint Photographic Experts Group (JPEG)	.jpg	Image	Compressed image file format, which loses a great deal of the image information.
Portable Network Graphic (PNG)	.png	Image	Image file format designed specifically for Internet use; highly compressed, but has high fidelity to the original image.
Wave	.wav	Audio	Microsoft audio format; uncompressed, and best used for short clips and sound effects.
Motion Pictures Expert Group (MPEG)	.mp3, .mp4, others	Audio and video	Compressed audio and video file format; good for long animations, video, and music.

Goodheart-Willcox Publisher

Figure 14-11. These are some of the file types commonly used in multimedia applications.

To play the sound clip, use the play( ) method:

```
sndMove.play( );
```

Image

The JavaFX controls for displaying images are the image and the image view controls. The ***image control*** holds the reference to the image file as an Image object. The ***image view control*** displays an Image object. Supported image formats are BMP, GIF, JPEG, and PNG.

To add an image, load the image file name into a JavaFX image control. Then, set that Image object as the property for an image view control. Take care, the image view control does not take the name of an image as its argument. It is looking for an Image object. The constructors are listed in the API in **Figure 14-12.** To use the image and image view controls, import these two classes:

```
import javafx.scene.image.Image;
import javafx.scene.image.ImageView;
```

Image and image view controls inherit from the Node class. Therefore, the properties are similar to other nodes in the scene. Their locations can be set by the setLayoutX and setLayoutY methods. Using the pattern for adding controls to the stage makes new controls easier to incorporate.

Specific to image views are image sizing methods. To specify the height of the image and maintain the original width, use the setFitHeight( ) method. Correspondingly, there is also a setFitWidth( ) method. These methods are described in **Figure 14-13.** To resize the images proportionally, set the preserveRatio property to true. If both "set fit" methods are used, the image may be distorted. However, if the preserveRatio property is true, Java works to get the best fit within a rectangle defined by the new height and width without distorting the image. If neither "set fit" method is used, the image is displayed with its own height and width properties.

For information on adding larger audio files, including music, search the Oracle Docs website for the API for the MediaPlayer object.

Constructor	Explanation
Image(String URL)	Creates a new Image object using the file located by a URL.
ImageView( )	Creates a new ImageView object.
ImageView(Image image)	Creates a new ImageView object using the specified Image object.
ImageView(String URL)	Creates a new ImageView object using the image located by a URL.

Goodheart-Willcox Publisher

Figure 14-12. These are the constructors for Image and ImageView objects.

Method	Explanation
setFitHeight(double)	Sets the value of the fitHeight property in an ImageView object, scaling the width of the image.
setFitWidth(double)	Sets the value of the fitWidth property in an ImageView object, scaling the height of the image.
setPreserveRatio(boolean)	Scales the image when resizing; true does not distort the image.

Goodheart-Willcox Publisher

Figure 14-13. These methods are used to control the size of an image.

Images can be added to JavaFX buttons as well. To create a button that displays an image, use this constructor from **Figure 14-3**:

```
Button( String text, Node image )
```

This code creates a new Button object with the specified text and image for its label.

The following example creates two frogs to animate for a game called Race to the Top. One frog is placed on the left of the stage, the other on the right, as shown in **Figure 14-14.** In addition, a button is placed using image file as well. The sizes of the frogs depend on the "set fit" methods. The button is automatically resized to accommodate the dimension of the image. All three frogs are based on the same image file, which is in the same folder as the application. Using the URL class to locate the image is optional here. We could use just the Image constructor that accepts a String. The list of nodes is defined as effects. Use the imv prefix for the ImageView objects.

```
// create frogs
URL resource = getClass( ).getResource( "frog.png" );

Image imgFrog = new Image( resource.toString( ) );

ImageView imvFrogL = new ImageView( imgFrog );
imvFrogL.setLayoutX( 40 );
imvFrogL.setLayoutY( 100 );
imvFrogL.setPreserveRatio( false );
imvFrogL.setFitHeight( 30 ); // width is not changed

ImageView imvFrogR = new ImageView( imgFrog );
imvFrogR.setLayoutX( 280 );
imvFrogR.setLayoutY( 100 );
imvFrogR.setPreserveRatio( true );
imvFrogR.setFitWidth( 100 ); // height is automatically changed

// create button
ImageView imvFrog = new ImageView( imgFrog );
btnPlayAgain = new Button( "Play", imvFrog );
effects.addAll( imvFrogL, imvFrogR, btnPlayAgain );
```

Goodheart-Willcox Publisher

Figure 14-14. An image file is placed on the stage in two separate instances. It is also used on the button.

Animation

Simple animation can be accomplished by changing the layout properties for the *x* and *y* coordinates of an image on the stage. Rolling dice to move the characters in the racing game created in Hands-On Example 12.3.1 is an example of this type of animation. To move the frogs shown in **Figure 14-14,** use a random-number generator to roll a die and then change the layoutX or layoutY property of the image.

Advanced animation is composed of changing images in an image control to suggest movement. For advanced information on JavaFX animations, search the Oracle Docs website for JavaFX animation and animation timer.

HANDS-ON EXAMPLE 14.2.3

Using Multimedia Controls 🔗

In this exercise, you will write an application to make a frog hop when a button is pressed. Before beginning this exercise, download the chapter files from the student companion website. The MultimediaSnippet.java file will be used as a starting point. You will also need the CLICK.WAV and frog-sm.png files. Keep them in the same folder as the snippet file.

1. Launch jGRASP, and open the MultimediaSnippet.java file.
2. Locate comment 1, and follow the instructions to set up the URL Resource and AudioClip objects.

```
/***** 1. Add audio: create resource URL to load file. */
URL resource = getClass( ).getResource( "CLICK.WAV" );
AudioClip sndMove = new AudioClip( resource.toString( ) );
```

3. Locate comment 2, and follow the instructions to create the control for the hopping frog.

```
/***** 2. Create frog for hopping. */
resource = getClass( ).getResource( "frog.png" );
Image imgFrog = new Image( resource.toString( ) );
ImageView viewFrog = new ImageView( imgFrog );
effects.add( viewFrog );
viewFrog.setLayoutX( 40 );
viewFrog.setLayoutY( 40 );
```

4. Locate comment 3, and follow the instructions to create the control for the button with an icon.

```
/***** 3. Create hop button with frog icon. */
resource = getClass( ).getResource( "frog-sm.png" );
ImageView iconFrog = new ImageView( imgFrog );
Button btnHop = new Button( "Hop", iconFrog );
btnHop.setLayoutX( 105 );
btnHop.setLayoutY( 120 );
btnHop.setVisible( true );
effects.add( btnHop );
```

5. Locate comment 4, and follow the instructions to write the event handler for the button. When the button is clicked, move the frog 10 pixels to the right, and play the click sound. When frog reaches the edge of the stage, reset it.

```
/***** 4. Handle button press to move frog. */
btnHop.setOnAction(
    event -> {
        // move frog
        viewFrog.setLayoutX( viewFrog.getLayoutX( ) + 10 );
        // if it hops off screen, reset to start
        if ( viewFrog.getLayoutX( ) >= scene.getWidth( ) ) {
            viewFrog.setLayoutX( 40 );
        }
        // play a sound
        sndMove.play( );
    } // end Hop event handler
); // end btnHop.setOnAction
```

6. Compile and run the application. Move the frog across the stage using the button. You have created a multimedia application!

Goodheart-Willcox Publisher

Try It!

Continue working in the MultimediaSnippet.java file. Add a new event. Add a button to move the frog down the stage by a fixed number of pixels. Include an appropriate label on the button. When the frog reaches the bottom of the stage, reset its position.

Coding Conundrum

A programmer has created an application with a set of radio buttons and an image view control that is to move up the screen. However, the user can select several of the radio buttons at a time. Additionally, the image view control moves down the screen instead of up. It is a conundrum! What did the programmer do wrong for the radio buttons (not shown here) and the image view control? The code for the image view control is shown below.

```
73 // TryIt  handle button press to move frog up */
74 btnUp.setOnAction(
75   event -> {
76     // move frog
77     viewFrog.setLayoutY( viewFrog.getLayoutY( ) + 10 );
78     // if it hops off screen
79     if ( viewFrog.getLayoutY( ) >= scene.getHeight( ) ) {
80       viewFrog.setLayoutY( 40 );
81     }
82     // play a sound
83     sndMove.play( );
84   } // end event handler
85 ); // end btnUp.setOnAction
```

SECTION REVIEW 14.2

Check Your Understanding

1. Which control has a special property to hide the characters entered by the user?
2. Which accessor method retrieves the text input from a text field?
3. When a set of options are independent and may be selected or not, which control should be used?
4. What is a multimedia asset?
5. Which control holds the reference to an image file?

Build Your Vocabulary

As you progress through this course, develop a personal computer science glossary. This will help you build your vocabulary and prepare you for a career. Write a definition for each of the following terms and add it to your computer science glossary.

asset	image control	uniform resource locator (URL)
check box	image view control	vertical box layout
focus	password field	
horizontal box layout	radio button	

JavaFX Games

You now know enough to make a simple game using JavaFX controls. It is an industry standard for GUIs that a player never has to stop and think about what to do to advance in the game. The instructions and controls must be intuitive. Once you think of an idea and draw it out on graph paper, take your design to a few potential players and see if they can figure out how to play. Their input will be very valuable. It will save time and frustration by not waiting until the game is coded completely to find out the GUI is hard to use or the game is not fun.

Learning Goals

- Design the behaviors for a basic video game.
- Formulate behaviors of objects, controls, and methods within the context of a video game.
- Implement code for a video game based on an established design.

Term

avatar

Essential Question

How can the GUI improve or detract from gameplay?

Design the Game

There are four phases to developing a game: designing the game, designing the program, collecting and creating the assets, and implementing the design in code. Thinking and drawing are the activities that complete the first three phases. Programming is not required until the last phase.

This section discusses how to create a basic game in the genre of racing games. It is a two-player game. The object of the game is to move an avatar to the top of the screen before the other player gets there. The first one to reach the top is the winner. The avatars are frogs. An **avatar** is the visual representation of a player character. The background graphic is a lily pond. There is a label at the top with the name of the game. On both sides of the game name, there is room to say "Winner!" for whichever frog wins. **Figure 14-15** shows a sketch of the layout for the game and its controls.

Tapping a key rolls two dice. The frog jumps up that many pixels. Dice are displayed on the bottom of the screen so each player can see the rolls. A label on each side tells the player which key to tap to move his or her frog. A Play button starts the game and provides a Play Again feature. At the start of the game, the frogs jump down to the bottom of the screen and the keys become active.

The first frog to get to the top wins. Congratulate the winner and make the keys inactive. Incorporate audio on keystrokes and for the win scenario.

Design the Program

Now that you know what controls are needed, you can begin to set up the game program. You need to add the controls and other objects to play the game.

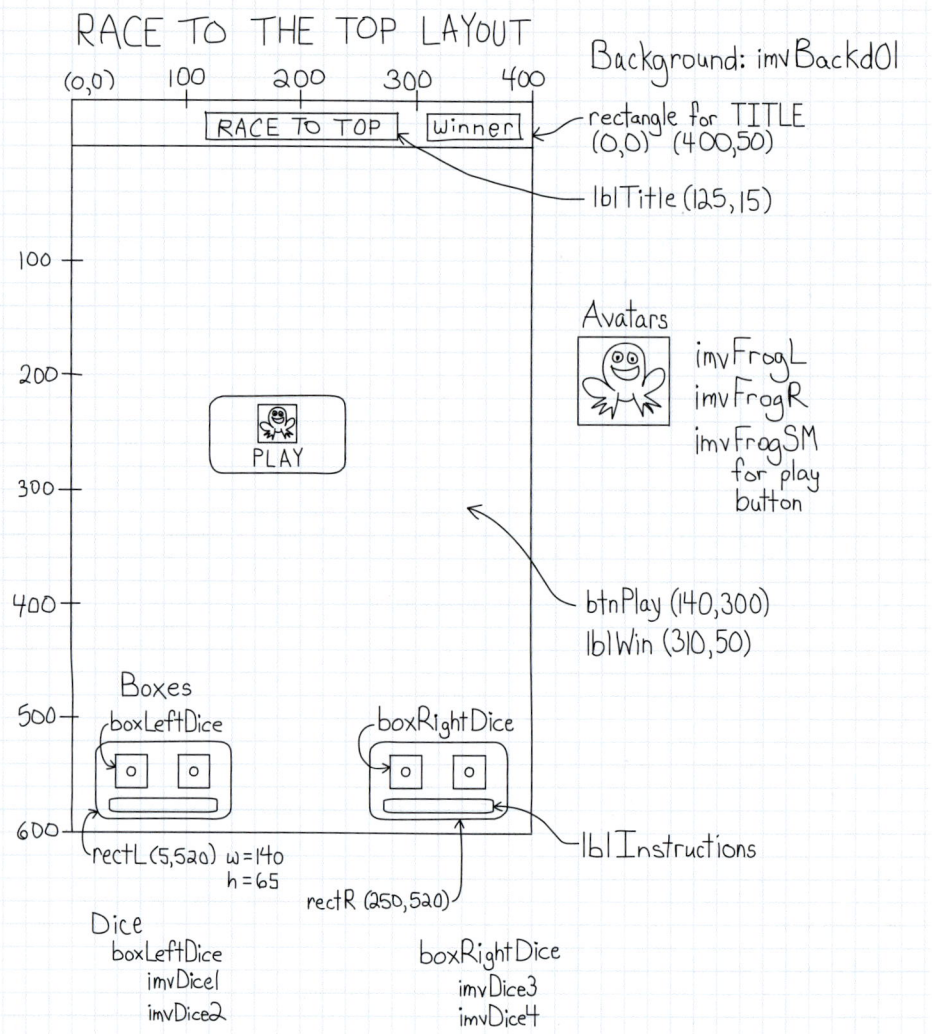

Goodheart-Willcox Publisher

Figure 14-15. This layout sketch shows the relative locations of the controls for a racing game.

Start by deciding on the identifiers for the controls. Refer to the layout sketch to do this. Remember, the program design is created on paper. Doing all this work on paper makes the coding so much easier.

Define Controls

A background image is needed along with a rectangle and a label to display the title. Use a prefix for each type of control to minimize confusion when you see an identifier. It will be easier to determine what that identifier represents with a meaningful prefix.

```
ImageView imvBackdrop; // image for background
Rectangle rectTitle; // rectangle for title
Label lblTitle; // label for title
```

The next layer of controls is the directions and dice images at the bottom of the stage. These elements consist of a rounded rectangle, label, and two dice images for each side. Appropriate names for these two rectangles are rectL and rectR. These names will serve as reminders that one is on the left of the stage and the other is on the right.

```
Rectangle rectL, rectR;
Label lblInstructions;
ImageView imvDice1, imvDice2, imvDice3, imvDice4;
```

A die has six faces. An image is required for each face. It makes sense to define an array to hold the six images for each die.

```
Image [ ] dieArray;
```

Each player has two dice. The dice will be displayed side-by-side for each player. The layout can be simplified by using a horizontal box to hold each pair of dice. Again, use meaningful names to help indicate the positions.

```
HBox boxLeftDice, boxRightDice;
```

The custom Die class created in Chapter 12 will be reused for this application. To use this class, a Die object needs to be created.

```
Die die;
```

At the end of the game, there will be a congratulation to the winner. This will be provided as text. To display the text, define a "Winner!" label.

```
Label lblTitle, lblWin
```

The game will play some sound effects. To do this, AudioClip objects are needed. A URL object is also needed to locate the path to the clips.

```
AudioClip sndMove, sndApplause;
URL resource;
```

After all the controls are defined, they need to be added. To add the controls, an observable list is defined.

```
ObservableList<Node> effects;
```

Finally, define the scene.

```
Scene scene;
```

Define Constants

It is important to avoid "magic numbers." To prevent "magic numbers" from appearing, several constants can be defined. Remember, the final keyword is used to define constants. The constants below define the play area on the stage.

```
// define constants
private final int BOTTOM = 480;  // starting y for avatars
private final int LEFT = 20;    // x for left avatar
private final int RIGHT = 310;  // x for right avatar
private final int TOP = 55;     // winning y coordinate for avatars
```

Define Custom Methods

Next, decide on the names for the custom methods and what they will accomplish. Start with setting up the stage. Define this first because the order in which controls are added to the stage is the order in which they are seen. Each control is added on top of the existing controls. A newly added control will cover or hide any controls in the same location.

```
setStage( );        // set the stage background image
setupDice( );    // set up dice and instructions
setupFrogs( );   // create frogs
setupAudio( );   // add audio
setupWinner( );    // create winner label
setupPlayButton( );   // set up Play/Play Again button
```

The setStage( ) method will add the background image, using a URL to set the path to the image, and locate it at coordinates (0, 50). Next, it will add an orange rectangle and set the title "Race to the Top" on top using font Tahoma, bold, 20 points. The label will be located at (125, 15). These controls should be added to the effects node in this order: background, rectangle, label.

The setupDice( ) method will instantiate an array of seven images and load the six die images found in the download folder into index 1 to 6. Then, it will set the properties for the four imvDice image views and add dieArray[1] as a default. It will instantiate and set properties for rectL and rectR. These are both rounded rectangles that are 140 × 65 and arc width heights and widths of 20. The rectangles are located at (5, 520) and (250, 520). The fill color is orange. The method will also instantiate and set up the two HBoxes for the dice. The imvDice1 and imvDice2 objects are added to boxLeftDice. The imvDice3 and imvDice4 objects are added to boxRightDice. Finally, the method will instantiate and set up the lblInstructions label. A single label is used with the text Tap A to roll dice\t\t\t\tTap L to roll dice. The label is located at (15, 560) in the font Arial bold, 15 points and color WHITE. These controls should be added to the stage in this order: rectangleS, boxes, instructions.

The setupFrogs( ) method will instantiate two frog image objects from a Frog class that you will create to extend the ImageView class. The arguments that Frog.java will accept are the x and y coordinates and the name of the image file. The constructor API is:

```
Frog( double x, double y, Image image )
```

The image to use is frog.png for the avatars. The images are added to two image views. The coordinates are set at (40, 0) and (300, 0) to start the frogs at the top of the stage. Add all to the stage. Order does not matter because these controls will not overlap.

The setupAudio( ) method will use the URL control to locate the path for two audio files: CLICK.WAV and APPLAUSE.WAV. The method will instantiate the two audio clip objects.

The setupWinner( ) method will instantiate a new label with the text "Winner!" This will be displayed over the winning avatar. The winning avatar will be determined by the event handler for the button. The label is located at (LEFT, 10) in the font Arial, bold, 20 point and color CORNFLOWERBLUE. The method sets the label to be invisible and adds it to the stage.

The setupPlayButton( ) method will set up the button. The button is dual purpose. It is used to start the game and restart the game to play again. The method will instantiate the button with the text "play" and the image view of the small frog for its icon. The button is located at (140, 300) and set to visible. Add it to the stage.

The btnPlay.setOnAction( ) method defines the event handler. The button event handler is added to the start method after all setup methods have been called. The event handler is where the code for the game action is placed. The only thing this event-handler method will do is call the startGame( ) method.

The startGame( ) method will start and restart the game to provide multiple plays without having to launch the game repeatedly. The actions in this method are to hide the winner label and move the two avatars to their play positions, which are (LEFT, BOTTOM) and (RIGHT, BOTTOM). It will hide the play button and change the button text to "Play Again" for the next use of the button. It also defines the event handlers for the two keys used to move the avatars. This method belongs to the scene because the keyboard is used and no other control on the stage is used.

FYI

The setupWinner( ) and the startGame( ) methods both must set the label invisible because the startGame( ) method is used for restart game as well.

The scene.setOnKeyReleased( ) method defines the event handler for the event of any key being released. The [A] and [L] keys are used in the game. So, the method gets the key code and then a switch statement is used to deal with each character, uppercase and lowercase. Next, the method will "roll the dice" to determine the number of pixels to move up the stage. It then shows the dice faces for the roll. If the avatar is at the TOP position, the method will play the applause audio clip, move the Winner label above the winner, set the label to be visible, and show the Play Again button. Stop game play by setting this method to null, which cancels the event handler from being called. Otherwise, if the avatar is not at the TOP position, play the click sound and adjust the *y* coordinate by the amount of the dice roll.

Implement the Design in Code

Now, it is time to write the code. This section walks you through the process of coding this game. Read through the explanations here. The code follows the design explained above. The following Hands-On Example provides the opportunity to work in a file already containing this code. All the game assets, the non-code items to add to the game, are provided for you to use in the following Hands-On Example.

There are many classes to import to make a game in JavaFX. In addition, you need to code the class for the game object: Frog.java. That is also part of the following Hands-On Example. Below are the classes to import.

```
/*  JavaFX Game - GUI controls
    your name here
*/

import java.util.*;
import javafx.application.Application;
import javafx.collections.ObservableList;
import javafx.scene.*;
import javafx.scene.effect.*;
import javafx.scene.shape.*;
import javafx.scene.text.*;
import javafx.scene.paint.*;
import javafx.stage.Stage;
import javafx.event.ActionEvent;
import javafx.event.EventHandler;
import javafx.scene.input.KeyEvent;

// add GUI elements
import javafx.scene.control.*;
import javafx.scene.image.*;
import javafx.scene.layout.HBox;

// add media elements
import javafx.scene.media.AudioClip;
import java.net.URL;
```

Once the required classes are imported, it is time to create the game class and the main method. The name of the game is Race to the Top. Name the class RaceToTheTopGame, and save the file under the same name. Apply what you have learned to understand this code:

```
public class RaceToTheTopGame extends Application {


    @Override
    public void start( Stage mainStage ) {
```

(Continued)

```
        Group root = new Group( ); // create a Group container
        scene = new Scene( root, 400, 600 ); // create a Scene


        mainStage.setTitle( "Race to the Top" ); // change the title
        mainStage.setScene( scene ); // add the scene to the stage
        mainStage.show( ); // display what is on the stage

    } // end start method

    public static void main( String [ ] args ) {
        launch( args ); /* Application method that launches
                the JavaFX runtime and your JavaFX application */
    } // end main method
} // end game
```

The next step is to define the controls as instance variables at the top of the program after the definition of the class and before the start( ) method. They are defined here so they will be available to all the methods in the program. Declare these instance variables to be private so no other code can use them except this game, but any method within this game can access their contents. This is an easy step because you identified all needed variables in the design above.

```
// define but do not instantiate controls
private AudioClip sndMove, sndApplause;
private Button btnPlay;
private Die die;
private Frog frogL, frogR;
private HBox boxLeftDice, boxRightDice;
private Image [ ] dieArray;
private ImageView imvBackdrop;
private ImageView imvDice1, imvDice2, imvDice3, imvDice4;
private Label lblTitle, lblWin, lblInstructionL, lblInstructionR;
private ObservableList<Node> effects;
private Rectangle rectTitle, rectL, rectR;
private Scene scene;
private URL resource;
// define constants for game object ( Frog ) locations
private final int BOTTOM = 480;
private final int LEFT = 20;
private final int RIGHT = 310;
private final int TOP = 55;
```

It is now time to add the calls to the setup methods designed above in the start method after the scene is instantiated. This is the streamlining of the code. Notice that there are six method calls below that clarify the big picture for the program. This makes it easy for someone new to the code to understand what is going on without having to wade through lots of detailed coding. At the end of the method calls, it is time to handle the Play/Play Again button-press event.

```
/***** get the list of current nodes in the group */
effects = root.getChildren( );

// call the setup methods
setStage( );        // set the stage
setupDice( );       // set up dice and instructions
setupFrogs( );      // create frogs
setupAudio( );      // add audio
setupWinner( );     // create winner label
setupPlayButton( ); // set up Play/Play Again button

// handle button press
btnPlay.setOnAction(
```

(Continued)

```
        event -> {
            startGame( );
        } // end play play/again event handler
    ); // end btnPlay.setOnAction
```

After the end of the main method, add each definition of the startup methods. Start with implementing the setStage method following the setStage method designed above. Make it easy on yourself and the next coders by defining each method in the order in which is it called.

```
// custom methods
public void setStage( ) {
    // add background image
    imvBackdrop = new ImageView( );
    resource = getClass( ).getResource( "background.png" );
    Image imgBackground = new Image( resource.toString( ) );
    imvBackdrop.setImage( imgBackground );
    imvBackdrop.setLayoutY( 50 );

    // add Title
    rectTitle = new Rectangle( 0, 0, 400, 50 );
    rectTitle.setFill( Color.ORANGE );
    lblTitle = new Label( "Race to the Top" );
    lblTitle.setFont( Font.font( "tahoma", FontWeight.BOLD, 20 ) );
    lblTitle.setLayoutX( 125 );
    lblTitle.setLayoutY( 15 );

    effects.addAll( imgBackdrop, rectTitle, lblTitle );

} // end setStage method
```

Now implement the setupDice method following the setupDice method design written out earlier. This goes after the setStage method and before the end game brace because it concerns adding more elements to the stage.

```
public void setupDice( ) {
    // initialize array of images of die for easy access
    dieArray = new Image[7];
    dieArray[1] = new Image( "1.jpg" );
    dieArray[2] = new Image( "2.jpg" );
    dieArray[3] = new Image( "3.jpg" );
    dieArray[4] = new Image( "4.jpg" );
    dieArray[5] = new Image( "5.jpg" );
    dieArray[6] = new Image( "6.jpg" );

    imvDice1 = new ImageView( );
    imvDice1.setImage( dieArray[1] );
    imvDice2 = new ImageView( );
    imvDice2.setImage( dieArray[1] );
    imvDice3 = new ImageView( );
    imvDice3.setImage( dieArray[1] );
    imvDice4 = new ImageView( );
    imvDice4.setImage( dieArray[1] );

    // set up rectangles to display the dice
    // on both sides of the stage
    rectL = new Rectangle( 5, 520, 140, 65 );
    rectL.setArcWidth( 20 );
    rectL.setArcHeight( 20 );
    rectL.setFill( Color.ORANGE );
    rectR = new Rectangle( 250, 520, 140, 65 );
    rectR.setArcWidth( 20 );
    rectR.setArcHeight( 20 );
    rectR.setFill( Color.ORANGE );
```

FYI

Some companies have a rule to put the custom methods in alphabetical order. It will be important to understand the company culture when you are working in the IT field.

(Continued)

```
        boxLeftDice = new HBox( 10 );
        boxLeftDice.getChildren( ).add( imvDice1 );
        boxLeftDice.getChildren( ).add( imvDice2 );
        boxLeftDice.setLayoutX( 35 );
        boxLeftDice.setLayoutY( 530 );

        boxRightDice = new HBox( 10 );
        boxRightDice.getChildren( ).add( imvDice3 );
        boxRightDice.getChildren( ).add( imvDice4 );
        boxRightDice.setLayoutX( 280 );
        boxRightDice.setLayoutY( 530 );

        lblInstructionL = new Label( "Tap A to roll dice" );
        lblInstructionR = new Label( "Tap L to roll dice" );
        lblInstructionL.setLayoutX( 15 );
        lblInstructionL.setLayoutY( 560 );
        lblInstructionL.setFont
            ( Font.font( "arial", FontWeight.BOLD, 15 ) );
        lblInstructionL.setTextFill( Color.WHITE );
        lblInstructionR.setLayoutX( 260 );
        lblInstructionR.setLayoutY( 560 );
        lblInstructionR.setFont
            ( Font.font( "arial", FontWeight.BOLD, 15 ) );
        lblInstructionR.setTextFill( Color.WHITE );

        effects.addAll( rectL, rectR, boxLeftDice, boxRightDice,
            lblInstructionL, lblInstructionR );
    } // end set up dice
```

At this time, set up the avatars. Follow the setupFrogs method design written out earlier. This is very short because of the Frog class you will write. Add it after the setupDice method before the end game brace.

```
    public void setupFrogs( ) {
        Image imgFrog = new Image( "frog.png" );
        frogL = new Frog( 40, 0, imgFrog );
        frogR = new Frog( 300, 0, imgFrog );

        effects.addAll( frogL, frogR);
    } // end setupFrogs
```

Next, implement the setupAudio method. Follow the setupAudio method design written out earlier. This goes after the setupFrogs method and before the end game brace.

```
    public void setupAudio( ) {

        resource = getClass( ).getResource( "CLICK.WAV" );
        sndMove = new AudioClip( resource.toString( ) );

        resource = getClass( ).getResource( "APPLAUSE.WAV" );
        sndApplause = new AudioClip( resource.toString( ) );
    } // end setupAudio
```

When one of the frogs reaches the top, the winner needs to be congratulated. The setupWinner method adds a label to the stage. Implement the Winner label as designed above. Add this code after the setupAudio method and before the end game brace.

```
    public void setupWinner( ) {

        lblWin = new Label( "Winner!" );
        lblWin.setFont( Font.font( "arial", FontWeight.BOLD, 20 ) );
        lblWin.setTextFill( Color.CORNFLOWERBLUE );
```

(Continued)

```
    lblWin.setLayoutX( LEFT );
    lblWin.setLayoutY( 10 );

    lblWin.setVisible( false );
    effects.add( lblWin );

} // end setupWinner label
```

This code would be extremely difficult to implement without planning. Clearly, if you had not spent the time and energy to specify what happens in the method and had not selected meaningful identifiers, how would you have known what statements to write? Designing on paper makes it much easier to enter the code. All that remains at this point is to write the Frog class and add the setupPlayButton button. These tasks are covered in the Hands-On Example.

HANDS-ON EXAMPLE 14.3.1

Finishing the Functionality of a Video Game ↗

In this exercise, you will write a Frog class from scratch and add the Play button to the game. Before beginning this exercise, download the chapter files from the student companion website. The RaceToTheTopSnippet.java file contains the code described in the text. The assets you will need are also provided: background image, frog images, dice images, audio files, and the Die class from a previous chapter.

1. Launch jGRASP, and begin a new, blank Java file.

2. Add a comment at the top of the file to describe the class.

   ```
   /** Frog class
      Define game objects for the Race to the Top game
   */
   ```

3. Add a statement to import the javaFX.scene.image package.

   ```
   import javafx.scene.image.*;
   ```

4. This Frog class extends the ImageView class to take advantage of the properties and methods already written in that class. Add the class name statement.

   ```
   public class Frog extends ImageView {

   } // end Frog class
   ```

5. Apply what you have learned about objects to code the Frog constructor. The arguments we want to capture are the location of the Frog object on the stage and which image to use. Insert this code inside the class statement.

   ```
   public Frog( double x, double y, Image image ) {
       setImage( image );
       this.setX( x );
       this.setY( y );
   } // end Frog constructor
   ```

6. Save the file as Frog.java in the same folder as the downloaded items.

7. Compile the class. It will not execute because there is no main method. Close the file.

8. Open the RaceToTheTopSnippet.java file. Compile and run it to verify it includes the work of this section.

9. Rename the class as RaceToTheTopGame, and save the file under the new name.

10. Locate comment 1 in the start method, and follow the instructions to set up the event handler for the button. When a player clicks the Play button, the application will call the startGame method. Notice that the last line in this entry starts with a right-hand parenthesis. It is the closing parenthesis to the btnPlay.setOnAction method.

```
/***** 1. Handle button press. */
btnPlay.setOnAction(
    event -> {
        startGame( );
    } // end play play/again event handler
); // end btnPlay.setOnAction
```

11. Locate comment 2 near the end of the game file. There is a definition of the setupPlayButton method, but no code has been entered. This method sets the properties for the button and adds the button to the stage. Add the code to instantiate the button, set its location, make it visible, and add it to the stage.

```
/***** 2. Create Play/Play Again button and add to stage. */
btnPlay = new Button( "Play", imvFrogSM );
btnPlay.setLayoutX( 140 );
btnPlay.setLayoutY( 300 );
btnPlay.setVisible( true );
effects.add( btnPlay ); // calls start game method
} // end set up Play button
```

12. Up to now, you have been adding controls to the scene. Now, you are going to react to the player. Locate comment 3 below where you just entered the code for comment 2. There is a definition of the startGame method, but no code has been entered. When a player presses the Play/Play Again button, this method is called. The first things to do are hide the Winner label, move the frogs to the bottom of the stage, hide the play buttons, and change the button text to Play Again for the next time it is used. Enter the code into the existing method definition.

```
// Hide Winner!
lblWin.setVisible( false );

// reseat frogs
imvFrogL.setLayoutY( BOTTOM );
imvFrogR.setLayoutY( BOTTOM );

// hide Play / Play Again button
btnPlay.setVisible( false );
btnPlay.setText( "Play Again" );
```

13. Within this method is the key-release event handler. For the rest of the game, this event handler will swing into action whenever *any* key is released. So, you need code to react to only an [A] or [L] key. Jobs to do are to roll dice to move the frog, show the results of the dice roll, check for a win and congratulate, and end this round. To end the round, set this handler to null. Now keys will not generate the event and pressing the keys will have no effect. Continue where you left off in step 12, and enter the code to initialize key-release event handler.

```
// init key handler
scene.setOnKeyReleased(
    event -> {

        // get the keystroke character
        String keyCode = event.getCode( ).getName( ).toLowerCase( );
        // reuse Die class from Chapter 12
        die = new Die( );
        // vars for showing dice and moving frogs
        int die1 = 1;
        int die2 = 1;
        int diceRoll = 2;

        if ( event.getEventType( ) == KeyEvent.KEY_RELEASED ){

            // handle user keystrokes: A for player1, L for player 2
            switch ( keyCode ) {
                case "a":
                    die1 = die.rollDie( );
```

(Continued)

```
                      die2 = die.rollDie( );
                      diceRoll = die1 + die2;

                      // show images for dice
                      imvDice1.setImage( dieArray[die1] );
                      imvDice2.setImage( dieArray[die2] );

                      // check for a win
                      if ( frogL.getY( ) <= TOP ) {

                          // left frog wins
                          sndApplause.play();

                          // stop playing
                          btnPlay.setVisible( true );
                          scene.setOnKeyReleased( null );

                      } else {
                          // play sound and move left frog
                          sndMove.play();
                          frogL.setY( frogL.getY( ) - diceRoll );
                      }

                  break;
                  case "l":
                      die1 = die.rollDie( );
                      die2 = die.rollDie( );
                      diceRoll = die1 + die2;

                      // show images for dice
                      imvDice3.setImage( dieArray[die1] );
                      imvDice4.setImage( dieArray[die2] );
                      // check for a win
                      if ( frogR.getY( ) <= TOP ) {

                          // right frog wins
                          sndApplause.play( );

                          // stop playing
                          btnPlay.setVisible( true );
                          scene.setOnKeyReleased( null );

                      } else {
                          // play sound and move right frog
                          sndMove.play();
                          frogR.setY( frogR.getY( ) - diceRoll );
                      }
                  break;
                  } // end switch
              } // end if key released
          } // end new Key Event
      ); // end scene.setOnKeyReleased method
  } // end startGame()
```

14. Do a visual check to verify there is an end game closing brace (and only one). Compile the application. Fix any errors that occur.

15. Run the application, and play the game.

Try It!

Continue working in the RaceToTheTop.java file. Implement the Winner label according to the design described in the text. In the setOnKeyReleased event, move the label to the LEFT or RIGHT position, depending on who is the winner, and make it visible. This should occur for each key event after the comment that either frog wins.

SECTION REVIEW 14.3

Check Your Understanding

1. What is an avatar?
2. In which phase of game development does brainstorming the game and making drawings of the layout take place?
3. Why is the program designed on paper instead of directly writing code?
4. How do you stop an event handler from continuing to execute?
5. In which phase of game development is code entered into the Java file?

Build Your Vocabulary

As you progress through this course, develop a personal computer science glossary. This will help you build your vocabulary and prepare you for a career. Write a definition for the following term and add it to your computer science glossary.

avatar

Science and Java

Influenza Vaccinations

Influenza is a respiratory infection caused by a virus that can be deadly. An influenza pandemic results when the virus spreads on a worldwide scale and infects a large portion of the population of the planet. These pandemics happen infrequently. There have been only nine pandemics in the last 300 years. The most recent one was in 2009. The 1918 influenza pandemic was the worst in recorded history. It is estimated to be responsible for the deaths of approximately 50–100 million people worldwide.

Library of Congress

Influenza vaccines, also known as flu shots, contain cultures of the virus that cause an immune response in the recipient. The body develops a defense based on this limited exposure to the virus and can fight the virus if exposed to it again. Early influenza vaccines contained live viruses. In the 1940s, the US military developed the first vaccines containing dead viruses. They were used on military personnel during World War II.

While the effectiveness of the vaccine varies from year to year, it typically provides modest to high protection against influenza. A person who has received a flu vaccination is less likely to contract the disease and better able to fight the disease if he or she does contract it. The United States Centers for Disease Control and Prevention (CDC) estimates that vaccination against influenza reduces sickness, medical visits, hospitalizations, and deaths. According to the CDC, influenza vaccination is the primary method for preventing influenza and its severe complications.

The CDC advises everyone to get a flu shot to prevent the disease and from having epidemics become pandemics. However, some people choose not to be vaccinated for various reasons.

Assignment

Before beginning this activity, download the chapter files from the student companion website. You are a researcher with your local health clinic. You have been tasked with developing a JavaFX animation that compares the likelihood of developing the flu for a person who has been vaccinated and one who has not. Here are some suggestions. Display two people images at the bottom of the screen. One has been vaccinated, the other has not. Display two virus images at the top of the screen. Move each virus downward a random number of steps. Use a random number between 1 and 10 for the person with no vaccine and between 1 and 5 for the person with a flu shot. Include a button for the user to advance the animation. When the virus infects (touches) a person, display "You have the flu!" Submit your design and code to your instructor. Use all four phases of game development to create the application.

Cooperative Coding

Sleep Game

Sometimes teens spend so many hours on social media that they begin to lose valuable sleep. This sleep loss can lead to moodiness, poor grades, and overeating. Lack of sleep can also intensify any existing problems like depression, anxiety, and attention deficit disorder (ADD).

Marcos Mesa Sam Wordley/Shutterstock.com

Researchers at the Wales Institute for Social and Economic Research, Data, and Methods published an article in the *Journal of Youth Studies*. They surveyed 900 teens between the ages of 12 and 15 about their social media use and its impact on sleep. The results of their research were alarming. One-fifth of the teens said they almost always wake up during the night and log in to social media. The teens reported feeling tired all the time and being less happy than teens whose sleep was uninterrupted. The study also revealed that girls were significantly more likely to engage in this behavior than boys.

Assignment

Your team is tasked with building a sleep game that brings awareness to the issue of sleep loss due to social media use. The overall activity should have a player try to get a good night's sleep. Every time the player pressed the [Z] key, the move should take the player closer to eight hours. However, setting up a random-number selection including negative numbers as well as positive ones will make the goal harder to obtain. Some positive numbers could represent good nutrition, maintaining the same bedtime, no social media at night, and later school start time. Negative numbers could represent texting at night, lack of physical activity, and overeating. These influences should display when the key is pressed.

Research the causes of lack of sleep as well as factors to improve sleep habits. Follow the three-step process that begins by designing the application on paper. Collect the assets appropriate for your application, which may include images and audio files. Finally, code the game by defining the JavaFX controls needed, creating the methods to set them up, adding controls to the scene, and writing event handlers. Capture the player's name and offer congratulations when the player reaches eight hours of sleep. Make adjustments to the game so it is possible to reach the goal.

Chapter Summary

Section 14.1: JavaFX Graphical User Interfaces

- A graphical user interface (GUI) is a screen that displays visual options with which the user can interact, such as buttons and text input, and it is important to evaluate the user experience (UX) when designing a GUI to present the best UX.
- To include control objects in a GUI, import the JavaFX control class, define an object of the JavaFX control class as a node, set the properties for the JavaFX control object, add the object to the list of nodes in the scene, and then write an event handler, if appropriate.

Section 14.2: GUI Input

- A common use of a text field is a log-in screen that includes entry for a username and password, and the text entered can be accessed with the getText accessor method.
- A radio button is a control presented in a group in which only one control can be selected, while a check box is a control presented in a group in which all, some, or none of the controls can be selected. Radio buttons need to be added to a ToggleGroup to ensure that only one button can be selected at a time.
- Assets for multimedia applications are external files such as images or sounds, and a URL object contains the full path to the file; an image control holds the reference to an image file as an Image object and an image view control displays an Image object.

Section 14.3: JavaFX Games

- The four phases to developing a game are: designing the game, designing the program, collecting and creating the assets, and implementing the design in code; the first three phases do not involve programming.
- Once the game design is set, formulate the behaviors of controls, constants, and methods on paper.
- Implementing the game design in code involves entering the code that was formulated based on the behaviors outlined in the second design phase into the Java application.

Chapter 14 Test

Multiple Choice

Select the best response.

1. What describes the feeling a user gets from interacting with a website or computer application?
   A. GUI
   B. UE
   C. UI
   D. UX

2. Which is the proper order of events for setting up the JavaFX controls?
   A. Add the control to the scene, define the control, set the control's layout properties.
   B. Define the control, set the control's layout properties, add the control to the scene.
   C. Set the control's layout properties, define the control, add the control to the scene.
   D. Set the control's layout properties, add the control to the scene, define the control.

3. Button, RadioButton, and CheckBox are all examples of:
   A. GUI control classes
   B. custom methods
   C. main stage arguments
   D. launch code

4. Which of the following groups of file extensions represents multimedia supported by JavaFX?
   A. JPG, GIF, JAR
   B. MP4, MV9, WAV
   C. GIF, DOC, TXT
   D. PNG, WAV, BMP

5. Which of the following is *not* a phase in developing a game?
   A. obtaining a contract to develop the game
   B. designing the game
   C. designing the program
   D. implementing the design in code

Completion

Complete the following sentences with the correct word(s).

6. A(n) _____ is a screen that displays visual options with which the user can interact.

7. The code executed when an event is generated is the _____.

8. A(n) _____ provides a side-by-side arrangement of controls.

9. To have the application find an asset, use the _____ class.

10. A computer is required in the _____ phase of developing a video game.

Matching

Match the correct term with its definition.

   A. UI controls

   B. VBox vbLogin = new Vbox( );

   C. URL object

   D. btnLogin.setOnAction(

   E. ToggleGroup group = new ToggleGroup( );

11. Setting up a layout box for a group of controls.

12. Beginning of an event handler.

13. Necessary for grouping radio buttons so only one may be selected.

14. Sets the path for each asset used in an application.

15. Text fields, labels, radio buttons, and check boxes.

Application and Extension of Knowledge

Before beginning these activities, download the chapter files from the student companion website. The JavaFXStartFile.java file will be used as a starting point for each of these activities.

1. Launch a web browser, navigate to a search engine, and search for Oracle Docs JavaFX Tooltip. Extend your knowledge of JavaFX controls by learning how to add a tooltip to an application. Then, launch jGRASP, open the JavaFXStartFile.java file, change the class name to ToolTip, and save the file under the new name. Implement the example given in Oracle Docs for the Hover Over Me button. Draw a layout, define controls, set properties, enter the code, and test the application.

2. Launch jGRASP, open the JavaFXStartFile.java file, change the class name to RadioButtonsNotGrouped, and save the file under the new name. Implement a set of radio buttons in which any number of the radio buttons can be selected or unselected at a given time. How do you achieve this result?

3. Launch jGRASP, open the JavaFXStartFile.java file, change the class name to KeepingScore, and save the file under the new name. Create a game with two players: the computer and a human player. For the human player, set up a button that rolls an eight-sided die. For the computer, get a random number between 1 and 8 each time the player rolls. Determine if the player or computer has rolled the greater number, and add 1 to the score of the winner. No one receives a point if it is a draw.

4. Launch jGRASP, open the JavaFXStartFile.java file, change the class name to AudioOnOff, and save the file under the new name. Implement a check box for the option to turn sound on or off for a game. Implement a Play Sound Button that plays a sound if the check box is checked. The button should not play the sound when the check box is unchecked. Draw a layout, define controls, set properties, enter the code, and test the application.

5. Investigate a timeline animation to move a shape across the stage one step at a time. Start by looking at the code in the Turtle Sprite.java file. Then, launch a web browser, navigate to a search engine, and search for Oracle Docs JavaFX timeline. Finally, launch jGRASP, open the JavaFXStartFile.java file, change the class name to JavaFXAnimation, and save the file under the new name. Design an application that moves a shape, such as a rectangle, across the stage one step at a time.

Online Activities

Complete the following activities, which will help you learn, practice, and expand your knowledge and skills.

Vocabulary. Practice vocabulary for this chapter using the e-flash cards, matching activity, and vocabulary game until you are able to recognize their meanings.

Communication Skills

Reading. Identify the main idea of this chapter as well as the key supporting details. Analyze the development of the main idea over the course of the chapter. How do details shape or refine the presentation of the main idea?

Writing. Writing is an academic skill applied each day in both personal and professional lives. Using standard English, write a paragraph about why writing is considered an academic skill. How do you think writing skills will help you in your career?

Speaking. Beginning a career calls for an individual to be able to stay motivated during a job search. Prepare a one- to two-minute speech you might deliver to a friend who is becoming discouraged while searching for a job. Deliver the speech to a classmate using note cards if necessary. Practice correct pronunciation and grammar.

Listening. *Empathic listening* occurs when you attempt to put yourself in the speaker's place and understand how the speaker feels. How can you show empathic listening when a classmate is asking for your feedback or opinion on a situation he or she is sharing with you?

Portfolio Development

College and Career Readiness

Technical Skills. Your portfolio must showcase the technical skills you have. Are you exceptionally good at working with computers? Do you have a talent for creating videos? Technical skills are very important for succeeding in school or at work.

1. Create a Microsoft Word document that describes the technical skills you have acquired. Use the heading Technical Skills and your name. Describe the skill, your level of competence, and any other information that will showcase your skill level. Save the document file.

2. Update your master spreadsheet.

CTSOs

Proper Attire. Some CTSOs require appropriate business attire from all entrants and those attending the competition. This requirement is in keeping with the mission of CTSOs: to prepare students for professional careers. To be sure the attire you have chosen to wear at the competition is in accordance with event requirements, complete the following activities.

1. Visit the organization's website and look for the most current dress code.

2. The dress code requirements are very detailed and gender specific. Some CTSOs may require a chapter blazer to be worn during the competition.

3. Complete a dress rehearsal when practicing for your event. Are you comfortable in the clothes you have chosen? Do you present a professional appearance?

4. In addition to the kinds of clothes you can wear, be sure the clothes are clean and pressed. You do not want to undermine your appearance or event performance with wrinkled clothes that may distract judges.

5. Make sure your hair is neat and worn in a conservative style. If you are a male, you should be freshly shaven and any facial hair neatly trimmed. Again, you do not want anything about your appearance to detract from your performance.

6. As far in advance of the event as is possible, share your clothing choice with your organization's advisor to make sure you will be appropriately dressed.

Careers in Computer Programming

Sections

Our world is digital. Phones, vehicles, financial institutions, hospitals—most aspects of modern life—cannot operate without computer code. Coding skills, or computer programming, are becoming core requirements for many well-paying jobs. Coding skills are in demand across a broad range of careers in all facets of government and business. The ability not only to use, but also to program, software is often required of business people as well as engineers and scientists.

This chapter looks at the job market demand for coding in order to highlight the wide range of employer demand for these skills. It emphasizes the preparation necessary for employment and the variety of opportunities that learning to code can open for students. It was not long ago that coding was seen by many people as an activity purely for a small segment of the population. Coding was associated with tech enthusiasts who spoke their own language. Public perception has definitely changed.

College and Career Readiness

Reading Prep

Before reading this chapter, read the Coding Conundrum feature in this chapter. Use it to help you focus on the most important concepts as you read the chapter.

While studying, look for the activity icon for:

- Vocabulary terms with e-flash cards and matching activities.
- Starter files for hands-on examples and other exercises.

These activities can be accessed at
www.g-wlearning.com/informationtechnology/1773

Chapter Glossary

Bureau of Labor Statistics (BLS): Principal federal agency responsible for measuring labor market activity, working conditions, and price changes in the economy; part of the Department of Labor.

C#: Programming language developed by Microsoft to complete its suite of programs called .NET; one of the primary languages for programming in Windows.

C++: Compiled programming language for use in system programming, desktop applications, and e-commerce servers.

career clusters: Sixteen groups of occupational and career specialties that share common knowledge and skills.

career plan: List of steps on a time line to reach career goals.

certification: Professional status earned by an individual after passing an exam focused on a specific body of knowledge.

formal education: Education received in a school, college, or university.

goal: Something a person wants to achieve in a specified time period.

hybrid job: Occupation requiring a combination of programming skills and industry-specific skills.

informational interviewing: Strategy used to interview a professional to ask for advice and direction rather than for a job opportunity.

internship: Short-term position with a sponsoring organization that provides an opportunity to gain on-the-job experience in a certain field of study or occupation.

JavaScript: Programming language used often with Internet browsers, game development, PDFs, and mobile and desktop applications; easy to learn and ideal for beginners.

long-term goal: Goal that will take a period of time greater than one year to achieve.

networking: Talking with people you know and then making new contacts to establish relationships that can help you achieve your goals.

portfolio: Collection of examples showing your qualifications, skills, and talents that support your career or personal goals.

postsecondary education: Education achieved after high school.

Python: General-use programming language known for its readability, straight-forward coding, and ease of learning.

résumé: Document that highlights a person's career goals, education, work history, and professional accomplishments.

short-term goal: Goal that can be achieved in less than one year.

Structured Query Language (SQL): Declarative language used for database management; often pronounced *sequel*.

Benefits of Careers in Coding

After completing an introduction to programming using the Java language, a student might be wondering if choosing a career in computer programming is something to be considered. There are many benefits from learning to code and making this a career goal. Coding has become a critical career skill. Employers have shown their willingness to pay a premium for the work of employees with coding and programming skills.

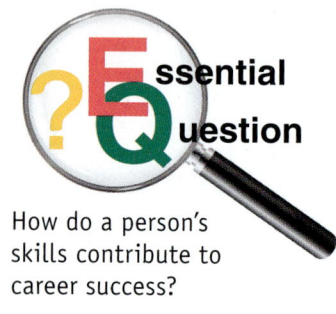

Essential Question

How do a person's skills contribute to career success?

Learning Goals

- Discuss benefits of a career in programming.
- Identify programming languages recommended to learn.

Terms

Bureau of Labor Statistics (BLS)
C#
C++
hybrid jobs

JavaScript
Python
Structured Query Language (SQL)

Discussion of Benefits

With technology driving the everyday lives of most Americans, a new kind of hybrid job has emerged. Blending technology with other fields is becoming more common in the workplace. Specifically, a *hybrid job* is an occupation requiring a combination of programming skills and industry-specific skills. These jobs are commonly found in design, data analysis, and marketing. The benefits of learning to code can be surprisingly wide-ranging. Here are a few of the ways learning to code can be beneficial:

- great earning potential
- strong demand for coding-related jobs
- proficiency in problem-solving
- career flexibility

Great Earning Potential

One of the strongest and most obvious draws of learning to code is the earning potential for coding and programming professionals. The *Bureau of Labor Statistics (BLS)* of the US Department of Labor is the principal federal agency responsible for measuring labor market activity, working conditions, and price changes in the economy. The BLS tracks salary and other important workforce information for an extensive variety of careers. According to the BLS, its mission is to collect, analyze, and disseminate essential economic information to support public and private decision-making. It is an independent statistical agency, so it can serve a wide range of communities by providing accurate, objective, relevant, timely, and accessible information.

Figure 15-1 shows the data from 2017 for programming-related jobs. Keep in mind that the average annual earnings for *all* occupations in 2017 was $50,620. Careers that involve some programming, coding, or scripting skills usually come with above-average salaries. Employment of computer and information technology occupations is projected to grow 13 percent from 2016 to 2026, which is faster than the average for all occupations. These occupations are projected to add about 557,100 new jobs.

Social networking, blogging, mass media, and online shopping are just some of the areas that have seen vast improvements thanks to developments in computers. Demand for IT workers will stem from greater emphasis on cloud computing, the collection and storage of big data, and information security.

Strong Demand for Coding-Related Jobs

The national average annual salary for all jobs is projected to grow at 7.4 percent according to the BLS. **Figure 15-2** shows the BLS projections for the coding and programming-related professions shown in **Figure 15-1.** Three out of five of these occupations exceed the national average.

What happened to computer programmers? The jobs are not disappearing. Analysts think computer programming skills will be blended into other related in-demand tech jobs. They will be part of the hybrid positions. An increase in hybrid positions has caused there to be fewer job postings for strictly programming and more opportunities that combine programming with a field-specific job.

Proficiency in Problem-Solving

If one would like to solve life's problems analytically rather than emotionally, learning to code is a positive step. If a problem arises, programmers are more capable of coming up with a solution that follows logical steps. When using an

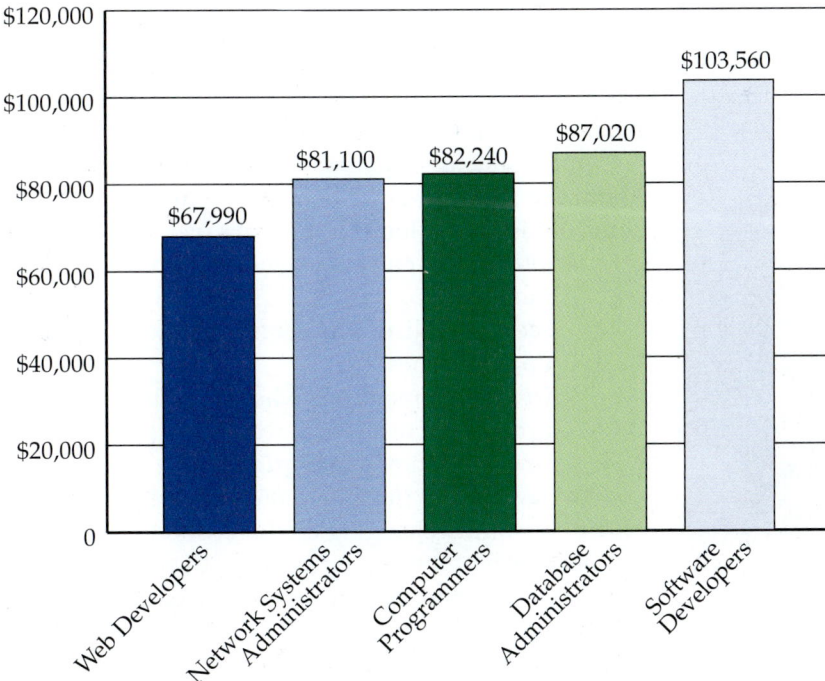

Goodheart-Willcox Publisher

Figure 15-1. These are the average annual salaries for several coding and programming careers for 2017, as reported by the Bureau of Labor Statistics.

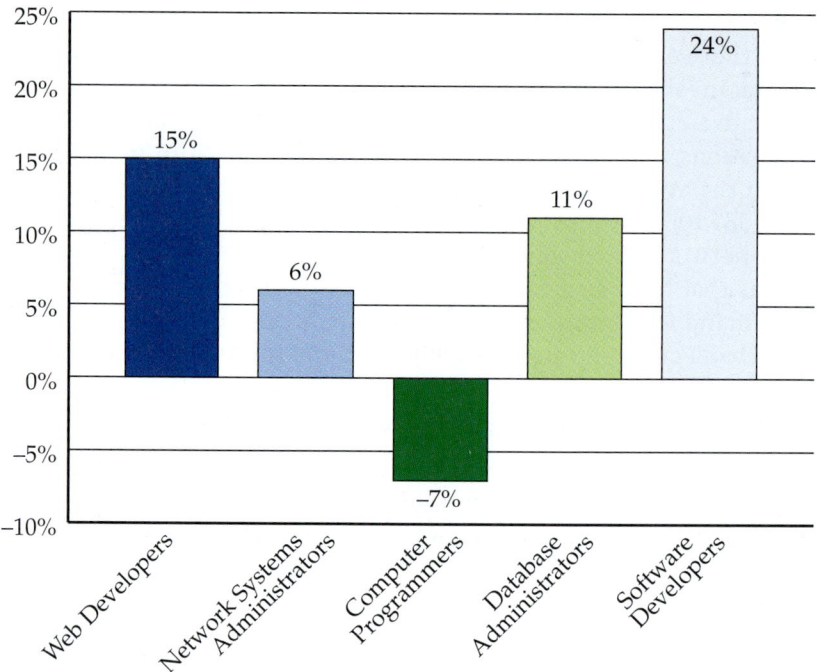

Figure 15-2. These are the Bureau of Labor Statistics projections for change in number of coding and programming jobs for 2016 to 2026.

algorithm on a computer, guidelines have already been set up. If one has a relationship or emotional issue, take a step back and use logic. Complex situations can be broken down in a collection of smaller tasks. This methodical approach to problems, in general, can help a programmer in all facets of life.

Career Flexibility

Many of the jobs of tomorrow will definitely include coding. Futurists have tried to predict them. The amount of data businesses will grow 50-fold over the next decade. As data becomes cheaper, faster, and more commonplace, the nature of work will begin to change as well.

According to Thomas Frey, a noted futurist author, many changes will cause new jobs to materialize. Here are some of his predictions for new jobs over the next 10 years:

- alternative-currency bankers who use electronic money
- seed capitalists who get funding for high-risk start-up companies
- waste data managers who streamline data storage
- urban agriculturalists who grow food in cities and towns, as shown in **Figure 15-3**

To have career flexibility, one needs to take advantage of the constantly changing digital economy. Even if you do not become a programmer, learning to code can give you a greater understanding of what is involved in programming. Coding increases communication skills so even a new coder is ready to interact with professionals in the field.

Figure 15-3. Urban agriculturalists grow food in cities and towns instead of on traditional rural farms.

HANDS-ON EXAMPLE 15.1.1

Accessing the Bureau of Labor Statistics

In this exercise, you will become familiar with the Bureau of Labor Statistics website. The BLS is a governmental agency and open to public use. Any user can find interesting information there.

1. Launch a web browser, and navigate to the BLS website (www.bls.gov).
2. Locate the menu bar, and click **Subjects>Inflation & Prices>Consumer Price Index**.
3. In your own words, define the Consumer Price Index (CPI). Do not directly copy the description on the website.
4. What is the current increase or decrease in the CPI?
5. Look at the chart for 12-month change by category. Which category has the most upward change in the past year? Include this with your description of the CPI and submit it to your instructor.
6. Locate the link for local and regional CPI. Click the link, and then choose your state and your city or a location close to it. Share one current fact that the BLS website displayed.

Programming Languages to Know

The programming languages you know can help advance your career as a computer programmer. The jobs a programmer can obtain and excel in depend on proficiency in certain programming languages. This list of six programming languages from the BLS gives some clarity on the languages most rewarded by employers. The BLS lists the number of computer programmer jobs as 294,900.

- SQL
- Java
- JavaScript
- Python
- C++
- C#

SQL

Structured Query Language (SQL) is a declarative language used for database management. SQL is often pronounced *sequel*. Common database software includes MySQL, SQL Server, Oracle, Microsoft Access, and IBM Db2. The programmers for these applications did not create a query or report-writing module. Instead, they all incorporate SQL into the application.

Java

Java is a general-purpose language that runs on many platforms. It is object-oriented and shares syntax with C and C++. It used to develop many website applications. As you know from studying this text, it is challenging to learn.

JavaScript

JavaScript is a programming language used often with Internet browsers, game development, PDFs, and mobile and desktop applications. It is easy to learn and ideal for beginners.

Python

Python is a general-use programming language known for its readability, straight-forward coding, and ease of learning. It is used by Google and YouTube.

C++

C++ is a compiled programming language for use in system programming. It is also used in desktop applications and e-commerce servers. C++ was created to improve on C, but is well-known for having a long learning curve.

C#

C# is a programming language developed by Microsoft to complete its suite of languages called .NET. It is one of the primary languages for programming in Windows. C#, which is pronounced *see-sharp*, resembles C++ and Java while trying to be more simplified and flexible. It plays a vital role in programming for the Internet, so it will be around for a while.

SECTION REVIEW 15.1

Check Your Understanding

1. What is a hybrid job?
2. Which federal agency measures labor market activity, working conditions, and price changes in the economy?
3. Why does the federal government list a projected decrease in programming jobs over the next ten years?
4. Which recommended programming language is used for database management?
5. Which recommended programming language was developed by Microsoft to complete the .NET suite of languages?

Build Your Vocabulary

As you progress through this course, develop a personal computer science glossary. This will help you build your vocabulary and prepare you for a career. Write a definition for each of the following terms and add it to your computer science glossary.

Bureau of Labor Statistics (BLS)	hybrid jobs	Python
C#	JavaScript	Structured Query Language (SQL)
C++		

Preparation for Careers in Coding

Preparation for a career in computer programming, and information technology in general, requires a specific set of skills. A candidate for this field must have formal training. In addition, various technical certifications are looked on favorably by employers. For example, Cisco certification in networking is a measure of proven success.

Most employers hiring in this area require a college degree, some technical certification, or some related experience. Some like to have all three. If a high school student had a chance to experience an introduction to Java course, that student has a big head start in preparing for a career in information technology. Proficiency in Java is a marketable skill that can be used to get a job in the IT field as well as help to learn other languages.

Learning Goals

- Develop a career plan.
- Create a results-oriented résumé and portfolio.

Terms

career clusters
career plan
certification
formal education
goal
informational interviewing
internship

long-term goal
networking
portfolio
postsecondary education
résumé
short-term goal

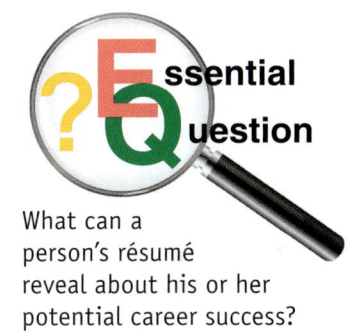

Essential Question

What can a person's résumé reveal about his or her potential career success?

Career Plan for IT

You have learned much about programming in this course. It has been challenging, but so enjoyable when your programs ran successfully. Planning for your career can be exciting, too. Your career choice will direct many other decisions. It will affect your educational choices and where you live.

A *career plan* is a list of steps on a time line to reach each of your career goals. It can be called a *postsecondary plan.* A career plan should include options for education. This may be four-year colleges, two-year colleges, and technical schools. The career plan should also address current job opportunities in your career of interest. **Figure 15-4** illustrates action items for a career plan.

Researching IT Career Information

Several resources exist to help research a career. People in IT rarely have the same job for more than a couple years. Sometimes the person finds a new, more-challenging, better-paying opportunity. In other cases, the person stays with the same company or organization, but the job changes because of new technology.

Action Items for a Career Plan: IT Specialist		
	Extracurricular and Volunteer Activities	**Work Experience**	**Education and Training**
During High School	• Volunteer in school IT lab or seek out co-op opportunities • Perform computer service and repair for nonprofit organizations	Work as an intern at a local electronics or computer-repair business	• Take classes in IT along with required coursework • Participate in career and technical student organizations (CTSOs)
During College	• Help student, nonprofit, or local groups identify the best solutions for their computing and electronic needs	Work as a part-time service or help-desk technician	• Follow the bachelor degree path for information technology
After College	• Obtain certification in your specialized area • Attend local IT professional and chamber of commerce events	Work as a technician	• Participate in appropriate professional development opportunities • Consider obtaining a master degree in computer science

Goodheart-Willcox Publisher

Figure 15-4. This is an example of a career plan.

An example of this occurred when the city of Baltimore passed a law requiring all police officers to wear body cameras. An employee who had the job of creating, delivering, and storing subpoenas on a database found her job had changed and she now had to discover how and where to store all the bodycam data.

Internet Research

The Internet is a good place to start. The Bureau of Labor Statistics tracks all jobs including jobs in IT. Also, most employers have websites that describe jobs they are currently seeking to fill and the qualifications needed. Additionally, most postsecondary schools have websites that provide career information. Governmental jobs are advertised on www.usa.gov.

The Occupational Information Network (O*NET) is a valuable resource for career information. O*NET Online (www.onetonline.org) is the most comprehensive database of occupational information. It was also created by the US Department of Labor and is regularly updated. This website contains data on salary, growth, openings, education requirements, skills and abilities, work tasks, and related occupations for more than 1,000 careers.

The *career clusters* are 16 groups of occupational and career specialties that share common knowledge and skills. Career clusters are centered on related career fields. The O*NET database can be searched by career cluster. Search for information technology.

Career Handbooks

Career handbooks offer a great place to begin researching specific careers, their industries, and area of the country or world in which these industries thrive. The Bureau of Labor Statistics publishes the *Occupational Outlook Handbook*. This handbook describes the training and education needed for various jobs. It provides up-to-date information about careers, industries, employment trends, and salaries. Information on IT careers are listed under the Computer and Information Technology occupational group.

Networking

Some open jobs are never advertised, especially in the IT field. Eighty percent of jobs are found by networking. *Networking* means talking with people you know and then making new contacts to establish relationships that can help you achieve your goals. Most employers would much rather hire someone who comes recommended by someone they already know. Asking family and friends for contacts in your field of choice is a good way to start. Then, get in touch with these professionals and ask if they know of any open jobs or employers who are hiring.

Informational Interviews

Informational interviews can give you unique insight into a career. *Informational interviewing* is a strategy used to interview a professional to ask for advice and direction rather than for a job opportunity. The purpose is to get a sense of what it is like to work in that profession. Be as professional and polite as possible. You may also find out who is hiring.

Setting SMART Goals

Another step in the career-planning process is to set goals. A *goal* is something a person wants to achieve in a specified time period. There are two types of goals: short term and long term. A *short-term goal* is one that can be achieved in less than one year. An example of a short-term goal may be getting an after-school job for the fall semester. A *long-term goal* is one that will take a period of time greater than one year to achieve. An example of a long-term goal is to attend college to earn a four-year degree.

Goals setting is the process of deciding what a person wants to achieve. Goals must be based on personal desires. Well-defined career goals follow the SMART goals model, as shown in **Figure 15-5**. *SMART goals* are specific, measurable, attainable, realistic, and timely.

Specific. A career goal should be specifically defined and stated. For example, "I want to have a career" is not a specific goal. Instead, you might say, "I want to have a career as an IT manager." When the goal is specific, it is easier to track progress.

Measurable. It is important to measure your progress so you know when you have reached your goal. For example, "I want to earn a bachelor degree in

FYI

Always follow up with your contact after an interview. Send a thank-you message to show appreciation for his or her time.

SMART Goals

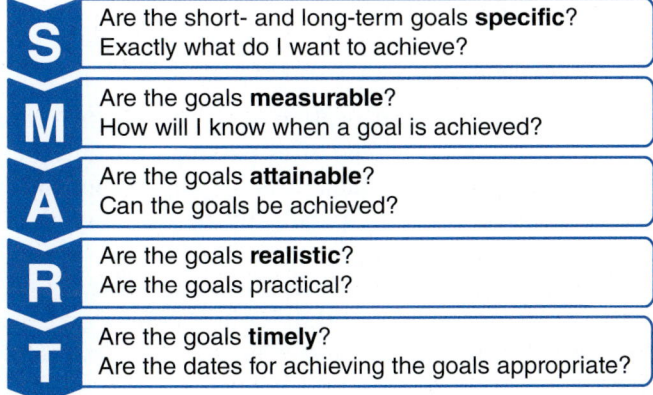

Goodheart-Willcox Publisher

Figure 15-5. The SMART goal model can help a person create well-defined career goals.

computer programming" is a measurable goal. When you earn the degree, you will know the goal was accomplished.

Attainable. Goals need to be attainable. For example, "I want to be a director of IT when I graduate from college." This is not reasonable for that point in a person's career. Gaining work experience is necessary before obtaining a management position. This goal becomes more attainable when coupled with a plan to gain the necessary aptitudes, skills, and experience required for a job position.

Realistic. Goals must be realistic. Obtaining a position as an IT manager may be practical with proper planning. It is not realistic for a new college graduate. Finding an entry-level position as a technician or system administrator and working your way up to IT manager over a period of years makes this a realistic goal.

Timely. A goal should have a starting point and an ending point. Setting a time frame to achieve a goal is the step most often overlooked. An end date can help you stay on track. For example, you may want to be a lead programmer by the time you are 35 years old. Aiming to get the experience and education to achieve this position by a specific age will help you remain motivated to reach the goal on time.

Discovering Educational, Certification, Training, and Experience Requirements

Your educational needs will depend on your career interest and goals. Some careers require a high school diploma followed by technical training or a bachelor degree. Others require graduate work, such as a master degree. Additionally, most IT careers require professional certification. Early planning on what is required can help you make decisions about your education.

Education

Formal education is the education received in a school, college, or university. *Postsecondary education* is any education achieved after high school, such as a technical certificate or college degree. Most IT careers require a college degree. Jobs higher on the career ladder often require additional formal education. This is called graduate, postgraduate, or continuing education.

Two-year and four-year colleges offer degrees in information technology. These programs may be found in computer science or management information systems departments. There are also licensed technical schools whose coursework concentrates on a particular IT career. For example, some technical schools focus on networking or web development. These schools grant certifications rather than diplomas or degrees.

Professional Certification

There are many types of certification available. *Certification* is a professional status earned by an individual after passing an exam focused on a specific body of knowledge. The individual usually prepares for the exam by taking classes either online or face-to-face specifically on the material tested. Certification exams may be given online or in person.

Certification programs are helpful to employers who are looking for a standardized way to assess a candidate's skill level. Having a certificate might give the candidate an extra advantage to gaining an interview. Some jobs *require* certification.

Some certifications must be renewed on a regular basis. For example, many certifications sponsored by Microsoft are only valid for a specific version of the

FYI

Certification may enhance a college degree, but it is not an alternative to a degree.

software. When the next version is released, another exam must be taken to be certified for the update. Other certification programs require regular continuing education (CE) classes to ensure individuals are current with the latest professional information.

Training and Experience

Most employers want candidates who have some real-world work experience. Taking advantage of a part-time position in the IT field is a good way to start. Many colleges hire students to handle technical support for their call centers. Others allow IT students to assist IT instructors in setting up and maintaining the computer labs. Some professors ask IT students to assist them with helping students in their classes.

Work experience can also be gained with an internship. An *internship* is a short-term position with a sponsoring organization that provides an opportunity to gain on-the-job experience in a certain field of study or occupation. Internships can be paid or unpaid. Regardless, they are a chance to gain valuable work experience. Many students take internships over summers while completing a degree.

Lifelong Learning

Lifelong learning is the voluntary attainment of knowledge throughout life. It typically refers to adults who are learning for the sake of learning in a variety of situations. Lifelong learning often relates to hobbies and interests, including art, cooking, foreign languages, outdoor recreation, and physical fitness.

There are numerous benefits to lifelong learning. The primary benefit is the knowledge gained, or learning for learning's sake. Learning more allows you to achieve goals you have established outside of your career or education and improve your self-esteem. Additional benefits include maintaining communication and social skills, expanding your interests, and mental acuity.

Hands-On Example 15.2.1

Investigating Certification

In this exercise, you will investigate software and hardware vendors and professional organizations that offer certification. Becoming certified in an area related to your career can be beneficial.

1. Launch a web browser, and navigate to the Microsoft website (www.microsoft.com).
2. Use the site's search function, and search for IT certification MCP to locate articles on certification.
3. Read several articles to see what types of certifications are available.
4. Navigate to the Cisco website (www.cisco.com).
5. Use the site's search function, and search for IT certification to locate articles on certification. Cisco offers several levels of network certification. What are the levels?
6. Investigate each level to see what is needed to pass the exam.
7. Navigate to the CompTIA website (www.comptia.org). CompTIA is a professional organization.
8. Click the **Certifications** link or select the certifications site options if offered in a pop-up window. Review the information provided on certification, training, testing, careers, and continuing education.
9. Prepare a paragraph on which of the certifications offered by these three organizations is most interesting to you. Share your opinion with your classmates and instructor.

Résumé and Portfolio

When seeking professional employment, it is almost always necessary to compose a résumé. A *résumé* is a document that highlights a person's career goals, education, work history, and professional accomplishments. **Figure 15-6** shows an example of a résumé for a high school student. The purpose of a résumé is to show you have what the employer seeks in a job candidate and to influence the reader into requesting an interview with you. You may have several versions of your résumé, each highlighting particular skills. Most job searches are conducted online through job boards. An electronic copy of your résumé must be included in your profile on these job boards.

In many cases, you will also need to have a portfolio to showcase your work. A *portfolio* is a collection of examples showing your qualifications, skills, and talents that support your career or personal goals. Portfolios are common for careers in which something tangible is created, such as a web page or online animation.

Creating a Résumé

The first impression most employers will have of you is your résumé. A potential employer will typically only spend 15 seconds reading the résumé before a decision is made whether or not to set up an interview. Therefore, a résumé should be neat and professional. It must be well-written and error-free.

When creating a résumé, use proper formatting and styles that are easy to read. Select an appropriate typeface, size, and formatting. A simple typeface in 11-point or 12-point type size is the best. The Internet will provide many examples of résumés you can review as a basis for your own résumé. If you hire a professional résumé writer, be sure that writer is experienced in the IT field. Ask to see samples of his or her work. A professional writer may cost between $250 and $400 depending on what you ask to be done.

A résumé should begin with your contact information followed by a career objective. Next, list professional accomplishments followed by work experience. Conclude by giving your educational experience and any certifications you have obtained.

Once the résumé is written, proofread it several times. Ask others to read it for errors as well. Potential employers may have to review hundreds of résumés for a single position. Be the one remembered for your skills and qualifications, not for the misspellings or grammatical mistakes you made.

Contact Information

The contact information should include your name, address, phone number, and e-mail address. It is okay to omit the street address, especially if you are posting on a job board. However, do not omit the city, state, and ZIP code. Some job boards have algorithms that automatically look for candidates in certain ZIP codes. Make sure the e-mail address is appropriate and professional. If needed, set up a new e-mail address.

Objective

The Objective section states what kind of position you are seeking. Use some of the specific terms that appear in the employer's job description. This shows the employer you are looking for a specific position, not just randomly sending out résumés. This section may be labeled Objective, Career Objective, or Career Goal.

Figure 15-7 shows an example of an objective section for someone seeking an IT position. Notice that the background qualifications are mentioned,

FYI

Many employers will check your social media presence before deciding to interview a candidate. Be sure your social media posts are all professional in nature.

FYI

The e-mail addresses XtremeMuscles@gmail.com and polkadot.momma@ yahoo.com are *not* professional e-mail addresses!

Robert Jefferies
123 Eastwood Terrace
Saratoga Springs, NY 60123
123-555-9715
rjefferies@e-mail.edu

OBJECTIVE
A mature and responsible high school student seeks an entry-level job as a computer repair technician.

EXPERIENCE
Saratoga Springs City Online Newspaper, Saratoga Springs, NY
September 2016 to present
Computer Support
- Maintained various computers for the newspaper.
- Assisted with network setup.
- Set up new computer stations.
- Installed new software as needed.

Hunter High School, Saratoga Springs, NY
September 2015 to September 2016
Student Volunteer
- Monitored the computer lab.
- Performed troubleshooting on lab computers.
- Installed software updates as needed.
- Performed routine software maintenance.

EDUCATION
Hunter High School, Saratoga Springs, NY
Expected graduation date: May 2017
Relevant coursework: Principles of Information Technology

HONORS
- Hunter High School Honor Roll, 8 quarters
- Winner: Student Troubleshooting Contest, 2014–2016

PUBLICATIONS
- Saratoga Springs City Online Newspaper
- Saratoga Springs City Calendar 2015

Figure 15-6. This is an example of a properly formatted résumé.

OBJECTIVE
Seeking a challenging position in information technology where broad knowledge of software applications, operating systems, hardware, and networks and a degree in applied information technology can be combined with skills in communication, customer relations, troubleshooting, and attention-to-detail to make the XYZ Company profitable.

Goodheart-Willcox Publisher

Figure 15-7. This is an example of an objective statement for a résumé.

including IT subjects and skills mastered. The candidate also states he or she wants to contribute to the profitability of the company. If the hiring organization is a nonprofit organization, instead of referencing profitability, you could say *to improve the quality of service to XYZ agency.*

Professional Accomplishments

The Professional Accomplishments section lists specific things you have done. These may be projects you completed, papers you published, or awards you have received. Think about your work and educational experience in terms of small successes. These accomplishments should have a beginning and an end. They should be projects or activities in which you had a measurable role.

Begin each statement with an active verb. **Figure 15-8** shows some examples that might be included on the résumé for a recent college graduate. Notice how this candidate has chosen to set the action verbs in a bold typeface to emphasize them. Each accomplishment also has a result. Potential employers like to see results.

Work History

The Work History section lists all places where you have worked. For someone just beginning in the IT field, this will be short. Include volunteer jobs as well as paid positions. Jobs that are not professional in nature, such as lifeguard or restaurant server, are often omitted unless some relevant experience was gained. Be prepared to explain any gaps in your work history. This section may be labeled Work History, Work Experience, or Experience. **Figure 15-9** shows an example of a work history section.

PROFESSIONAL ACCOMPLISHMENTS
Built a computer from individual components. Results: Gained hands-on knowledge of hardware and problem solving; began small, profitable business in computer repair.

Mastered coding in Visual Basic.NET, Java, and C++ as well as database software. Results: Became proficient in programming and database usage.

Goodheart-Willcox Publisher

Figure 15-8. This is an example of a Professional Accomplishments section for a résumé.

WORK HISTORY
University of Baltimore, Baltimore, MD (10/2019 to 8/2021)
Help Desk Analyst: Duties included building positive relations with faculty and students; solving hardware, Microsoft Office, PeopleSoft, and Internet problems.

Goodheart-Willcox Publisher

Figure 15-9. This is an example of a Work History or Experience section for a résumé.

Education and Certifications

The Education and Certification section lists degrees, where they were obtained, and the areas of concentration. The most recent should be first. Also include any certificates you have and the organizations that granted them. List any awards received in school or from the community. An inventory of various hardware and software expertise should also appear here. **Figure 15-10** shows an example of an Education and Certifications section.

Creating a Portfolio

If you have created web pages or video games, hard-copy prints should be included in a portfolio. Color representations of these projects are better. If you are creating an electronic portfolio, include the fully functional digital version. If you have completed any programming projects, these should be represented by screen captures and documentation.

Some common elements included in a portfolio are photocopies of certificates of accomplishment and diplomas. Samples of work, letters of recommendation, and any documents showing a talent or skill appropriate for the position should also be included.

Bring your portfolio to the interview. Do not submit it with your résumé and application unless specifically requested to do so.

EDUCATION AND CERTIFICATIONS
B.S., University of Baltimore, Baltimore, MD, Applied Information Technology, 2019

Eagle Scout Award, Boy Scouts of America, Baltimore Area Council, 2014

- PC Software: Java, C++, Cold Fusion, HTML, Microsoft Office, SQL, Windows Server
- Operating Systems: Windows 10, Linux

Goodheart-Willcox Publisher

Figure 15-10. This is an example of an Education and Certifications section for a résumé.

HANDS-ON EXAMPLE 15.2.2

Submitting a Résumé

In this exercise, you will investigate open positions posted on online job boards. In addition to networking, online job searches are the primary way most people find jobs.

1. Launch a web browser, and navigate to a job search site, such as Monster (www.monster.com) or Indeed (www.indeed.com).
2. Use the site's search function, and enter the key words computer support technician. Use your ZIP code to narrow the search to your area.
3. Read a few of the postings and assess what the job requirements are.
4. Investigate how to upload a résumé and apply for one of the jobs.
5. Write a paragraph explaining how to upload a résumé.

SECTION REVIEW 15.2

Check Your Understanding

1. Another name for a career plan is a(n) _____.
2. What is the publication by the Bureau of Labor Statistics that describes the training and education needed for various jobs?
3. How long do employers typically spend reviewing a résumé before deciding whether or not to set up an interview with the candidate?
4. Which section of a résumé lists specific things you have done using action verbs?
5. What is the term for a collection of samples of work, letters of recommendation, and any documents showing a talent or skill appropriate for a job?

Build Your Vocabulary

As you progress through this course, develop a personal computer science glossary. This will help you build your vocabulary and prepare you for a career. Write a definition for each of the following terms and add it to your computer science glossary.

career clusters	informational interviewing	portfolio
career plan	internship	postsecondary education
certification	long-term goal	résumé
formal education	networking	short-term goal
goal		

Cooperative Coding

Bureau of Labor Statistics Speaker

The Bureau of Labor Statistics (BLS) was created in 1880 by the Department of the Interior. In 1886, it published its first annual report, which was on industrial depressions. The BLS currently employs about 2,400 workers. Roughly half of those workers are economists and statisticians. The BLS provides many resources. One of those resources is providing experts to be speakers at conferences. The mission statement of BLS is:

> to collect, analyze, and disseminate essential economic information to support public and private decision making. As an independent statistical agency, BLS serves its diverse user communities by providing products and services that are accurate, objective, relevant, timely, and accessible.

speaker should have a wide range of knowledge of careers and have relevant information for young people planning their educations for their futures. You have been asked to select an expert from the Bureau of Labor Statistics because of its mission.

Navigate to the BLS website (www.bls.gov). Click **Home**>**BLS Speakers Available** from the menu at the top of the page. A sliding menu is displayed in the middle of the page. Click the right arrow to move through the slides to see what topics are available. Discuss each topic with your team. On each slide, there is a link that can be clicked for more information.

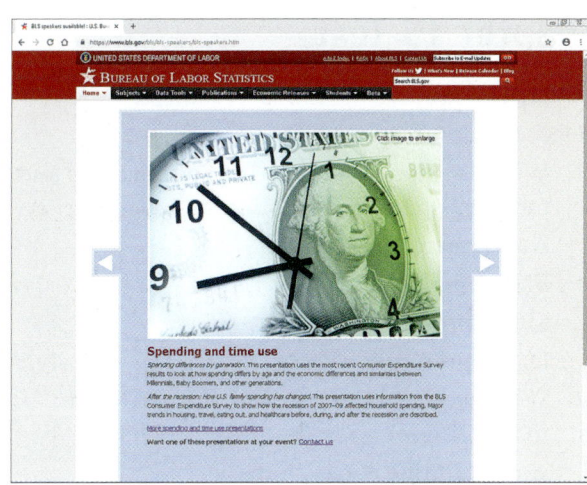

Goodheart-Willcox Publisher

Monkey Business Images/Shutterstock.com

Assignment

You and your team are in charge of planning a career fair to include about one hundred attendees from your school. The fair will include stations from a wide range of industries and professions as well as a keynote speaker. Your team has been asked to obtain a speaker for the keynote address that will set the tone for the event. The keynote

Each member should choose a primary and secondary option from the available subjects. Then, as a group, choose a single topic. Write a short justification for your committee's choice based on the relevance of the topic and the background of the audience.

1. What was the process your group used to agree on a single topic?
2. How would you go about requesting a speaker from the BLS?

Chapter Summary

Section 15.1: Benefits of Careers in Coding
- Learning to code can provide many career benefits, including great earning potential, strong demand for coding-related jobs, proficiency in problem-solving, and career flexibility.
- Programming languages recommended to know by the Bureau of Labor Statistics include SQL, Java, JavaScript, Python, C++, and C#.

Section 15.2: Preparation for Careers in Coding
- When creating a career plan, research information on your selected career, including educational, certification, training, and experience requirements.
- A résumé highlights a person's career goals, education, work history, and professional accomplishments, while a portfolio is a collection of examples showing your qualifications, skills, and talents that support your career or personal goals.

Chapter 15 Test

Multiple Choice

Select the best response.

1. What is a position requiring coding skills and industry-specific qualifications called?
   A. partnership
   B. hybrid
   C. corporation
   D. programmer

2. What would the future job of waste data manager do?
   A. streamline data storage
   B. use electronic money
   C. get funding for high-risk start-up companies
   D. grow food in cities and towns

3. Which programming language is used to question databases and make reports?
   A. Java
   B. JavaScript
   C. Python
   D. SQL

4. What is a professional status earned by an individual after passing an exam focused on a specific body of knowledge?
   A. certification
   B. diploma
   C. bachelor degree
   D. master degree

5. Which section of a résumé lists specific things you have done?
   A. Objective
   B. Professional Accomplishments
   C. Work History
   D. Education and Certifications

Completion

Complete the following sentences with the correct word(s).

6. Which governmental agency is responsible for measuring labor market activity, working conditions, and price changes in the economy?

7. _____ is a language developed by Microsoft to complete its suite of languages called .NET.

8. If a SMART goal has a starting and ending date, the goal is considered a(n) _____ goal.

9. _____ interviewing is a strategy used to meet with a professional to ask for advice and direction.

10. The first impression most employers will have of you is your _____.

Matching

Match the correct term with its definition.

   A. portfolio
   B. networking
   C. résumé
   D. Bureau of Labor Statistics
   E. career flexibility

11. Summary of accomplishments and education.
12. Benefit of being able to code.
13. Examples of projects.
14. How most IT jobs are found.
15. US governmental agency that tracks all aspects of jobs.

Application and Extension of Knowledge

1. Using the Bureau of Labor Statistics website (www.bls.gov), investigate several careers in the IT field, and note the median salary of each. Enter the data into a spreadsheet, and create a chart that shows the relationship of each career and its salary.

2. Find a posting for an IT job that interests you on Monster, CareerBuilder, or any other career site. Create a résumé using the guidelines in this chapter. For education and certifications you have not earned yet, use future dates for when you think you will secure them.

3. Using the career you selected in #2, investigate the educational requirements for the position. Create an education plan that will lead you to the career. Include at least two options for schools. Be sure to include estimates on the cost of completing your education.

4. Select an advanced position in the IT field that you might like to have after ten years in the field. Create a career plan to meet this goal. Follow the guidelines in this chapter.

5. Investigate the programming languages listed in this chapter as well as other languages. Identify one language other than Java you would like to learn. Write a one-page summary comparing this language to Java and explaining why you would like to learn it in addition to Java.

Online Activities

Complete the following activities, which will help you learn, practice, and expand your knowledge and skills.

Vocabulary. Practice vocabulary for this chapter using the e-flash cards, matching activity, and vocabulary game until you are able to recognize their meanings.

Communication Skills

Reading. Reading word by word is slow, and it lessens both concentration and the ability to connect concepts to form meaning. Active readers read groups of words, or *phrases,* rather than individual words. Practice reading this chapter phrase by phrase rather than word by word. How does this affect your understanding of the chapter?

Writing. Schools often have a code of conduct that applies to expected behavior for students. Your school might call it an honor code, behavior policy, or another name. Write a one-paragraph summary of your school's code of conduct as it applies to student behavior. Then, write a paragraph to describe the impact of this code on the school community as a whole.

Speaking. The way you communicate with others will have a lot to do with the success of the relationships you build. Create a speech that will introduce you to a counselor at a local college. The counselor should be a person you have never met. Deliver the speech to your class. How did the style, words, phrases, and tone used influence the way the audience responded to the speech?

Listening. *Reflective listening* occurs when the listener shows an understanding of what was said. Engage in a conversation with a classmate about a topic covered in this chapter. After the conversation, restate what your classmate said. How much did you remember?

Portfolio Development

College and Career Readiness

Organizing Your Portfolio. You have collected various items for your portfolio and tracked them in your master spreadsheet. Now is the time to organize the contents. Review the items and select the ones you want to include in your final portfolio. There may be documents you decide not to use. Next, create a flowchart to lay out the organization for your portfolio. Your instructor may have specific guidelines for you to follow.

1. Review the documents you have collected. Select the items you want to include in your portfolio.

2. Check the quality of each item in your folders. Make sure the documents you scanned are clear. Do a final check of the documents you created to make sure they are high quality in form and format.

3. Create the flowchart. Revise it until you have an order that is appropriate for the purpose of the portfolio.

CTSOs

Event Preparation. No matter what competitive events you participate in for a CTSO, you have to be well organized and prepared. Study the content exhaustively before the event and make sure all tools have been secured and travel arrangements have been made. Confirming details well in advance will decrease stress and leave you free to concentrate on the event itself. To prepare for a competition, complete the following activities.

1. Pack appropriate clothing, including comfortable shoes and professional attire.

2. Prepare all technological resources, including anything that you might need for preparation or competition. Double-check to make sure any electronic presentation material is saved in a format that is compatible with the machines available to you at the event.

3. If the event calls for visuals, have them prepared in advance, packed, and ready to take with you.

4. Bring registration materials, including a valid form of identification.

5. Bring study materials you have used to study for the event. If note cards are acceptable when making a presentation, make sure your notes are complete and easy to read. Have a back-up set in case of an emergency.

6. At least two weeks before the competition, create a checklist of what you need for the event. Include every detail down to a pencil or pen. Then, use this checklist before you go into the presentation so you do not forget anything.

Computing and Society

Sections

When you go online or log into any network of computers, you have a responsibility to the other users, and they have a responsibility to you. Users agree to abide by basic principles of right and wrong. Users agree to observe civility and show respect for others' ideas. The Internet is an extension of society, a new dimension of the world around us. The same standards of societal behavior people use in public apply to actions and behaviors online. Users agree to not participate in cyberbullying, exploitation, and theft. They also agree to abide by licensing agreements for software and hardware. Unfortunately, as a computer user, you must be aware of and know how to counter others who do not conform to society's ethics.

College and Career Readiness

Reading Prep

Before reading this chapter, turn each heading into a review question. As you read each section, see if you can answer the question you created.

While studying, look for the activity icon for:

- Vocabulary terms with e-flash cards and matching activities.
- Starter files for hands-on examples and other exercises.

These activities can be accessed at www.g-wlearning.com/informationtechnology/1773

Chapter Glossary 🔗

acceptable use policy (AUP): Set of rules that explains what is and is not acceptable use of school- or company-owned equipment and networks.

antivirus software: Detects and removes malicious software from a computer; also known as cyberdefense software.

blue skimming: Using a Bluetooth-enabled card skimmer to steal data.

computer virus: Computer code carried in another program that can replicate itself in order to corrupt or otherwise harm either data files or the software used to process these files.

cookies: Small text files that websites plant on the computer user's hard drive when a user initially visits the website.

copyright: Acknowledges the ownership of a work and specifies that only the owner has the right to sell the work, use it, or give permission for someone else to sell or use it.

cyberbullying: Harassment that takes place using electronic technology.

data breach: Security violation in which sensitive, protected, or confidential data is copied, transmitted, viewed, stolen, or used by an individual unauthorized to do so.

data vandalism: Manipulation or destruction of data found in cyberspace.

digital citizenship: Standard of appropriate behavior when using technology to communicate.

dumpster diving: Digging through the trash to locate information.

ethics: Principles of right and wrong.

hacking: Activity performed by computer programmers to break into e-mails, websites, computer systems, and files of other computer users.

identity theft: Stealing someone's personal information and using that information to commit theft or fraud.

infringement: Use of copyrighted material without permission.

intellectual property: Something that comes from a person's mind, such as an idea, invention, or process.

licensing agreement: Contract that gives one party permission to market, produce, or use the product or service owned by another party.

malware: Software that intentionally performs actions to disrupt the operation of a computer system, collect private information, or otherwise harm the computer or user.

open source: Software that has had its source code made available to the public at no charge and with no restrictions.

patent: Gives a person or company the right to be the sole producer of a product for a defined period of time.

phishing: Attempt to get private information by appearing as a harmless request.

piracy: Unethical and illegal copying or downloading of software, files, or other protected material.

plagiarism: Claiming another person's material as your own.

social engineering: Using various social methods to obtain information on an individual.

trademark: Protects taglines, slogans, names, symbols, and any unique method to identify a product or company.

ransomware: Blocks the user's access to programs until the user pays to unlock them.

scareware: Software designed to cause enough anxiety so the computer user leaps at the chance to take corrective action.

spyware: Program that secretly collects a user's data and behavior.

Trojan horse: Program that invites the user to run it while concealing malicious code that will be executed.

worm: Stand-alone computer program, not hidden in another program, that replicates itself in order to spread to other computers.

Computing and Ethics

Making good moral decisions is vital to our computerized society. Respecting others' intellectual property is part of that idea. These philosophies are especially important when using the Internet. Understanding licensing agreements for someone else's creation is necessary for managing all that is available on the Internet. Maintaining your own privacy and respecting the privacy of all other users is another facet of dealing with a digitized way of life.

Cyberspace allows users to act without anyone watching them. It is imperative that users make good decisions and do the right thing even with lack of supervision. For example, cyberbullying is an immoral behavior in cyberspace. There are sometimes legal liabilities for this online behavior, but there are always ethical responsibilities.

Essential Question

What implications does digital citizenship have for society as a whole?

Learning Goals

- Propose ethical actions related to intellectual property and activities in cyberspace.
- Discuss current and future issues of data privacy.

Terms

acceptable use policy (AUP)	intellectual property
copyright	licensing agreement
cyberbullying	open source
data breach	patent
digital citizenship	piracy
ethics	plagiarism
infringement	trademark

Ethical Responsibilities

Ethics are principles of right and wrong. Ethical use of the Internet involves not breaking the law. It is also tied to respecting the intellectual property of others. Acting ethically means doing what most individuals and social groups believe to be morally correct and good for society.

Digital citizenship is one aspect of ethics. *Digital citizenship* is the standard of appropriate behavior when using technology to communicate. Many schools and companies provide an acceptable use policy to guide students and employees on expectations for ethical use of computer resources. An *acceptable use policy (AUP)* is a set of rules that explains what is and is not acceptable use of school- or company-owned equipment and networks. This document should describe policies on copyrights, fair use, file sharing, online privacy, cyberpredators, and cyberbullying.

Intellectual Property

Intellectual property is something that comes from a person's mind, such as an idea, invention, or process. This includes written words as well as software. Computer code is intellectual property.

Plagiarism is claiming another person's material as your own. This is unethical and illegal. If you copy the work of someone else without permission, you are committing plagiarism. Crediting the source does not give you permission to use the material.

Copying whole sentences and paragraphs and using them as your own without attribution is unethical. However, consulting articles can be an effective first step while researching a topic. The Internet has made it easy to find information. It has made discovering plagiarism easier, too. There are many resources that can be used to check for plagiarism.

Laws protect individuals from the unauthorized and unethical use of intellectual property. This is done through copyrights, patents, and trademarks.

A *copyright* acknowledges the ownership of a work and specifies that only the owner has the right to sell the work, use it, or give permission for someone else to sell or use it. All original material is automatically copyrighted as soon as it is in a tangible form. An idea cannot be copyrighted.

Copyright laws cover all original work whether it is in print, on the Internet, or in any form. Copyrighted material may be indicated by the © symbol or the statement "copyright by." Lack of the symbol or statement does *not* mean the material is not copyrighted.

Any use of copyrighted material without permission is called *infringement.* Some examples of copyright infringement include: copying DVDs and selling them, printing copies of a book without permission, and distributing commercial music on the Internet.

A *patent* gives a person or company the right to be the sole producer of a product for a defined period of time. Once a patent expires, others can produce the product, as shown in **Figure 16-1.** Patents protect an invention that is functional or mechanical.

A *trademark* protects taglines, slogans, names, symbols, and any unique method to identify a product or company. A service mark is similar to a trademark, but it identifies a service rather than a product. Trademarks and service marks do not protect a work or product.

Intellectual laws provide protection to companies, organizations, and individuals who wish to develop innovations. Without this protection, the cost of innovation may be too great to pursue the idea or product. However, it may be argued that intellectual laws limit what can be developed to compete with protected material. This may be seen as having a harmful effect on innovation.

Although a copyright may be registered with the US Copyright Office at the Library of Congress, it does not need to be registered.

Hxdbzxy/Shutterstock.com

Figure 16-1. When LEGO toys were created, they were patented. Once the last patent expired in 1989, other companies could produce interlocking-brick toys. LEGO, however, remains a registered trademark of The LEGO Group.

HANDS-ON EXAMPLE 16.1.1

Locating Information's Source

The Internet is a great resource for locating information. It can be used to find material to support your original work or to locate the source of information.

1. Launch a web browser, and navigate to a search engine.

2. Enter the following quotation into the search box exactly as shown, including punctuation. Be careful not to mistype or misspell any words.

```
We're the richest economy in the history of the world. For the majority of Americans
not to get the benefits of this extraordinarily prosperous economy, there's something
fundamentally wrong.
```

3. Count the number of websites on the first page of results that contain these exact words in order.

4. Identify the media in which the original quote appeared.

5. Identify the author and original date of the quote. Be prepared to discuss your responses.

Licensing

Piracy is the unethical and illegal copying or downloading of software, files, or other protected material. Examples of protected material include games, images, movies, and music. Piracy is a form of stealing and carries a heavy penalty, including fines and incarceration. Imagine spending months or years developing a new game. For example, the most recent Spiderman game took four years to complete. The investment is costly in time and money. The developers want a financial return on the work. It is their creation and they own it. When somebody engages in piracy of that work, it is stealing from the developers.

As a rule, judges dispense tough sentences for software piracy. Here is an example. The owner of a popular website was sentenced in federal court to six years in prison because he and his coconspirators sold pirated copies of Adobe, Autodesk, and Macromedia—more than $4.1 million of copyrighted software. This theft resulted in nearly $20 million in losses to the software owners. At the time of sentencing, this was the longest prison term ever handed down in a software piracy case. The criminal was also ordered to forfeit the proceeds of his illegal conduct, pay restitution of more than $4.1 million, and perform 50 hours of community service.

How do you get to play a game or use software? The answer is licensing agreements. A *licensing agreement* is a contract that gives one party permission to market, produce, or use the product or service owned by another party. The agreement grants a license usually in return for a fee or royalty payment. When purchasing and installing software, including games, you agree to follow the terms of a license. For example, the cost of Microsoft Office 365 personal version is about $70 for one year. A different example is a photo supplier, such as Shutterstock or Getty Images. These companies charge a fee and, in return, grant you a license to use the image you download.

Not all software has usage restrictions. The term *open source* applies to software that has had its source code made available to the public at no charge and with no restrictions. LibreOffice and GIMP (GNU Image Manipulation Program) are examples of open-source software. To use the software, you must accept the terms of a license, but the legal terms differ dramatically from proprietary licenses. A Creative Commons (CC) license is one of several public licenses that allows for free distribution. A CC license is used when an author wants to give other people the right to share, use, and enhance a work.

FYI

Linux is an example of an open-source operating system.

HANDS-ON EXAMPLE 16.1.2

Investigating Licensing Agreements

Licensing agreements are drawn up to preserve the intellectual property of programs. This includes software for databases, word processing, accounting, sales management, billing, and even games. Developers spend months and years on their creations and should be compensated for them. Download the chapter files from the student companion website. Then, open the file Ch16SoftwareLicensingAgreement.doc, which is a sample licensing agreement. Read the agreement, and then answer the following questions.

1. Who is the licensor?
2. What is the name of the software?
3. How many PCs can it run on?
4. Is the software allowed to be used in a different location?
5. What date does the agreement expire?
6. How much does the software cost?
7. How long must the users maintain confidentiality about the code?

Cyberbullying

It is important to understand the far-reaching impact of your online actions. You may post something that is hurtful to someone else. However, once a message or image is posted, it can be difficult to remove it.

Cyberbullying is harassment that takes place using electronic technology. It includes using social media, text messages, or e-mails to harass or scare a person with hurtful words or pictures. A victim of cyberbullying cannot be physically seen or touched by the bully. However, this does not mean the person cannot be harmed by the actions. Cyberbullying is unethical and can be prosecuted. Several states have passed antibullying laws.

Cyberbullying is especially dangerous because of its relentlessness. Those who are cyberbullied are often also physically bullied. However, cyberbullying goes on twenty-four hours a day, seven days a week. It can reach those being bullied when they are alone without support.

The federal government created the www.StopBullying.gov website to educate people about this problem, as shown in **Figure 16-2.** According to this website, nine percent of students in grades 6 through 12 experienced cyberbullying during the school year studied. Additionally, 15 percent of high school students (grades 9 through 12) were electronically bullied in the past year.

FYI

Cyberbullying is not a joking matter. If you suspect that someone is being cyberbullied, report it.

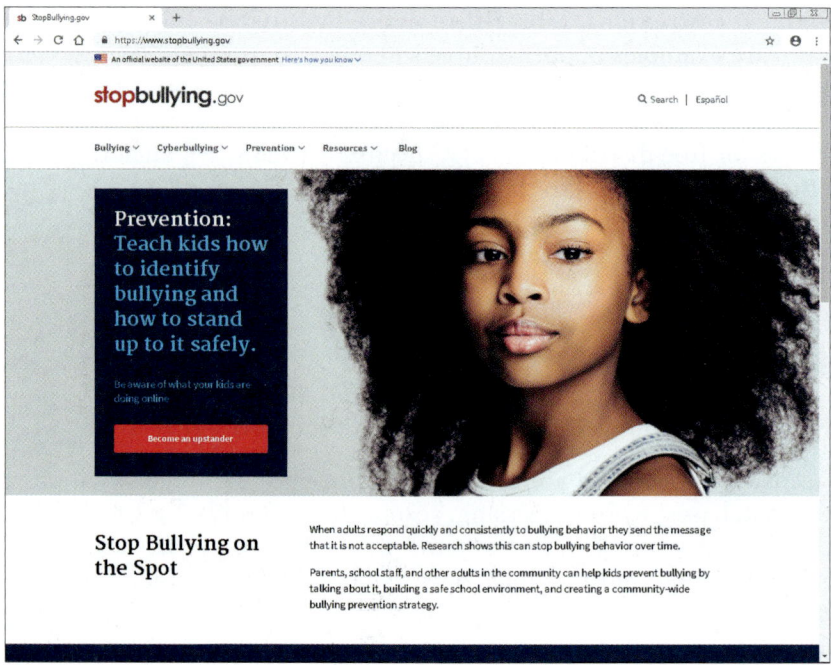

Figure 16-2. The StopBullying.gov website provides information from various governmental agencies on bullying and cyberbullying, including how to prevent and respond to bullying. The website is managed by the US Department of Health and Human Services.

HANDS-ON EXAMPLE 16.1.3

Locating Cyberbullying Information

It is important to understand what cyberbullying is and the negative impact it can have. Protect yourself from cyberbullying, and help to protect others.

1. Launch a web browser, and navigate to the www.StopBullying.gov website.
2. Using the website menu, click **Cyberbullying** and then **What is cyberbullying**, or use the search function to search for what is cyberbullying.
3. Read the information provided on the web page.
4. Using the menu on the left, click **Cyberbullying Tactics**, or use the search function to search for cyberbullying tactics.
5. Read the common tactics of cyberbullying. Have you experienced any of these or seen someone else experience them?
6. Using the menu on the left of the web page, click **Prevent Cyberbullying**, or use the search function to search for prevent cyberbullying.
7. Read the warning signs of cyberbullying provided on the page. Have you seen any of these signs in friends or other people you know?

Current and Future Issues of Data Privacy

Computers and the use of the Internet have changed many things in today's society. This digital life has revolutionized how people learn, shop, eat, search for jobs, and find their way around. People surf the web on a daily basis for information, directions, entertainment, and many other things. Now you can pay bills, order food, and buy groceries online. Other activities include checking e-mail and trading stocks. The police and the FBI have even used the Internet to help capture and convict criminals. There are many benefits, but also increased responsibilities.

Current Issues

Privacy of personal digital information is a great concern. Every time you purchase something online, the vendor places your information in a database. Your location and your phone calls are tracked in a database by your service provider, as shown in **Figure 16-3.** The names of people to whom you send e-mail, the websites you visit, and your search terms are all part of your Internet history. Companies large and small as well as governmental agencies have a lot of information about you. This data is available to anybody who can hack into the computer systems that store the data. As defined by the federal government, a *data breach* is a security violation in which sensitive, protected, or confidential data is copied, transmitted, viewed, stolen, or used by an individual unauthorized to do so.

Monkey Business Images/Shutterstock.com

Figure 16-3. The location of your smartphone can be tracked and stored as data, which means there is a history of everywhere you have taken your phone.

Be careful what information you post online. Do not overshare information. There are unethical cybersurfers who troll the Internet looking for personal data to steal and sell to others. Even if a photograph appears innocent, it may contain a clue to a password. For example, the name of your school may be obvious in the image. If you use this as part of your password, you have given hackers a hint.

There are four major companies that compete for our attention online: Google, Amazon, Facebook, and Twitter. These companies have developed algorithms with a goal of trying to keep users on their sites. At Stanford University, there is a Persuasive Technology Lab that investigates the techniques technology companies use to maintain your attention. For example, Snapchat is used by many teenagers. Snapchat created a log feature to tell a user how many continuous days he or she sent and received texts from another user. Some users are so addicted to this practice they give their passwords to friends if they are going to be "phoneless" for a day. Another example of persuading users to stay on a website is the way YouTube automatically starts playing the next episode or video. YouTube is trying to keep the user on the site.

This persuasive activity is done to influence users to buy services or products. However, there may be a positive use of this persuasive and monitoring ability. An ethical use might be to collectively develop algorithms for the benefit of society. For example, collective attention could be used to alleviate poverty, affect positive activities related to the environment, challenge social injustices, or combat loneliness.

FYI

Data breaches can expose credit card information. Always check monthly statements to verify purchases.

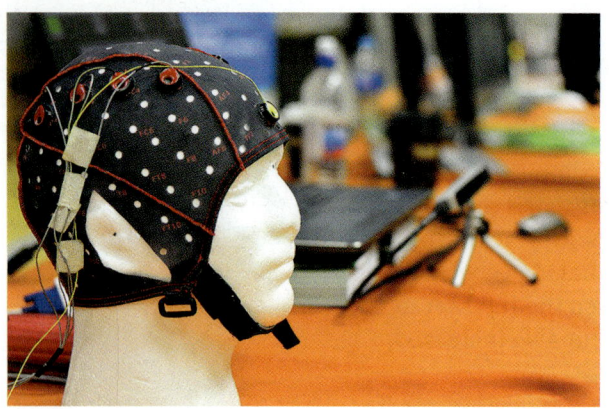

Figure 16-4. This cap contains electrodes that will sense electrical activity in the brain of the wearer.

Future Issues

According to the Center for Humane Technology, smartphones and the Internet are so essential to our daily lives, we have become a captive audience. Billions of people are connected to these tools, so people must feel there is an enormous benefit. Technology companies are able to have a direct channel to persuade users 24 hours a day, seven days a week. Additionally, no other media has such an accurate, personalized profile of what people text, click, and watch.

Your thoughts should be private. What if technology were able to tell what you were thinking? Brain-decoding technology is in its infancy, but is used in China. Using portable EEG monitors, high-speed train engineers in Shanghai are tracked constantly at work for attention-to-task and their emotional state. Electroencephalography (EEG) is the measure and mapping of electrical transmissions in the brain. It is the basis for brain-decoding technology. Sensors are placed on the head that can detect electrical activity in the brain, as shown in **Figure 16-4.** Brain-decoding technology raises concern over the right to freedom and privacy of thought.

SECTION 16.1 REVIEW

Check Your Understanding

1. _____ are principles of right and wrong.
2. What is an acceptable use policy (AUP)?
3. LibreOffice and GIMP are examples of _____ software.
4. What is a data breach?
5. How is brain-decoding technology being used on high-speed trains in China?

Build Your Vocabulary

As you progress through this course, develop a personal computer science glossary. This will help you build your vocabulary and prepare you for a career. Write a definition for each of the following terms and add it to your computer science glossary.

acceptable use policy (AUP)	ethics	patent
copyright	infringement	piracy
cyberbullying	intellectual property	plagiarism
data breach	licensing agreement	trademark
digital citizenship	open source	

Computing and Security

There are many types of threats to computer systems and computer users. Protection of digital data will continue to require a combination of technologically advanced hardware, frequently updated software, and safe practices and procedures. Ultimately, it is the individual computer user who is responsible for cybersecurity. It is not a duty to be trusted to the latest software update or the network administrator. Regardless of the equipment and antivirus programs, the actions of computer users are largely responsible for failures to protect computer data. This section discusses how to prevent computer threats and how to protect stored data.

Learning Goals

- Analyze threats to computers and users.
- Describe methods for protecting data.

Terms

antivirus software	phishing
blue skimming	ransomware
computer virus	scareware
cookies	spyware
data vandalism	Trojan horse
hacking	worm
malware	

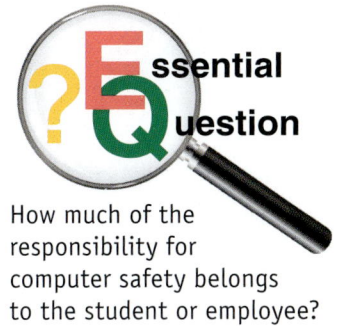

Essential Question

How much of the responsibility for computer safety belongs to the student or employee?

Analyzing Threats

There are many types of threats to computer systems and computer users. Malware most frequently finds its way into a computer by way of an executable code hidden in another program. Often the user does not find out about the damage until long after a successful attack. Malware attacks may occur when the user downloads seemingly harmless data from the Internet or transfers files from a shared flash drive. Other threats originate from phishing, data vandalism, cookies, and computer hacking.

Malware

Malware is software that intentionally performs actions to disrupt the operation of a computer system, collect private information, or otherwise harm the computer or user. The word *malware* is an abbreviated form of "malicious software," meaning software that intends to harm. It is a broad category of harmful software including:

- computer viruses;
- computer worms;
- Trojan horses;
- spyware; and
- scareware and ransomware.

A *computer virus* consists of computer code carried in another program that can replicate itself in order to corrupt or otherwise harm either data files or the software used to process these files. A computer worm is similar to a virus. However, a *worm* is a stand-alone computer program, not hidden in another program, that replicates itself in order to spread to other computers. A *Trojan horse* is a program that invites the user to run it while concealing malicious code that will be executed.

Spyware is a program that secretly collects a user's data and behavior. Spyware may activate a webcam, log keystrokes, collect login passwords, collect bank and credit card information, monitor Internet habits, or create tailored pop-up ads. Spyware is commonly used for advertising. When used this way, it is usually called *adware.* Best practices for avoiding spyware are given in **Figure 16-5.**

Similar to spyware are skimmers that steal credit or debit card data at gas pumps, ATMs, and payment terminals. The hacker inserts a hardware device on the outside or inside of a pay terminal. Then, credit or debit card information is intercepted during the payment transaction and transmitted to the hacker via Bluetooth technology. Using a Bluetooth-enabled skimmer to steal data is called *blue skimming.*

Scareware is software designed to cause enough anxiety so the computer user leaps at the chance to take corrective action. An example is a message that warns the user of a virus infection, then offers a free antivirus scan or a toll-free telephone number to receive help. However, the "scan" actually installs malicious software or the telephone "helper" asks to take over your computer. *Ransomware* is similar to scareware, but blocks the user's access to programs until the user pays to unlock them.

Cookies

Cookies are not malware, but can present security concerns. *Cookies* are small text files that websites plant on the computer user's hard drive when a user initially visits the website. They are used to identify users and often prepare customized web pages. On many e-commerce sites, cookies are required to keep track of items in the shopping cart. Many password-protected websites also require cookies to keep the user logged in.

Activity	Action and Reason
Pop-up dialog boxes	Do not click links within pop-up dialog boxes. Just clicking within the dialog box or a "close" button within the window may result in spyware being installed. Instead, close the pop-up dialog box by clicking on the standard close button (X) in the upper-right corner of the title bar.
Pop-up windows on websites	These are windows created using HTML or other website-formatting language. They can be recognized because they do not look like standard dialog boxes generated by the operating system. Often, all parts of a pop-up window, including what appears to be a standard close button (X), will activate spyware. Try pressing the [Esc] key to close the window or close the browser window.
Unexpected dialog boxes	Be suspicious of unexpected dialog boxes that ask whether you want to run a particular program or perform another type of task. Close the dialog box by clicking the standard close button (X) in the title bar.
Links offering antispyware software	These links may actually install the spyware it claims to be eliminating. Only install antispyware from the developer's website, not from a third-party site.

Goodheart-Willcox Publisher

Figure 16-5. Good practices for safe use of the Internet and the World Wide Web.

Cookies are also used to target the user with advertisements. For example, a user may search the Internet for "tours to Bali." The search engine places a cookie on the user's computer containing this information. Then, the user goes to a website for the daily news, and ads for tours to Bali show up in a sidebar. The news site has used the cookie from the search engine to target the ad to the user. By selling cookies, the search engine generates revenue. While this use of cookies is not a threat to the computer, it does represent collecting information about the user.

Phishing

Phishing is an attempt to get private information by appearing as a harmless request. For example, a user may get a telephone call that he or she has won a special prize or vacation. The offer states all that is necessary is to fill out a survey. However, the survey requests personal, sensitive information. The scam tries to obtain the target's full name, year of birth, credit card number, or Social Security number. If the request is online, the scammers might try to ask for those facts plus a physical address and phone number.

Data Vandalism

Data vandalism is the manipulation or destruction of data found in cyberspace. It is unethical and can be illegal, as shown in **Figure 16-6.** For example, a hacker may break into a school computer database and alter grades. Commercial competitors may engage in data vandalism, such as hacking into the website of a competing business to change the URLs to hyperlinks for invalid addresses.

Computer Hacking

Hacking is an activity performed by computer programmers to break into e-mails, websites, computer systems, and files of other computer users. Hacking

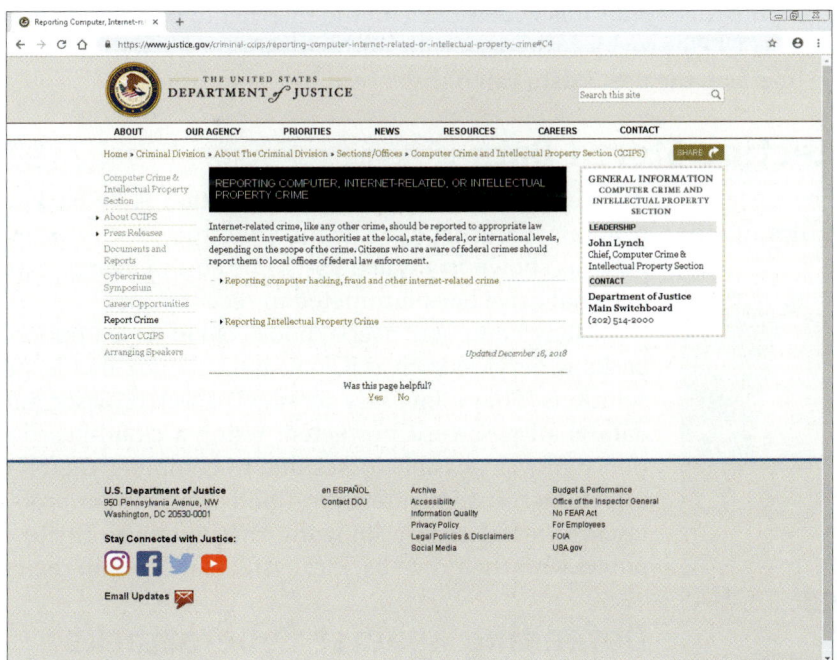

Goodheart-Willcox Publisher

Figure 16-6. The Department of Justice maintains a website for reporting computer crimes, including hacking and data vandalism.

is often unethical and illegal. However, there are legitimate hackers as well. Many companies hire hackers to find faults in their own computer systems. Then the faults can be fixed before they are exploited.

Protecting Data

Becoming aware of threats is the first step in securing data. The Internet is a network of connected devices all over the world. Large-scale coordination occurs among many different machines across multiple paths every time a web page is opened or an e-mail is sent. The domain name system (DNS) is used to identify every device on the Internet. Using the DNS, devices can look up other Internet protocol (IP) addresses. This allows end-to-end communication between devices. The advantage of the Internet is a seemingly unlimited worldwide communication. When it comes to privacy, however, this extensive exchange of data is the main disadvantage.

Internet Security Protocols

Internet protocols tell computers, modems, routers, and networks how to communicate with each other. Protocols also provide instructions on how to verify and handle information being received and transmitted. Internet security protocols are especially important because they reduce or eliminate malicious uses of the Internet. Two important Internet security protocols are SSH and HTTPS.

SSH

The secure shell (SSH) is a network protocol that secures data communication and remote command execution. It is used to authenticate the client and server machines and to establish a secure channel between them.

HTTPS

The secure hypertext transfer protocol (HTTPS) secures communication over computer networks. It is widely used on the Internet to provide secure connections. HTTPS provides the user with authentication of a website and web server. In effect, the user's data can only be read by the website.

Protecting Stored Data

The easiest way to protect information from corruption is by backing up those files in other locations. External hard drives are often used for this purpose, as shown in **Figure 16-7.** The cost of a large-capacity external drive has plummeted in recent years.

Many computer users choose cloud-based options for backups. Services such as iCloud and Dropbox are file-storage solutions. They allow easy access via a web browser, and the data is all password protected. Using a cloud-based solution removes possible losses due to computer thefts, virus infections, ransomware, fires, floods, and similar problems. Users can feel secure knowing that massive off-site computers with their own backup systems will retain their files.

Defending against Cyberattacks

In the rapidly changing world of cyberattacks, it is a very good idea to run antivirus software. *Antivirus software* detects and removes malicious software from a computer.

Pressmaster/Shutterstock.com

Figure 16-7. A common way to protect data is to create regular backups on an external hard drive.

It is also called cyberdefense software. Most actively prevent infections by scanning incoming data for malware.

If for any reason a virus cannot be removed, it will be quarantined. This means the virus is isolated from the rest of the files. Once in quarantine, the virus cannot infect anything else.

Cyberthreats evolve rapidly. Therefore, the antivirus software must be regularly updated. In most cases, the software will check for updates at set intervals, such as once a day or once a week.

FYI

Some antivirus software is available as freeware or open-source software. Others are for-purchase only.

HANDS-ON EXAMPLE 16.2.1

Investigating Antivirus Software

Antivirus software is an important part of cybersecurity. Before purchasing antivirus software, it is a good idea to research the different utility programs that are available.

1. Launch a web browser, and navigate to a search engine.
2. Enter these keywords: best antivirus software ratings.
3. In the search results, look for a link from a computer magazine, such as PC Magazine or PC World, that is a current review of antivirus software. A link to Consumer Reports that provides a review is also a good choice.
4. Click the link, and review the current ratings and prices. Ratings and prices will vary by source. Most software is usually between $50 and $100 annually. Some software offers a free version with basic features and a premium version with enhanced features for a price.
5. Select an antivirus software you feel would best meet your needs. Write one paragraph justifying your choice.

SECTION REVIEW 16.2

Check Your Understanding

1. _____ is software that intentionally performs actions to disrupt the operation of a computer system, collect private information, or otherwise harm the computer or user.
2. What are small text files used by websites to identify users and often prepare customized web pages?
3. When is hacking legitimate?
4. Of what are SSH and HTTPS examples?
5. What does it mean to quarantine a virus?

Build Your Vocabulary

As you progress through this course, develop a personal computer science glossary. This will help you build your vocabulary and prepare you for a career. Write a definition for each of the following terms and add it to your computer science glossary.

antivirus software	hacking	scareware
blue skimming	malware	spyware
computer virus	phishing	Trojan horse
cookies	ransomware	worm
data vandalism		

Safe Computing

The Internet has opened a wide range of information and activities. You can educate yourself about any topic. You can follow content from those you admire. You can easily learn about current events. Social media, in particular, allows users to build relationships with friends, colleagues, mentors, role models, and others. It increases the visibility of businesses tailored to individual needs. Users can connect at any time. Social media can help you bond with others before, during, and after an in-person event, such as a convention, conference, or meeting. People can get to know you prior to meeting you and you can share mutual expectations. Then, the face-to-face meeting will be much more valuable.

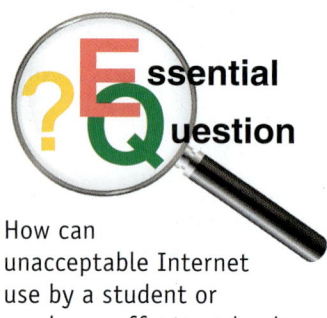

?Essential Question

How can unacceptable Internet use by a student or employee affect a school or company as a whole?

Learning Goals

- Evaluate websites for reliability of security.
- Formulate practices to safeguard personal information on social media.

Terms

dumpster diving
identity theft

social engineering

Determining Reliable Websites

FYI

Hackers can mask e-mail addresses to appear legitimate.

There are no iron-clad guarantees for knowing which websites can be reliably used without picking up a virus. Some of the most well-known and secure websites have been hacked. However, there are a number of ways to greatly reduce the risks of downloading a virus.

The safest way to navigate to a site is to manually enter the address into your web browser. Clicking a link to the site on another web page could send you to a fake site.

If the link is in an e-mail from an unknown sender, do not click the link. Delete the e-mail. Inserting links in e-mail is a very common method for getting people to go to virus-infected sites.

Another trick by hackers is to send fake e-mail from the address of a person you know with one line and a link. It often says "Hi, thought you would like this." Without clicking on the link or verifying with your contact, delete this e-mail, too.

URL

http://www.twitter.com

Protocol
Third-level domain
Second-level domain
Top-level domain

Goodheart-Willcox Publisher

Figure 16-8. A web address consists of several subparts. The second-level domain is what most people think of when talking about a website.

Thorough Inspection

The top-level domain is the last part of an Internet address, such as .com, as shown in **Figure 16-8.** When receiving e-mail or surfing the web, check this part of the address to see if it is a typically reliable domain. Governmental domains (.gov) and educational domains (.edu) are usually safe and reliable. These domains are not available to the general public to use, as shown in **Figure 16-9.** However, many educational institutions provide their students with .edu e-mail addresses or web space. Student accounts are often not as closely monitored as the school's main sites. Therefore, they may present the risk of malware.

Look at the structure and appearance of the website. If it is very basic and poorly designed, it may pose a risk. The site may be passing along malware intentionally or it may contain files infected due to carelessness. Look for misspellings and grammatical errors, broken links, and elements that overlap or are otherwise improperly formatted. If the site is complex and appears to have been built by dedicated programmers, it is more likely to be a safe site, although this is not guaranteed.

Before going to a site, enter the site name into a search engine along with the word "complaints." If there are problems with the site, there will likely be many complaints.

Many risky websites are designed to have a very professional appearance. Just because a site appears well-developed does not mean it is safe.

Safe File Downloads

In the rapidly changing cyberwar, any file could be potentially unsafe. There are some questions you can ask to help decide if the download is likely safe or unsafe.

- Have you successfully used the site before?
- Is the website a governmental site?
- Does the site name or invitation to download look suspicious?
- Does the site or file have a digital logo or signature?
- Does your antivirus software warn about the site or the file?
- Do you have sufficient time to download the file without interruption?
- Can the file be downloaded to a separate drive and tested there before installing it on your main computer?

Be particularly wary of certain file types, such as those with these file extensions: .bat, .com, .exe, .pif, and .scr. These files are executable files that run programs, which may be malware.

Carefully check the source to make sure you trust it before downloading the file. Remember, a file can only have one file name extension. If a file appears to have two file name extensions, such as readme.txt.exe, only the last one is the actual extension. This could be an attempt to make the user think the file is safe, but it likely contains malware.

Be Social Media Savvy

Identity theft is an illegal act that involves stealing someone's personal information and using that information to commit theft or fraud. Unfortunately,

Top-Level Domain	Members	Example
com	Commercial entities	g-w.com
gov	US governmental agencies	whitehouse.gov
edu	US colleges and universities	stanford.edu
org	Organizations	internetsociety.org
net	Open domain, anyone can register	sourceforge.net

Goodheart-Willcox Publisher

Figure 16-9. Common top-level domains and examples of who may use them.

identity theft is rampant on the Internet. Most credit card companies will compensate victims of fraud. However, the victim still pays a huge price in time and aggravation while receiving new credit cards, notifying banks, and notifying vendors.

Millions of people fall victim to fraud each year. The Federal Trade Commission (FTC) estimates that as many as nine million Americans have their identities stolen each year. The impact can be financially and emotionally devastating. The FTC maintains a website for reporting identity theft (www.identitytheft.gov), as shown in **Figure 16-10.** It also provides steps for recovering from identity theft.

Computer users should protect their identities when visiting websites. Most legitimate websites include a privacy statement, typically found at the bottom of the home page. Here are some additional steps each Internet user should take to minimize the threat of identity fraud.

- Never click any links in unsolicited e-mails.
- Do not set social media profiles to public.
- Immediately delete messages from suspicious senders.
- Fill in only required fields in online forms and consider using an alternate spelling of your name.
- Carefully inspect the terms of service for websites.
- Be sure to understand the purpose of any check box relating to sharing your information.
- Ensure that the lock symbol is shown in the browser's status bar and look to see that the address begins with https.

A common method for stealing someone's identity is social engineering. *Social engineering* is using various social methods to obtain information on an individual, such as searching social media platforms, dumpster diving, or even phone calls. *Dumpster diving* is digging through the trash to locate information, such as addresses, account numbers, and prescreened credit or insurance offers.

FYI

Most websites track visitors as they navigate through cyberspace.

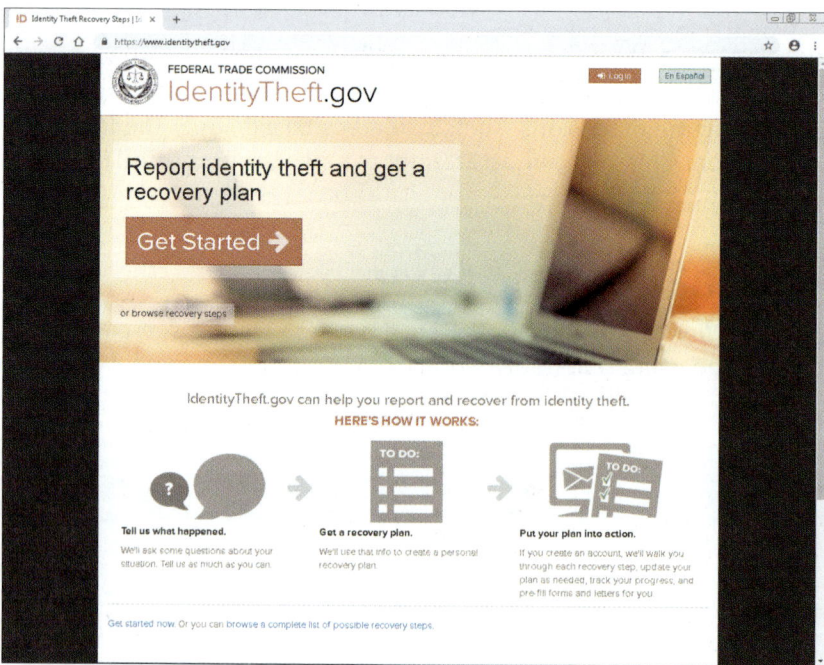

Goodheart-Willcox Publisher

Figure 16-10. If you have been a victim of identity theft, report it to the Federal Trade Commission.

Make sure your social-networking profiles are set to private. Social media websites, such as Facebook and Twitter, should not be allowed to display any personally identifiable information. Check the security settings for each account. Configure all accounts to control your status, who can gain access to information, who can post, and who can share account information with others. Use each site's search function to find out how to change security settings.

Be careful what information you post online. Do not overshare information. Also, give consideration to which photographs you post. What is shown in the photograph may be used to identify you. However, electronic images may also contain unseen information that identifies where and when the photograph was taken.

Consider opting out of pre-screened offers of credit and insurance. You can opt out for five years or permanently. To opt out, call 1-888-567-8688 or visit www.optoutprescreen.com.

HANDS-ON EXAMPLE 16.3.1

Preventing Identity Theft

Identity theft is a real online problem, but it is not limited to the Internet. The FTC provides information on how to protect yourself from identity theft and what to do if you are a victim.

1. Launch a web browser, and navigate to the FTC's consumer website (www.consumer.ftc.gov).
2. Using the site's menu, click **Privacy, Identity & Online Security**>**Identity Theft**. Alternately, search the site for identity theft.
3. Locate the article titled How to Keep Your Personal Information Secure, and open it.
4. Read the information.
5. Identify steps to take to keep your personal information secure both offline and online.

SECTION REVIEW 16.3

Check Your Understanding

1. Governmental websites are usually trustworthy. What is the top-level domain for a governmental website?
2. Dumpster diving is an example of a(n) _____ technique.
3. _____ is an illegal act that involves stealing someone's personal information and using that information to commit theft or fraud.
4. What is the address for the website maintained by the Federal Trade Commission for reporting identity theft?
5. What should your social networking profiles be set to in order to keep personally identifiable information from being shared?

Build Your Vocabulary

As you progress through this course, develop a personal computer science glossary. This will help you build your vocabulary and prepare you for a career. Write a definition for each of the following terms and add it to your computer science glossary.

dumpster diving   identity theft   social engineering

Cooperative Coding

Tech-Support Scams

In March 2019, a federal judge issued a restraining order against a company in Utah that provided technical support for computer users. The Federal Trade Commission (FTC) alleges the company deceived consumers to gain access to their computers. Then, the company convinced the consumers their computers were infected with malware and needed costly repairs and other services that were not needed. Most of the victims were older consumers. Older adults are more likely than younger people to lose money to tech-support scams.

Verbaska/Shutterstock.com

Assignment

There is a retirement community close to your school. As a school-outreach program, you and your team have been asked to give a talk to seniors who live in the retirement community. The presentation is to be about ways to avoid tech-support scams and what to do if a person is a victim. Include these topics in your presentation:

- spotting and avoiding tech-support scams
- what to do if you think there is a problem with your computer
- what to do if you were the victim of a scam
- how to report a tech-support scam

Use the FTC website (www.ftc.gov) as a resource. Using the site's search function, search for how to spot, avoid, and report tech-support scams. This site contains several resources on this topic. As a group, review the information on the site and plan a presentation. Each member should be responsible for one part of the presentation. You may decide to prepare the presentation as a slideshow or as a verbal presentation.

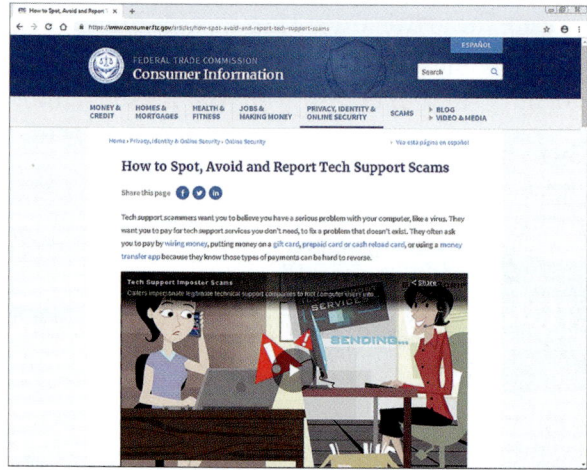

Goodheart-Willcox Publisher

1. Which resources on the FTC website did your team use as reference?
2. What format did your team decide to use for the presentation, and why?

Chapter Summary

Section 16.1: Computing and Ethics

- Ethics are principles of right and wrong and should be used to guide how intellectual property is handled, including copyrights, patents, and trademarks, as well as adhering to licensing agreements and not engaging in cyberbullying.
- Current issues of privacy of personal information include online presence, phone calls, and location monitoring among others, while in the future, privacy concerns may additionally be related to such things as monitoring brain activity.

Section 16.2: Computing and Security

- There are many types of malware that intentionally performs actions to disrupt the operation of a computer system, collect private information, or otherwise harm the computer or user; cookies, phishing, data vandalism, and computer hacking are other concerns related to computer security.
- Two important Internet security protocols are SSH and HTTPS, and antivirus software can be used to detect and remove malware, but the easiest way to protect data from corruption is to make regular backup copies.

Section 16.3: Safe Computing

- Steps can be taken to help determine if a website is reliable, from a thorough examination of the site and its address to checking for correct file extensions of downloaded files.
- Identity theft is an illegal act that involves stealing someone's personal information and using that information to commit theft or fraud, which is commonly conducted through social engineering, and the FTC maintains a website for reporting identity theft; setting social media profiles to be private and carefully choosing what to post can help protect against identity theft.

Chapter 16 Test

Multiple Choice

Select the best response.

1. Which of the following is considered plagiarism?
   A. Referencing an article in your term paper.
   B. Copying paragraphs from an online article and claiming them as your own.
   C. Copying music from a CD and giving it away.
   D. Producing an item without a proper licensing agreement.

2. What would copying and selling unlicensed copies of a DVD be considered?
   A. infringement
   B. plagiarism
   C. identity theft
   D. fair use

3. Which of the following is not malware?
   A. worms
   B. computer viruses
   C. cookies
   D. Trojan horses

4. A(n) _____ is a device that enables hackers to steal credit card information at gas pumps and ATMs.
   A. skimmer
   B. hacker
   C. plagiarizer
   D. black box

5. Which of these file extensions should not raise much concern after a download?
   A. .txt
   B. .exe
   C. .scr
   D. .com

Completion

Complete the following sentences with the correct word(s).

6. _____ is something that comes from a person's mind, such as an idea, invention, or process.

7. _____ is harassment that takes place using electronic technology.

8. When a computer is infected with _____, the perpetrators want money to unlock it.

9. Stealing someone's personal information and using it to commit fraud is _____.

10. _____ is using various social methods to obtain information on an individual.

Matching

Match the correct term with its definition.
   A. licensing agreement
   B. antivirus software
   C. cyberbullying
   D. intellectual property
   E. advice on reliable websites

11. Harassment that takes place using electronic technology.

12. Is it a governmental site?

13. Contracts that give one party permission to market something owned by another party.

14. Prevents malware infection.

15. Game, invention, or program.

Application and Extension of Knowledge

1. Choose a well-known business to investigate to see if the company has any trademarks or patents. Visit the corporate website and look for a copyright statement. Based on your findings, prepare for a class discussion to describe how this company protects its intellectual property.

2. Research the laws that relate to Internet security. When were they created? What are their purposes? Next, research laws regulating hacking and browser hijacking. Summarize what you learned about Internet security and ethical conduct.

3. Identify five applications of passwords in your personal life. These may be your cell phone, your computer, online gaming websites, social media sites, or anything that requires a password. Research the password requirements for each. Identify which characters are allowed, if there is a minimum length, and any other requirement. Make a table or chart to compare the requirements. Then, locate three different websites that check the strength of a password and use the sites to check the strength of your passwords.

4. The federal government and the White House have undertaken many cybersecurity policy initiatives. The White House website (www.whitehouse.gov) contains information about cybersecurity. Using the site's search function, search for foreign policy cybersecurity. Locate an article on foreign policy related to cybersecurity. Read the article, then write a summary of the government's objectives.

5. The United States Department of Justice website (www.justice.gov) can be used for reporting computer and intellectual-property crimes. Using the site's search function, search for reporting intellectual-property crime. Locate the guide for reporting intellectual-property crime, and read the pamphlet. Write a summary of what a person or organization should do if they are a victim of an intellectual-property crime.

Online Activities

Complete the following activities, which will help you learn, practice, and expand your knowledge and skills.

Vocabulary. Practice vocabulary for this chapter using the e-flash cards, matching activity, and vocabulary game until you are able to recognize their meanings.

Communication Skills

Reading. When engaging in active reading, it is important to relate what you are reading to your *prior knowledge,* which is what you already know or have already experienced. This helps you understand and form judgments about what you are reading. Select one of the sections of this chapter to read again. Assess whether your prior knowledge helped you understand the content.

Writing. *Compare* means to look at two things to find what is the same between them. *Contrast* means to look at two things in order to find what is different between them. Write one paragraph comparing working part-time while in school to not working while in school. Write another paragraph contrasting these scenarios. Develop and strengthen your ideas as needed by revising and editing your writing.

Speaking. Most people in the United States act as responsible and contributing citizens. How can a person demonstrate social and ethical responsibility in a digital society? Can you think of ways that are not discussed in this chapter? Share your opinions with the class.

Listening. *Deliberative listening* is listening to decide the quality or validity of what is being said. Listen to a classmate who is trying to persuade you. To *persuade* is to convince a person to take a course of action or adopt a viewpoint. Cite which points the person made that would convince you to do what he or she wants you to do.

Portfolio Development

College and Career Readiness

Presenting Your Portfolio. You have organized the components of the portfolio. Now, you will create the final product. Start with a flowchart to create the order of your documents. After you have sorted through the documents you want to include, print a copy of each. Next, prepare a table of contents for the items. This will help the person reviewing the portfolio understand its contents.

Your instructor may have examples of print and digital portfolios you can review for ideas. There may be an occasion where a print portfolio is required rather than a digital one. The organizational processes are similar. Search the Internet for articles about how to organize a print or digital portfolio.

1. Review the documents you have collected. Select the items you want to include in your portfolio. Make copies of certificates, diplomas, and other important documents. Keep the originals in a safe place.
2. Create the slideshow, web pages, or other medium for presenting your digital portfolio.
3. View the completed digital portfolio to check its appearance.
4. Place physical items in a binder, folder, or other container.
5. Present the portfolio to your instructor, counselor, or other person who can give constructive feedback.
6. Review the feedback you received. Make necessary adjustments and revisions.

CTSOs

Day of the Event. You have practiced all year for this CTSO competition, and now you are ready. Whether it is for an objective test, written test, report, or presentation, you have done your homework and are ready to shine. To prepare for the day of the event, complete the following activities.

1. Get plenty of sleep the night before the event so you are rested and ready to go.
2. Use your event checklist before you go to the presentation so you do not forget any materials needed for the event.
3. Find the room where the competition will take place and arrive early. If you are late and the door is closed, you will be disqualified.
4. If you are making a presentation before a panel of judges, practice what you are going to say when you are called on to speak. State your name, your school, and any other information requested. Be confident, smile, and make eye contact with the judges.
5. When the event is finished, thank the judges for their time.

Installing Java

Java is an object-oriented programming language. It uses a combination of a compiler and an interpreter to run programs. To write Java programs, download and install three software systems:
- Java Development Kit (JDK)
- JavaFX software development kit (SDK)
- integrated development environment (IDE)

The Java Development Kit contains the Java compiler, the Java Virtual Machine (JVM), other software tools, and the Java Class Library. The JavaFX software development kit contains software tools for creating graphics.

An integrated development environment provides a user-friendly interface to the compiler and JVM. Many fine IDEs are available for writing and running Java programs. This textbook uses jGRASP because it is easy to use for writing many small programs.

Installing the Java Development Kit

This textbook uses the open-source version of the Java Development Kit. This version of the software is free to use under the GNU General Public License. To install the JDK:

1. Launch a web browser, and navigate to the jdk.java.net website.
2. Click the link for the latest "ready for use" JDK.
3. Download the version or "build" that fits your operating system. Do *not* select a beta or "early access" version. Those are still being tested and may contain errors.
4. After the download is complete, unzip the file, and move the unzipped folder to an easily accessible location on your computer. Write down the path to this folder.

A word about Java versions: as of this writing, the latest version is JDK 13. Oracle updates Java every six months. Most of the changes in the new versions do not apply to the sections of Java used in this textbook. The version you install might have a higher version number, but the programs you will write will still run.

Installing JavaFX

To run graphics programs, you will need to install Open JavaFX. This software development kit is also open source and free to use. To install the SDK for JavaFX:

1. Launch a web browser, and navigate to the openjfx.io website.
2. Locate and click the **Download** link. This will take you to the Gluon website where you can download the appropriate JavaFX SDK for your operating system.
3. After the download is complete, unzip the file, and move the unzipped folder to an easily accessible location on your computer. It is recommended

putting JavaFX in the same folder in which you put the Java JDK. Write down the path to this folder.

Installing the IDE

The jGRASP IDE was developed at Auburn University. It is free to use and is ideal for students who will be writing many small programs. To download and install the IDE:

1. Visit the jGRASP website (www.jgrasp.org).
2. Locate and click the download link, and follow the instructions for your operating system.

After jGRASP is installed, before writing any Java programs, jGRASP needs to know where Java and JavaFX are located. To inform jGRASP where these programs are located:

1. Launch jGRASP.
2. Click on **Settings>PATH/CLASSPATH>Workspace** on the jGRASP menu bar. The **Settings for Workspace** dialog box is displayed.
3. Click the **PATH** tab and then the **PATHS** subtab.
4. Click the **New** button. The **New PATH** dialog box is displayed.
5. Click the **Browse** button, navigate to the bin subfolder under the main folder where you saved the Java JDK on your computer, and click the **Choose** button. Finally, click the **OK** button in the **New PATH** dialog box. When you finish, the window should look as shown, but your specific folder location should appear.

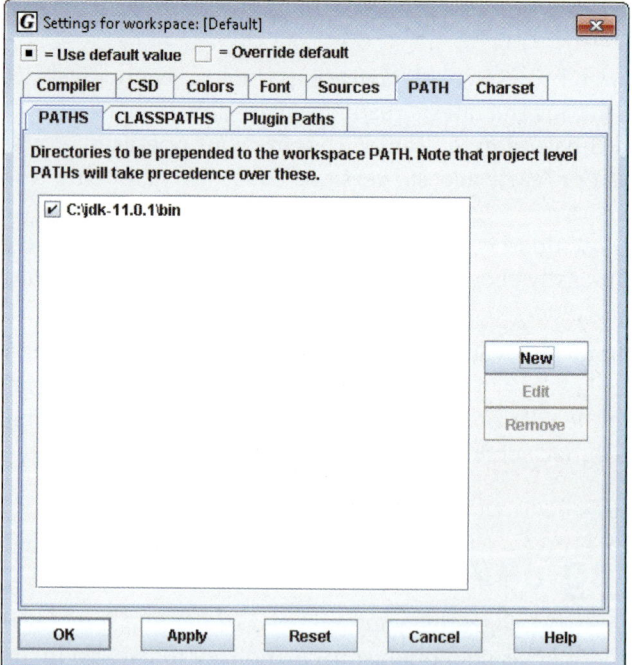

Goodheart-Willcox Publisher

6. Click the **OK** button to close the **Settings for Workspace** dialog box. Now you need to identify the location of JavaFX.
7. Click **Settings>Compiler Settings>Workspace** on the jGRASP menu bar. The **Settings for Workspace** dialog box is displayed.
8. Click the **Compiler** tab and then the **Flags/Args** subtab.

9. In the **FLAGS or ARGS** column, unselect the radio button next to **Compile**. The square will be empty (not filled in) when unselected. Next, click in the text box to the left, and enter:

```
--module-path C:\Users\javafx-sdk-11\lib --add-modules=
   javafx.controls --add-modules=javafx.media
```

but enter your exact path for C:\Users\javafx-sdk-11\lib. Be sure to enter two hyphens before module-path and also before add-modules.

10. In the FLAGS2 or ARGS2 column, unselect the radio button next to **Run**. Then, click in the text box and enter the following, again substituting your exact path for C:\Users\javafx-sdk-11\lib.

```
--module-path C:\Users\javafx-sdk-11\lib --add-modules=javafx.
controls --add-modules=javafx.media
```

11. When you finish, the window should look as shown, but your specific folder location should appear. Click the **OK** button to close the dialog box. Now you need to identify the version of Unicode.

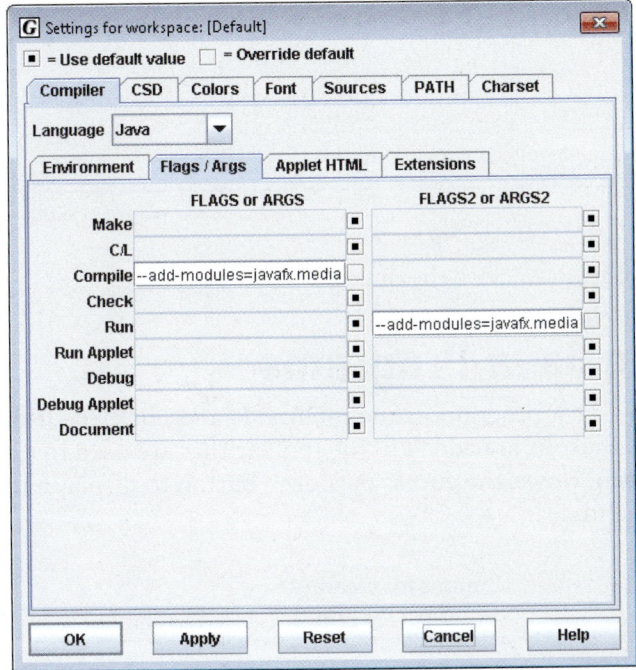

Goodheart-Willcox Publisher

12. Click **Settings>Compiler Settings>Workspace** on the jGRASP menu bar. The **Settings for Workspace** dialog box is displayed.

13. Click the **Charset** tab.

14. For all three drop-down lists, unselect the radio button (square), and select UTF-8 in the drop-down list, as shown.

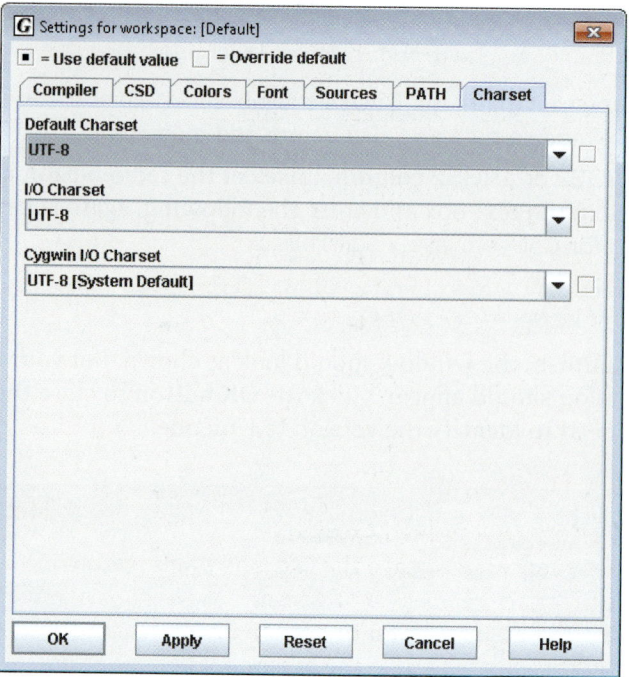

Goodheart-Willcox Publisher

15. Click the **OK** button to close the dialog box.

Running Java Programs

To create a Java program, launch jGRASP, and click **File**>**New**>**Java** from the menu bar. Buttons are added to the toolbar that are used to create Java programs, as shown. Hover the cursor over each button to display a tooltip indicating the command.

Buttons for creating
Java programs

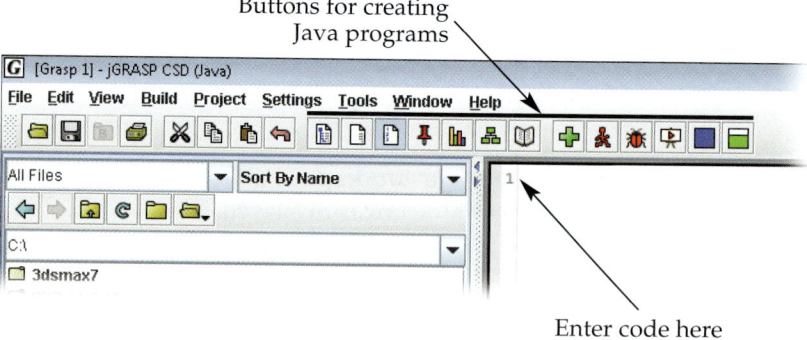

Enter code here

Goodheart-Willcox Publisher

The general steps to write and run a Java program are:

1. Enter the Java code into the source code editor window.
2. Click **File**>**Save As** to save the code.

Compile file

Find and run
the main method
or applet

3. Compile the program by clicking the **Compile file** button on the toolbar. Compiler messages will be displayed in the **Compile Messages** tab at the bottom of the jGRASP window.
4. Execute the program by clicking the **Find and run the main method or applet** button on the toolbar. The program output will be displayed in the **Run I/O** tab at the bottom of the jGRASP window. If it is a graphic program, the graphics will appear in a separate window.

Try an example. This will provide a test of your installation. Launch jGRASP. Click **File>New>Java** on the menu bar. The following code will be explained in the textbook as you learn how to program in Java. For now, simply enter it into the source code editor. Make sure you enter it exactly as listed here.

```java
//  Test Program
public class InstallationTestProgram {

    public static void main( String [ ] args ) {

        System.out.println( "Java is installed!" );
    }
}
```

Save the file by clicking **File>Save As** on the menu bar. In the save-as dialog box, navigate to where you wish to save the file, and click the **OK** button. Be sure to use the default file name suggested. To compile and execute the program, click the **Find and run the main method or applet** button on the toolbar. This output should appear in the **Run I/O** tab at the bottom of the jGRASP window:

```
Java is installed!
```

If errors occur, verify you have correctly entered the code. Also, verify you set the correct paths in the compiler options.

Formatting Output

Java provides a method in the String class for formatting the output: String.format( ). To use this method, both a format string and a list of values to format are required. This format string and list of values to print are called arguments.

Formatted output requires a format string and a list of variables, constants, or expressions. The creation of the string format variable follows this pattern:

```
String s = String.format( "text and specifiers", list of values );
```

A format specifier begins with a percent sign (%) and ends with a converter. The general syntax for specifiers is:

```
%[argument_index$][flags][width][.precision]converter
```

- % is the indication that a specifier is starting
- argument_index$ is the number of the location of the variable in the list of values, with the first value being 1; this is optional, but is useful if an argument is going to be used more than once in the output string
- flags are additional formatting; see the table that follows
- width is the number of spaces to fill with the value
- precision is the number of decimal places to display
- converter is the data type of the variable to be formatted

The arguments are variables separated by commas. The specifiers describe how they are to be formatted. Remember to print the formatted String using the System.out.print( ) or System.out.println( ) method. For example:

```
String s = String.format( "A common approximation for PI "
+ "is%5.2f.", Math.PI );
System.out.println( s );
// outputs: A common approximation for PI is 3.14.
```

String.format( ) Specifiers, Flags, and Escape Sequences		
Purpose	**Code**	**Explanation**
Floating-point conversion	%#.#f	Specify number of characters wide with the first # and number of decimal places with the second #
Decimal-integer conversion	%#d	Specify number of characters wide with #
String conversion	%#s	Specify field width with #
Percent conversion	%%	Display the percent sign (%)
Line-separator conversion	%n	Go to the next line
Left-justified flag	%-	Causes the number to be left-justified; default alignment is right-justified
Zero-padded flag	%0	Fill empty characters in field with zeros (0)
Comma-separator flag	%,	Insert a comma every three digits

Continued

String.format( ) Specifiers, Flags, and Escape Sequences		
Purpose	**Code**	**Explanation**
Sign flag	%+	Show the sign of a positive or negative number
Parentheses flag	%(	Show negative numbers in parentheses
Escape strings	\n	Go to the next line; note that %n is the preferred usage
	\t	Insert a tab

Responding to Java Error Messages

Below are the most common error messages a new programmer will likely see. If you receive any of these messages, check your code for the conditions mentioned for each error. Compiler errors appear when you attempt to compile the application. Runtime exceptions can occur when the program is running.

Compiler Messages

class, interface or enum expected

- This error is usually caused by too many braces. Check that your braces are matched; that is, you have one closing brace (}) for each opening brace ({).

else without if

- Make sure that you do *not* have a semicolon after the if condition.
- Check that your braces are matched; that is, you have an opening and an ending brace around the statements that should execute if the condition is true and also an opening and an ending brace around the statements that should execute if the condition if false.

<identifier> expected

- This error is often a side effect of some other error condition, such as missing a semicolon at the end of a statement or missing an opening brace for a block.

illegal start of type

- This error is usually caused by too many braces. Check that your braces are matched; that is, you have one closing brace (}) for each opening brace ({).

Not a statement

- This error occurs most likely because of a punctuation error. The compiler cannot determine what the statement is. Carefully inspect the statement for proper syntax and punctuation.

')' expected

- This error occurs if the code has more left parentheses than right parentheses. Check that the code has the same number of left parentheses as right parentheses.

reached end of file while parsing

- This error is usually caused by a missing closing brace. Check that every opening brace ({) has a matching closing brace (}), especially for main and for the class.

';' expected

- This error is the result of a statement without its terminating semicolon. The line number in the message is often the line *after* the missing semicolon.

cannot find symbol...

- If the symbol is a variable, check for a typo, and remember that uppercase and lowercase count.
- If the symbol is a class name, check that you have imported the class at the top of the program.
- If the symbol is a method, check that you have parentheses after the method name.
- Check that the variable name is not out of scope.

unclosed comment

- This error is caused if the end of a block comment (*/) is either missing or incorrect, perhaps as /* instead of */.

unclosed string literal

- This error occurs if the ending quotation mark (") is missing from a String or if the String starts on one line and ends on another line. Strings must start and end on the same line.

Runtime Exceptions

"No main methods, JavaFX Applications, applets, or MIDlets found in file"

- Check that the main method is spelled correctly:

```
public static void main( String [ ] args )
```

- Check that you are not attempting to execute a custom class that has no main method.

ArithmeticException: / by zero

- This exception occurs when an integer is divided by 0. One way to avoid this exception is to check before dividing to make sure the denominator is not 0.

ArrayIndexOutOfBoundsException

- This exception is reported when an array index is less than 0 or equal to or greater than the length of the array. A common cause of this exception is to use <= instead of < in a for loop condition. For example:

```
index <= arrayName.length // incorrect
```

instead of

```
index < arrayName.length  // correct
```

FileNotFoundException

- This exception is reported when the Scanner constructor cannot find the file. Check if the file name sent to the File constructor is correctly spelled and that the file is in the same folder as the .java file.

InputMismatchException

- This exception occurs when the user's input does not match the data type that a Scanner method is attempting to read. For example, the program attempts to input an integer using the nextInt method, but the user enters a double or a String.

NoSuchElementException

- This exception occurs when an attempt is made to read data from a file after the end of the file has been reached.

NullPointerException

- This exception occurs if a null object reference is used to call a method. A reference is null when it is defined, but does not yet point to an object. For example:

```
String word;    // word is null
```

StringIndexOutOfBoundsException

- This exception is reported when a String index is less than 0 or equal to or greater than the length of the String. A common cause of this exception is to use <= instead of < in a for loop condition. For example:

```
index <= stringName.length( )  // incorrect
```

instead of

```
index < stringName.length( )   // correct
```

Using Turtle Graphics

The Turtle class is a custom class written by the authors that can be used to draw shapes. You do this by telling the Turtle object to move forward, move backward, or turn. The turtle has a drawing pen. You can tell the turtle to lift up or to put down the pen. If the pen is down, the turtle leaves a mark as it moves. If the pen is up, the turtle does not leave a mark.

The Turtle class uses four .java files, which are provided on the student companion website (www.g-wlearning.com/informationtechnology/1773). The files are Turtle.java, Sprite.java, SpriteAnimator.java, and another Java file that will be the application. The SpriteApplication.java file is provided as a starting point for an application. To create a drawing using this starter file, change the name of the program to reflect your application, then save the file under the new name. All files should be put in the same folder. Compile the files in this order:

1. SpriteAnimator.java
2. Sprite.java
3. Turtle.java
4. the file you will use for your drawing

As written, the SpriteApplication.java file creates a window that is 800 pixels wide and 500 pixels high. These sizes can be changed by changing the values of the windowWidth and windowHeight variables near the top of the SpriteApplication.java file. To give the drawing a title, change the value of the windowTitle variable near the top of the SpriteApplication.java file. All drawing code should be written in the buildScript method.

The first step in the buildScript method is to instantiate a Turtle object. The Turtle constructor has this API:

```
Turtle( Group root, double startXLocation, double startYLocation )
```

In the SpriteApplication.java file, the root contains the Turtle and the drawings. The second and third arguments give the x and y locations in the window where the Turtle object should first appear.

When a Turtle object starts, its direction (or heading) is up and the pen is down. To create drawings, use the methods of the Turtle class shown in the following table.

Method	Explanation
void forward(double pixels)	Moves the turtle forward by the number of pixels.
void backward(double pixels)	Moves the turtle backward by the number of pixels.
void turnRight( )	Turns the turtle 90 degrees to the right.
void turnRight(double degrees)	Turns the turtle right by the number of degrees.
void turnLeft( )	Turns the turtle 90 degrees to the left.
void turnLeft(double degrees)	Turns the turtle left by the number of degrees.

Continued

Method	Explanation
void moveTo(double x, double y)	Moves the turtle to the *x, y* location. The turtle's direction does not change.
void setDrawingColor(Color color)	Sets the color the turtle will use to mark its path. The specified color will continue to be the drawing color until this method is called with a different color.
void penDown( )	Sets down the pen. The turtle leaves a mark when the pen is down.
void penUp( )	Lifts up the pen. The turtle does not leave a mark when the pen is up.
void hide( )	Hides the turtle so it does not appear in the window. Any marks the turtle has made are still displayed.
void show( )	Shows the turtle.
void setSpeed(double speed)	Sets the drawing speed. These static constants can be used to set the speed: Turtle.SLOW, Turtle.MEDIUM, Turtle.FAST, and Turtle.NO_ANIMATION.
double getX( )	Returns the *x* position of the turtle.
double getY( )	Returns the *y* position of the turtle.
boolean isPenUp( )	Returns true if the pen is up. Returns false if the pen is down.
boolean isPenDown( )	Returns true if the pen is down. Returns false if the pen is up.
double getHeading( )	Returns the direction the turtle is facing in degrees. Facing up is 0 and the degrees increase clockwise to a maximum of 359 degrees.
void clear( )	Removes all lines drawn by the turtle.
Color getDrawingColor( )	Returns the current drawing color.

GLOSSARY

A

abstraction: Generalization of patterns found in problem solving. (1)

acceptable use policy (AUP): Set of rules that explains what is and is not acceptable use of school- or company-owned equipment and networks. (16)

access modifier: Specifies the scope of a class, method, or instance variable across an application. (12)

accessor method: Allows a client to view the data in an object by returning the value of an instance variable; also called a *getter.* (12)

actual parameter: Value sent to the method from the client when the method is called. (12)

algorithm: Sequence of steps used to solve a problem. (1)

alpha value: Amount of transparency. (7)

antivirus software: Detects and removes malicious software from a computer; also known as cyberdefense software. (16)

append: Keep the data in an existing file and add new data to the end of the file. (11)

application programming interface (API): Documentation on how to create objects of a given class, what methods are available, and how to call those methods. (6)

arc: Section of an ellipse. (7)

array: Ordered sequence of data of the same data type. (13)

asset: External file used for multimedia. (14)

avatar: Visual representation of a player character. (14)

B

bar graph: Consists of side-by-side lines and is commonly used to compare individual values. (13)

base: How many numbers are used in a number system. (2)

binary number system: Positional number system that has a base of two. (2)

binary operator: Combines only two values at a time. (5)

block: Subsection of code; begins with an open brace (curly bracket) and ends with a closing brace. (3)

block comment: Comment that starts with /* and continues over multiple lines until ended by */. (3)

blue skimming: Using a Bluetooth-enabled card skimmer to steal data. (16)

buffer: Storage area in memory. (6)

bug: Anomaly or error in a program. (5)

Bureau of Labor Statistics (BLS): Principal federal agency responsible for measuring labor market activity, working conditions, and price changes in the economy; part of the Department of Labor. (15)

button: Control used to trigger an action when the user clicks the control. (14)

bytecode: Set of instructions composed of compact numeric codes that is a blend of compiling and interpreting. (2)

C

C#: Programming language developed by Microsoft to complete its suite of programs called .NET; one of the primary languages for programming in Windows. (15)

C++: Compiled programming language for use in system programming, desktop applications, and e-commerce servers. (15)

call: To use a method. (4)

career clusters: Sixteen groups of occupational and career specialties that share common knowledge and skills. (15)

career plan: List of steps on a time line to reach career goals. (15)

Note: The number in parentheses following each definition indicates the chapter in which the term can be found.

case: Each choice for a switch statement. (8)

casting: Creation of a temporary value for the duration of an operation; also *type casting*. (5)

catching an exception: Exception handler receives the exception object from the runtime system. (11)

central processing unit (CPU): Fetches coded instructions, decodes them, and then runs or executes them. (2)

certification: Professional status earned by an individual after passing an exam focused on a specific body of knowledge. (15)

check box: Control presented in a group in which all, some, or none of the controls can be selected. (14)

checked exception: One that must be handled or acknowledged by the application. (11)

child class: New class that inherits methods and data from a parent class. (7)

circle: Closed curve with each point on the curve the same distance from a center point; type of ellipse. (7)

class: Contains the instructions for creating objects. (2) Unit of code that works together; every program is made up of one or more classes. (3)

class scope: Means a variable can be used throughout an entire class. (12)

client: Application that instantiates objects and calls methods of another class. (12)

COBOL: Stands for common business oriented language; a programming language written for business applications and one of the first programming languages that did not require the use of binary numbers for instructions. (1)

comment: Text that the compiler ignores; intended to communicate information to the programmer. (3)

compiled: Conversion of code written in a high level programming language into machine code *before* it is run. (2)

compiler error: Error in Java syntax detected by the compiler. (3)

computational thinking: Thinking like a computer operates. (1)

computer virus: Computer code carried in another program that can replicate itself in order to corrupt or otherwise harm either data files or the software used to process these files. (16)

concatenation operator: Joins data values and String literals into one message for output; represented by the plus sign (+). (4)

condition: Boolean expression. (8)

condition-controlled loop: Loop that repeats until a boolean expression (condition) tests false; while loop. (9)

constant: Value that cannot change during the execution of the program. (4)

constructor: Special method in a class that creates (instantiates) an object, making sure the object has valid values for all its data. (6)

control flow statement: Statement that causes different statements to be executed based on certain conditions. (8)

cookies: Small text files that websites plant on the computer user's hard drive when a user initially visits the website. (16)

copyright: Acknowledges the ownership of a work and specifies that only the owner has the right to sell the work, use it, or give permission for someone else to sell or use it. (16)

counter-controlled loop: Loop regulated by a counter; for loop. (9)

cyberbullying: Harassment that takes place using electronic technology. (16)

D

data breach: Security violation in which sensitive, protected, or confidential data is copied, transmitted, viewed, stolen, or used by an individual unauthorized to do so. (16)

data type: Format in which the data will be stored and the size the variable will be given in memory. (4)

data vandalism: Manipulation or destruction of data found in cyberspace. (16)

debug: Remove bugs from a program. (5)

decomposition: Breaking a problem into small, doable steps. (1)

decrement operator: Subtracts 1 from the value of a specified variable. (5)

default constructor: Constructor that takes no arguments and assigns default values to instance variables. (12)

digital citizenship: Standard of appropriate behavior when using technology to communicate. (16)

divide equals operator: Divides the value of a specified variable by a given amount. (5)

do/while loop: Loop that iterates a set of statements while a boolean expression is true. (9)

dot notation: Placing a period (.) between the object or class name and the method name when calling a method; also used between the object or class name of a static constant or public variable. (6)

dumpster diving: Digging through the trash to locate information. (16)

E

echo: To say back what was input so it can be verified the input was correct. (5)

element: Each data item in an array. (13)

empty String: String object that has been created, but contains no characters. (10)

equality operator: Tests if two variables or expressions evaluate to the same value. (8)

equals method: Method of the String class provided to perform an equivalent function to determine if the characters of the Strings are the same. (8)

equalsIgnoreCase method: Compares the characters in Strings considering uppercase and lowercase letters to be equal. (8)

escape character: Indicates the following character has special meaning. (5)

ethics: Principles of right and wrong. (16)

event: Object created as a result of user action. (14)

event handler: Code executed when an event is generated. (14)

exception: Error detected by a method from which it cannot recover. (4)

exception handler: Code that tells the runtime system what to do when a specific exception occurs. (11)

explicit type casting: Occurs when the programmer tells the compiler to convert a value to a different type, which applies for that operation only. (5)

expression: Combination of operators, numbers, constants, and variables that results in a single value. (5)

extends: Creates a child class while inheriting methods and data from a parent class. (7)

F

file pointer: Mechanism that keeps track of the next data to read in the external file. (9)

focus: Indicates which control is active to accept input. (14)

font: Set of characters of a typeface in one specific style and size. (7)

for loop: Loop that iterates a block of code a predetermined number of times. (9)

formal education: Education received in a school, college, or university. (15)

formal parameter: Parameter name listed in the method header. (12)

G

getter: Allows a client to view the data in an object by returning the value of an instance variable; also called an *accessor method.* (12)

goal: Something a person wants to achieve in a specified time period. (15)

graphical user interface (GUI): Screen displays visual options with which the user can interact. (14)

group: Virtual container within a scene that holds nodes in the JavaFX application. (7)

H

hacking: Activity performed by computer programmers to break into e-mails, websites, computer systems, and files of other computer users. (16)

hard-coding: Defining a variable value in the code. (4)

hash code: Sequence of characters that uniquely identifies an object; usually consists of the class name followed by @ and a hexadecimal address. (12)

hexadecimal notation: Positional number system that has a base of 16. (2)

high-level language: Programming language that has a limited set of recognizable English words. (2)

horizontal box layout: Provides a side-by-side arrangement of the controls. (14)

hue: Mixture of the color contributions of the base colors. (7)

hybrid job: Occupation requiring a combination of programming skills and industry-specific skills. (15)

I

identifier: Name the programmer chooses for a variable in the program; also used as names for programs and for any methods the program defines. (4)

identity theft: Stealing someone's personal information and using that information to commit theft or fraud. (16)

if-then statement: Selection statement. (8)

image control: Holds the reference to an image file as an Image object. (14)

image view control: Displays an Image object. (14)

immutable: Unable to be changed. (6)

implicit type casting: Java automatically promotes the operands. (5)

increment operator: Adds 1 to the value of a specified variable. (5)

index: Number representing the relative position of a character in a String; first index is always 0. (10)

infinite loop: Loop that executes forever (never ends); results from poor programming. (9)

informational interviewing: Strategy used to interview a professional to ask for advice and direction rather than for a job opportunity. (15)

infringement: Use of copyrighted material without permission. (16)

inherit: To take on characteristics of a parent class. (12)

initialization: Code that gives the counter in the loop its beginning value. (9)

input: Data that is typed, scanned, or otherwise sent to a computer system. (2)

input validation: Process of ensuring information entered into a program is acceptable. (11)

instance: An object of a class. (6)

instance variable: Indicates each object (instance) of the class will have that data and the values of the data will vary from object to object. (12)

instantiation: Creation of an object from a class. (2)

instruction set: Complex actions developed using the basic functions of calculation and decision. (1)

integrated development environment (IDE): Application that provides a source code editor and an interface to write, compile, and run programs. (3)

intellectual property: Something that comes from a person's mind, such as an idea, invention, or process. (16)

internship: Short-term position with a sponsoring organization that provides an opportunity to gain on-the-job experience in a certain field of study or occupation. (15)

interpreted: Conversion of code written in a high-level programming language into machine code *as* it is run. (2)

is-a relationship: Object of a subclass is also an object of the superclass. (12)

iteration: Each repetition of the statements in a loop. (9)

J

Java: Object-oriented computer programming language with features that minimize confusion and errors and allow programmers to produce code that operates on a wide variety of platforms. (1)

Java Development Kit (JDK): Set of tools and code for writing Java programs. (3)

Java Virtual Machine: Software that interprets Java code into machine code at runtime. (2)

JavaScript: Programming language used often with Internet browsers, game development, PDFs, and mobile and desktop applications; easy to learn and ideal for beginners. (15)

K

keyword: Code or word that has special meaning in the Java language. (3)

L

label: Noneditable text used to provide information to the user. (14)

Lambda expression: Shortcut from the full event handler statement. (14)

licensing agreement: Contract that gives one party permission to market, produce, or use the product or service owned by another party. (16)

line: Straight segment between two endpoints. (7)

line comment: Comment that begins with // and continues to the end of the line; does not extend beyond the line. (3)

literal value: Textual representation of a value. (4)

logic error: Error in the program's algorithm or in the implementation of the algorithm that causes incorrect results; not detected by the compiler. (3)

logical operator: Allows the combination of Boolean expressions into more complex expressions. (8)

long-term goal: Goal that will take a period of time greater than one year to achieve. (15)

loop: Block of code, or statements, that is repeated as many times as required. (9)

loop body: Consists of zero, one, or more valid Java statements that are repeated within a loop. (9)

loop counter: Variable defined in the initialization; used to keep track of the number of iterations of the loop. (9)

looping condition: Requirement for repeating statements in a loop. (9)

M

machine language: Code that the CPU uses consisting of 0s and 1s. (2)

malware: Software that intentionally performs actions to disrupt the operation of a computer system, collect private information, or otherwise harm the computer or user. (16)

memory: The part of the computer that stores information for immediate processing. (2)

method: Behaviors of an object. (2)

minus equals operator: Decreases the value of a specified variable by a given amount. (5)

mixed-type arithmetic: Use of an operator with two differently typed variables. (5)

mod equals operator: Divides the value of a specified variable by a given amount and stores the remainder. (5)

modulus operator: Finds the remainder in division. (5)

mutator method: Allows a client to change the values of the data in an object; also called a *setter.* (12)

N

nested loop: One loop placed inside another loop. (9)

nested selection: One selection placed inside another selection statement. (8)

networking: Talking with people you know and then making new contacts to establish relationships that can help you achieve your goals. (15)

node: Object in a scene. (7)

null: Keyword indicating an object reference has no value. (6)

O

object: Self-contained unit of data and code to manage that data. (1)

object-oriented coding: Writing code in a programming language that encapsulates code and data into objects. (1)

object reference: Name of the object; holds a location in memory used to find the object. (6)

observable list: Master list of the nodes in the scene for updating the display automatically when the nodes change. (7)

open source: Software that has had its source code made available to the public at no charge and with no restrictions. (16)

opening a file: Coding the File and Scanner constructors to read an external file. (9)

operand: Each value in an operation. (5)

operating system (OS): Specific set of software persistently stored on the computer that manages all of the devices as well as locates instructions and provides them to the CPU. (2)

operator precedence: Order in which operations are performed in Java. (5)

output: Data that has been digitized into a useful format and provided to the user. (2)

overloaded constructor: Has the same class name as another constructor, but takes a different number or data type of arguments. (12)

overloaded method: Two or more methods that share a name but take different arguments. (6)

P

package: Collection of related classes in one location. (6)

package access: Means the item is available to any method or application in the same package or folder. (12)

paper check: Manually writing out a description of the sequence of instructions to be executed and then analyzing it for proper logic. (9)

parameter: Any argument sent by the client to the method. (12)

parent class: Class that can be extended to create new classes with the same methods and fields. (7)

password field: Text field that has a special property to hide the characters entered. (14)

patent: Gives a person or company the right to be the sole producer of a product for a defined period of time. (16)

pattern recognition: Identifying parts of one problem that are replicated within a problem or in other problems. (1)

peripheral: External computer device not critical to basic operation. (2)

phishing: Attempt to get private information by appearing as a harmless request. (16)

piracy: Unethical and illegal copying or downloading of software, files, or other protected material. (16)

pixel: Point on the display that has an x and y location and a color; picture element. (6)

plagiarism: Claiming another person's material as your own. (16)

platform: Computer environment depending on a unique instruction set. (1)

plus equals operator: Increases the value of a specified variable by a given amount. (5)

portability: Ability of a program to run on multiple computers. (3)

portfolio: Collection of examples showing your qualifications, skills, and talents that support your career or personal goals. (15)

postsecondary education: Education achieved after high school. (15)

primitive data type: One of eight data types built into Java. (4)

prompt: Message displayed to the user indicating what to input or what action to take. (4)

pseudorandom number: Number that belongs to a set of a sequence of numbers long enough to appear to be random, but is not. (6)

Python: General-use programming language known for its readability, straight-forward coding, and ease of learning. (15)

R

radio button: Control presented in a group in which only one control can be selected. (14)

ransomware: Blocks the user's access to programs until the user pays to unlock them. (16)

raster: Pixel-based graphic in which each individual pixel holds its own color value and is available for modification. (7)

real number: Number with a fractional part. (4)

relational operator: Compares two numeric expressions to see if one is greater, lesser, or both are equal. (8)

résumé: Document that highlights a person's career goals, education, work history, and professional accomplishments. (15)

root: Always the first node. (7)

runtime engine: Software that interprets bytecode into machine code. (2)

runtime error: Error detected by the JVM while the program is executing; not found by the compiler. (3)

S

scale: To multiply each value by a set amount. (13)

scareware: Software designed to cause enough anxiety so the computer user leaps at the chance to take corrective action. (16)

scene: Controls the layout of all items in a group. (7)

scene graph: Organizes all the effects in a scene. (7)

scope: Block of code in which a variable is defined; a variable cannot be used outside its scope. (8)

search key: Value to find in a search. (13)

selection statement: Allows the program to choose a set of statements to execute based on a certain condition at the time the selection statement is processed. (8)

sequential search: Algorithm for searching an array by comparing the search key to each element in order. (13)

setter: Allows a client to change the values of the data in an object; also called a *mutator method.* (12)

shape: Any two-dimensional object defined in geometry. (7)

shortcut concatenation operator: Appends a String or other data type to an existing String; represented in Java with the symbol +=. (10)

shortcut operator: Streamlines common expressions to simplify the code and make it easy to read. (5)

short-term goal: Goal that can be achieved in less than one year. (15)

social engineering: Using various social methods to obtain information on an individual. (16)

software development life cycle: Process that provides structure to the production and maintenance of software; software is written, installed, used, and maintained. (1)

sort: To arrange elements in an array in some order based on their values. (13)

spyware: Program that secretly collects a user's data and behavior. (16)

stack trace: History of the methods called before the error occurred. (11)

stage: Window created by the start method where the graphical output is displayed. (7)

statement: Performs the work of the program; can extend over multiple lines and ends with a semicolon. (3)

static method: Called with a class name instead of an object reference. (6)

storage: Where information is kept by the computer so that it can be saved to view, play, or be reused. (2)

string literal: Sequence of characters enclosed in quotation marks. (4)

Structured Query Language (SQL): Declarative language used for database management; often pronounced *sequel.* (15)

subclass: Class that inherits methods and data by extending a superclass. (12)

substring: String composed of a sequence of zero to all the characters in a String in the original order. (10)

superclass: Original class from which another class inherits. (12)

switch statement: Type of selection statement that tests the value of a variable and selects statements based on any number of choices. (8)

syntax: Rules of grammar that specify how code is to be written in a programming language. (2)

T

test plan: Outlines a systematic approach to evaluating the success of a program. (5)

test set: Combination of sample inputs and outputs for a single test. (5)

this object reference: Special reference to the object on which the method should operate. (12)

throwing an exception: Sending an exception object to the runtime system. (11)

times equals operator: Multiplies the value of a specified variable by a given amount. (5)

token: Unit of characters collected via tokenizing. (6)

tokenizing: Skipping white space and collecting non-white-space characters. (6)

trademark: Protects taglines, slogans, names, symbols, and any unique method to identify a product or company. (16)

traverse: Visit each character in a String in order one at a time. (10)

Trojan horse: Program that invites the user to run it while concealing malicious code that will be executed. (16)

true block: Set of statements to execute when the condition of an *if-then* statement is true. (8)

truth table: Shows the truth-value of a compound selection statement for each component in the statement. (8)

truth-value: Whether a Boolean statement is true or false; used to determine branching to execute different code segments. (8)

typeface: Single design of text characters and punctuation. (7)

U

unary operator: Operator in which only one expression is used. (8)

unchecked exception: One that is reported by the RunTimeException class and does not need to be handled by the programmer. (11)

Unicode encoding system: Encoding standard for representing international text characters. (2)

uniform resource locator (URL): Address that points to a specific document or other resource on a computer network. (14)

user error: Incorrect action by the user. (3)

user experience (UX): Feeling users get from interacting with a website or computer application. (14)

user interface (UI): How the user enters data. (2)

V

variable: Memory location defined by name and data type whose value is able to be changed during program execution. (4)

vector: Graphic composed of shapes based on mathematical statements. (7)

vertical box layout: Aligns controls from top to bottom. (14)

W

while loop: Loop that iterates a set of statements while a boolean expression is true; statements may not be executed at all if the tested condition is false when the loop begins. (9)

white-space character: A space, tab, or new line character; used as a separator. (3)

worm: Stand-alone computer program, not hidden in another program, that replicates itself in order to spread to other computers. (16)

wrapper class: Encloses, or "wraps," a primitive type into an object, which can then be used to call methods. (10)

INDEX

A

absolute values, 157
abstraction, 8
acceptable use policy (AUP), 458
access modifiers, 327
accessor methods, 337–340
actual parameters, 333
addition operator, 101
AI (artificial intelligence), 4, 47
algorithms
    defined, 9
    expression of, 9
    logic errors in, 65
    most elegant solution and, 9
    perfect squares example, 9–10
    writing, 8–9
alpha value, 179
ALU (arithmetic logic unit), 110
American Standard Code for Information Interchange (ASCII), 43–44
Analytical Engine, 13
animation, 416
antivirus software, 468–469
append, defined, 314
Application class, 166
application programming interface (API), 134
applications classes, 326
application software (apps), 16, 33
arc shapes, 187, 189, 191
argument index, 120
arguments, 120, 135
ArithmeticException, 305
arithmetic logic unit (ALU), 110
arithmetic operators, 100–105
ArrayIndexOutOfBoundsException, 367
arrays, 360
    creating, 362–366
    element-by-element processing, 367, 370–371
    element default values in, 363
    elements, accessing, 364
    filling, 368
    finding average of, 368–369
    finding minimum and maximum values, 369
    graphing, 371–372, 375–376
    naming, 362
    of objects, 376–381
    printing, 368
    runtime errors with, 367
    searching, 384–387, 389
    size of, 364
    sorting, 387, 389
Arrays class, 387
artificial intelligence (AI), 4, 47
ASCII (American Standard Code for Information Interchange), 43–44
assets, 414
asterisk character, 101
audio devices, 29
audio files, 414–415
AUP (acceptable use policy), 458
avatars, 419

B

Babbage, Charles, 13
backslash character, 123
backups, 468
bar graphs, 371
base, defined, 39
BASIC (Beginners All-purpose Symbolic Instruction Code), 35–36
binary number system, 40, 42
binary operators, 100–103
bits, 40
block comments, 56
blocks, 56
blue skimming, 466
boolean data type, 77, 204
Boolean expressions, 204
boundary values, 125–126
brain-decoding technology, 464
buffers, 147
bugs, 126
Bureau of Labor Statistics (BLS), 438, 441, 453
buttons, 399, 408, 411–413
bytecode, 36
byte data type, 76
bytes, 40

C

C#, 442
C++, 442
calculations, humans *versus* computers, 5
calculators, 4, 41
call, defined, 87
calling methods. *See* methods
CamelCase, 55, 80

Note: Page numbers followed by *f* indicate figures.